Fascinating Life Sciences

This interdisciplinary series brings together the most essential and captivating topics in the life sciences. They range from the plant sciences to zoology, from the microbiome to macrobiome, and from basic biology to biotechnology. The series not only highlights fascinating research; it also discusses major challenges associated with the life sciences and related disciplines and outlines future research directions. Individual volumes provide in-depth information, are richly illustrated with photographs, illustrations, and maps, and feature suggestions for further reading or glossaries where appropriate.

Interested researchers in all areas of the life sciences, as well as biology enthusiasts, will find the series' interdisciplinary focus and highly readable volumes especially appealing.

Wen-San Huang

Reptiles in the Changing Climate

Species and Ecosystems in Flux on Orchid Island

Springer

Wen-San Huang
Director Office
National Museum of Natural Science (NMNS)
Taichung, Taiwan

ISSN 2509-6745 ISSN 2509-6753 (electronic)
Fascinating Life Sciences
ISBN 978-981-92-1886-8 ISBN 978-981-92-1887-5 (eBook)
https://doi.org/10.1007/978-981-92-1887-5

Cover caption: The familiar Taiji (Yin–Yang) symbol–black (yin) and white (yang) with opposite dots–represents the coexistence of complementary forces. In this cover, a reptile egg–eating kukri snake forms the dividing line. On the right (yang), eggs are rarely eaten, allowing species such as the long-tailed skink to hatch. On the left (yin), eggs are more easily consumed; however, the snake's territorial behavior still allows green sea turtle eggs to survive. Originating in ancient China, yin and yang describe opposing yet interdependent forces. This cover reflects their balance, emphasizing that ecological stability depends on harmony among species, climate, and environment. (Illustration by Ms. Hsun-Yu Pong and Ms. Shu Chin Yeh.)

This Springer imprint is published by the registered company Springer Nature Singapore Pte Ltd.
The registered company address is: 152 Beach Road, #21-01/04 Gateway East, Singapore 189721, Singapore

Preface

I have conducted research on Orchid Island for nearly three decades (1997–present), during which time I have published more than forty papers. These span the spectrum from fundamental ecological data collection to the impacts of climate change on island biota. These papers employ techniques from ecology, physiology, molecular biology, behavioral science, and climate change to comprehensively examine the challenges faced by the island's reptile fauna. I endeavor to provide a thorough overview of how the primary reptile populations on this single island are experiencing widespread decline and even extinction due to the impacts of climate change. This field remains largely under-documented. Consequently, this volume synthesises extensive findings from my long-term research, compiling them for the benefit of future scholars.

Frankly, without comprehensive ecological data research (such as reproductive cycles), there would be no basis for conceptualizing the subsequent impacts of climate change on island species; without the discovery of Kukri sanke's foraging behavior, the later connections within the food chain would never have been established. Thus, to truly grasp the full panorama, one must painstakingly piece together the puzzle bit by bit, only then revealing the whole picture. I was once derided as a stamp collector: merely capable of drudgery and piecemeal work. Thankfully, this labourer has finally been able to write a book.

Reaching Orchid Island necessitates travel by small aircraft, which often experiences considerable turbulence due to crosswinds. I am frequently asked why I risk life and limb to conduct research on this remote island, especially since ecological studies are not currently mainstream scientific pursuits and seem to attract little attention. Frankly, it's a good question. Whenever posed, I invariably counter: Have you ever been in love? When you love someone, you charge ahead without hesitation, committed to standing by them through thick and thin. That is precisely how I feel about my research into Orchid Island's reptiles.

The book's narrative approach—"one theory, five major breakthroughs, a 30-year chain of evidence"—draws on long-term book into reptiles on Taiwan's Orchid Island. All 40-plus papers have been published in renowned international journals. I believe that the novel theory and robust data presented in this book represent a significant contribution to advancing research in ecological evolution toward new discoveries. The greatest original contribution of this book is the establishment of a cross-scale chain of evidence in ecological evolution, linking individual behavioral decisions all the way to ecosystem fate. Furthermore, it raises a new scientific question, establishes a new theoretical framework, and transforms the way we understand the natural world. The previous book has largely been confined to a single scale: Studies of individual behavior focus solely on behavior; life history studies focus solely on reproductive strategies; studies of populations focus solely on species interactions; and books on climate change focus solely on environmental effects. My book on Orchid Island, however, is the first to systematically link these scales together.

The academic positioning of this book: Ecology and evolutionary biology have long faced a core challenge: how to link adaptive behavior at the individual level with stability at the ecosystem scale. Over the past few decades, most research has examined individual behavior, life history strategies, interspecific interactions, food web structure, and the effects of global change in isolation. However, there has been a lack of a theoretical framework capable of integrating these different scales; consequently, it has not been possible to fully explain how ecosystems respond to environmental change, or why some ecosystems remain resilient while others collapse. Using Orchid Island as a natural laboratory, I have established a long-term research system spanning over 30 years. Through studies of lizards, snakes, green sea turtles, weevils, and invasive species, I have gradually developed a new conceptual framework: the "resource-driven eco-evolutionary resilience theory." Its central tenet is that the fate of an ecosystem is not directly determined by climate change or species abundance, but rather by how key resources transmit their influence across different scales through behavioral decisions, biological interactions, and life history evolution.

In other words, the stability or collapse of an ecosystem is the cumulative result of the decisions made by countless individuals in response to changes in resources. The findings presented in this book reveal the following five major breakthroughs: 1. Predation risk drives life history evolution: starting from individual decisions; 2. resources shape social organization: from individuals to populations; 3. interspecific interactions reorganize community structure: from populations to communities; 4. marine subsidies maintain ecosystem resilience: from assemblages to ecosystems; and 5. climate change determines ecosystem fate through resource networks: from ecosystems to global change. This book provides the first direct evidence that global change can cause cascading ecosystem collapses through the disappearance of key resources and establishes a crucial empirical foundation for the climate-mediated resource collapse hypothesis. Therefore, by synthesizing the above research findings, this book proposes the "resource-driven eco-evolutionary resilience theory."

Taichung, Taiwan Wen-San Huang

Acknowledgments

For more than 28 years, my work has been done with the cooperation of many people. In particular, lizards or snakes are very difficult to catch by traps or other means; two or more people must work together to catch them by chasing or encircling. My colleague, Mr. Cheng-Shiung Chang, played an important role in this. The images that come to my mind only show two people using laundry bags to catch lizards in ditches, seaside reefs, piles of fallen leaves in the mountains, on manzanita or on the roadside cement, and so on. Therefore, I am very grateful to Mr. Chang for his uncomplaining assistance and help over a long period of time. I would also like to thank my outstanding graduate students and postdocs for their help, who were always ready to provide great ideas to improve my initial wild ideas. I would like to express my gratitude to the students who have been working on this project for a long time: Jung-Ya Hsu, Chen-Pan Liao, Hui-Yun Tseng, Jhan-Wei Lin, Chi-Chien Kuo, Chun-Chia Chou, Yu-Wei Ko, Hui-Ying Yeh, Hao Hsu, Min-Hao Hsu, Chang-Yi Lee, Lo-Yen Chang, Shiang-Yu Huang, Chen-Song Chang, Yi-Ying Chen, Yu-De Zhang, Wei-Hong Lin, Tsang-Pon Wang, Hao-Chen Tang, Shi-Ping Huang, Tai-En Ou.

I would also like to thank Hidetoshi Ota, who was at the beginning of my research, for his encouragement and paper suggestions. I would also like to thank my Cornell PhD committee members, Harry Greene, Kelly Zamudio, Kraig Adler, and especially Harry Greene, who was so busy that I would always stop by his office to ask for help. Harry is very sweet and my lifelong research idol, and David Pike, Rulon Clark, and Andrea Manica have also given me a lot of help with my papers. The hand-drawn drawings in this book were all done by Ms. Hsun-Yu Pong, who finally completed all the reptiles on Orchid Island after years of effort. All the scientific drawings in this book were re-modified by Chen-Pan Liao with great effort. I am also very grateful to Min-Hao Hsu for proofreading the text and ensuring the accuracy of the references. I must also thank Mr Chung-Wei Yu for providing the splendid reptile photographs in Chap. 1.

When the book was first written, it was a bit loose, unclear, and there were too many repetitions, so thanks to Min-Hao Hsu for his help with proofreading and word-by-word editing to make the book what it is now. I also thank my wife, Annie Chia-Ling Wang, my daughter, Betty Shi-Ping Huang, and my son, Mason Chen-Wei Huang. With their love and support, my life is complete and peaceful.

Contents

An Introduction to Orchid Island and the Significance of Studying Island Biogeography and Climate Change

1

Abstract

This chapter shall briefly introduce island biogeography, climate change, the ecological outlook for reptiles on small islands, the cornerstone of research—fundamental ecological studies, the importance of long-term ecological research, animal behavioral studies, and the origins of reptile ecology research on Orchid Island. It shall cover Orchid Island's geographical location, geological history, cultural heritage, and its flora and fauna resources, alongside the distribution of reptile species inhabiting the island.

Keywords

Orchid Island · Reptiles · Ecology · Behavior · Location · Geology · Climate change

1.1 Introduction

Orchid Island is located near Taiwan and the Philippines, and its biota is naturally closely related to those of these two islands. Therefore, the scope of research cannot be separated from issues such as island biogeography, which has evolved over a long period of time, and recent climate change.

1.2 Island Biogeography

The term "island biogeography" was coined by ecologists Robert MacArthur and Edward Wilson in the 1960s. Based on Wilson's decades of research on ants, they proposed an important model aimed at predicting the number of species that would emerge on a new island. This theory was initially developed to explain the

W.-S. Huang, *Reptiles in the Changing Climate*, Fascinating Life Sciences,
https://doi.org/10.1007/978-981-92-1887-5_1

species-area relationship on islands and the extinction rate patterns associated with island distance and size. Since islands represent isolation mechanisms, any isolated areas can be categorized under the term "island." For example, isolated ecosystems now encompass mountain peaks, fragmented forests, and even natural habitats isolated due to human land development.

In undisturbed island environments, species abundance is determined by immigration and extinction. Once a species successfully colonizes an island, its extinction rate is influenced by island size; this can be described using the species-area curve or effect. Larger islands have larger habitat areas and a greater variety of habitats. Larger habitat sizes reduce the probability of extinction caused by random events. Habitat diversity increases the number of species that can survive after migration. Over time, the two forces of extinction and migration offset each other, leading to a balanced level of species richness. Therefore, factors such as the distance to the nearest "neighbor" and the mainland, isolation time, island size, and habitat size all influence island species diversity. However, the above-mentioned processes are the result of natural selection under conditions of no human interference and occur on a slow timescale. Once humans forcibly intervene or alter the environment, the rate of species extinction accelerates. For example, climate change.

1.3 Climate Change

The primary impacts of climate change on island ecosystems include: rising sea levels submerging coastal lands, frequent extreme weather events leading to habitat loss and biodiversity decline; altered temperature and rainfall patterns affecting species survival and reproduction, such as increased invasive species threats to native species, particularly exacerbated by the vulnerability of island ecosystems. The impacts on species are briefly summarized as follows:

1. Sea-level rise and habitat loss: In low-lying islands, rising sea levels directly submerge coastal areas, reducing biological habitats and even leading to the complete disappearance of islands. For example, from 1997 to 2022, the beach on Orchid Island receded by 11 meters, and the depth of the sand was reduced to only 40 centimeters. This caused sea turtles to be unwilling to lay eggs even when they came ashore because the sand was too shallow, which had a huge impact on the reptiles on Orchid Island (Chap. 14).
2. Increased extreme weather events: Climate change-induced extreme weather events, such as typhoons, heavy rains, and droughts, pose threats to the survival of island species. For example, as mentioned in Chap. 14 of this book, frequent and intense typhoons can cause beach erosion or a decrease in sand, leading to the disappearance of sea turtle nesting sites.
3. Impact on species survival and reproduction: Climate change-induced temperature increases may exceed the tolerance limits of animals or their offspring, affecting their survival and reproduction. For example, as mentioned in

Chap. 13 of this book, the long-tailed skink refuses to lay eggs in high-temperature concrete burrows because its eggs cannot survive temperatures exceeding 40 °C.

4. Invasion of invasive species: Climate change may alter the ecological environment of islands, creating conditions for invasive species to establish themselves. Invasive species may compete with native species for resources, prey on them, or spread diseases, posing a threat to island biodiversity. For example, the recent invasion of the multi-lined skink in southern Taiwan has led to a decline in native lizard populations. The multi-lined skink has also recently invaded Orchid Island, and it is likely to pose a threat to local native species.
5. Vulnerability of island ecosystems: Island ecosystems are typically more isolated, with relatively fewer species and a higher proportion of endemic species. Especially vulnerable islands are most susceptible to significant impacts from climate change, primarily due to their simpler biological composition, low connectivity, and short food chains (Chap. 3). Therefore, if any link in the food chain is disrupted by climate change, it can lead to extremely severe consequences (Chap. 14) or even species extinction. For example, the kukri snake, which primarily feeds on sea turtle eggs, has shifted to preying on the eggs of other lizards, causing a severe decline in most lizard populations and even leading to the extinction of some species.

1.3.1 Prospects of Reptile Ecology in a Small Island

Since I joined the museum in 1992, I have been conducting ecological research on various groups of reptiles. Since most reptiles are a group of species that are not easy to study, and many species do not make sounds, direct visual observation is required to find reptiles. To date, there is still a lack of understanding of the ecological habits of many species. In view of this, our research focuses on establishing basic ecological data on species, such as food habits, reproductive biology, and habitat use, and then extends to the study of animal behavior, operational experiments, and long-term observations of species activity patterns in the field to investigate the survival of species in response to changes in environmental factors, and then to the effects of environmental changes (e.g., global warming) on them. The taxa studied include Scincidae, Gekkonidae, Lacertidae, Agamidae, and several snake species.

The more we study, the more we realize that nature's arrangement is so incredible. A reptile-egg-eating snake can manipulate the ecology, population changes, behavioral evolution, and even life history adaptations of all reptiles to achieve maximum fitness. All of these end results would not have occurred without the reptile-egg-eating snake's own peculiar territorial behavior—females would have fought and defeated males for food. The following is a brief description of the process and results of several experiments. Detailed results can be found in subsequent chapters.

1.3.2 The Cornerstone of Research: Basic Ecological Research

All in-depth studies must be based on basic knowledge and understanding of the species, and only through field observations can we find good problems. Although Taiwan is located in a subtropical region, there are many high mountains on the island with altitudes over 1000 m. I wanted to understand the life history of reptiles at different altitudes on the island. From 1992 to 1996, I conducted studies on *Takydromus hsuehshanensis, Sphenomorphus taiwanensis, Diploderma brevipes,* and *Plestiodon elegans* in the Wushe and Hohan Mountains. By capturing and measuring external morphological features and by studying anatomical and reproductive sections, the reproductive cycles and differences between males and females of each species were understood, and basic ecological data of the species were established.

1.3.3 Importance of Long-Term Ecological Studies

After about 5 years of continuous research on high-altitude species, I began to search for a suitable tropical habitat for our research after understanding the food habits, reproductive cycles, activity times, and habitat use of high-altitude species in Taiwan, and in 1997, I visited Orchid Island for the first time and began a fascinating research journey from there. The island is a typical tropical island with only about 5000 inhabitants and no serious human development, thus preserving its pristine nature. Through continuous and stable surveys every month, a total of 18 species of terrestrial reptiles are known, including five species of snakes (*Elaphe carinata, Oligodon formosanus, Viridovipera stejnegeri, Psammodynastes pulverulentus,* and *Ramphotyphlops braminus*), and 13 species of lizards (*Eutropis longicaudata, E. multicarinata, E. cumingi, Sphenomorphus incognitus, Diploderma swinhonis, Gekko kikuchii, Gekko hokouensis, Hemidactylus frenatus, Hemiphyllodactylus typus, Lepidodactylus lugubris, L. yami, Emioa atrocostata,* and *Takydromus sauteri*). Some of these species, such as *E. multicarinata, E. cumingi, G. kikuchii,* and *L. yami*, are of the Philippine lineage, making Orchid Island even more special.

Because of the special characteristics of the species and environment on Orchid Island, I chose to conduct a long-term and in-depth study here. In the past 25 years of monthly surveys, in addition to the publication of a new recorded species of Kusu's southern lizard (*E. cumingi*) in 2000, eight other species of lizards in Orchid Island were also studied. I also used morphological measurements, stomach contents analysis, and reproductive sections to understand the male and female dimorphism, food habits, and reproductive cycles of each species to establish the biology of tropical species.

The special features of Orchid Island are not only the terrestrial reptiles. There are only two places in Taiwan where green sea turtle (*Chelonia mydas*) come ashore to lay eggs, and Orchid Island is one of them. From May to October every year, green sea turtles come ashore at night to lay eggs in Tungching Bay and Little Badai Bay on Orchid Island. The eggs of the turtles are an excellent food for the Kukri snake, which is a snake that feeds on the soft-shell eggs of reptiles and has a stable population in Orchid Island. In my study, I found an interesting phenomenon: after the green sea turtle laid their eggs, the Kukri snake would arrive at the spawning site within 3–7 h, and through morphological measurements, I found that a very high percentage of kukri snake in Orchid Island had tails cut. The results attracted me to investigate the interactions between male and female Kukri snakes, and the research results became the first study in the world to identify the territorial behavior of snakes. In order to compete for food resources, kukri snake will drive out competitors of the same species, and the fighting process can easily result in the loss of part of its tail. This results in males reaching the eggs earlier than females, thus contributing to our interest in observing and subsequently discovering territorial behavior.

The importance of long-term ecological studies is not only to gain insight into the life history and habits of species, but also to understand the effects of changes in environmental factors on animals through long-term research data. Global climate change is currently the hottest topic, but very few studies have been conducted on species from tropical regions. Reptiles are ectotherms, and in addition to the direct impact of climate change on the adults themselves, the eggs they lay are also greatly affected by the external environment. My studies in Orchid Island show that many long-tailed skink lay their eggs in concrete holes adapted to artificial structures such as retaining walls. The cement structure not only provides an environment for the lizard to lay eggs but also maintains a higher temperature inside the cement cave compared to the natural environment in the wild, resulting in shorter hatching time and higher hatching rate, larger hatchlings, and higher subsequent survival rate. However, the long-term research data provided a surprising finding: due to the increase in ambient temperature, the high temperature of the cement cave is no longer suitable for the long-tailed southern lizard! For reptiles, an increase in temperature within the appropriate temperature range is beneficial for egg incubation, but beyond the upper limit of incubation temperature, hatching rates begin to decline. Climate change causes temperature increases in both concrete culverts and natural habitats, but the rate of temperature increase in concrete culverts is three times higher than that in natural habitats. This became the commonly known [ecological trap].

1.3.4 Animal Behavior Studies

In the field, I sometimes observed very special phenomena that I did not understand at first. During a survey on Orchid Island, I was once walking on the road to the weather station when something suddenly fell down from the slope above us. At that time, I didn't know why the long-tailed skink attacked the kukri snake, but I thought it was very special. Later on, I found a long-tailed skink attacking a kukri snake in a concrete hole again, which led me to associate this behavior and to test the meaning of this behavior. The long-tailed skink that attacked the kukri snake had eggs in the concrete hole, so I tentatively speculated that such an attack might be the result of parental protection of the eggs. Since only about 140 species (about 3%) of reptiles have parental care behavior, parental care behavior can be divided into several levels. Some species provide care by simply staying near the eggs and not actively defending them, some raise the temperature of incubating eggs, and a few other species attack predators to improve egg survival. I found that the long-tailed skink in concrete holes attacked the kukri snake. Whether such behavior represents parental care behavior of the long-tailed skink needs to be verified in operational trials.

Many studies suggest that parental care behavior may be a cost to the mother lizard itself, reducing the survival rate or subsequent fecundity of her offspring, so most reptiles do not exhibit this behavior unless it has a greater benefit to the future fitness of the species, i.e., the eggs or hatchlings that are cared for have a higher survival rate. Do long-tailed skinks with eggs in concrete holes attack the kukri snake more frequently than lizards without eggs? And does this behavior increase the survival rate of eggs? Can long-tailed skinks distinguish egg predators? I conducted a test in a concrete retaining wall hole in Orchid Island and placed a kukri snake into the concrete hole with and without eggs. The hatching rate of protected eggs was significantly higher than that of unprotected eggs, indicating that parental care has a significant effect on the survival rate of offspring. In our experiments, I also placed the stinky snake in a concrete hole to test whether the lizard could recognize its natural enemies with different hunting pressure. The results of the test showed that all the lizards would run away when they encountered the stinky snake, regardless of whether they had eggs or not. In addition, egg-caring lizards were oblivious to other non-threatening lizards of the same species, such as the Swinhone tree lizard and scaly fly lizard, and did not show any behavioral response. All the results provide a clear message that long-tailed skinks have different behavioral responses to different kinds of invaders.

It was mentioned earlier that parental care behavior may cause the parent to spend less time foraging, resulting in lower subsequent survival or fecundity of the parent, but such a hypothesis requires long-term field data and is not easy to prove. In my study of long-tailed skinks, I found that the length of time that females spent protecting their eggs varied, so I wanted to find out if the length of time spent protecting eggs affected the survival rate and fecundity of females. Through 4 years of intensive testing and research, I obtained different results: the duration of egg guarding did not affect the subsequent condition of female lizards. However, it is not clear why the length of time spent nursing eggs affects the length of time spent nursing eggs. Since the survival rate of eggs from uncared-for parental lizards is low, we used different frequencies of natural enemy disturbance to demonstrate whether natural enemies are a factor affecting the duration of egg care in long-tailed skinks, and high frequencies of disturbance by the kukri snake did induce long-tailed skinks to extend parental care to ensure better egg survival. Of course, if the frequency of disturbance by the kukri snake is too high, the female will eat her own eggs, which is commonly known as filial cannibalism. These studies will be discussed in detail in the following chapters.

1.4 Location, Geological Age, Culture, and Flora and Fauna in Orchid Island

Orchid Island has evolved through different names in different periods. According to Min-Lin Hong (Hong, 1978), the name "Red Head"was changed to "Orchid Island" by the Civil Affairs Office of the Taiwan Provincial Office of the Chief Executive in the Republic of China (R.O.C.) in 1947 because the previous name "Red Head" was easily associated with the fear of poisoning by red insects, and because the island is rich in the endemic species of five-leaved butterfly orchids, and so the name "Orchid Island" has been used ever since.

1.4.1 Location of Orchid Island

Orchid Island is located in the western Pacific Ocean, southeast of Taiwan and south of Green Island, facing the Bashi Strait to the south and the Bataan Islands of the Philippines. The northernmost island of the Bataan Islands, Yami Island, is 99 kilometers away from Little Orchid Island. It is between 121 degrees 5 min East and 22 degrees North. It has an area of 48.3892 square kilometers (Fig. 1.1).

Fig. 1.1 Panoramic view of Orchid Island. (Photo https://flic.kr/p/2jdUTD4 released by the Office of the President, Taiwan (ROC) is licensed under CC BY 2.0 https://creativecommons.org/licenses/by/2.0/. Colors and geometry are changed)

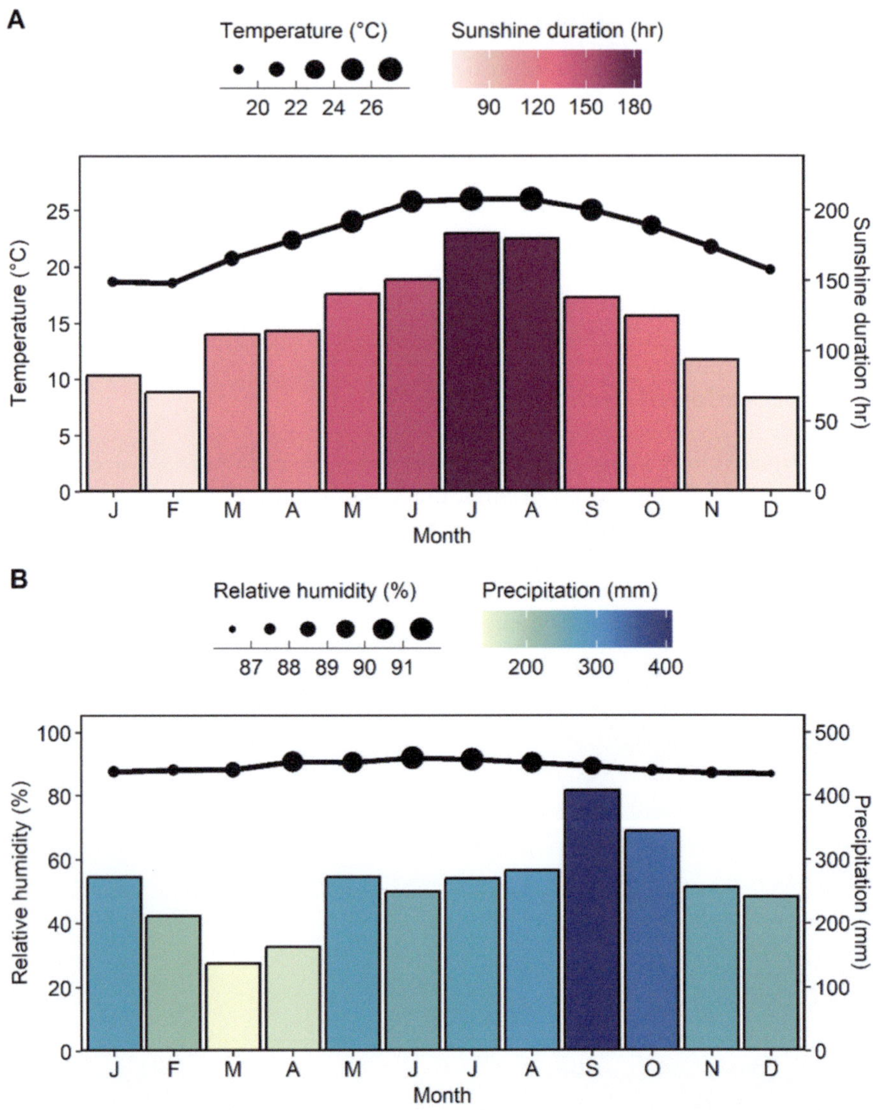

Fig. 1.2 Annual variations in sunshine duration, mean maximum air temperature (**a**), precipitation and relative humidity (**b**) during 1992–2002 (using monthly averages) at Orchid Island, Taitung County, Taiwan (22°03′N, 121°33′E). (Data source: Orchid Island Weather Station, Central Weather Bureau, Taipei, Taiwan)

The climate is tropical rainforest, with an annual rainfall of over 3000 mm and 209 days of rain per year (Fig. 1.2). There used to be seven tribes in Orchid Island, but now there are five villages: Yayo Village, Hontao Village, Dongqing Village, Yuzen Village, and Langtao Village. The total population is about 5000 people.

1.4.2 Tao (Dawu) Culture

According to the prehistoric archaeological sites on Orchid Island, the Dawu (formerly known as the Yami) may have lived on Orchid Island for hundreds or even thousands of years, while the Dawu are linguistically and culturally derived from the Bataan Islands in the north of the Philippines, and are members of the larger South Island language group.

Because of its special geographical location, Orchid Island is a major stepping stone for human, animal, and plant migration in Southeast Asia, and is rich in Dawu culture and rare flora and fauna, which is valuable for academic research on indigenous origins and culture, paleogeography, paleobiogeography and biology in Taiwan, the Ryukyus and the South Seas. During the Japanese occupation, the Japanese government officially designated Orchid Island as a research area in 1895, maintaining the traditional living conditions of the local inhabitants and forbidding foreigners from colonizing or developing the area.

Fishing is the most important traditional livelihood on Orchid Island, and the related calendar, rituals, and artistic expressions are all inextricably linked to fishing activities. The Dawu tribe is an egalitarian social organization (there is no so-called chief or "chieftain"), and there is no culture of brewing wine or the custom of hunting the head out of the grass. [cited from Tao (Orchid Island)–Wikimedia web page].

In terms of linguistic classification, the islanders belong to the same South Island language group as the aboriginal people of Taiwan, and the local oral history, material civilization, geographic space, and maritime navigation knowledge of Orchid Island tend to lean toward the southern island of Bataan Itbayat. In recent years, the Dawu people have made many trips to the Bataan Islands in search of their roots, confirming their cultural affinity. The Dawu people have developed a unique social culture on Orchid Island in response to the natural environment, and are different from other aboriginal hunting cultures in Taiwan in terms of production patterns. The Dawu people divide the year into three seasons by the activity of catching flying fish, and develop a yearly calendar unique to Orchid Island, which also leads to the development of related social organizations (fishing groups), taboos, and craft techniques (Cited from Orchid Island Township Office, web page).

The origins of the Tao tribe are widely discussed, but the origins of the Tao tribe can be broadly divided into:

1. Alpine origin: The only ancestors left in Orchid Island after 9 years of overseas flooding.
2. Coastal origin: The ancestors came from overseas to live here on a wooden box ship.
3. Overseas origin: Women from overseas floating in wooden crates united with local men to reproduce.

These myths have been passed down through the tribes, but the common theme is that the Dawu people come from two systems, "stone" and "bamboo", and that men and women from both systems married each other to produce a healthy Dawu tribe.

The production work of the Dawu people is divided on the basis of gender. The men's production activities are mainly fishing for fish in the sea, cultivating taro fields in the fields, and cutting down trees in the forests to build houses and board boats. Women are responsible for taro fields, housework, and weaving.

The Dawu people broadly divide their food into two major categories, namely the main food kanen and the side food yakan, the former being simply called "rice" by the Dawu people and the latter "vegetables"; usually, the main daily task of the women is to "take rice" mangerp so kanen, while the men "find vegetables" manersavat.

1.4.3 Agriculture

The crops cultivated are yams, taro, sweet potatoes, chestnuts, ramie, and rattan; fruits and vegetables include watermelon, betel nut, bananas, coconuts, and longan. Water taro is the main agricultural product and is a staple of daily life. Water taro (Fig. 1.3) is also an important ritual item in ceremonial events, called ceremonial taro. Millet karayi is usually given as a gift to friends and relatives during ceremonial occasions.

Fig. 1.3 Water taro fields near the seafront and villages. (Photo by Wen-San Huang)

1.4.4 Fishery

Traditionally, a man's social status is determined by his skill in fishing and hunting, and the value of fishing and hunting activities exceeds the value of agricultural production. There are two types of fishing in the sea: fishing for flying fish and fishing for deep-water fish. For example, in the first month (paneneb), fishing is done by night lighting, and only one net is used to catch flying fish, and no other boat group is allowed to fish in this month. In the second month (pikokaod), fishing is done in the same way as in the previous month. The same as the previous month, but members of other groups can participate in their own group fishing or shore fishing. In the third and fourth months, each crew can go to another crew to fish by preparing a torch and net (vanaka), which is called "night torch lure flying fish fishing method."

Deepwater fishing, in contrast, does not follow a fixed method. It may occur during the day and then continue into the night with additional fishing or the catching of seabirds. This combined approach enables fishermen to harvest a wide variety of marine life.

1.4.5 Livestock

The main livestock raised by the Dawu people are chickens, pigs, and goats. Chickens and pigs are raised in a free-range manner, and the people only raise chickens for sacrifice and rarely eat eggs or kill them. Although goats are the property mark of the clan, like pigs, they are regarded as revered animals and are often grazed with the mark of ownership cut out of their ears.

The special culture of Orchid Island also includes the attitude toward wild animals. For example, when we rented a house in the village, the owner warned us not to bring the sand from the beach into the house. The reason for this is that the green sea turtles are considered by the Orchid Islanders to be an unlucky thing, and the sand they crawl on is a taboo to the Lanyu (Mandarin Pronunciation; Orchid Island) Islanders. When the children in Orchid Island see the lizards we collect, they call them Malagou, which means thief, so they don't catch wild animals at all. This also means that the wildlife in Orchid Island is not bothered by the locals. As for the attitude toward plants, they also try to maintain the integrity of the forest, and only cut down a very small number of large trees for use as a plank boat. Therefore, the natural ecological environment of Orchid Island has basically remained natural for hundreds of years, which is an excellent field laboratory for scholars who study reptiles.

1.4.6 Geological Age of Orchid Island

Taiwan is located at the intersection of plates, so the earth's crust is quite unstable. When magma erupts from cracks in the ground, volcanic groups such as Datun and Keelung (mainland Taiwan) are formed on land. They are all concrete evidence of volcanic and crustal movements under the sea in Taiwan. However, although these islands are also formed by volcanic lava, their formation dates and methods are

Fig. 1.4 Beautiful mountain terrain sculpted by millions of years of waves. (Photo by Wen-San Huang)

different. The main component of volcanic islands formed by explosive eruptions is andesite, a rock found mainly in volcanic island arcs in the Pacific Rim, and the lava is mixed with volcanic clastic rock to form a raised island on the surface. The area of this island is usually small, and the overall shape is conical; the terrain on the island is uneven, not as flat as Penghu. It is mainly formed by recent magmatic activities within 2 million years. The volcanic activities in the eastern coast of Taiwan, Orchid Island, and the Miocene-Pliocene of Green Island have created excellent geological landscapes, such as the Lanyu Dragonhead, Red-Head, Tank, and other volcanic agglomerates or volcanic tuffs (Chen et al., 1994).

The sculpting of the mountains and the sea over the past 2 million years has created many beautiful landscapes on the island (Fig. 1.4).

1.4.7 Flora and Fauna of Orchid Island

According to the survey report collection, there are about 1049 species of vascular plants, 21 species of Mammalia, 489 species of Aves, 30 species of Reptilia, 3 species of Amphibia, and 884 species of Fish (5 species of Chondrichthyes and 779 species of Actinopterygii) in Orchid Island (Lee, 2010). Among the 30 species of reptiles in Orchid Island, 19 species of reptiles are more common, including 9 species of lizards: long-tailed lizard(*Eutropis longicaudata*), Dory's lizard (*E. multicarinata*), caly fly lizard (*Sphenomorphus incognitus*), coastal island lizard (*Emoia atrocostata*), Swinhoe's tree lizard (*Diploderma swinhonis*), lead gecko (*Gekko hokouensis*), Lanyu gecko (*Gekko kikuchii*), common gecko (*Hemidactylus*

Fig. 1.5 The 27 species of reptile on Orchid Island. (**a**) *Eutropis longicaudata*, (**b**) *E. multicarinata*, (**c**) *Sphenomorphus incognitus*, (**d**) *Emioa atrocostata*, (**e**) *Diploderma swinhonis*, (**f**) *Gekko hokouensis*, (**g**) *Gekko kikuchii*, (**h**) *Hemidactylus frenatus*, (**i**) *Takydromus sauteri*), (**j**) *E. cumingi*, (**k**) *Lepidodactylus yami*, (**l**) *L. lugubris*, (**m**) *Hemiphyllodactylus typus*, (**n**) *Oligodon formosanus*, (**o**) *Elaphe carinata*, (**p**) *Psammodynastes pulverulentus*, (**q**) *Viridovipera stejnegeri stejnegeri*, (**r**) *Ramphotyphlops braminus*, (**s**) *Amphiesma stolata*, (**t**) *Calamaria pavimentata*, (**u**) *Chelonia mydas*, (**u**) *Eretmochelys imbricata*, (**w**) *Lepidochelys olivacea*, (**x**) *Laticauda colubrina*, (**y**) *L. laticaudata*, (**z**) *L. semifasciata*, and (**Γ**) *Emydocephalus ijimae*. (S: Photo by Wen-San Huang; V and W photo courtesy of Hung-Ming Chen; the other pictures photo courtesy of Chung-Wei Yu)

Fig. 1.5 (continued)

Fig. 1.5 (continued)

frenatus), and Sauter grass lizard(*Takydromus sauteri*) (Fig. 1.5a–i). The four less common lizards include Cu′s skink (*E. cumingi*), Yami scaled toe gecko (*Lepidodactylus yami*), scaled toe gecko (*L. lugubris*), and the half-leaf toed gecko (*Hemiphyllodactylus typus*) (Fig. 1.5j–m). The seven species of terrestrial snakes include kukri snake (*Oligodon formosanus*), the stinky snake (*Elaphe carinata*), Mock viper (*Psammodynastes pulverulentus*), Green bamboo viper (*Viridovipera stejnegeri stejnegeri*), blind snake (*Ramphotyphlops braminus*), flowering snake (*Amphiesma stolata*), and wire snake (*Calamaria pavimentata*) (Fig. 1.5n–t). The three sea turtles include the green sea turtle (*Chelonia mydas*), hawksbill turtle (*Eretmochelys imbricata*), and (Olive ridley sea turtle) (Fig. 1.5u–w). The four sea snakes are yellow-lipped green-spotted sea snake(*Laticauda colubrina*), black-lipped green-spotted sea snake (*L. laticaudata*), broad-tailed green-spotted sea snake(*L. semifasciata*), and Iijima's sea snake(*Emydocephalus ijimae*) (Fig. 1.5x–Γ). There are three less common species, including sea snakes such as Ripley's sea snake(*Hydrophis ornatus*), black-headed sea snake (*H. melanocephalus*), and black-backed sea snake(*Pelamis platura*). It is likely that these three species of sea snakes were brought to Orchid Island by ocean-going fishing vessels rather than their place of origin, so I have not provided photos here. Because all animals are considered ominous and cursed by the Dawu people, locals do not harass or capture wild animals. As a result, the reptile diversity of the area has been preserved intact, making it an important island for ecology and evolution studies.

This book is a synthesis of my research on common reptiles in Orchid Island from 1997 to the present, with the hope of understanding the operation and future of reptiles in Orchid Island through basic ecology, ecological behavior, evolution and climate change. The next chapter will explore where the reptiles of Orchid Island came from and how they crossed the harsh ocean environment to reach this small island.

References

Chen, C. H., Liu, T. K., Yang, T. Y., & Chen, Y. G. (1994). *Explanatory text of the geologic map of Taiwan, scale 1:50000, sheet 71, Lanyu*. National Taiwan University.

Hong, M. L. (1978). The history of the investigation of the Lanyu Island by the officials of the Taitung office in the twenty-third year of the Guangxu period. *Taiwan Literature, 29*, 1–15.

Lee, C. T. (2010). *Survey and evaluation of natural ecosystems in Orchid Island*. Marine National Park Headquarters Commissioning Program.

When and How Animals Are Dispersed to Orchid Island?

2

Abstract

Orchid Island surfaced about 2 million years ago (Chap. 1), which means that all creatures reached Orchid Island within this time, including all reptiles. Since the reptiles of Orchid Island came from Taiwan and the Philippines, when and how did they actually come? This chapter focuses on these two questions, mainly on molecular DNA studies and physiological tolerances. In addition, this chapter does not study how the Philippine lizards crossed the ocean to Orchid Island, but will use the example of weevils in both places to explain the mechanism of crossing the ocean.

Keywords

Dispersal · Phylogeography · Genetic divergence · Biogeography · Kuroshio current · Salinity tolerance · Lizard · Weevil

While Taiwan is home to the majority of the 30 species of reptiles on Orchid Island, the terrestrial *Eutropis multicarinata, E. cumingi, Lepidodactylus yami,* and *Gekko kikuchii* come from the Philippine Islands. The terrestrial species of *Gekko hokouensis, Hemidactylus frenatus, Diploderma swinhonis, E. longicaudata, Sphenomorphus incognitus,* and *Takydromus sauteri, Ramphotyphlops braminus, Viridovipera stejnegeri, Psammodynastes pulverulentus, Elaphe carinata,* and *Oligodon formosanus, Hemiphyllodactylus typus, L. lugubris, Amphiesma stolata*, and *Calamaria pavimentata* all come from the mainland Taiwan. The pelagic reptiles include the green sea turtle (*Chelonia mydas*), hawksbill turtle (*Eretmochelys imbricata*), loggerhead sea turtle (*Lepidochelys olivacea*), yellow-lipped green-spotted sea snake (*Laticauda colubrina*), black-lipped green-spotted sea snake (*L. laticaudata*), and *L. semifasciata*, *Emydocephalus ijimae, Hydrophis ornatus, H. melanocephalus, Pelamis platura*, and *Emioa atrocostata*. The Lanyu

W.-S. Huang, *Reptiles in the Changing Climate*, Fascinating Life Sciences,
https://doi.org/10.1007/978-981-92-1887-5_2

gecko (*Gekko kikuchii*) was previously considered to be endemic to Orchid Island, but recent studies have shown that it is also distributed in the Philippine Islands, except that the first-named type is located on Orchid Island (Brown et al., 2013), so the Lanyu gecko is also thought to come from the Philippines.

In the Analects of Confucius, the phrase "It is a pleasure to have friends who come from afar" is used. It is generally interpreted as: Isn't it pleasant to have friends who share the same interests and come from afar? It is often used to express welcome to friends who come from far away.

But at the animal level, it is not so simple. There may be a lot of habitats at first for different species to occupy, but when the habitats are saturated, later animals will compete for habitat with the same or similar species occupying the habitat. However, it is not easy for animals to reach Orchid Island, because even during the Great Ice Age, mainland Taiwan was never connected to Orchid Island, because the water around Orchid Island was nearly 1000 meters deep, so animals had to be able to pass the oceanic test to reach Orchid Island.

However, typhoons may accelerate the spread of Taiwanese animals to Orchid Island. Although most typhoons blow from the east or southeast toward Orchid Island to mainland Taiwan, there are also many typhoons formed in the South China Sea that blow from Taiwan to Orchid Island due to certain factors. For example, the Typhoon Tempin in 2012. If there are animals on a dead tree or leaf, the typhoon may speed up the landing on Orchid Island. Tempin was the typhoon with the most unusual movement path that year. Typhoon No. 14 was formed at 0000 UTC on August 19 in the eastern part of the Philippines, and first moved northward and then westward to attack Taiwan. It made its first landfall in Pingtung, Taiwan and entered the South China Sea in the early morning of August 24. During the period in the South China Sea, the tropical storm Bolaven No. 15 had a Fujiwhara effect.

The second time at 2:00 on the 28th passed through the sea off the Taiwan Strait, and finally accelerated northward along the eastern sea of Taiwan. The Central Weather Bureau issued a land-based typhoon warning for the second time.

The total duration of the warning was up to 1 week, and the total number of warnings issued reached 54, which is the second-longest typhoon since 2000. They both passed through the sea between Orchid Island and Green Island.

2.1 Where and When Did It Come from? An Example of the Swinhoe's Tree Lizard from Taiwan to Orchid Island

Due to their geographical isolation and relative youth, islands are regarded as natural laboratories for evolutionary studies, where the historical processes of species diversity can be seen (Emerson, 2002; Losos & Ricklefs, 2009). According to several authors (Bermingham & Moritz, 1998; Emerson, 2002; Avise, 2009; Borges & Hortal, 2009; Puppo et al., 2016; Gabrielli et al., 2020), the phylogeography of the island enables biologists to comprehend how previous geographic history has shaped diversity, leading to evolutionary differentiation within species lineages. Intraspecific phylogeographic analyses are a crucial tool for investigating occult historical processes of population diversification in order to further seek the level of genetic diversity and connectivity of a lineage's intra-island distribution (Bermingham & Moritz, 1998; Emerson, 2002). Such research has revealed patterns of population growth, gene flow, and diverse origins by reconstructing intraspecific isolated populations with various historical groups and diversity (Holland & Hadfield, 2002; Dufresnes et al., 2020; Spadavecchia et al., 2021; Withers et al., 2021).

Because of its distinctive geological history, the island of Taiwan (hereinafter referred to as Taiwan) is a particularly interesting location for phylogeographic studies of terrestrial species. With a land area of 35,900 km^2, this large island is a continental island located 130 km from the eastern edge of the Eurasian continent (Huang et al., 2014; Ali, 2018). During the glacial period, a land bridge connecting Taiwan and the mainland allowed the majority of terrestrial vertebrates from Taiwan to colonize this area (Li et al., 2006; Jing et al., 2007; Huang et al., 2014; Qu et al., 2015). Recent research, however, indicates that some species colonized the area before land bridges were built (e.g., Shinohara et al.; 2014; Lv et al., 2018), and that they also came from the Ryukyu and Luzon island arcs in addition to the mainland (Komaki et al., 2017; Yang et al., 2018) and the Luzon island arc (Shang, 2001; Esselstyn & Oliveros, 2010; Tseng et al., 2018). Taiwan is geographically complex because, in contrast to the type of continental islands connected to passive continental margins (Sibuet & Hsu, 2004; Ali, 2018; Huang et al., 2018), it arises from orogenic margins at the boundary of the radiating plate. Proto-Taiwan was created by the rapid oblique convergence of the Luzon arc on the Philippine Sea plate and the Asian continental margin in the late Miocene (Teng, 1990; Ramsey et al., 2007). Around 6.5 million years ago (mya), in the Late Miocene, Taiwan was gradually raised above sea level as a result of the collision between the Asian continent and the Philippine Sea plate (Ramsey et al., 2007; Ali, 2018; Huang et al., 2018). During 3–1 mya, Taiwan underwent successive orogenic movements that resulted in the formation of the Central Range, a north-south trending mountain range in Taiwan that is approximately 4 km above sea level (Ramsey et al., 2007, Ali, 2018, Huang et al., 2018). In addition, the collision triggered ongoing volcanism in the Northern Luzon Arc, close to Orchid Island and Green Island, two islands in the southeast of Taiwan (Yang et al., 1996; Huang et al., 2018). As a result, strains from various colonial pathways have adapted to complicated geographic processes and varying altitudes that are dissimilar to their native habitats (Ali, 2018; He et al., 2018).

According to several studies (Warren et al., 2003; Borges & Hortal, 2009; Parent, 2012; Gabrielli et al., 2020), the degree of habitat heterogeneity and isolation is what drives intra-island diversity. Taiwan's unusual geographic history and environment have caused species that are only found there to become isolated, diverge, and evolve to a high degree of diversity (He et al., 2018). The divergence within some strains in Taiwan has been the subject of numerous studies (e.g., You et al., 2015; Wu et al., 2016; Kurita et al., 2017). Taiwan has been the subject of numerous studies aimed at elucidating the genetic history of intra-island diversity. Collectively, these studies suggest that species retreat to more southerly locations or lower elevation refuges, causing periodic expansion during past glacial periods (Lin et al., 2002; Yeh et al., 2004; Cheng et al., 2005; Lee et al., 2006; Lin et al., 2012; Weng et al., 2016; Tsai et al., 2021), or north-south oriented central mountain ranges and west-east oriented rivers are the main topographic barriers causing divergence (Yeh et al., 2004; Shih et al., 2006; Lin et al., 2012; Tseng et al., 2015). The phylogeographic histories of populations at the population level have, however, received relatively little attention, and refuges and barriers differ across lineages.

To investigate the possibility of intra-island diversity, we looked at the population genetic history of the Swinhoe's tree lizard, *Diploderma swinhonis*. The genus *Diploderma* evolved in Taiwan with five endemic species originating from southern China (Ota, 1991; Wang et al., 2019). The distribution of *D. swinhonis* extends throughout Taiwan, Orchid Island, and Green Island (Ota, 1991; Huang, 2007). According to Yang et al. (2018), the divergence between *D. swinhonis* and its sister species occurred 10 mya ago. The intraspecific phylogeny of *D. swinhonis* remains unclear, although (Ota, 1997) inferred that the Central Range isolates the eastern and western populations. Based on unpublished studies and field observations, there is heterogeneity in morphology, ornamentation, and behavior among different populations of *D. swinhonis* (Shang, 2001). Therefore, understanding their genetic diversity becomes an interesting issue. Coupled with its mainland origin, it is worthwhile to examine the phylogeography and population genetics of this species in Taiwan to understand how it dispersed throughout the terrestrial and offshore islands after arriving in Taiwan.

The purpose of this study is to illustrate *D. swinhonis*, internal differentiation and genetic structure. and to use intraspecific phylogeographic analysis to retrace its evolutionary history. We concentrate on the following issues: (1) Is there a clear differentiation of the population? (2) How did the species disperse and diverge within Taiwan and the two islands (Fig. 2.1), and was its genetic structure related to geological events and geographic barriers, or did it periodically expand during past glacial periods? In response to the first query, we predict that *D. swinhonis* diverged through geographic isolation, based on phylogenetic reports of Taiwanese reptiles. The phylogeographic population history of *D. swinhonis* allows us to propose a hypothesis of intraspecific divergence within the Taiwan region.

2.1.1 Phylogenetic Tree

The resulting alignments were edited to create the fully aligned length of 1122 bp Cytb, 1185 bp 16S, 1236 bp BACH1, and 1310 bp RAG1. The mitochondrial sequences presented four main clades in Taiwan Island (Fig. 2.2). The ML tree and BI tree showed that *D. swinhonis* formed a monophyletic lineage (ML bootstrap values = 100; BI posterior probability = 1). The *Diploderma swinhonis* diverged into four clades with well-supported values (ML bootstrap values ≥70% and BI posterior probability ≥0.95), except for one weekly support value in the BI tree. Four clades represented four monophyletic groups: western, southern, southeastern, and eastern clades. The distribution of the western clade (W) covered from the north of Tainan to the north of Lanyang Creek in Yilan and the eastern clade (E) distributed

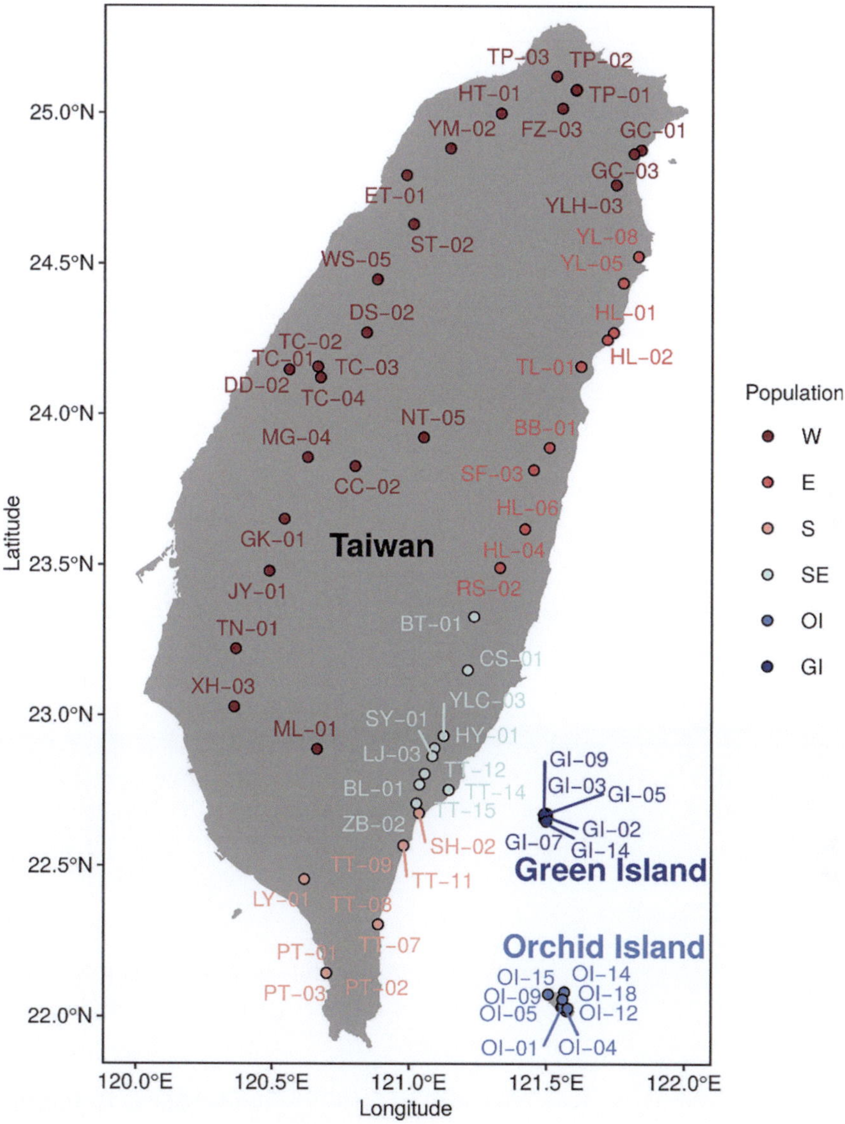

Fig. 2.1 Map of sampling localities on Taiwan Island, Green Island, and Orchid Island. Points correspond to sampling localities, colored according to populations. *W* western population, *SE* southeastern population, *E* eastern population, *S* southern population, *OI* group in Orchid Island, *GI* group in Green Island. (Redrawn from Chang (2022), unpublished master's thesis, National Chung Hsing University, Taichung, Taiwan)

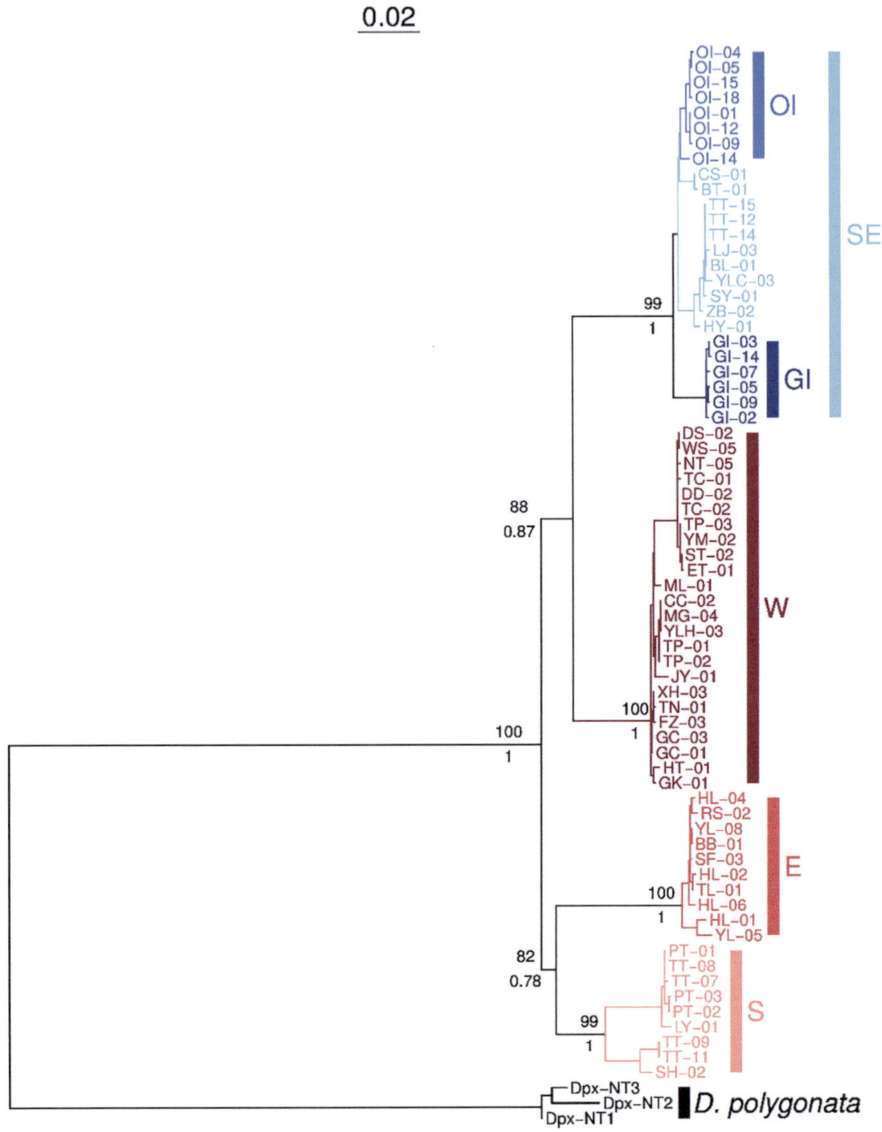

Fig. 2.2 Maximum likelihood phylogenetic tree for *D. swinhonis* based on the mitochondrial dataset (concatenated Cytb and 16S sequences). *D. polygonata* were used as outgroups from GeneBank and a collected sample. Bootstrap values from maximum likelihood analysis and posterior probabilities from Bayesian inference analysis were provided above and below the branches, respectively. *W* western population, *SE* southeastern population, *E* eastern population, *S* southern population, *OI* group in Orchid Island, *GI* group in Green Island. (Redrawn from Chang (2022), unpublished master's thesis, National Chung Hsing University, Taichung, Taiwan)

from the south of Lanyang Creek to Ruisui of Hualien. The southeastern clade (SE) and the southern clade (S) are bounded by Zhiben Creek in Taitung. Two insular islands, Orchid Island and Green Island, both were monophyletic groups belonging to the SE clade. The phylogenetic relationships of four clades indicated that the W clade was a sister group to the SE clade and the E clade was a sister group to the S clade, forming the western-southeastern (W-SE) and the eastern-southern clade (E-S), respectively. The weekly support value in the BI tree was on the divergence between S and E (BI posterior probability = 0.78). The two insular islands both formed the monophyletic groups belonging to the SE clade with well-supported values.

Compared to mitochondrial sequences, the concatenated sequences comprising 4 gene sequences, showed similar topology in addition to weak support value on the divergence between S and E in the BI tree. However, the phylogeny based on nuclear sequences demonstrated a contrary topology. All of the support values in the two analyses were poor support or were of no use (ML bootstrap values <70% and BI posterior probability support values <0.80). The phylogeny had a mixture of resolved nodes and polytomies. Thus, we determined that the mitochondrial ML tree was used to present our phylogeny recording with BI posterior probability (Fig. 2.2).

2.1.2 Population Structures

The summary statistics for the different K values are provided in Fig. 2.3. Cluster analysis suggested the four populations in our data ($K = 4$) and gave the population memberships for each individual that were consistent with the four clades in our phylogeny results. Among the four populations, we found that the haplotype diversity was different from nucleotide diversity, and the haplotype diversity varied from mitochondrial and nuclear datasets (Table 2.1). For the nucleotide diversity, the

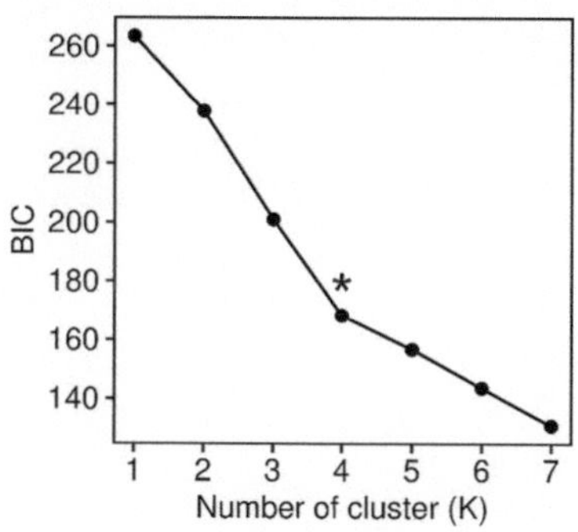

Fig. 2.3 The value of the BIC for the retained model (*) and the different values of K. The Bayesian Information Criterion (BIC) indicates the optimal number of clusters ($K = 4$) for clustering analysis. Optimal cluster is determined using Ward's clustering method (Jombart et al., 2010). (Redrawn from Chang (2022), unpublished master's thesis, National Chung Hsing University, Taichung, Taiwan)

Table 2.1 Measures of population genetic diversity for mitochondrial and nuclear sequences from populations. The sample sizes (n), variable sites (VA), number of haplotypes (H), haplotype diversity (h), nucleotide diversity (π), Tajima's D, and Fu's Fs are shown (SD = standard deviation). Panel A, grouping by gene; panel B, grouping by populations based on mitochondrial dataset; panel C, grouping by population based on nuclear dataset

	Gene/group	*n*	VA	$\pi \pm$ SD	H	h ± SD	Tajima's D	Fu's Fs
(A)	Cytb	68	243	0.0601 ± 0.0019	48	0.977 ± 0.004	1.424	15.561
	16S	69	121	0.0301 ± 0.0006	39	0.960 ± 0.007	1.292	8.168
	BACH1	65	50	0.0039 ± 0.0002	56	0.966 ± 0.006	−1.465	−34.434
	RAG1	68	86	0.0053 ± 0.0002	63	0.964 ± 0.007	−1.818	−33.479
(B)	W	24	54	0.0065 ± 0.0002	17	0.950 ± 0.011	0.760	3.057
	SE	25	93	0.0113 ± 0.0005	22	0.967 ± 0.000	0.782	3.875
	E	10	50	0.0057 ± 0.0012	10	0.932 ± 0.001	−0.287	2.538
	S	9	89	0.0160 ± 0.0026	7	0.889 ± 0.039	1.591	13.204
(C)	W	23	40	0.0032 ± 0.0001	39	0.989 ± 0.008	−0.383	−30.962
	SE	24	45	0.0037 ± 0.0002	34	0.957 ± 0.020	−0.270	−15.860
	E	10	28	0.0025 ± 0.0003	12	0.879 ± 0.065	−0.746	−1.693
	S	9	57	0.0042 ± 0.0005	17	0.993 ± 0.021	−1.512	−7.087

W western population, *SE* southeastern population, *E* eastern population relatively low diversities were found in all nuclear genes, compared to mitochondrial genes. The highest nucleotide diversity was found in the southern population and the eastern and western populations contained lower nucleotide diversity than the southeastern populations in the two datasets. While the highest haplotype diversity was also found in the southern population from the nuclear dataset, the mitochondrial dataset presented the lowest haplotype diversity. Haplotype diversity was higher in the western population despite having low nucleotide diversity, especially in the nuclear dataset. The eastern population contained low nucleotide and haplotype diversity, both in mitochondrial and nuclear datasets.

The p-distances between populations were calculated from the mitochondrial dataset, indicating the base differences per site from averaging over all sequence pairs between populations from 5.45% to 7.03% (Table 2.2). The F_{ST} values suggested the gene flows among populations were from 0.62 to 0.75 in the mitochondrial dataset and were from 0.09 to 0.43 in the nuclear dataset (Table 2.3). The high F_{ST} values meant four populations have been highly isolated in the mitochondrial dataset. Different from the mitochondrial dataset, the relatively low F_{ST} values demonstrated gene sharing among populations in the nuclear dataset. Notably, the southern and southeastern populations had high levels of gene flow ($F_{ST} = 0.09$) compared to other populations in the nuclear dataset.

The mitochondrial haplotype networks showed that the phylogenetic relationships in the Cytb gene (Fig. 2.4) were somewhat different from 16S rRNA (Fig. 2.5), and were different from the nuclear haplotypes (Figs. 2.6 and 2.7). No haplotype was shared among locations of the populations for the mitochondrial dataset. The phylogenetic relationships of Cytb were highly similar to our phylogeny for the mitochondrial dataset, whereas the 16S haplotypes presented the difference in that the western population was closely related to the southern population. We also identified the haplotype locations of two insular islands from the SE population. Both Orchid Island and Green Island had not shared mitochondrial haplotypes

Table 2.2 Pairwise p-distances among populations based on mitochondrial dataset. The number of base differences per site from averaging over all sequence pairs between groups is shown. Standard error estimates are shown above the diagonal

	W	E	SE	S
W	–	0.0047	0.0044	0.0045
E	0.0645	–	0.0050	0.0045
SE	0.0545	0.0703	–	0.0045
S	0.0581	0.0585	0.0624	–

W western population, *SE* southeastern population, *E* eastern population

Table 2.3 Population pairwise F_{ST} for mitochondrial dataset. Mitochondrial and nuclear F_{ST} are shown below and above the diagonal, respectively

	W	SE	E	S
W	0	0.37158	0.47363	0.24282
SE	0.67251	0	0.31559	0.09543
E	0.75039	0.73691	0	0.30525
S	0.62459	0.62085	0.67715	0

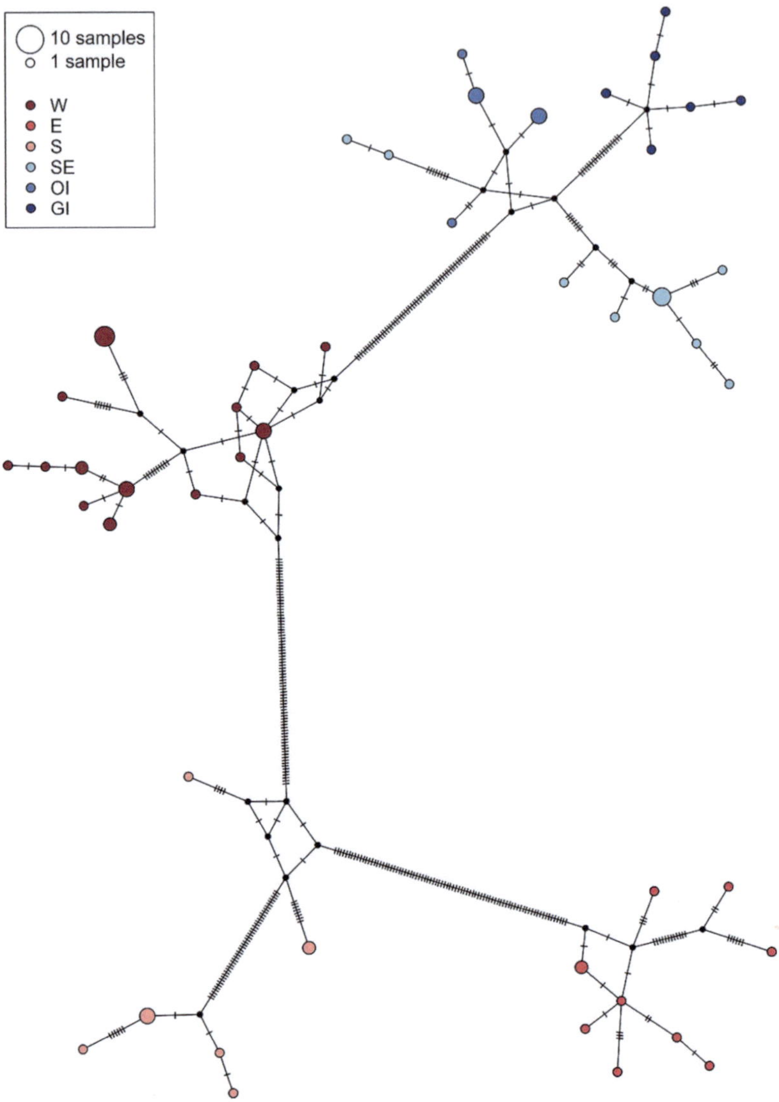

Fig. 2.4 Haplotype network of *D. swinhonis* using a mitochondrial gene, Cytb-phased sequences, including 68 individuals and 136 sequences. Black dots and lines indicate haplotypes of absent ancestors and nucleotide substitutions. Sizes and colors correspond to the number of sequences and population localities. *W* western population, *SE* southeastern population, *E* eastern population, *S* southern population, *OI* group in Orchid Island, *GI* group in Green Island. (Redrawn from Chang (2022), unpublished master's thesis, National Chung Hsing University, Taichung, Taiwan)

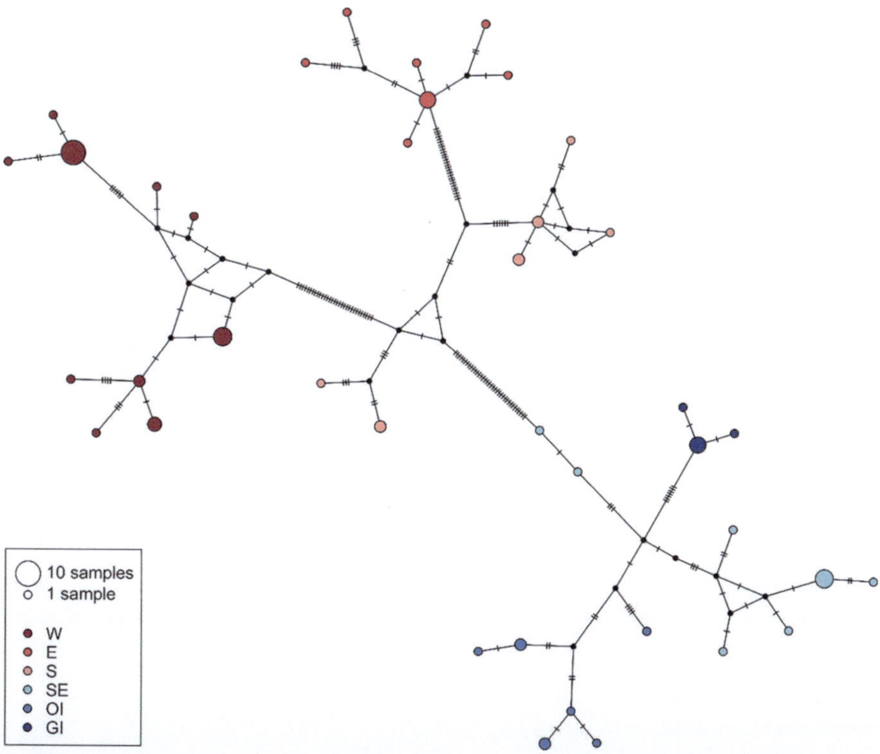

Fig. 2.5 Haplotype network of *D. swinhonis* using a mitochondrial gene, 16S-phased sequences, including 69 individuals and 138 sequences. Black dots and lines indicate haplotypes of absent ancestors and nucleotide substitutions. Sizes and colors correspond to the number of sequences and population localities, respectively. *W* western population, *SE* southeastern population, *E* eastern population, *S* southern population, *OI* group in Orchid Island, *GI*: group in Green Island. (Redrawn from Chang (2022), unpublished master's thesis, National Chung Hsing University, Taichung, Taiwan)

among locations. Compared to mitochondrial networks, the nuclear haplotype relationships among populations were relatively unclear. For the nuclear dataset, there were haplotypes shared among locations, including insular islands, for both BACH1 and RAG1.

2.1.3 Divergence Times

Based on the mitochondrial dataset, the *D. swinhonis* phylogeny of BEAST v2.6.6 was used in the same BI and ML trees to estimate the timing of recent common ancestry and population divergence, including mean nodal age and 95% maximum posterior density (Table 2.4 and Fig. 2.8). The first lineage divergence is thought to have occurred around 5.41 mya, when the W-SE clade split off from the E-S clade

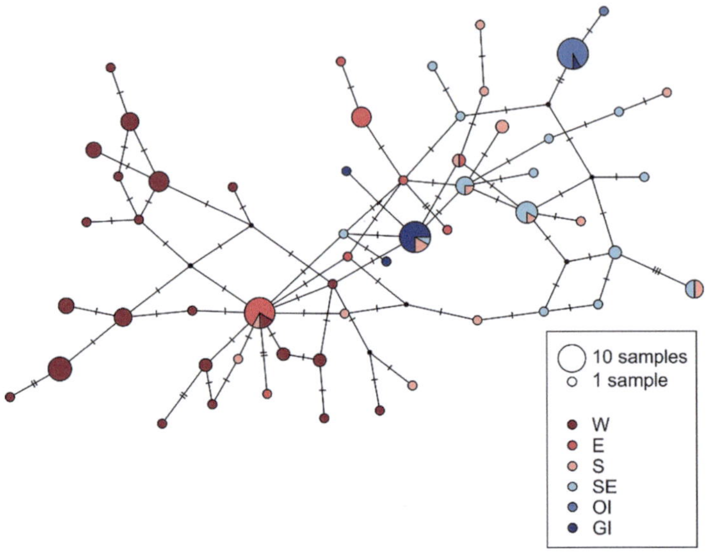

Fig. 2.6 Haplotype network of *D. swinhonis* using a nuclear gene, BACH1-phased sequences, including 66 individuals and 132 sequences. Black dots and lines indicate haplotypes of absent ancestors and nucleotide substitutions. The number of sharing haplotypes present by pie charts. Sizes and colors correspond to the number of sequences and population localities. *W* western population, *SE* southeastern population, *E* eastern population, *S* southern population, *OI* group in Orchid Island, *GI* group in Green Island. (Redrawn from Chang (2022), unpublished master's thesis, National Chung Hsing University, Taichung, Taiwan)

during the early Pliocene, swinhonis. It is believed that the divergence between the eastern and southern populations and the divergence between the western and southeastern populations (4.05 mya) are comparable (4.24 mya). The estimated ages of the four populations' most recent common ancestors, with the southern, southeastern, eastern, and western populations listed in chronological order from the most ancient to the most recent, ranged between 2.6 and 1.62 mya. The Green Island date departs first from the Southeast population for both islands (2.38 mya). The OI group's common ancestor was around 1.19 mya old during the late Pleistocene.

2.1.4 Models of Past Phylogeographic History

The genetic results of *D. swinhonis* suggest that the mitochondrial dataset differs historically between the western, eastern, southeastern, southern, and two islands. In contrast, the nuclear dataset suggests that the four populations share nuclear haplotypes (Figs. 2.6 and 2.7) and relatively high levels of gene flow among them (Table 2.3). According to the afore mentioned data, barriers to mitochondrial gene flow have emerged and nuclear gene flow patterns between populations have changed. So, in order to determine the scenario that best accounts for the obstacles to gene flow variation in historical processes within species, we tested a variety of

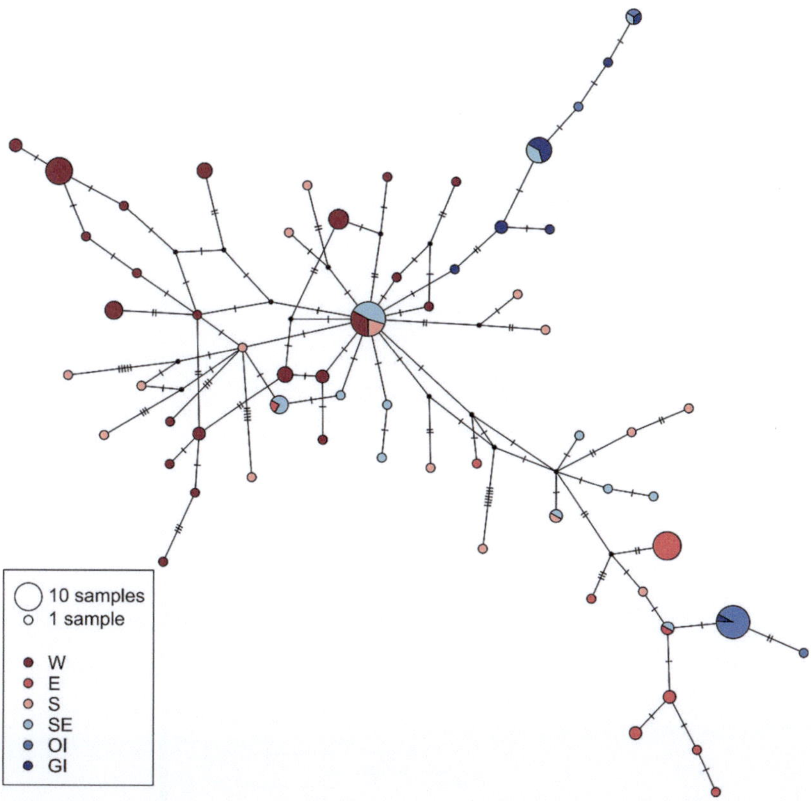

Fig. 2.7 Haplotype network of *D. swinhonis* using a nuclear gene, RAG1-phased sequences, including 69 individuals and 138 sequences. Black dots and lines indicate haplotypes of absent ancestors and nucleotide substitutions. The number of sharing haplotypes present by pie charts. Sizes and colors correspond to the number of sequences and population localities. *W* western population, *SE* southeastern population, *E* eastern population, *S* southern population, *OI* group in Orchid Island, *GI* group in Green Island. (Redrawn from Chang (2022), unpublished master's thesis, National Chung Hsing University, Taichung, Taiwan)

Table 2.4 Mean node age estimates (mya) and 95% highest posterior densities (HPD). The estimated date (mya) and inferred using BEAST based on Bayesian inference analysis

MRCA of clade	Node ages	95% HPD
All *D. swinhonis*	5.41	3.46, 7.95
W-SE clade	4.05	2.42, 6.06
S-E clade	4.24	2.45, 6.52
Western population	1.62	0.82, 2.75
Southeastern population	2.38	1.35, 3.66
Eastern population	1.85	0.83, 3.24
Southern population	2.6	1.32, 4.30

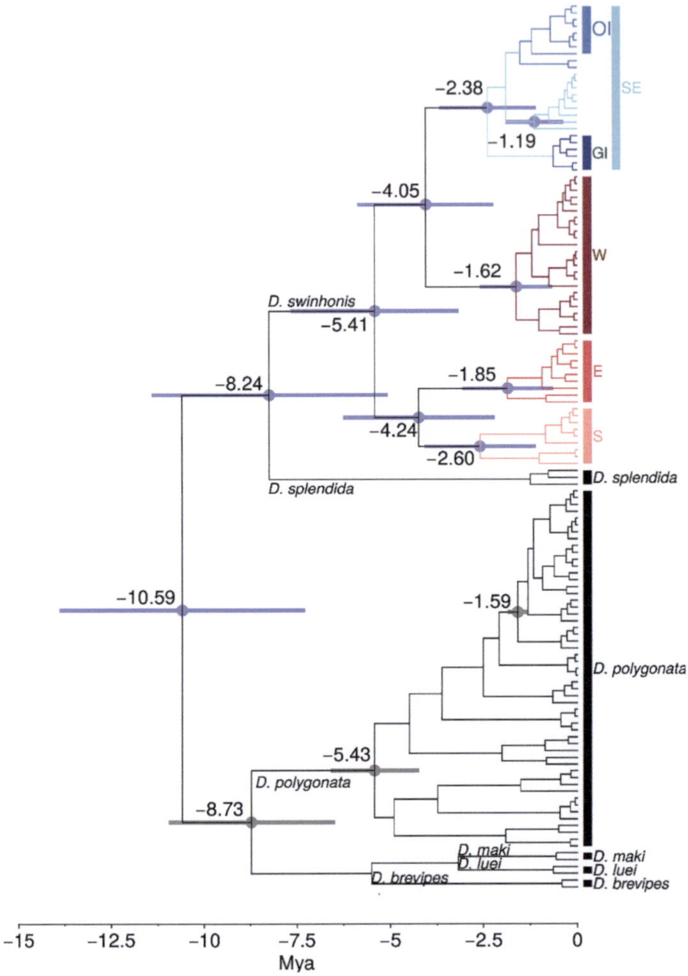

Fig. 2.8 Estimated time of the divergence among mitochondrial populations and additional *Diploderma* species for outgroups. Numbers indicate the expected times of divergences. Gray error bars indicate the HPDs of times of divergences for calibration nodes. Blue error bars indicate the HDPs of estimated times of divergences. Colors correspond to population localities. *W* western population, *SE* southeastern population, *E* eastern population, *S* southern population, *OI* group in Orchid Island, *GI* group in Green Island. (Redrawn from Chang (2022), unpublished master's thesis, National Chung Hsing University, Taichung, Taiwan)

models that included biogeographic processes as well as the potential for past diffusion. However, genetic variability appears to be relatively low and contains polymorphism in phylogeny, according to our findings from nuclear genes. We used the mitochondrial dataset to infer the time of differentiation and the subsequent tests rather than the combined dataset.

In order to compare the biogeographic processes and the likelihood of diffusion over time, the best model with Akaike Information Criterion (AIC) supported by the mitochondrial dataset was proposed using BioGeoBEARS phylogenetic statistics (Table 2.5) show that the likelihoods assigned to the data by both models are the same. Three similar hypothetical scenarios with relatively low AICs were found for scenarios simulating the probability of diffusion in various time periods (H1, H2, and H3 in Table 2.5). These scenarios point to two distinct periods in the history of phylogeography and indicate that the likelihood of diffusion has changed. First, a pattern between sites (H1 and H2) at 6.5–5 mya is proposed, in which there are no barriers between them or barriers have emerged in the central mountain range (H10). Second, it is proposed that barriers isolated from all sites appear at 5-0 mya (H1 and H10) or 4-0 mya (H2).

2.1.5 Discussion

Based on mitochondrial phylogeny, cluster analysis, and network reconstruction of the population structure, we identified four populations in Taiwan. The phylogenetic relationships and distributions of the four populations within Taiwan display unheard-of patterns in our mitochondrial results. To determine the geographic events occurring at the time of population divergence, we tested various diffusion probability models using historical processes. We tested different diffusion probability models with historical processes to infer the geographic events occurring at the time of population divergence. The best models identified by the AIC indicated that three similar models were most appropriate for our mitochondrial data (Table 2.5). Based on these three models and the timing of differentiation, we reconstructed the phylogeographic history of *D. swinhonis* as follows. The first divergence between the W-SE clade and the S-E clade occurred at 5.41 mya according to the divergence time of the mitochondrial sequences (Fig. 2.8). These models present gene flow between all populations during 6.5–5 mya or between populations adjacent to the original Central Range (Table 2.5). Rarely, there any obvious geographical barriers between the two clades. For example, the shortest distance between the sampling sites of these two clades of eastern and southeastern populations is only about 3.7 km, separated by a stream with a maximum width of 200 m. In addition, other boundaries between these two clades are not clearly visible. In addition, the other boundaries of these two clades lack geographic barriers. Therefore, we believe that the first divergence event is not associated with a geologic event. Despite the lack of evidence for this divergence event in our study, we propose the following hypotheses. First, the strain of *D. swinhonis* arrived in the original Taiwan area and diverged into two clades due to adaptation to different environmental features, without geographic separation. Possible factors such as habitat preference or changes in reproductive behavior, resulted in reproductive barriers. Second, *D. swinhonis* diverged into two groups before arriving in Taiwan. Although studies have generally reported that the island of Taiwan had a history of 6.5 million years after its appearance above sea level, few geologists believe that Taiwan may have

Table 2.5 Akaike information criterion (AIC) from BioGeoBEARS model. The AIC was used to compare different possible models and determine which model is the best fit for our mitochondrial data. The AIC of the three optimal models (*) is the smallest, and the difference between the AICs of the three models is not more than 3

	Relative likelihood (AIC) based on biogeographic model					
Models of past history	DEC	DEC + J	DIVALIKE*	DIVALIKE+J*	BAYAREALIKE	BAYAREALIKE+J
H0	72.91	73.53	64.26	66.25	89.33	83.19
*H1	56.29	58.27	48.21	50.21	81.99	66.47
*H2	56.88	55.31	50.9	51.04	78.97	56.25
H3	61.06	57.22	54.84	53.04	79.05	57.67
H4	64.21	59.52	57.61	55.82	78.94	60.4
H5	60.74	61.45	48.29	50.41	80.55	70.57
H6	76.76	78.05	63.3	65.27	88.96	85.3
H7	66.04	61.92	58.34	58.17	80.92	65.93
H8	62.41	62.48	56.65	56.54	76.4	64.66
H9	59.92	60.02	54.73	55.51	76.43	63.16
*H10	60.32	59.98	51.63	52.78	73.51	66.11
H11	73.72	75.16	63.13	65.11	88.83	85.27
H12	65.06	62.28	57.67	57.72	80.4	65.98
H13	61.74	60.53	55.58	55.8	80.87	64.92
H14	57.99	58.31	52.92	54.19	75.88	63.59
H15	71.53	71.75	62.77	64.74	88.47	80.33
H16	63.32	56.99	56.46	54.93	79.03	59.58
H17	60.41	56.88	53.97	52.56	76.02	57.29
H18	71.15	71.64	62.82	64.79	87.48	80.34
H19	63.57	58.92	56.95	55.22	79.38	59.83
H20	71.15	71.49	62.97	64.92	87.4	79.97

appeared only about 3 mya. Therefore, we cannot completely rule out the possibility that divergence occurred prior to the arrival of Taiwan. The two clades may have evolved different environmental preferences and occupied different localities before spreading to Taiwan. Finally, because genetic barriers evolved first, the boundary between the two clades did not exist as a geographical barrier and showed an unprecedented pattern in modern Taiwan.

Secondary divergence occurred at about 4 mya, forming four populations (Fig. 2.8). The S-E clade diverged into southern and eastern populations at 4.24 mya, and then the W-SE clade diverged into western and southeastern populations at 4.05 mya (Fig. 2.8). Our hypothetical model suggests that genetic or geographic barriers between these four populations existed after 5–4 mya. Here, we first discuss the divergence between the western and southeastern populations. We suggest that the divergence between western and southeastern populations is isolated from the central mountain range. Successive mountain-building movements have caused the mountain belts in Taiwan to rise. This geographic isolation has led to species differentiation that is not adapted to higher altitudes (e.g. Shih et al., 2006; Tseng et al., 2015; Yeh et al., 2004), e.g., studies of *Plestiodon chinensis* in China have shown differentiation between western and eastern Taiwan as different subspecies (Kurita et al., 2017). Based on the rate of rise of the Central Range (0.65–1.28 mm/year) (Huang et al., 2018), the height of the mountain belt at 5 mya (650–1280 m) is higher than the distribution height of *D. swinhonis* (Shang et al., 2009). Therefore, the north–south trending central mountain range should have served as a barrier between the western and eastern strains during the early historical stage in Taiwan.

For the divergence between the southern and eastern populations, we proposed two hypotheses. The first is the invasive expansion of the southeastern populations. BioGeoBEARS analysis shows that our data best fit the biogeographic model, and we can infer no long-distance dispersal variation events (Table 2.5). The current distribution of the populations indicated that the S-E clade occupied the eastern and southern parts of Taiwan. Due to the distribution of the southeastern population, there is no contact between the eastern and southern populations. In addition, the S-E clade diverged into two populations at the same time as the estimated divergence date of the W-SE clade. We hypothesize that the distribution of the S-E clade in two different locations caused contact difficulties between the two populations, ultimately leading to isolation and divergence. Second, divergence arose from the specific environmental preferences of the two major clades. Similar to the first hypothesis of clade divergence, the S-E clade prefers specific environmental features only in these two locations, and the S-E clade spreads from its ancestral range to the southern and eastern parts of Taiwan. This divergence event can be investigated in the future, i.e., whether the current distribution is due to adaptation of the two major clades to specific environmental preferences, and further research is needed to investigate sufficient evidence of trait and habitat variation between the S-E and W-SE clades.

Two island states, Orchid Island and Green Island, have also contributed to independent evolutionary lineages and increased diversity (Shang, 2001; Esselstyn & Oliveros, 2010; Tseng et al., 2018). Although genetic diversity trends were not significantly higher in the southeastern populations than in other populations (Table 2.1), mitochondrial evolutionary trees showed that mitochondria on Green Island and OI differentiated into monophyletic groups at 2.38 mya and 1.19 mya, respectively (Fig. 2.8). The order of origin of the monophyletic groups on these two islands differs from the order of island occurrence. Typically, the initial colonization events occurred on the old volcanic islands that first appeared above sea level (Cowie & Holland, 2006). Orchid Island appeared earlier than Green Island (Yang et al., 1996), but our results show that the order of dispersal and differentiation to the islands was opposite to the order of island occurrence. Since there was no land bridge connecting these two islands to Taiwan during the past glacial period, and their divergence predates the past glacial period, the explanation of colonization by land bridge is not supported. The explanation we offer here is that the spreading events took place according to the sequence of their proximity to Taiwan. During the formation of these two islands, the Luzon arc to which they belonged has been approaching Taiwan in a northwesterly direction since the collision of the Philippine Sea plate s (Yang et al., 1996; Sibuet & Hsu, 2004). We suggest that cross-sea dispersal is related to relative distances, so that the southern population first dispersed to Green Island, which is relatively close to Taiwan (Fig. 2.8), before dispersing a second time to Orchid Island, and thus the southeastern population dispersed twice across the sea to both islands as the Luzon arc approached. We also revealed a discordance between the mitochondrial and nuclear sequences of *D. swinhonis*. Mitochondrial-nuclear discordance can be defined as a distinct pattern of apparent divergence between these two genetic marker types (Toews & Brelsford, 2012). As we mentioned above, the geographical distance between the southern and southeastern populations is very short. The mitochondrial FST between these two populations (0.73) is greater than the nuclear FST (0.09) (Table 2.3). The interpopulation FSTs indicate a considerable level of gene flow between these two populations with respect to nuclear genes, even though the mitochondrial results of these two populations are already deeply divergent in 5.41 mya (Fig. 2.8). In addition to the relative levels of gene flow shown by the FST, this difference can also be seen in the haplotype network. We have a high degree of discrepancy between the results of the two datasets (Figs. 2.4, 2.5, 2.6, and 2.7). The mitochondrial network shows four differentiated populations and two independent branches of the island (Figs. 2.4 and 2.5). There were no common haplotypes in the two mitochondrial genes, 16S and Cytb (Figs. 2.4 and 2.5). In contrast, all four populations and islands share nuclear gene haplotypes (Figs. 2.6 and 2.7).

Discordance between mitochondrial and nuclear genes has been identified in previous biogeographic studies (Toews & Brelsford, 2012; Dai et al., 2013; Dufresnes et al., 2020; Streicher & Day, 2020). Studies have suggested several possible evolutionary factors for mitochondrial-nuclear discordance: there is the spread of gender bias (Peters et al., 2012; Dai et al., 2013; Phuong et al., 2017), asymmetric introgression, or gene flow (Phuong et al., 2017; Ivanov et al., 2018; Dufresnes et al., 2020; Semenov et al., 2021), incomplete lineage sorting (Firneno Jr. et al. 2020), nuclear mitochondrial pseudogenes (Song et al., 2008; Song et al., 2014), and male genes (Hedtke & Hillis, 2011). Our findings suggest that mitochondrial genes have a stronger phylogenetic structure, whereas nuclear genes have a higher rate of gene flow (Figs. 2.4, 2.5, 2.6, and 2.7), which represents asymmetric gene flow that occurred abruptly in the past or is now occurring gradually (Toews & Brelsford, 2012). Based on the above studies, we hypothesized the following possible hypotheses.

The first is male-biased diffusion (Peters et al., 2012; Toews & Brelsford, 2012; Dai et al., 2013), as male-biased gene flow is usually characterized by weak or nucleus-free genetic structures (Peters et al., 2012). Asymmetric gene flow is more likely to occur due to climatic instability (Dufresnes et al., 2020; Phuong et al., 2017). Thus, nuclear gene flow is likely to arise due to periodic glaciation, resulting in complex secondary contacts and current mitochondrial–nuclear discordance. This mechanism will depend on further behavioral studies to confirm. The second hypothesis is that mitochondrial genes show strong frequency differences in different parts of the gene flow (Toews & Brelsford, 2012).

The second hypothesis is that mito-nuclear discordance was driven by incomplete lineage sorting (ILS). ILS inferred from discordance among different gene trees from independent loci (Joly et al., 2009) usually resulted from variant effective population sizes and substitution rates between mitochondrial and nuclear genes (Joly et al., 2009, Bernardo et al., 2019, Firneno Jr. et al. 2020). Our nuclear sequences show the polymorphisms and nuclear haplotypes shared among populations (Figs. 2.6 and 2.7), which have been regarded as evidence of recent divergence because of the retention of ancestral polymorphisms (Peters et al., 2007). Consequently, there is the probability that nuclear DNA has not yet achieved monophyletic groups in *D. swinhonis*, and the ILS may contribute to discordance between.

mitochondrial and nuclear genes.

In conclusion, the phylogenetic history of *D. swinhonis* tree lizards reveals three main factors that contribute to intraspecific diversity. First, geographic barriers have led to genetic divergence. Since *D. swinhonis* arrived in Taiwan during the uplift above sea level in the original Taiwan area, its diversification process is related to the history of geomorphological changes in Taiwan. During the geographic event, the uplift of the central mountain range at 4–5 mya showed elevational differences that isolated the gene flow between the east and west lineages, and the marine isolation of the island also led to separate groups in the last two million years. Second, the complex geological history of Taiwan has led to environmental diversity. We hypothesize that environmental diversity may have contributed to the differentiation of lineages without geographical barriers due to specific local adaptations. Which characteristics of lineages are appropriate will need to be studied in the future. Third, we found a mitochondrial-nuclear discordance in *D. swinhonis*. We hypothesize that a second contact of nuclear genes resulted from a male-biased expansion during the last degeneration. Selection for mitochondrial genes may also have contributed to deep differentiation and strong structure. The complex geological history of Taiwan, where plates converge, has driven environmental features that have produced intraspecific divergence in *D. swinhonis*. The result also shows that *D. swinhonis* dispersed to Lanyu around 1.2 mya (Fig. 2.8).

2.2 How Long Did It Take the Weevil to Get from the Philippines to Orchid Island?

This study investigates the dispersal patterns of the weevils from the Babuyan and Bataan Islands in the northern Philippines, Orchid Island and Green Island in Taiwan, and Ishigaki Island and Iriomote Island in the southern part of Ryukyu, Japan. These islands are home to a population of *Pachyrhynchus orbifer* complex, which exhibits a wide range of markings, including globular markings, green, pink, or yellow stripes, and even no markings at all. Whether the similarity and relatedness of the markings between these individuals on various islands was significant was one of the questions we were interested in answering.

The islands, which stretch from Luzon to Yaeyama, are separated by a very deep trench, and they weren't joined when the sea level was lower during the last ice age. The most likely scenario is that they sank to various islands as they floated through the ocean. For instance, the larvae of the striped weevil on Orchid Island feed on the trunk of the checkerboard foot, which serves as its host plant. The fruit of the sea drifting plant known as checkerboard foot can float in the water and then spread to various islands. The ball-backed weevil can float to various islands with the chess-board foot if it can attach itself to it. Additionally, the globe-backed weevil's ability to drift and disperse at sea may be accelerated by the Kuroshio current's rapid northward flow along Luzon.

Ball-back weevil's (*P. jitanasaius) ability to travel across the sea.*

The ball-backed weevil is a group of non-flying insects that is primarily found on one island in the Philippines, including, of course, Orchid Island and Green Island in Taiwan. This is due to the healing of wing sheaths and the degeneration of the hind wings.

The past evolutionary history of species dispersal patterns among oceanic islands (when they dispersed to different places) has been studied extensively using DNA information, but it is difficult to comprehend the true dispersal mechanisms (how they dispersed, by birds, sea drift, typhoons, etc.). We've discovered through previous DNA research that the globe-back has moved between islands a lot over the last million years. The checkerboard foot, a sea drifting plant, is the herbivore of the striped bulbous weevil, and the checkerboard foot's eggs and larvae develop there.

We were able to conduct salinity tolerance experiments using eggs, larvae, and adults thanks to artificial propagation, which Jiemin Jiang helped with. Fortunately, the Kuroshio Ocean Cultural and Educational Foundation's "Kuroshio 101 Drift Project" allowed us to drift on Kuroshio with larvae and board feet, truly in the wild. We were able to test whether ball-back dispersion could occur during the Kuroshio tide. The findings show that the ball-back weevil's eggs and larvae are more tolerant of salinity and more likely to disperse across the sea at these two stages, but the adults are not as likely to do so.

Without the aid of other animals or floating objects, terrestrial species, particularly non-agile species (those that cannot fly or swim), cannot cross the ocean. However, it can be inferred from genetic data that the flightless weevil was able to travel great distances to colonize far-off islands. Here, we use captive *P. jitanasaius* to examine three different aspects of larval survival: (i) the physiological tolerance of weevils (egg, larval, and adult stages) to various salinities; (ii) the survival of larvae in a laboratory simulation of a marine environment; and (iii) the survival of larvae in field experiments in the ocean using fish poison tree fruits floating in the Pacific Kuroshio. While we discovered that larval survival in seawater was lower than in freshwater, if any larvae survived after 7 days of submersion, some of them eventually developed into adults during subsequent rearing. No adult survived for more than 2 days, regardless of salinity. After floating in salt water for 6 days in the lab and 6 days in Kuroshio, only 2 of the 18 larvae were able to survive in fruit. The results of this study offer the first empirical proof that *P. jitanasaius* larvae can survive "drifting" in ocean currents and that their eggs and larvae have the highest likelihood of crossing ocean barriers. This trait may aid in the transoceanic dispersal of these wingless insects and help to determine where they are found and how they form new species on islands in the western Pacific.

For species with little capacity for dispersal, the ocean frequently acts as a formidable geographic barrier (Whittaker & Fernández-Palacios, 2007). Important biogeographic study topics include species colonization processes on marine islands and evolutionary patterns (Cowie & Holland, 2006). The colonization histories of various species on such islands have been studied using molecular evidence since the development of molecular genetics (Raxworthy et al., 2002; Vences et al., 2003; Gillespie, 2004; Cowie & Holland, 2008; Harbaugh et al., 2009; Claridge et al., 2017; Machado et al., 2017). However, there is little knowledge about long-distance dispersal mechanisms and natural transport routes across marine barriers for the majority of island species with restricted mobility.

The remoteness of the island and the colonists' ability to disperse are two important factors that affect the success of island colonization in the ocean (Simberloff & Wilson, 1969; de Queiroz, 2005; Gillespie et al., 2012). Few colonists have the ability to reach remote islands and successfully establish populations there (Schrago & Russo, 2003). Long dispersal distances of more than 1000 km are possible for migrant species, such as birds and butterflies (Brower, 1996; Hake et al., 2001). However, it is challenging to spread non-susceptible terrestrial organisms to islands (Blaustein et al., 1994; Vences et al., 2003). The movement of non-susceptible species by wind and animals, as well as drifting on floating debris, such as tree trunks, is a potential means of dispersal at sea (Gillespie et al., 2012; Incagnone et al., 2014). On debris floating on the ocean surface, many organisms have been discovered, including terrestrial insects (Heatwole & Levins, 1972; Peck, 1994; Thiel & Gutow, 2005). Following significant tsunamis, insects have been observed drifting thousands of kilometers in the Pacific Ocean over a six-year period (Carlton et al., 2017).

Ocean currents are massive, continuous, steady movements of seawater that have an impact on the spread and colonization of island organisms (de Queiroz, 2005, Gillespie et al., 2012). One of the main currents in the western Pacific Ocean is called Kuroshio. It starts north of the equator and primarily flows north from the Philippine island of Luzon to the main islands of Japan via the Babuyan, Bataan, and Ryukyu Islands. A variety of organisms, such as seagrasses [e.g., seagrasses; (Kuo et al.,

2006)] and the genetic structure of their populations [e.g., *Plestiodon* skinks (Kurita & Hikida, 2014) and *Diploderma* tree lizards (Yang et al., 2018)]. For *Pachyrhynchus* weevils in the Taiwan-Luzon volcanic belt, Kuroshio may have carried floating plant parts between islands, thus facilitating their dispersal (Tseng et al., 2018).

There are a number of potential "screens" that could prevent species from dispersing and establishing during transoceanic colonization, including salinity tolerance (Coulson et al., 2002) and environmental changes (climate change, temperature). Survival of species that float in the ocean depends on their ability to tolerate salinity (Hart et al., 1991). Some invertebrates, including terrestrial cryptostigmatic collembola (Coulson et al., 2002) and insects (Kobayashi et al., 2014), have had their salt tolerance examined. *Pachyrhynchus* weevils have been shown to be capable of long-distance dispersal along the Taiwan-Luzon volcanic belt in a previous study (Tseng et al., 2018) . However, the extent to which *Pachyrhynchus* weevils can survive in seawater remains unclear.

Pachyrhynchus weevils (Order Coleoptera, Family Curculionidae) is a group of colorful beetles found mainly on islands in Southeast Asia, including the South Ryukyu Islands (Iriomote and Ishigaki Islands), the Green and Orchid Island of Taiwan, and the Philippines (Schultze, 1923; Starr & Wang, 1992). *Pachyrhynchus* weevils are flightless insects with fused elytra and atrophied hindwings. These weevils had a cavity beneath their fused elytra that could store air, according to a previous experiment, indicating that they could float on water for at least 6 h (Schultze, 1923). Additionally, they are wood-boring insects, and the stems or fruits of their host plants are where their larvae and pupae reside (Kayashima et al., 1940; Ôshiro, 1991, Hsu et al., 2017). *Pachyrhynchus* weevils may be dispersed throughout the ocean as adults drifting on floating objects or as eggs, larvae, and pupae inside the stems and fruits of their host plants, according to their morphological adaptations and wood-boring life history (Tseng et al., 2018).

In this chapter, we investigated the possibility that during Kuroshio, *Pachyrhynchus* weevils could migrate to different islands in the ocean and disperse there. We specifically hypothesized that (i) *Pachyrhynchus* weevil adults can survive in seawater better than their eggs and larvae because their hard exoskeleton is waterproof and an air cavity under their fused dorsal spine allows them to float on the ocean surface. (ii) Eggs and larvae of *Pachyrhynchus* weevils that inhabit the stems and fruits of their host plants, such as *Barringtonia asiatica*, can survive oceanic drifts in Kuroshio because of the large box-like fruits of *B. asiatica*, which have a thick layer of sponge-like fibers that are extremely waterproof and buoyant, helping to isolate the weevils from seawater. To test these two hypotheses, we used laboratory-reared *Pachyrhynchus jitanasaius* to examine (i) the hatching rate and egg length of *P. jitanasaius* eggs, and the survival of *P. jitanasaius* larvae and adults under low (freshwater), medium (brackish), and high (seawater) salinity conditions in the laboratory; (ii) the survival of *P. jitanasaius* larvae survival rates; (ii) the survival of *P. jitanasaius* larvae placed in stems of *B. asiatica* in a laboratory simulated marine environment; and (iii) the survival of *P. jitanasaius* larvae placed in fruit of *B. asiatica* drifting in the Kuroshio tide between Orchid Island and Yaeyama Island, offshore of Taiwan in the western Pacific Ocean (Fig. 2.9).

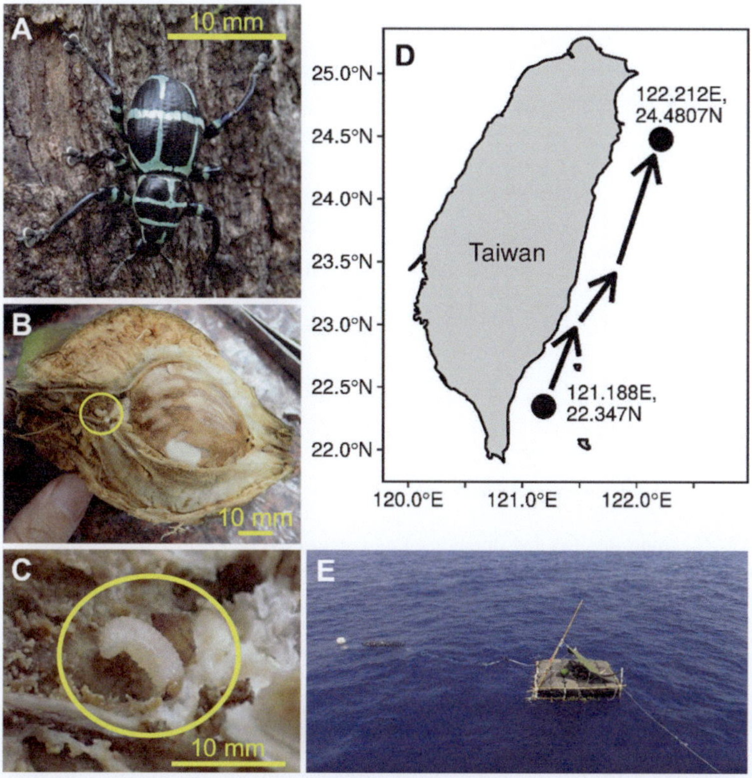

Fig. 2.9 (**a**) *Pachyrhynchus jitanasaius* inhabit the tree trunk of the ironwood tree (*Casuarina equisetifolia*) on Green Island. (**b**) *P. jitanasaius* larvae (indicated by a circle) inhabit the fruit of the fish poison tree (*Barringtonia asiatica*). (**c**) The circle indicates the fifth to sixth instar larvae of *P. jitanasaius*. (**d**) The map of the Taiwan and nearby islands shows the rafting route from Dawu, Taitung, to Guishan Island, Yilan. (**e**) The fruits with larvae were fixed with fishing nets and tied to a plastic platform made of 36 floating buoys and bamboos. (Adapted from Yeh et al. (2018), Journal of Experimental Biology, with permission. Photos rearranged and modified. (**d**) Redrawn.)

2.2.1 Salinity Tolerance of Eggs, Larvae, and Adults

In this study, logistic regression showed no difference in hatching rates of *P. jitanasaius* eggs between the three salinity treatments for the same immersion time (marginal likelihood ratio test, $P = 0.461$). In addition, hatching rates between the two immersion periods (7 and 14 days) under the same salinity conditions showed no difference (marginal likelihood ratio test, $P = 0.895$) (Fig. 2.10a). In all three salinity conditions, some larvae hatched while immersed. We found no difference in the mean duration of the egg stage (9–10 days) between the salinity treatment (Type III F test, $F_{3,565} = 0.0362$, $P = 0.991$) and the immersion period (Type III F test, $F_{2,25} = 0.242$, $P = 0.090$) (Fig. 2.10b). We also confirmed the homogeneity of the variance of the egg stage between treatments (Brown-Forsythe

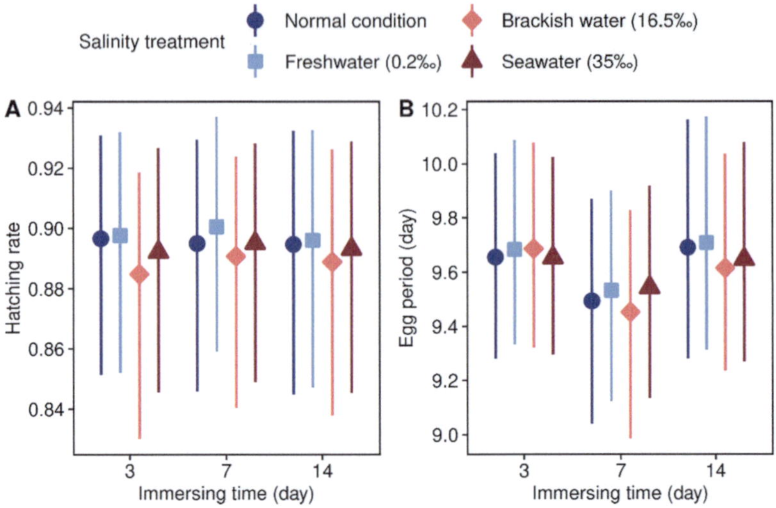

Fig. 2.10 Hatching rate (**a**) and duration of egg stages (**b**) of *Pachyrhynchus jitanasaius* eggs under different salinity conditions and immersion times. (Redrawnfrom Yeh et al. (2018), Journal of Experimental Biology, with permission)

test, $F_{11,541} = 0.0362$, $P = 0.825$). These results indicate that hatching rates and egg duration were similar regardless of immersion duration or salinity.

Larval survival was measured during three periods: during the immersion experiment (from day 1 to day 7) (Fig. 2.11a), after immersion (from day 7 to adulthood) (Fig. 2.11b), and from initial immersion to adulthood (day 1 to adulthood) (Fig. 2.11c). The survival rates of *P. jitanasaius* larvae after 7 days of immersion were 100%, 68%, 75%, and 25% for the control, freshwater, brackish water, and seawater groups, respectively (Fig. 2.11a). Body size (i.e., head width) of larvae showed no correlation with survival rate. There were significant differences in larval survival among the four salinity treatments (marginal likelihood ratio test: $\chi^2 = 22.666$; DF $= 3$; $P < 0.0001$). In addition, subsequent multiple comparisons revealed significant differences in hazard ratios between the control and other treatments (freshwater, brackish water, and seawater), suggesting that water immersion affected larval survival. Similarly, there was a significant difference in hazard ratios between the seawater and freshwater or brackish water treatments. However, there was no significant difference in hazard ratios between freshwater and brackish water treatments. The comparison of the different salinity treatments showed that salinity affected the survival rate of larvae. On the other hand, in the second period considered (after immersion, from day 7 to adulthood), there were no significant differences in survival rates between the four salinity treatments (marginal likelihood ratio test: $\chi^2 = 2.016$; DF $= 3$; $P = 0.569$; see Fig. 2.11b). Thus, despite the sensitivity of the larvae to seawater, a small fraction that survived immersion could grow to adulthood (19 (100%), 13 (68%), 15 (75%), and 5 (25%) surviving individuals for control, freshwater, brackish water, and seawater, respectively) (Fig. 2.11c).

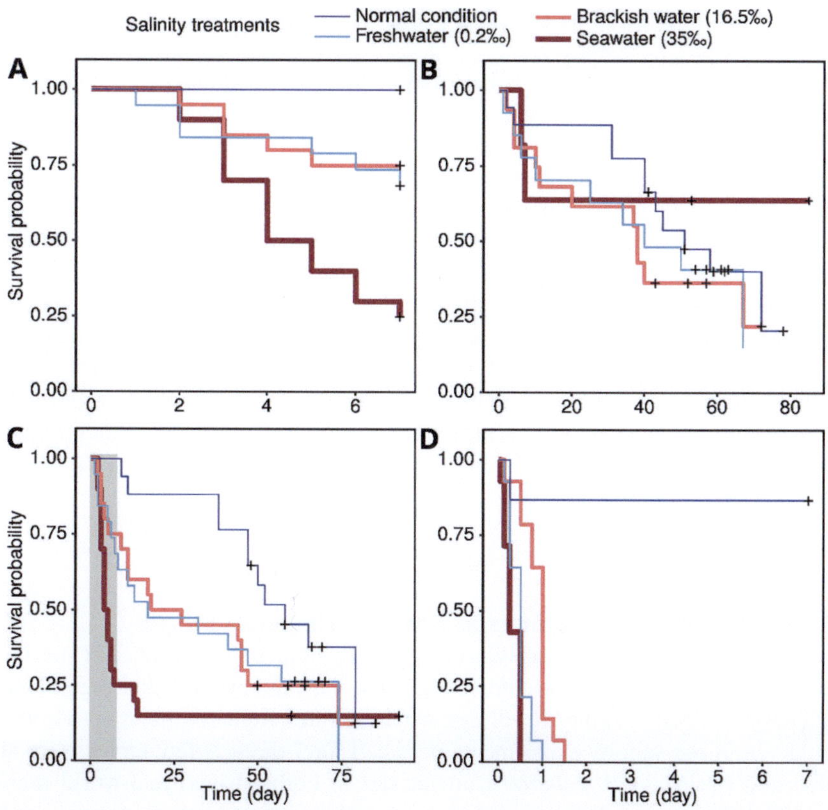

Fig. 2.11 Survival probabilities of *Pachyrhynchus jitanasaius* under three salinity conditions over time. (**a**) Survival probabilities of the larvae from immersion day 1 to day 7, (**b**) survival probabilities of the larvae after day 7 of immersion to adult emergence, (**c**) survival probabilities of the larvae from day 1 of immersion to adult emergence, and (**d**) survival probabilities of the adults from day 1 of immersion to day 7. (Redrawnfrom Yeh et al. (2018), Journal of Experimental Biology, with permission)

Under all three salinity conditions, *P. jitanasaius* adults survived for less than 40 h (1.66 days) and 70% and 95% of the individuals died before 12 h and 24 h (0.5–1 day), respectively (Fig. 2.11d). Adult survival rates differed significantly among the three salinities and control treatments (marginal likelihood ratio test: $\chi^2 = 44.128$; DF = 3; $P < 0.0001$). In addition, subsequent multiple comparisons showed that the hazard ratios for the control and the three salinity treatments were significantly different, with water immersion reducing adult survival. The hazard ratios for freshwater versus brackish and seawater treatments were significantly different, indicating that salinity affected adult survival. However, there was no significant difference in the hazard ratios between brackish water and seawater treatments. These results indicate that *P. jitanasaius* adults cannot survive in water regardless of salinity, and the higher the salinity, the lower the survival rate of adults.

2.2.2 Survival Rate in the Drift Experiment

In this study, four of the 22 fruits used in the drifting experiment in the ocean were lost due to wave action. There was a significant difference between the two experiments and the control treatment $\chi^2 = 35.902$; DF = 2; $P < 0.0001$). Multiple comparisons showed significant differences in survival probabilities between the control and drifting experiments and between the control and simulated ocean treatments, but no differences were found between the drifting and simulated experiments (Fig. 2.12). Larval survival rates were lower in both the drift and simulation groups than in the control group. Most importantly, two of the 18 *P. jitanasaius* larvae and two of the 22 larvae survived in the drift and simulation experiments, respectively. These four larvae emerged as adults and subsequently survived for over a month in the laboratory, suggesting that a small proportion of *P. jitanasaius* larvae drifting in B. asiatica fruit could survive the oceanic journey of Kuroshio for 6 days and at least ~300 km, despite a survival rate of ~10%.

2.2.3 Discussion

2.2.3.1 Egg and Larval Stages as Possible Dispersal Stages

Our research shows that *P. jitanasaius* adults have a much lower chance of surviving in seawater than do their eggs and larvae. Our analysis revealed that regardless of immersion time and salinity level, egg hatching rates were comparable (84–93%). The larvae in our salinity experiments persisted for 7 days. A few larvae were also added and placed in *B. asiatica* fruit thrived in both lab and real-world marine environments. Two adult *Pachyrhynchus* were able to float on water for at least 6 h in a previous experiment; after 12 h, one weevil had died, but the other was still alive (Schultze, 1923). Similar findings were found in our study, which found that survival rates dropped in seawater and that 70% of adult weevils survived for less than 12 hours.

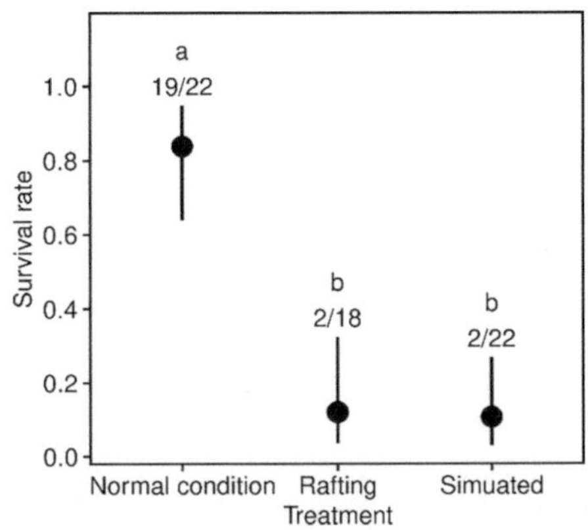

Fig. 2.12 Survival rates of *Pachyrhynchus jitanasaius* larvae in simulated and rafting experiments. Numbers above the bars indicate the number of larvae that survived (left) and the total number of larvae (right)

2.2.3.2 Physiological Mechanisms of Seawater Tolerance in Different Life Cycle Stages

A physiological adaptation called osmoregulation is used in various environments. The osmoregulation strategies of insects, particularly Coleoptera, are not well understood (Nicolson, 1980; Riddle, 1986; Naidu, 2001; Pallarés et al., 2015). Most studies on osmoregulation have focused on desiccation in terrestrial beetles (Nicolson, 1980; Naidu, 2001, 2006) and few studies have examined salt tolerance in aquatic (Pallarés et al., 2015) or terrestrial coleopteran insects. Beetles that show osmosis regulation can maintain hemolymph osmosis concentrations in different saline environments (Pallarés et al., 2015). Our research offers the first illustration of salinity tolerance in flightless weevils, which can colonize other islands by swimming across the ocean as eggs and larvae.

Some invertebrate species' eggs have developed unique characteristics that enable them to adapt to environments with high salinities. For instance, the crystalline structure of the eggs of the rod-shaped insect *Megacrania alpheus* allows them to survive in seawater for 150 days (Kobayashi et al., 2014). Our research leads us to hypothesize that, like rod insects, *P. jitanasaius* eggs may be protected from seawater by their structural characteristics (Shu et al., 2015). Some aquatic invertebrates exhibit osmotic adaptation when submerged in freshwater or hypersaline water during the larval stage. It was discovered that the peculiar rectal segment of mosquito larvae that feed on polar algae regulates the ion concentration and osmotic pressure of the hemolymph (Clark et al., 2004). Due to effective osmoregulation, marine butterfly larvae (*Philanisus plebeius*) can survive salinities ranging from 90 mM/l (15% seawater) to 900 mM/l (150% seawater)(Leader, 1972). To find out whether *P. jitanasaius* larvae use particular osmoregulatory strategies to deal with the difficulties of seawater salinity, more physiological research is required.

According to theory, adult beetles have an impermeable cuticle that acts as a defense against environmental osmotic pressure. However, *Pachyrhynchus* larvae and eggs exhibit better salinity tolerance even though the cuticle of adults may be thicker and less permeable. Even though they do survive longer in freshwater than in saltwater, most adult *Pachyrhynchus* cannot survive underwater for longer than 24 h. *Hygrotus salinarius* showed varying salinity tolerances throughout its life cycle, with adults and larvae exhibiting various osmotic pressure regulation mechanisms in environments with high salinities (Tones, 1978). The ability of *Pachyrhynchus* weevils to tolerate salinity is demonstrated here, and it offers a potential system for further research into the mechanisms of osmotic regulation in flightless insects.

Additionally, our study offers the first concrete proof that *P. jitanasaius* larvae can withstand marine drift inside of fruit *B. asiatica* (fish poison tree). Typhoons, floods, and other natural disasters have a significant impact on the coastal habitats of tropical islands (Hirsh & Marler, 2002) and influence the distribution of plants (Walker, 1991; Gannon & Willig, 2008) and animals such as sticklebacks (Kobayashi et al., 2014) and iguanas (Censky et al., 1998).

B. asiatica is a host plant for *P. sonani* and can be found on Orchid Island (Hsu et al., 2017). *B. asiatica* is a salt-tolerant plant that is found in large numbers along tropical coasts in the western Pacific Ocean and Indian Ocean from eastern Sanjiba to Taiwan (Mojica & Micor, 2007). The outermost layer of *B. asiatica* fruits is substantial and water-resistant (Quigley et al., 2014). The fruit floats because of the air pockets in the sponge-like middle layer. The seeds are safeguarded by the thick, firm inner layer. The coconut-like hard and spongy layer can float on the surface for more than 2 years (Tsou & Mori, 2002; Thiel & Gutow, 2005; Quigley et al., 2014). Therefore, the fruits, stems, and roots of *B. asiatica* (host plant) can keep separate the eggs and larvae of *Pachyrhynchus* from seawater while providing a continuous food source (Thiel & Gutow, 2005).

2.2.3.3 Ocean Currents, Host Plant Characteristics, and Physiological Tolerance All Influence *Pachyrhynchus* Weevil Dispersal and Distribution

Currents are crucial in determining the biogeography of island biodiversity, and oceanic drift is a significant mechanism of species dispersal (Fraser et al., 2011). Animals of different taxonomic groups, including terrestrial invertebrates and vertebrates, that use oceanic drift for dispersal were found on marine floats between islands in the Puerto Rican Bank (Heatwole & Levins, 1972) and the Galapagos Islands (Peck, 1994). The floats likely came from the coastlines of nearby islands. The swift Kuroshio, traveling north along the Taiwan-Luzon volcanic belt, carries live eggs and larvae enmeshed in drifting *B. asiatica* fruit, offering a potential carrier for *Pachyrhynchus* weevil dispersal among remote islands in the area. This study offers the first empirical evidence in support of this theory: *Pachyrhynchus* weevils can cross marine barriers due to their physiological tolerance of seawater, the traits of their host plant, *B. asiatica*, and swift currents (the Kuroshio Current). They can then use this combination of factors to enhance their dispersal and colonization and spread to other islands.

2.3 How to Resist the Salinity of the Sea Crossing from Taiwan to Orchid Island

Terrestrial organisms can travel from one land mass to another through aquatic dispersal. The capacity and likelihood of this dispersal are ecologically relevant to how organisms are distributed on land and in islands, which in turn affects biogeography and biodiversity in terms of evolution (Calsbeek & Smith, 2003; Vidal et al., 2008; Nikula et al., 2013). The only way for organisms to spread between land masses is by crossing the ocean, not by using land bridges, according to molecular and fossil evidence (Glor et al., 2005; Vidal et al., 2008; Yang et al., 2018). This suggests that over-water dispersal is an important strategy for organisms. Although many introduced species have used human traffic to cross the ocean in the Anthropocene, species that artificially invade new areas can still naturally spread to nearby areas (Betancur-R et al., 2011). For instance, Norway rats spread to nearby islands through natural drift or swimming after being introduced to islands around the world via artificial boats (Russell et al., 2010; Tabak et al., 2014). Thus, overwater dispersal is not only important for understanding the distribution of native species but also plays a substantial role in determining the extent to which invasive species can reach.

Extreme salinity conditions and critical water loss are the main physical challenges for terrestrial organisms as they drift across the ocean (Kobayashi et al., 2014; Hopkins & Brodie, 2015; Yeh et al., 2018). High salinity environments are stressful for organisms. Salinity tolerance varies among species (Hart et al., 2012; Jud et al., 2014; Kobayashi et al., 2014); this variation affects species' chances of surviving oceanic drift, which in turn causes differences in their capacity for global dispersal at the species level, which further affects species distribution (Jud et al., 2014; Josefson, 2016). Studies on the relationship between this physiological capacity and dispersal, however, have primarily concentrated on plants, freshwater animals, and arthropods (e.g., Guja et al., 2010; Esser & Cumberlidge, 2011; Yeh et al., 2018). There hasn't been much research done on how terrestrial vertebrates move around. Among reptiles, some species occur in saline wetlands or on the coast and even survive there for a long time (Neill, 1958; Ellis, 1981; Koleva et al., 2017), suggesting that they can tolerate higher salinity environments. However, most previous studies on salinity tolerance have focused on marine or estuarine species (Ellis, 1981; Davenport & Macedo, 1990; Lillywhite et al., 2009). In the few studies on freshwater and terrestrial species, salinity treatments were usually oral or injected (Templeton et al., 1972; Hart et al., 2012), while treatments in contact with saline water, the most relevant situation faced by drifting individuals, were rarely performed in the previous literature. Furthermore, no study has ever connected the distribution of species with salinity tolerance. Salinity tolerance in direct contact with water and its relationship to distribution, a crucial ability to cross the ocean, have thus far not been fully understood.

For organisms drifting over the ocean, dehydration is the biggest problem and a major contributor to salinity tolerance. In addition to losing water through respiration, reptiles also lose water primarily through their skin (Dawson et al., 1966; Prange & Schmidt-Nielsen, 1969). Skin water loss is related to climate adaptation to the habitat environment (Baeyens & Rountree, 1983; Roberts & Lillywhite, 1983; Guillon et al., 2014), with species living in dry and hot habitats losing water more slowly than species living in wet and cold habitats (Dmi'el et al., 1997; Guillon et al., 2014). Additionally, lower water loss rates suggest that the species can hold onto water for longer periods of time, which may postpone the onset of dehydration-related death (Heatwole & Veron, 1977; Hillman & Gorman, 1977). Thus, species with lower water loss rates are more likely to survive the overwater drift period than species that lose water more quickly. However, in the case of vertebrates crossing the ocean, the relationship between water loss and survival is rarely reported.

There are two distinct developmental stages for crossing the ocean for the majority of reptiles: the individual stage and the egg stage. The individual stage of swimming, floating, or drifting across the ocean is well documented right now (Censky et al., 1998; Calsbeek & Smith, 2003; Gerlach et al., 2006; Borden, 2007). Small reptiles are more likely to cross the ocean by floating or drifting than are larger species that can swim far distances (Schoener & Schoener, 1984; Calsbeek

& Smith, 2003). Some small species of geckos are thought to spread through artificial boats or natural drifting (Snyder, 1917; Fisher, 1997). In *Anolis*, it can float on seawater for a short period of time (Schoener & Schoener, 1984), dispersing among the islands that make up their natural range (Calsbeek & Smith, 2003; Glor et al., 2005). On the other hand, it's likely that during the egg stage, over-water dispersal takes place (Snyder, 1917; Holden et al., 2014). The type of eggshell may be a major determinant of the salinity tolerance and water loss capacity of reptile eggs. Reptile eggs come in two main varieties: hard-shelled eggs and eggs with a parchment shell (Packard & DeMarco, 1991). The former is more flexible and permeable to water exchange because they have thinner shells and less calcium than the latter. Parchment paper-shell eggs are very sensitive to environmental humidity (Packard et al., 1979), whereas hard-shell eggs have a dense hard shell that limits the exchange of water and materials from the environment (Packard et al., 1982). As a result, hard-shell eggs ought to be more tolerant of seawater. Some gecko studies have shown that eggs can tolerate seawater well after immersion treatment (Brown & Alcala, 1957; Brown & Duffy, 1992). Only *Anolis sagrei* has reported successful hatching of parchment-shell eggs after submersion in seawater (Losos et al., 2003).

Currently, a body of water separates Taiwan and nearby islands from the Asian continent (at a depth of about 70 m). However, during the last glacial periods of 1.55 Ma, Taiwan's main island and the smaller islands between Taiwan and China were joined to the Asian continent, and between 26.5 and 18 ka, when the sea level fell 135 m (Fig. 2.13a). Taiwan, as well as these western islands, is therefore a continental island. Via land bridges, terrestrial species can disperse in Taiwan and on the Asian continent. Contrarily, oceanic trenches along the nearby eastern coast of Taiwan have rapidly deepened the eastern part of Taiwan's ocean (to depths of over 1000 m), making it historically impossible for the eastern islands of Taiwan to connect to any island or continent (Fig. 2.13a). Because of this, the eastern islands are oceanic and can only be reached by water diffusion by the species. An understanding of interspecific salinity variation can show how the current distribution of species was created when combined with the historical context of these geographic connections across Taiwan and these islands.

To determine the potential for overwater dispersal of the small lizard species from Taiwan and nearby islands, we looked at the salinity heterogeneity of these species in this study. We then looked at the current distribution of these species and their salinity tolerance in a historical geological context. For this study, six species of small lizards were chosen, including two introduced species and four native species (Table 2.6 and Fig. 2.13b–g). In order to simulate extreme drift conditions with direct contact with water, we conducted semi-submersion experiments on these six species (*P. elegans, E. longicaudata, E. multifasciata, D. swinhonis, H. frenatus,* and *A. sagrei*) and their eggs (only eggs of *E. longicaudata*, *D. swinhonis* and *H. frenatus* are chosen). In order to determine the likelihood of natural transoceanic dispersal, we first used semi-submersion experiments to measure the water loss and survival/hatching rates of these tiny lizards and their eggs. Second, we looked at the distribution of these six species and the relationship between these shifts in salinity

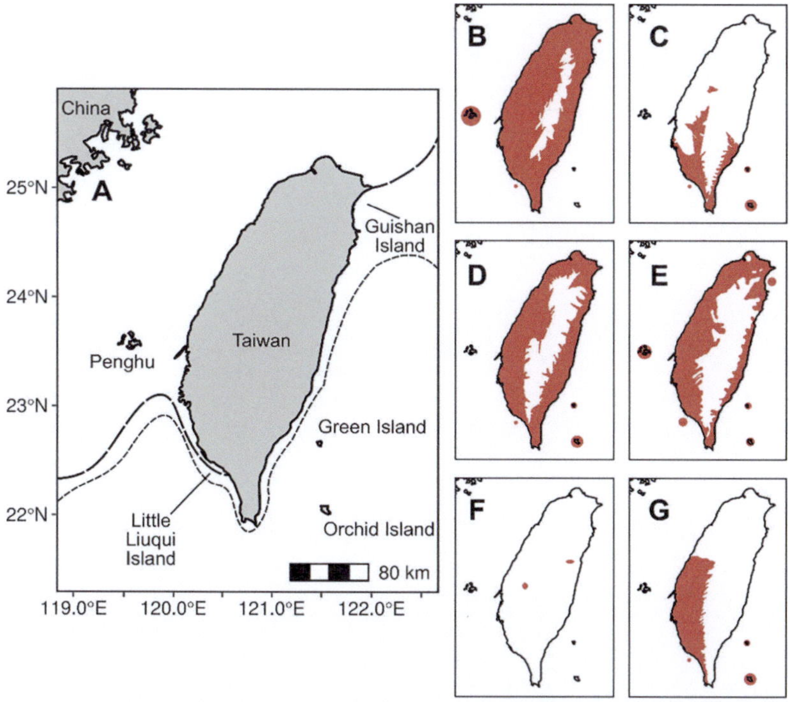

Fig. 2.13 The historical land connection with the Asian continent in Taiwan and current distribution of the six lizard species. The long dashed line represents the coastline during the Last Glacial Maximum 26.5–18 ka, and the short dashed line represents the coastline before 1.55 Ma (**a**). The red areas represent the distribution of *P. elegans* (**b**), *E. longicaudata* (**c**), *D. swinhonis* (**d**), *H. frenatus* (**e**), *A. sagrei* (**f**), and *E. multifasciata* (**g**). (Redrawn from Hsu (2019), master's thesis, National Chung Hsing University, Taichung, Taiwan)

tolerance, as well as the historical background of geographic association in Taiwan and the nearby islands. Finally, for the sake of conservation, we evaluated the possibility of these two introduced species dispersing.

2.3.1 Water Loss Rate of Lizards

P. elegans individuals had the highest total rate of water loss (RTWL) among the six species in the control group (Fig. 2.14a), demonstrating that this species rapidly loses water in air. In *E. longicaudata* and *E. multifasciata*, RTWLs were low in the control group, and there was no significant difference between the control and freshwater treatments (FW) (Fig. 2.14b, f). Meanwhile, in the other four species, the RTWLs of FW were significantly lower than those of the control group (Fig. 2.14a–e). In FW, RTWLs were similar in the other five species except for negative values in *D. swinhonis* (males: −0.069 ± 0.087, females: −0.032 ± 0.058; Fig. 2.14c), indicating that *D. swinhonis* drank water (males: 0.021–0.073, females: 0.019–

Table 2.6 Distributions of six lizard species in Taiwan and adjacent islands. The distance is the closest distance to Taiwan

	Taiwan	Penghu	Little Liuqiu Island	Guishan Island	Green Island	Orchid Island	Philippines
Distance from Taiwan	–	47 km, west	13 km, southwest	11 km, northeast	44 km, southeast	63 km, southeast	161 km, southeast
Plestiodon elegans	O	O	O	O	X	X	X
Eutropis longicaudata	O	X	O	X	O	O	X
Eutropis multifasciata	O	X	O	X	O	O	O
Diploderma swinhonis	O	X	O	X	O	O	X
Hemidactylus frenatus	O	O	O	O	O	O	O
Anolis sagrei	O	X	X	X	X	X	X

The direction is the relative position to Taiwan. The character O indicates that this species exists in that location, and the character X indicates no distribution.

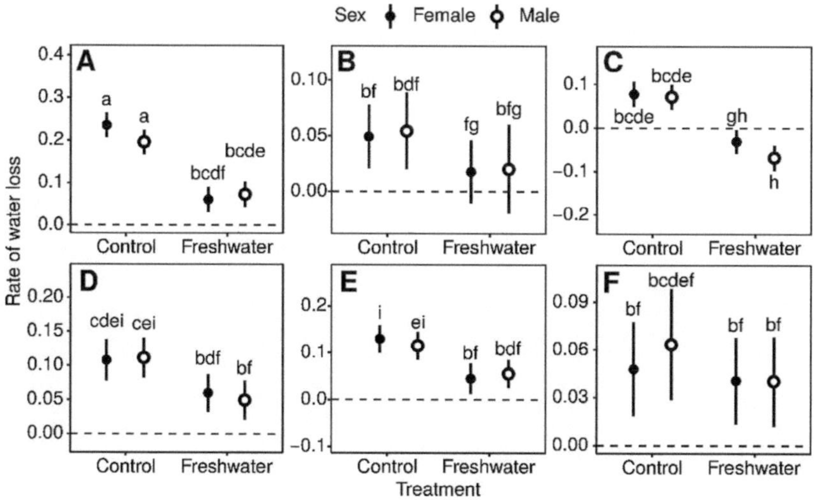

Fig. 2.14 Rates of total water loss (RTWLs) of six lizard species from Taiwan, *P. elegans* (**a**), *E. longicaudata* (**b**), *D. swinhonis* (**c**), *H. frenatus* (**d**), *A. sagrei* (**e**), and *E. multifasciata* (**f**). Circles and vertical bars indicate the post hoc means and 95% confidence intervals of RTWLs, respectively. Alphabet letters above the bars present the results of post hoc comparisons; two groups with the same letter indicate no significant difference. (Redrawn from Hsu et al. (2021), PLOS ONE, licensed under CC BY 4.0)

0.063; Fig. 2.14a–f). There was no difference in RTWLs between males and females in all treatments for all species. Since most of the individuals in the brine treatment (SW) died before the end of the treatment period, resulting in sparse data on RTWLs in this group, we did not include them for comparison in this study.

2.3.2 Survival Rate of Lizards

Survival rates (90–100%) for all species did not differ between the control group and FW (Fig. 2.15). On the contrary, the survival rate of SW for most species, except *E. longicaudata* and *E. multifasciata*, was significantly lower compared to the control group and FW (Fig. 2.15). When treated with SW, the survival rates of the four affected species were extremely low, indicating their low tolerance to saline water (Fig.2.15a–e). Meanwhile, the survival of two *Eutropis* species was not significantly affected by brine immersion (Figs. 2.15b, f). In all treatments, the sex of all species was not correlated with the survival rate.

2.3.3 Water Loss Rate of Eggs

In eggs of *E. longicaudata* and *D. swinhonis*, RTWLs were lower than zero in the control and FW (Fig. 2.16a, b), indicating that eggs from both treatments absorbed water during the experimental treatment. The RTWLs were significantly lower in

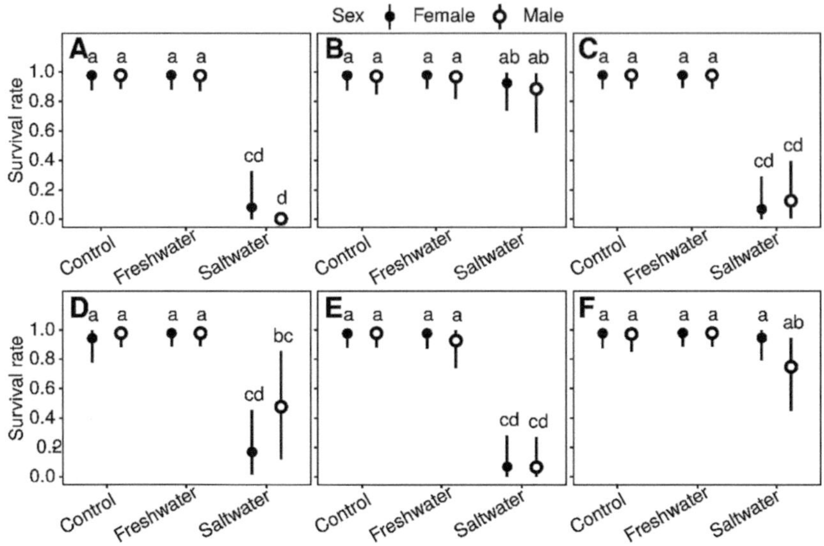

Fig. 2.15 Survival rates of six species of lizards in Taiwan, *P. elegans* (**a**), *E. longicaudata* (**b**), *D. swinhonis* (**c**), *H. frenatus* (**d**), *A. sagrei* (**e**), and *E. multifasciata* (**f**), on three consecutive days. The circles and bars indicate the post hoc mean and 95% confidence interval of survival rates, respectively. Letters above the bars indicate the results of post hoc comparisons; two groups with the same letter indicate no significant difference. (Redrawn from Hsu et al. (2021), PLOS ONE, licensed under CC BY 4.0)

FW, indicating that they absorbed more water than the control. Meanwhile, RTWLs in SW were higher for both parchment-shell egg-laying species than for the control and FW (Fig. 2.16a, b), indicating that eggs of these two species lost significant water in SW. In contrast, RTWLs for *H. frenatus* did not differ significantly among treatments and did not differ from zero, suggesting that none of the treatments affected water in the eggs and that little net water loss/absorption occurred in these hard-shelled eggs (Fig. 2.16c).

2.3.4 Hatching Rate of Eggs

In *E. longicaudata* and *D. swinhonis*, hatching rates were similar in the control and FW (Fig. 2.16d, e), but were significantly lower in SW, indicating that saline immersion did reduce hatching rates in both species. In contrast, in *H. frenatus*, hatching rates in the control (100%) were not different from those in FW (100%) and SW (100%; Fig. 2.16f), suggesting that no treatment affected the hatching rate of these hard-shelled eggs.

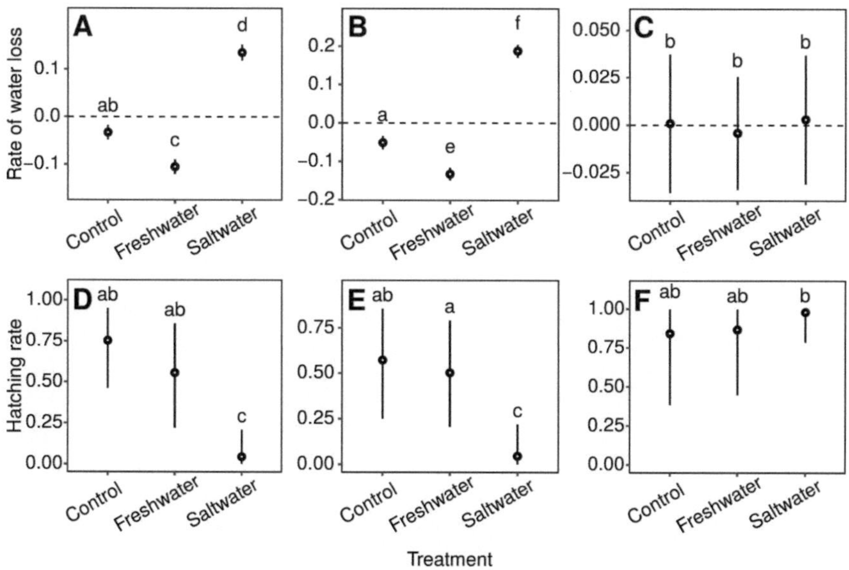

Fig. 2.16 Water loss rates (RTWLs, A-C) and hatching rates (D-F) for eggs of three lizard species from Taiwan. RTWLs for eggs of *E. longicaudata* (**a**), *D. swinhonis* (**b**), and *H. frenatus* (**c**). The hatching rates of eggs of *E. longicaudata* (**d**), *D. swinhonis* (**e**), and *H. frenatus* (**f**) over a three-day period. The circles and bars indicate the post-test mean and 95% confidence interval of RTWLs, respectively. Letters above the bars indicate the results of post hoc comparisons; identical letters in both groups indicate no significant differences. Post hoc comparisons of RTWLs and hatching rates were analyzed separately. (Redrawn from Hsu et al. (2021), PLOS ONE, licensed under CC BY 4.0)

2.3.5 Discussion

Our findings unequivocally show the diversity of lizard species' tolerance to salinity at the individual and oviposition stages, which may also be an indication of inter-specific transoceanic dispersal. Our findings primarily explain the distribution of these species on the islands surrounding Taiwan and how they got there by combining historical geology and the origin of the species (Fig. 2.17). This is the first time a semi-immersion experiment has been used to test a terrestrial vertebrate's tolerance to salinity before correlating this physiological trait with the species' distribution across islands.

Among the native species, *P. elegans* is less likely to cross the ocean, while *E. longicaudata* is clearly tolerant of SW, suggesting that the species is more capable of dispersal by oceanic drift. The other two native species, *D. swinhonis* and *H. frenatus,* were moderately tolerant of SW. Compared to *P. elegans*, they may have a better ability to cross the ocean if the floating period is shorter than 3 days. Among the introduced species, *E. multifasciata* has the potential to disperse to the oceans like the native *Eutropis* of Taiwan. In contrast, *A. sagrei* tolerates SW, and

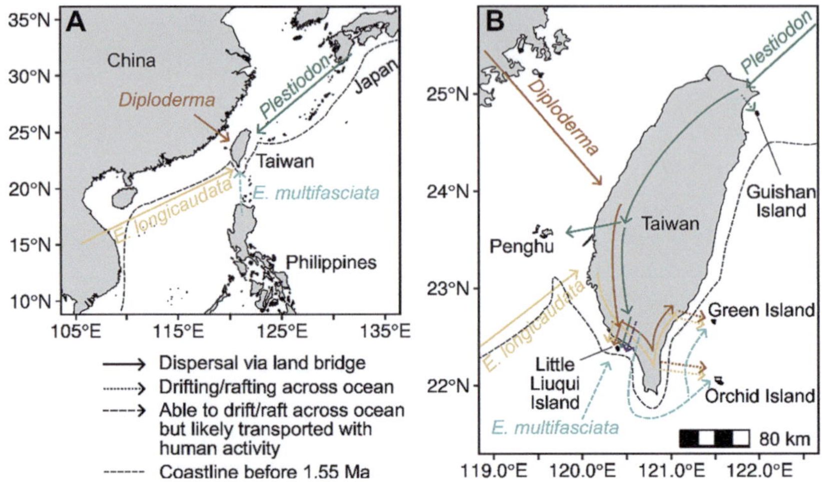

Fig. 2.17 Distribution routes of four species of lizards in Taiwan. (Redrawn from Hsu et al. (2021), PLOS ONE, licensed under CC BY 4.0)

they cannot float over the ocean for as long as the native *D. swinhonis* and *H. frenatus*. Water dispersal is still possible if they can reach land by drifting across the ocean in 2 days, e.g. through small straits, island series and archipelagos. This may be why *A. sagrei* has been observed entering the ocean and reaching neighboring islands (Calsbeek & Smith, 2003).

Differences in SW survival rates between species may be determined by differences in RTWL. Although we do not have these data because of the high mortality rate of SW, the RTWL values for the control group and the differences between the control group and FW still provide a glimpse of the differences in the amount of water loss between species and may occur in SW. Mechanistically, this difference in water loss may be due to differences in species body size (Kurita & Hikida, 2014; Karin et al., 2016; Wang et al., 2019), the amount of subcutaneous lipid content (Baeyens & Rountree, 1983; Roberts & Lillywhite, 1983; Karin et al., 2016), and the morphology of the scales (Cox & Cox, 2015). We do not know the lipid content among these species, but we did find that *E. longicaudata* and *E. multifasciata* had relatively larger scale sizes than the other species (personal observation). Also, body size was not a possible factor in this study, as it was not significant in the RTWL analysis. Evolutionarily, phylogeny (Dawson et al., 1966; Kurita & Hikida, 2014) and adaptation to habitat climate (species living in hotter and drier habitats lose water more slowly) (Guillon et al., 2014; Diniz-Filho et al., 2019) may also be associated with differences in RTWL. *Plestiodon* originated in the temperate regions of Northeast Asia (Heger & Fox, 1992); therefore, it is physiologically adapted to low temperatures, whereas other species originated in tropical regions (Kattan & Lillywhite, 1989; Peig & Green, 2009) with high temperatures. This long-term effect may result in a higher rate of water loss in *P. elegans* compared to other species.

In the egg experiments, *E. longicaudata* and *D. swinhonis* had the poorest egg tolerance SW. In contrast, the low RTWL and high hatching rate (regardless of treatment) of *H. frenatus* indicated that eggs of this species tolerated SW well. In previous studies, gecko eggs coped well with SW in both experiments without gas exchange (complete immersion) (Dmi'el, 1985) and in experiments with air contact (eggs surrounded by wet gauze) (Brown & Alcala, 1957), suggesting that these eggs may survive both floating and drifting on the ocean. The different tolerance of the eggs of these three species to SW may be due to the type of eggshell; the eggs of *E. longicaudata* and *D. swinhonis* were parchment-shelled, while those of *H. frenatus* were hard-shelled. The RTWL results of this study are consistent with previous studies showing that the permeability of the former is generally higher than that of the latter (Packard et al., 1982). This difference may largely determine the heterogeneity of egg tolerance to SW and ocean crossing ability.

The current study primarily reflects the current distribution of each species (Figs. 2.13b–g), and the dispersal history of some species can be inferred when combined with the historical topography surrounding Taiwan (Fig. 2.17). *P. elegans* in this study has a low salinity tolerance, which helps to explain why the species is only found on Penghu and Little Liuqiu Island in the western islands and not on Green Island or Orchid Island (Fig. 2.13b). Access to the oceanic islands requires overwater dispersal. This may further explain why *Plestiodon* species are distributed almost exclusively on the mainland and continental islands (Heger & Fox, 1992; Wegener et al., 2014). *P. elegans* diverged from a related species present in Japan and the Ryukyu Islands at about 4.23 Ma(Heger & Fox, 1992). At that time, the Ryukyu Islands and Taiwan were still connected to the East Asian mainland. The ancestral species probably spread southward from the Ryukyu Islands to Taiwan via land (Fig. 2.17a) and then across the ocean to the western islands (Fig. 2.17b).

E. longicaudata, tolerant to high salinity levels, suggests that the species has the ability to drift over long distances in the ocean, which explains the distribution of the species to the oceanic islands near Taiwan (Fig. 2.13c). The species spread to the oceanic islands by drifting or rafting after evolving in Southeast Asia at around 38 Ma (Peig & Green, 2009), probably reached Taiwan via land bridges while Taiwan was still connected to the Asian continent (Fig. 2.17a), and then dispersed to the oceanic islands by drifting or rafting (Fig. 2.17b). Currently, *E. longicaudata* does not occur in northern Taiwan due to the limitation of low winter temperatures. However, if the climate continues to change warming, it may spread northward in Taiwan and may cross the ocean to the northern islands near Taiwan.

According to our findings, *D. Swinhonis* are widely dispersed, including the two oceanic islands Green Island and Orchid Island, and they can only survive for 1 day in the ocean (Fig. 2.13d). The genus originated in China's mainland and likely spread via two routes: land bridges and water to these islands, where they can arrive in less than a day (Fig. 2.17a, b). The forces of climatic events, such as monsoons or typhoons, may reduce the drift time to these islands, even for the farthest Orchid Island, due to the close proximity of Taiwan to its neighboring islands (Table 2.6). Thus, their distribution on the western islands can be explained by land-bridge dispersal and water dispersal, while the distribution on the eastern islands is entirely from the transoceanic route.

A noteworthy conflict between the results and the current distribution occurs in the northern part of Guishan Island. Guishan Island emerged from the ocean about 7000 years ago (Bansal & Karanth, 2010) and was never connected to Taiwan by a land bridge. *P. elegans*, which has a low tolerance for salt water, occurs on this island (Fig. 2.13b), while *D. swinhonis*, which has a moderate tolerance for salt water, does not (Fig. 2.13d). In addition, *Diploderma* species were dispersed in the Ryukyu Islands by Kuroshio (Yang et al., 2018), but not in Katsura Island, which is also on the Kuroshio route. One explanation is that for both species, over-water dispersal is a rare event, and this only happened by chance for *P. elegans*. Alternatively, perhaps the microhabitat on this small island is suitable for *P. elegans* but not for *D. swinhonis*; the former occurs in dense shrubbery, while the latter requires forest. A similar situation exists in Penghu, located in Luqiao (Fig. 2.13a), where *P. elegans* is present but *D. swinhonis*, native to China, is not (Burkner, 2017) (Fig. 2.13b, d). The environment of Penghu may not be suitable for the long-term survival of *D. swinhonis* because there are too few trees for this arboreal species.

The most widespread distribution of *H. frenatus* on the islands is not consistent with its moderate saltwater tolerance, but it is consistent with the excellent tolerance of eggs to seawater in this study. It is difficult to know where this species originated because they are often dispersed through human activities (Brandley et al., 2012). In the present study, we show a high potential for dispersal through transoceanic dispersal at the egg stage, consistent with the speculations of previous studies (Snyder, 1917; Holden et al., 2014). This study's findings on the high SW tolerance of gecko eggs may help to explain the widespread distribution and low genetic differentiation of the species on land and in Southeast Asian islands (Kattan & Lillywhite, 1989).

In addition, we also suggest conservation issues for these two invasive lizard species; *E. multifasciata* has the potential to naturally disperse to any island and should be carefully monitored in the future. For *A. sagrei*, we can focus on islands near Taiwan to prevent natural overwater dispersal, as well as those islands where agricultural transport is flourishing. *E. multifasciata* is thought to have been introduced to Kaohsiung in southern Taiwan through artificial transport since 1994 (Chen et al., 2001), and they have now spread outward to central Taiwan and some neighboring islands (Fig. 2.13g). In our results, the high salinity tolerance of *E. multifasciata* showed that they were able to disperse from Taiwan and even from the Philippines to these islands via Kuroshio, as it could survive (>70%) for 3 days (Table 2.7; Fig. 2.15), and some individuals even survived for 5 days after semi-submersion treatment (unpublished data), exceeding the drift time from the Philippines to Taiwan (89.4 h; Table 2.6). However, we still believe that they were introduced by human activities (Fig. 2.17a, b) because they were recently found on these islands (Green Island in 2008; Orchid Island in 2017) and live densely in areas near ports and airports. More evidence is needed to confirm how the species arrived in Taiwan and on these islands. The point is that if the climate continues to warm, it is highly likely that they will settle on other islands near Taiwan in the future, even on the northern islands. On the other hand, *A. sagrei* is also thought to have been introduced to Taiwan by human transport(Carranza & Arnold, 2006). In our results,

Table 2.7 Accumulated survival rates in six lizard species in the saltwater treatment

Species	Sex	*N*	Survival rate (%)		
			Day 1	Day 2	Day 3
P. elegans	Male	10	0	0	0
	Female	10	10	0	0
E. longicaudata	Male	6	100	83	83
	Female	10	100	100	90
D. swinhonis	Male	10	90	10	10
	Female	10	70	0	0
H. frenatus	Male	10	60	40	30
	Female	10	80	40	10
E. multifasciata	Male	10	100	90	70
	Female	11	100	100	91
A. sagrei	Male	10	90	60	0
	Female	10	70	40	0

the moderate saline tolerance of *A. sagrei* suggests that it may drift and survive within one or 2 days, which is sufficient to reach the neighboring islands. Compared to *E. multifasciata*, *A. sagrei* is less likely to disperse naturally to other islands because of its low salinity tolerance and its current limited distribution in Taiwan. We propose to focus on human activities to prevent further invasion of this introduced species.

If you are unfamiliar with the basic ecology of reptiles on Orchid Island or lack accurate information, it will seriously affect the completeness of the subsequent chapters of this book. Therefore, the next chapter will present ecological data and reproductive biology data on the nine most common reptiles on Orchid Island.

References

Ali, J. R. (2018). Islands as biological substrates: Continental. *Journal of Biogeography, 45*, 1003–1018.

Avise, J. C. (2009). Phylogeography: Retrospect and prospect. *Journal of Biogeography, 36*, 3–15.

Baeyens, D. A., & Rountree, R. L. (1983). A comparative study of evaporative water loss and epidermal permeability in an arboreal snake, *Opheodrys aestivus*, and a semi-aquatic snake, *Nerodia rhombifera*. *Comparative Biochemistry and Physiology Part A: Physiology, 76*, 301–304.

Bansal, R., & Karanth, K. P. (2010). Molecular phylogeny of *Hemidactylus* geckos (Squamata: Gekkonidae) of the Indian subcontinent reveals a unique Indian radiation and an Indian origin of Asian house geckos. *Molecular Phylogenetics and Evolution, 57*, 459–465.

Bermingham, E., & Moritz, C. (1998). Comparative phylogeography: Concepts and applications. *Molecular Ecology, 7*, 367–369.

Bernardo, P. H., Sánchez-Ramírez, S., Sánchez-Pacheco, S. J., Álvarez-Castañeda, S. T., Aguilera-Miller, E. F., Mendez-de la Cruz, F. R., & Murphy, R. W. (2019). Extreme mito-nuclear discordance in a peninsular lizard: The role of drift, selection, and climate. *Heredity, 123*, 359–370.

Betancur-R, R., Hines, A., Acero, A., Ortí, P. G., Wilbur, A. E., & Freshwater, D. W. (2011). Reconstructing the lionfish invasion: Insights into greater Caribbean biogeography. *Journal of Biogeography, 38*, 1281–1293.

Blaustein, A. R., Wake, D. B., & Sousa, W. P. (1994). Amphibian declines: Judging stability, persistence, and susceptibility of populations to local and global extinctions. *Conservation Biology, 8*, 60–71.

Borden, R. O. S. S. (2007). *Varanus salvator* (Asian water monitor) migration. *Biawak, 1*, 84.

Borges, P. A. V., & Hortal, J. (2009). Time, area and isolation: Factors driving the diversification of Azorean arthropods. *Journal of Biogeography, 36*, 178–191.

Brandley, M. C., Ota, H., Hikida, T., et al. (2012). The phylogenetic systematics of blue-tailed skinks (*Plestiodon*) and the family Scincidae. *Zoological Journal of the Linnean Society, 165*, 163–189.

Brower, L. (1996). Monarch butterfly orientation: Missing pieces of a magnificent puzzle. *Journal of Experimental Biology, 199*, 93–103.

Brown, W. C., & Alcala, A. C. (1957). Viability of lizard eggs exposed to sea water. *Copeia, 1957*, 39–41.

Brown, S. G., & Duffy, P. K. (1992). The effects of egg-laying site, temperature, and salt water on incubation time and hatching success in the gecko *Lepidodactylus lugubris*. *Journal of Herpetology, 26*, 510–513.

Brown, R. M., Siler, C. D., Oliveros, C. H., Welton, L. J., Rock, A., Rodriguez, J., Jose, E., & Diesmos, A. C. (2013). The amphibians and reptiles of Luzon Island, Philippines, VIII: The herpetofauna of Cagayan and Isabela provinces, northern Sierra Madre Mountain range. *Zookeys, 266*, 1–120.

Burkner, P.-C. (2017). An R package for Bayesian multilevel models using Stan. *Journal of Statistical Software, 80*, 1–28.

Calsbeek, R., & Smith, T. B. (2003). Ocean currents mediate evolution in Island lizards. *Nature, 426*, 332.

Carlton, J. T., Chapman, J. W., Geller, J. B., Miller, J. A., Carlton, D. A., McCuller, M. I., Treneman, N. C., Steves, B. P., & Ruiz, G. M. (2017). Tsunami-driven rafting: Transoceanic species dispersal and implications for marine biogeography. *Science, 357*, 1402–1406.

Carranza, S., & Arnold, E. N. (2006). Systematics, biogeography, and evolution of *Hemidactylus* geckos (Reptilia: Gekkonidae) elucidated using mitochondrial DNA sequences. *Molecular Phylogenetics and Evolution, 38*, 531–545.

Censky, E. J., Hodge, K., & Dudley, J. (1998). Over-water dispersal of lizards due to hurricanes. *Nature, 395*, 556.

Chang, L. Y. (2022). *Intraspecific phylogeography of the Swinhoe's tree lizard (Diploderma swinhonis): Complicated geographical history arose the unusual processes of genetic divergence within island* (Master's thesis). National Chung Hsing University, Taichung, Taiwan.

Chen, Y. G., Wu, W. S., Chen, C. H., & Liu, T. K. (2001). A date for volcanic eruption inferred from a siltstone xenolith. *Quaternary Science Reviews, 20*, 869–873.

Cheng, Y.-P., Hwang, S.-Y., & Lin, T.-P. (2005). Potential refugia in Taiwan revealed by the phylogeographical study of *Castanopsis carlesii* Hayata (Fagaceae). *Molecular Ecology, 14*, 2075–2085.

Claridge, E. M., Gillespie, R. G., Brewer, M. S., & Roderick, G. K. (2017). Stepping-stones across space and time: Repeated radiation of Pacific flightless broad-nosed weevils (Coleoptera: Curculionidae: Entiminae: Rhyncogonus). *Journal of Biogeography, 44*, 784–796.

Clark, T. M., Flis, B. J., & Remold, S. K. (2004). Differences in the effects of salinity on larval growth and developmental programs of a freshwater and a euryhaline mosquito species (Insecta: Diptera,Culicidae). *Journal of Experimental Biology, 207*, 2289–2295.

Coulson, S., Hodkinson, I., Webb, N., & Harrison, J. (2002). Survival of terrestrial soil-dwelling arthropods on and in seawater: Implications for trans-oceanic dispersal. *Functional Ecology, 16*, 353–356.

Cowie, R. H., & Holland, B. S. (2006). Dispersal is fundamental to biogeography and the evolution of biodiversity on oceanic islands. *Journal of Biogeography, 33*, 193–198.

Cowie, R. H., & Holland, B. S. (2008). Molecular biogeography and diversification of the endemic terrestrial fauna of the Hawaiian islands. *Philosophical Transactions of the Royal Society B: Biological Sciences, 363*, 3363–3376.

Cox, C. L., & Cox, R. M. (2015). Evolutionary shifts in habitat aridity predict evaporative water loss across squamate reptiles. *Evolution, 69*, 2507–2516.

Dai, C., Wang, W., & Lei, F. (2013). Multilocus phylogeography (mitochondrial, autosomal and Z-chromosomal loci) and genetic consequence of long-distance male dispersal in black-throated tits (*Aegithalos concinnus*). *Heredity, 110*, 457–465.

Davenport, J., & Macedo, E. A. (1990). Behavioural osmotic control in the euryhaline diamondback terrapin *Malaclemys terrapin*: Responses to low salinity and rainfall. *Journal of Zoology, 220*, 487–496.

Dawson, W. R., Shoemaker, V. H., & Licht, P. (1966). Evaporative water losses of some small Australian lizards. *Ecology, 47*, 589–594.

de Queiroz, A. (2005). The resurrection of oceanic dispersal in historical biogeography. *Trends in Ecology & Evolution, 20*, 68–73.

Diniz-Filho, J. A. F., Souza, K. S., Bini, L. M., Loyola, R., Dobrovolski, R., Rodrigues, J. F. M., et al. (2019). A macroecological approach to evolutionary rescue and adaptation to climate change. *Ecography, 42*, 1124–1141.

Dmi'el, R. (1985). Effect of body size and temperature on skin resistance to water loss in a desert snake. *Journal of Thermal Biology, 10*, 145–149.

Dmi'el, R., Perry, G., & Lazell, J. (1997). Evaporative water loss in nine insular populations of the lizard *Anolis cristatellus* group in the British Virgin Islands. *Biotropica, 29*, 111–116.

Dufresnes, C., Nicieza, A. G., Litvinchuk, S. N., Rodrigues, N., Jeffries, D. L., Vences, M., Perrin, N., & Martínez-Solano, Í. (2020). Are glacial refugia hotspots of speciation and cytonuclear discordances? Answers from the genomic phylogeography of Spanish common frogs. *Molecular Ecology, 29*, 986–1000.

Ellis, T. M. (1981). Tolerance of sea water by the American crocodile, *Crocodylus acutus*. *Journal of Herpetology, 15*, 187–192.

Emerson, B. (2002). Evolution on oceanic islands: Molecular phylogenetic approaches to understanding pattern and process. *Molecular Ecology, 11*, 951–966.

Esselstyn, J. A., & Oliveros, C. H. (2010). Colonization of The Philippines from Taiwan: A multilocus test of the biogeographic and phylogenetic relationships of isolated populations of shrews. *Journal of Biogeography, 37*, 1504–1514.

Esser, L. J., & Cumberlidge, N. (2011). Evidence that salt water may not be a barrier to the dispersal of Asian freshwater crabs. *The Raffles Bulletin of Zoology, 59*, 259–268.

Fisher, R. N. (1997). Dispersal and evolution of the Pacific Basin gekkonid lizards *Gehyra oceanica* and *Gehyra mutilata*. *Evolution, 51*, 906–921.

Fraser, C. I., Nikula, R., & Waters, J. M. (2011). Oceanic rafting by a coastal community. *Proceedings of the Royal Society B: Biological Sciences, 278*, 649–655.

Gabrielli, M., Nabholz, B., Leroy, T., Milá, B., & Thébaud, C. (2020). Within-Island diversification in a passerine bird. *Proceedings of the Royal Society B: Biological Sciences, 287*, 20192999.

Gannon, M. R., & Willig, M. R. (2008). Island in the storm: Disturbance ecology of plant-visiting bats on the hurricane-prone Island of Puerto Rico. In T. H. Fleming & P. A. Racey (Eds.), *Island bats: Evolution, ecology, and conservation* (p. 560). University of Chicago Press.

Gerlach, J., Muir, C., & Richmond, M. D. (2006). The first substantiated case of trans-oceanic tortoise dispersal. *Journal of Natural History, 40*, 2403–2408.

Gillespie, R. (2004). Community assembly through adaptive radiation in Hawaiian spiders. *Science, 303*, 356–359.

Gillespie, R. G., Baldwin, B. G., Waters, J. M., Fraser, C. I., Nikula, R., & Roderick, G. K. (2012). Long-distance dispersal: A framework for hypothesis testing. *Trends in Ecology & Evolution, 27*, 47–56.

Glor, R. E., Losos, J. B., & Larson, A. (2005). Out of Cuba: Overwater dispersal and speciation among lizards in the *Anolis carolinensis* subgroup. *Molecular Ecology, 14*, 2419–2432.

Guillon, M., Guiller, G., DeNardo, D. F., & Lourdais, O. (2014). Microclimate preferences correlate with contrasted evaporative water loss in parapatric vipers at their contact zone. *Canadian Journal of Zoology, 92*, 81–86.

Guja, L. K., Merritt, D. J., & Dixon, K. W. (2010). Buoyancy, salt tolerance and germination of coastal seeds: Implications for oceanic hydrochorous dispersal. *Functional Plant Biology, 37*, 1175–1186.

Hake, M., Kjellén, N., & Alerstam, T. (2001). Satellite tracking of Swedish ospreys *Pandion haliaetus*: Autumn migration routes and orientation. *Journal of Avian Biology, 32*, 47–56.

Harbaugh, D. T., Wagner, W. L., Allan, G. J., & Zimmer, E. A. (2009). The Hawaiian archipelago is a stepping stone for dispersal in the Pacific: An example from the plant genus *Melicope* (Rutaceae). *Journal of Biogeography, 36*, 230–241.

Hart, B. T., Bailey, P., Edwards, R., Hortle, K., James, K., McMahon, A., Meredith, C., & Swadling, K. (1991). A review of the salt sensitivity of the Australian freshwater biota. *Hydrobiologia, 210*(1), 105–144.

Hart, K. M., Schofield, P. J., & Gregoire, D. R. (2012). Experimentally derived salinity tolerance of hatchling Burmese pythons (*Python molurus bivittatus*) from the Everglades, Florida (USA). *Journal of Experimental Marine Biology and Ecology, 413*, 56–59.

He, J., Gao, Z., Su, Y., Lin, S., & Jiang, H. (2018). Geographical and temporal origins of terrestrial vertebrates endemic to Taiwan. *Journal of Biogeography, 45*, 2458–2470.

Heatwole, H., & Levins, R. (1972). Biogeography of the Puerto Rican Bank: Flotsam transport of terrestrial animals. *Ecology, 53*, 112–117.

Heatwole, H., & Veron, J. E. N. (1977). Vital limit and evaporative water loss in lizards (Reptilia, Lacertilia): A critique and new data. *Journal of Herpetology, 11*, 341–348.

Hedtke, S. M., & Hillis, D. M. (2011). The potential role of androgenesis in cytoplasmic–nuclear phylogenetic discordance. *Systematic Biology, 60*, 87–96.

Heger, N. A., & Fox, S. F. (1992). Viability of lizard (*Sceloporus undulatus*) eggs exposed to simulated flood conditions. *Journal of Herpetology, 26*, 338–341.

Hillman, S. S., & Gorman, G. C. (1977). Water loss, desiccation tolerance, and survival under desiccating conditions in 11 species of Caribbean *Anolis*. *Oecologia, 29*, 105–116.

Hirsh, H., & Marler, T. (2002). Damage and recovery of *Cycas micronesica* after typhoon Paka. *Biotropica, 34*, 598–602.

Holden, M., Girard, F., & Ineich, I. (2014). Eggs and hatchlings of *Hemiphyllodactylus typus* Bleeker, 1860 (Gekkonidae) on a small islet in New Caledonia. *Herpetology Notes, 7*, 509–513.

Holland, B. S., & Hadfield, M. G. (2002). Islands within an Island: Phylogeography and conservation genetics of the endangered Hawaiian tree snail *Achatinella mustelina*. *Molecular Ecology, 11*, 365–375.

Hopkins, G. R., & Brodie, E. D. (2015). Occurrence of amphibians in saline habitats: A review and evolutionary perspective. *Herpetological Monographs, 29*, 1–27.

Hsu, C. F., Tseng, H. Y., Hsiao, Y., & Ko, C. C. (2017). First record of the host plant and larvae of *Pachyrhynchus sonani* (Coleoptera: Curculionidae) on Lanyu Island, Taiwan. *Entomological Science, 20*, 288–291.

Hsu, M. H. (2019). *The possibilities of over-ocean dispersal in lizards of Taiwan and the adjacent islands* (Master's thesis). National Chung Hsing University, Taichung, Taiwan.

Hsu, M. H., Lin, J. W., Liao, C. P., Hsu, J. Y., & Huang, W. S. (2021). Trans-marine dispersal inferred from the saltwater tolerance of lizards from Taiwan. *PLOS ONE, 16*(2), e0247009.

Huang, W. S. (2007). Ecology and reproductive patterns of the agamid lizard, *Japalura swinhonis*, on an East Asian Island, with comments on the small clutch sizes of Island lizrds. *Zoological Science, 24*, 181–188.

Huang, C.-W., Lee, Y.-C., Lin, S.-M., & Wu, W.-L. (2014). Taxonomic revision of *Aegista subchinensis* (Möllendorff, 1884) (Stylommatophora, Bradybaenidae) and a description of a new species of Aegista from eastern Taiwan based on multilocus phylogeny and comparative morphology. *ZooKeys, 445*, 31–55.

Huang, C.-Y., Chen, W.-H., Wang, M.-H., Lin, C.-T., Yang, S., Li, X., Yu, M., Zhao, X., Yang, K.-M., Liu, C.-S., Hsieh, Y.-H., & Harris, R. (2018). Juxtaposed sequence stratigraphy, temporal-spatial variations of sedimentation and development of modern-forming forearc Lichi Mélange in North Luzon trough forearc basin onshore and offshore eastern Taiwan: An overview. *Earth-Science Reviews, 182*, 102–140.

Incagnone, G., Marrone, F., Barone, R., Robba, L., & Naselli-Flores, L. (2014). How do freshwater organisms cross the "dry ocean"? A review on passive dispersal and colonization processes with a special focus on temporary ponds. *Hydrobiologia, 750*, 103–123.

Ivanov, V., Lee, K. M., & Mutanen, M. (2018). Mitonuclear discordance in wolf spiders: Genomic evidence for species integrity and introgression. *Molecular Ecology, 27*, 1681–1695.

Jing, M., Yu, H.-T., Wu, S.-H., Wang, W., & Zheng, X. (2007). Phylogenetic relationships in genus *Niviventer* (Rodentia: Muridae) in China inferred from complete mitochondrial cytochrome b gene. *Molecular Phylogenetics and Evolution, 44*, 521–529.

Joly, S., McLenachan, P. A., & Lockhart, P. J. (2009). A statistical approach for distinguishing hybridization and incomplete lineage sorting. *The American Naturalist, 174*, E54–E70.

Jombart, T., Devillard, S., & Balloux, F. (2010). Discriminant analysis of principal components: A new method for the analysis of genetically structured populations. *BMC Genetics, 11*(1), 94.

Josefson, A. B. (2016). Species sorting of benthic invertebrates in a salinity gradient–importance of dispersal limitation. *PLoS One, 11*, e0168908.

Jud, Z. R., Nichols, P. K., & Layman, C. A. (2014). Broad salinity tolerance in the invasive lionfish *Pterois* spp. may facilitate estuarine colonization. *Environmental Biology of Fishes, 98*, 135–143.

Karin, B. R., Metallinou, M., Weinell, J. L., Jackman, T. R., & Bauer, A. M. (2016). Resolving the higher-order phylogenetic relationships of the circumtropical *Mabuya* group (Squamata: Scincidae): An out-of-Asia diversification. *Molecular Phylogenetics and Evolution, 102*, 220–232.

Kattan, G. H., & Lillywhite, H. B. (1989). Humidity acclimation and skin permeability in the lizard *Anolis carolinensis*. *Physiological Zoology, 62*, 593–606.

Kayashima, I., et al. (1940). Notes on a weevil (*Pachyrhynchus moniliferus* germ.) injurious to cacao-fruit in Phillippine Islands. *Transactions of the Natural History Society of Formosa, 30*, 126–129.

Kobayashi, S., Usui, R., Nomoto, K., Ushirokita, M., Denda, T., & Izawa, M. (2014). Does egg dispersal occur via the ocean in the stick insect *Megacrania tsudai* (Phasmida: Phasmatidae)? *Ecological Research, 29*, 1025–1032.

Koleva, V., Kornilev, Y., Telenchev, I., Lukanov, S., Hristova, B., & Natchev, N. (2017). Salt tolerance's toll: Prolonged exposure to saline water inflicts damage to the blood cells of dice snakes (*Natrix tessellata*). *Web Ecology, 17*, 1–7.

Komaki, S., Lin, S.-M., Nozawa, M., Oumi, S., Sumida, M., & Igawa, T. (2017). Fine-scale demographic processes resulting from multiple overseas colonization events of the Japanese stream tree frog, *Buergeria japonica*. *Journal of Biogeography, 44*, 1586–1597.

Kuo, J., Kanamoto, Z., Iizumi, H., Aioi, K., & Mukai, H. (2006). Seagrasses from the Nansei Islands, Southern Japanese archipelago: Species composition, distribution and biogeography. *Marine Ecology, 27*, 290–298.

Kurita, K., & Hikida, T. (2014). Divergence and long-distance overseas dispersals of Island populations of the Ryukyu five-lined skink, *Plestiodon marginatus* (Scincidae: Squamata), in the Ryukyu archipelago, Japan, as revealed by mitochondrial DNA phylogeography. *Zoological Science, 31*, 187–194.

Kurita, K., Nakamura, Y., Okamoto, T., Lin, S.-M., & Hikida, T. (2017). Taxonomic reassessment of two subspecies of Chinese skink in Taiwan based on morphological and molecular investigations (Squamata, Scincidae). *ZooKeys, 687*, 131–148.

Leader, J. P. (1972). Osmoregulation in the larva of the marine caddis fly, *Philanisus plebeius* (Walk.) (Trichoptera). *Journal of Experimental Biology, 57*, 821–838.
Lee, Y.-J., Hwang, S.-Y., Ho, K.-C., & Lin, T.-P. (2006). Source populations of *Quercus glauca* in the last glacial age in Taiwan revealed by nuclear microsatellite markers. *Journal of Heredity, 97*, 261–269.
Li, S.-H., Li, J.-W., Han, L.-X., Yao, C.-T., Shi, H., Lei, F.-M., & Yen, C. (2006). Species delimitation in the Hwamei *Garrulax canorus*. *Ibis, 148*, 698–706.
Lillywhite, H. B., Menon, J. G., Menon, G. K., Sheehy, C. M., 3rd, & Tu, M. C. (2009). Water exchange and permeability properties of the skin in three species of amphibious sea snakes (*Laticauda* spp.). *Journal of Experimental Biology, 212*, 1921–1929.
Lin, S.-M., Chen, C. A., & Lue, K.-Y. (2002). Molecular phylogeny and biogeography of the grass lizards genus *Takydromus* (Reptilia: Lacertidae) of East Asia. *Molecular Phylogenetics and Evolution, 22*, 276–288.
Lin, H.-D., Chen, Y.-R., & Lin, S.-M. (2012). Strict consistency between genetic and topographic landscapes of the brown tree frog (*Buergeria robusta*) in Taiwan. *Molecular Phylogenetics and Evolution, 62*, 251–262.
Losos, J. B., & Ricklefs, R. E. (2009). Adaptation and diversification on islands. *Nature, 457*, 830–836.
Losos, J. B., Schoener, T. W., & Spiller, D. A. (2003). Effect of immersion in seawater on egg survival in the lizard *Anolis sagrei*. *Oecologia, 137*, 360–362.
Lv, Y.-Y., He, K., Klaus, S., Brown, R. M., & Li, J.-T. (2018). A comprehensive phylogeny of the genus *Kurixalus* (Rhacophoridae, Anura) sheds light on the geographical range evolution of frilled swamp treefrogs. *Molecular Phylogenetics and Evolution, 121*, 224–232.
Machado, A., Rodríguez-Expósito, E., López, M., & Hernández, M. (2017). Phylogenetic analysis of the genus *Laparocerus*, with comments on colonisation and diversification in Macaronesia (Coleoptera, Curculionidae, Entiminae). *ZooKeys, 651*, 1–77.
Mojica, E.-R. E., & Micor, J. R. L. (2007). Bioactivity study of *Barringtonia asiatica* (Linnaeus) kurz. Seed aqueous extract in *Artemia salina*. *International Journal of Botany, 3*, 325–328.
Naidu, S. (2001). Water balance and osmoregulation in *Stenocara gracilipes*, a wax-blooming tenebrionid beetle from the Namib Desert. *Journal of Insect Physiology, 47*, 1429–1440.
Naidu, S. G. (2006). Haemolymph amino acid, sugar and glycerol levels in the Namib Desert tenebrionid *Physadesmia globosa* (Coleoptera: Tenebrionidae) during dehydration and rehydration. *European Journal of Entomology, 103*, 305.
Neill, W. T. (1958). The occurrence of amphibians and reptiles in saltwater areas, and a bibliography. *Bulletin of Marine Science, 8*, 1–97.
Nicolson, S. W. (1980). Water balance and osmoregulation in *Onymacris plana*, a tenebrionid beetle from the Namib desert. *Journal of Insect Physiology, 26*, 315–320.
Nikula, R., Spencer, H. G., & Waters, J. M. (2013). Passive rafting is a powerful driver of transoceanic gene flow. *Biology Letters, 9*, 20120821.
Ôshiro, Y. (1991). Studies on the Pachyrhynchus infernalis effecting *Mangifera indica* in Ishigaki Island. *Okinawa Agriculture Research Society, 26*, 19–23.
Ota, H. (1991). Taxonomic redefinition of *Japalura swinhonis* Günther (Agamidae: Squamata), with a description of a new subspecies of *J. polygonata* from Taiwan. *Herpetologica, 47*, 280–294.
Ota, H. (1997). Historical biogeographical implications in the variation and diversity of amphibians and reptiles in Taiwan. In K. Y. Lue & T. H. Chen (Eds.), *Proceedings of the symposium on the phylogeny, biogeography and conservation of fauna and flora of East Asian region* (pp. 75–86). National Science Council.
Packard, M. J., & DeMarco, V. G. (1991). Eggshell structure and formation in eggs of oviparous reptiles. In D. C. Deeming & M. W. J. Ferguson (Eds.), *Egg incubation: Its effects on embryonic development in birds and reptiles* (pp. 53–70). Cambridge University Press.
Packard, G. C., Taigen, T. L., Packard, M. J., & Shuman, R. D. (1979). Water-vapor conductance of testudinian and crocodilian eggs (class Reptilia). *Respiration Physiology, 38*, 1–10.

Packard, M. J., Packard, G. C., & Boardman, T. J. (1982). Structure of eggshells and water relations of reptilian eggs. *Herpetologica, 38*, 136–155.

Pallarés, S., Arribas, P., Bilton, D. T., Millán, A., & Velasco, J. (2015). The comparative osmoregulatory ability of two water beetle genera whose species span the fresh-hypersaline gradient in inland waters (Coleoptera: Dytiscidae, Hydrophilidae). *PLoS One, 10*, e0124299.

Parent, C. (2012). Biogeographical and ecological determinants of land snail diversification on islands. *American Malacological Bulletin, 30*, 207–215.

Peck, S. B. (1994). Aerial dispersal of insects between and to islands in the Galapagos Archipelago, Ecuador. *Annals of the Entomological Society of America, 87*, 218–224.

Peig, J., & Green, A. J. (2009). New perspectives for estimating body condition from mass/length data: The scaled mass index as an alternative method. *Oikos, 118*, 1883–1891.

Peters, J. L., Zhuravlev, Y., Fefelov, I., Logie, A., & Omland, K. E. (2007). Nuclear loci and coalescent methods support ancient hybridization as cause of mitochondrial paraphyly between gadwall and falcated duck (*Anas* spp.). *Evolution, 61*, 1992–2006.

Peters, J. L., Bolender, K. A., & Pearce, J. M. (2012). Behavioural vs. molecular sources of conflict between nuclear and mitochondrial DNA: The role of male-biased dispersal in a Holarctic Sea duck. *Molecular Ecology, 21*, 3562–3575.

Phuong, M. A., Bi, K., & Moritz, C. (2017). Range instability leads to cytonuclear discordance in a morphologically cryptic ground squirrel species complex. *Molecular Ecology, 26*, 4743–4755.

Prange, H. D., & Schmidt-Nielsen, K. (1969). Evaporative water loss in snakes. *Comparative Biochemistry and Physiology, 28*, 973–975.

Puppo, P., Curto, M., & Meimberg, H. (2016). Genetic structure of *Micromeria* (Lamiaceae) in Tenerife, the imprint of geological history and hybridization on within-Island diversification. *Ecology and Evolution, 6*, 3443–3460.

Qu, Y., Song, G., Gao, B., Quan, Q., Ericson, P. G. P., & Lei, F. (2015). The influence of geological events on the endemism of east Asian birds studied through comparative phylogeography. *Journal of Biogeography, 42*, 179–192.

Quigley, D. T. G., Gainey, P. A., & Dinsdale, A. (2014). First records of *Barringtonia asiatica* (Lecythidaceae) from UK waters and a review of north-western European records. *New Journal of Botany, 4*, 107–109.

Ramsey, L. A., Walker, R. T., & Jackson, J. (2007). Geomorphic constraints on the active tectonics of southern Taiwan. *Geophysical Journal International, 170*, 1357–1372.

Raxworthy, C. J., Forstner, M. R. J., & Nussbaum, R. A. (2002). Chameleon radiation by oceanic dispersal. *Nature, 415*, 784–787.

Riddle, W. A. (1986). Hemolymph osmoregulation in three species of beetles. *Comparative Biochemistry and Physiology Part A: Physiology, 83*, 619–626.

Roberts, J. B., & Lillywhite, H. B. (1983). Lipids and the permeability of epidermis from snakes. *Journal of Experimental Zoology, 228*, 1–9.

Russell, J. C., Miller, S. D., Harper, G. A., MacInnes, H. E., Wylie, M. J., & Fewster, R. M. (2010). Survivors or reinvaders? Using genetic assignment to identify invasive pests following eradication. *Biological Invasions, 12*, 1747–1757.

Schoener, A., & Schoener, T. W. (1984). Experiments on dispersal: Short-term floatation of insular anoles, with a review of similar abilities in other terrestrial animals. *Oecologia, 63*, 289–294.

Schrago, C. G., & Russo, C. A. (2003). Timing the origin of New World monkeys. *Molecular Biology and Evolution, 20*, 1620–1625.

Schultze, W. (1923). A monograph of the pachyrrhynchid group of the Brachyderinae, Curculionidae: Part I. The genus *Pachyrrhynchus* Germar. *The Philippine Journal of Science, 23*, 609–673.

Semenov, G. A., Linck, E., Enbody, E. D., Harris, R. B., Khaydarov, D. R., Alström, P., Andersson, L., & Taylor, S. A. (2021). Asymmetric introgression reveals the genetic architecture of a plumage trait. *Nature Communications, 12*, 1019.

Shang, G. (2001). *Natural portraits of lizards of Taiwan*. Big Tree Press.

Shang, G., Yang, Y. J., & Lee, P. H. (2009). *The atlas of amphibians and reptiles of Taiwan*. Taipei, Taiwan: Owl Publishing House.

Shih, H.-T., Hung, H.-C., Schubart, C. D., Chen, C. A., & Chang, H.-W. (2006). Intraspecific genetic diversity of the endemic freshwater crab *Candidiopotamon rathbunae* (Decapoda, Brachyura, Potamidae) reflects five million years of the geological history of Taiwan. *Journal of Biogeography, 33*, 980–989.

Shinohara, A., Kawada, S.-I., Son, N. T., Koshimoto, C., Endo, H., Can, D. N., & Suzuki, H. (2014). Molecular phylogeny of East and Southeast Asian fossorial moles (Lipotyphla, Talpidae). *Journal of Mammalogy, 95*, 455–466.

Shu, L., Suter, M. J.-F., & Räsänen, K. (2015). Evolution of egg coats: Linking molecular biology and ecology. *Molecular Ecology, 24*, 4052–4073.

Sibuet, J.-C., & Hsu, S.-K. (2004). How was Taiwan created? *Tectonophysics, 379*, 159–181.

Simberloff, D., & Wilson, E. O. (1969). Experimental zoogeography of islands: The colonization of empty islands. *Ecology, 50*, 278–296.

Snyder, J. O. (1917). Notes on Hawaiian lizards. *Proceedings of the United States National Museum, 54*, 19–25.

Song, H., Buhay, J. E., Whiting, M. F., & Crandall, K. A. (2008). Many species in one: DNA barcoding overestimates the number of species when nuclear mitochondrial pseudogenes are coamplified. *Proceedings of the National Academy of Sciences, 105*, 13486–13491.

Song, H., Moulton, M. J., & Whiting, M. F. (2014). Rampant nuclear insertion of mtDNA across diverse lineages within Orthoptera (Insecta). *PLoS One, 9*, e110508.

Spadavecchia, G., Chiocchio, A., Bisconti, R., & Canestrelli, D. (2021). Paso doble: A two-step late Pleistocene range expansion in the Tyrrhenian tree frog *Hyla sarda*. *Gene, 780*, 145489.

Starr, C. K., & Wang, H.-Y. (1992). Pachyrhynchine weevils (Coleoptera: Curculionidae) of the islands fringing Taiwan. *Journal of Taiwan Museum, 45*, 5–13.

Streicher, J. W., & Day, J. J. (2020). The toad's warts: Discordance creates bumpy expectations of mitochondrial-nuclear evolution between species. *Molecular Ecology, 29*, 3400–3402.

Tabak, M. A., Poncet, S., Passfield, K., Carling, M. D., & Martinez del Rio, C. (2014). The relationship between distance and genetic similarity among invasive rat populations in The Falkland Islands. *Conservation Genetics, 16*, 125–135.

Templeton, J. R., Murrish, D. E., Randall, E. M., & Mugaas, J. N. (1972). Salt and water balance in the desert iguana, *Dipsosaurus dorsalis*. *Zeitschrift für Vergleichende Physiologie, 76*, 255–269.

Teng, L. S. (1990). Geotectonic evolution of late Cenozoic arc-continent collision in Taiwan. *Tectonophysics, 183*, 57–76.

Thiel, M., & Gutow, L. (2005). The ecology of rafting in the marine environment. I. The floating substrata. *Oceanography and Marine Biology: An Annual Review, 42*, 181–264.

Toews, D. P. L., & Brelsford, A. (2012). The biogeography of mitochondrial and nuclear discordance in animals. *Molecular Ecology, 21*, 3907–3930.

Tones, P. I. (1978). Osmoregulation in adults and larvae of Hygrotus salinarius Wallis (Coleoptera, Dytiscidae). *Comparative Biochemistry and Physiology Part A: Physiology, 60*, 247–250.

Tsai, C.-L., Kubota, K., Pham, H.-T., & Yeh, W.-B. (2021). Ancestral haplotype retention and population expansion determine the complicated population genetic structure of the hilly lineage of *Neolucanus swinhoei* complex (Coleoptera, Lucanidae) on the subtropical Taiwan Island. *Insects, 12*, 227.

Tseng, S.-P., Wang, C.-J., Li, S.-H., & Lin, S.-M. (2015). Within-Island speciation with an exceptional case of distinct separation between two sibling lizard species divided by a narrow stream. *Molecular Phylogenetics and Evolution, 90*, 164–175.

Tseng, H.-Y., Huang, W.-S., Jeng, M.-L., Villanueva, R. J. T., Nuñeza, O. M., & Lin, C.-P. (2018). Complex inter-island colonization and peripatric founder speciation promote diversification of flightless *Pachyrhynchus* weevils in the Taiwan–Luzon volcanic belt. *Journal of Biogeography, 45*, 89–100.

Tsou, C.-H., & Mori, S. A. (2002). Seed coat anatomy and its relationship to seed dispersal in subfamily Lecythidoideae of the Lecythidaceae (the Brazil nut family). *Botanical Bulletin of Academia Sinica, 43*, 37–56.

Vences, M., Vieites, D. R., Glaw, F., Brinkmann, H., Kosuch, J., Veith, M., & Meyer, A. (2003). Multiple overseas dispersal in amphibians. *Proceedings of the Royal Society B: Biological Sciences, 270*, 2435–2442.

Vidal, N., Azvolinsky, A., Cruaud, C., & Hedges, S. B. (2008). Origin of tropical American burrowing reptiles by transatlantic rafting. *Biology Letters, 4*, 115–118.

Walker, L. R. (1991). Tree damage and recovery from hurricane Hugo in Luquillo experimental forest, Puerto Rico. *Biotropica, 23*, 379–385.

Wang, K., Che, J., Lin, S., Deepak, V., Aniruddha, D.-R., Jiang, K., Jin, J., Chen, H., & Siler, C. D. (2019). Multilocus phylogeny and revised classification for mountain dragons of the genus *Japalura sl* (Reptilia: Agamidae: Draconinae) from Asia. *Zoological Journal of the Linnean Society, 185*, 246–267.

Warren, B. H., Bermingham, E., Bowie, R. C. K., Prys-Jones, R. P., & Thébaud, C. (2003). Molecular phylogeography reveals Island colonization history and diversification of western Indian Ocean sunbirds (*Nectarinia*: Nectariniidae). *Molecular Phylogenetics and Evolution, 29*, 67–85.

Wegener, J. E., Gartner, G. E., & Losos, J. B. (2014). Lizard scales in an adaptive radiation: Variation in scale number follows climatic and structural habitat diversity in *Anolis* lizards. *Biological Journal of the Linnean Society, 113*, 570–579.

Weng, Y. M., Yang, M. M., & Yeh, W. B. (2016). A comparative phylogeographic study reveals discordant evolutionary histories of alpine ground beetles (Coleoptera, Carabidae). *Ecology and Evolution, 6*, 2061–2073.

Whittaker, R. J., & Fernández-Palacios, J. M. (2007). *Island biogeography: Ecology, evolution, and conservation* (2nd ed.). Oxford University Press.

Withers, S. J., Parsons, S., Hauber, M. E., Kendrick, A., & Lavery, S. D. (2021). Genetic divergence between isolated populations of the North Island New Zealand rifleman (*Acanthisitta chloris granti*) implicates ancient biogeographic impacts rather than recent habitat fragmentation. *Ecology and Evolution, 11*, 5998–6014.

Wu, S.-P., Huang, C.-C., Tsai, C.-L., Lin, T.-E., Jhang, J.-J., & Wu, S.-H. (2016). Systematic revision of the Taiwanese genus Kurixalus members with a description of two new endemic species (Anura, Rhacophoridae). *ZooKeys, 557*, 121–153.

Yang, T. F., Lee, T., Chen, C.-H., Cheng, S.-N., Knittel, U., Punongbayan, R. S., & Rasdas, A. R. (1996). A double Island arc between Taiwan and Luzon: Consequence of ridge subduction. *Tectonophysics, 258*, 85–101.

Yang, S. F., Komaki, S., Brown, R. M., & Lin, S. M. (2018). Riding the Kuroshio current: Stepping stone dispersal of the Okinawa tree lizard across the east Asian Island arc. *Journal of Biogeography, 45*, 37–50.

Yeh, W.-B., Chang, Y.-L., Lin, C.-H., Wu, F.-S., & Yang, J.-T. (2004). Genetic differentiation of *Loxoblemmus appendicularis* complex (Orthoptera: Gryllidae): Speciation through vicariant and glaciation events. *Annals of the Entomological Society of America, 97*, 613–623.

Yeh, H.-Y., Tseng, H.-Y., Lin, C.-P., Liao, C.-P., Hsu, J.-Y., & Huang, W.-S. (2018). Rafting on floating fruit is effective for oceanic dispersal of flightless weevils. *Journal of Experimental Biology, 221*, jeb190488.

You, C.-W., Poyarkov, N. A., Jr., & Lin, S.-M. (2015). Diversity of the snail-eating snakes *Pareas* (Serpentes, Pareatidae) from Taiwan. *Zoologica Scripta, 44*, 349–361.

Ecology and Reproductive Biology of Nine Species of Lizards on Orchid Island

3

Abstract

The basic ecological data in this chapter make a significant contribution to subsequent chapters. For example, the clutch size, reproductive period, and egg-laying sites of lizard species with elastic eggshells are the primary prey targets of reptile-egg-eating kukri snakes. In terms of snake predation, what kind of adaptations will they develop?

Five of the nine species of insectivorous lizards are active foragers, while four of them are sit-and-wait foragers. The distribution of habitat types and microhabitats varies greatly between species. Although the prey consumed is varied, there are big differences between species.

Keywords

Reproductive cycle · Microhabitat · Food habit · Clutch sizes · Lizards · Skinks · Gecko · Morphology

3.1 Introduction

Because islands are natural, simple, and isolated habitats, they provide an excellent environment for evolutionary ecologists to study individual adaptations as well as populations, assemblages, and ecosystems. For a long time, the diverse tropical island fauna has attracted researchers from a variety of disciplines. Even today, understanding the historical processes, current patterns, and mechanisms of island biodiversity is one of the main goals of evolutionary and community biology (Grant, 1998; Ota, 1999).

Reptiles can be important model organisms for various ecological and evolutionary studies because they are (1) moderately large for field observation, (2) moderately

W.-S. Huang, *Reptiles in the Changing Climate*, Fascinating Life Sciences,
https://doi.org/10.1007/978-981-92-1887-5_3

dispersing for studying biogeographic dispersal and isolation, and (3) relatively high in diversity for studying ecological relationships among species.

Some previous biogeographers believed that the fauna of Orchid Island is characterized by a higher proportion of Philippine elements compared to the main island of Taiwan (Kuroda, 1925; Kano, 1933). To date, four of the 16 native reptiles known from Orchid Island (*G. kikuchii, Lepidodactylus yami, Eutropis multicarinata, and E. cumingi*; Chap. 2) have been assigned to the Philippine element, while most of the remaining species are probably derived from the spread of Taiwan (Ota, 1991; Ota & Huang, 2000).

Most temperate oviparous lizards appear to exhibit spring gonadal development in both males and females, followed by courtship, mating, and oviposition (Fitch, 1970; Duvall et al., 1982), in contrast, tropic lizards typically breed continuously (Huang, 1997a). Although Taiwan is usually classified as a subtropical region, it does have high-altitude mountains at about 4000 m above sea level. Despite being categorized as a subtropical region, Taiwan does have high mountains that are about 4000 m above sea level. In the past, I showed that three Taiwanese high-elevation lizards, *Plestiodon* (previously *Eumeces*) *elegans*, *Sphenomorphus taiwanensis*, and *Takydromus hsueshanensis*, exhibit annual breeding patterns that are comparable to those of temperate lizards (Huang, 1996b, 1997a). Although low-altitude lizards in Taiwan have also been the subject of research (Cheng, 1987), lizards from tropical regions, such as those from Orchid Island, have not received as much attention.

According to general consensus, a species of lizard's reproductive cycle may be a result of simple phenotypic responses to its environment (e.g., ambient temperature, photoperiod, precipitation, resource dynamics) (Huang, 1998), implied phenotypic responses to its environment (Duvall et al., 1982; Duvall et al., 1993), the effects of systematic developmental constraints (James & Shine, 1985; Vitt, 1986; Ota & Lin, 1994), or a combination of these factors. Numerous species' reproductive cycles have been compared in an effort to determine the root causes of various reproductive patterns (e.g., Vitt, 1986). Although some species of Taiwanese lizards have been studied, few such attempts have been made to date (Cheng & Lin, 1977, 1978; Kato & Ota, 1994; Huang, 1996a, b, 1997a, b, 1998). Additionally, few studies have focused on predation factors, which may be crucial in controlling the timing of reproduction.

The length of the breeding season on islands is explained by two hypotheses (Stamps & Buechner, 1985). 1. Resource hypothesis: The first breeding date is anticipated to be similar to or earlier due to the higher resource levels in each territory of the island and terrestrial habitats. If the start of the breeding season is determined by the extreme climate, then a milder island climate should promote earlier breeding on the island. In contrast, the defense hypothesis: for island territorial owners, the defense hypothesis predicts that breeding will begin later. It implies that the abundant food in the territory will be consumed by the animals living there, that outsiders or conspecifics will eat the food, or that too much time will be spent engaging in aggressive behavior (Brown, 1975; Stamps & Buechner, 1985). These authors, however, did not take into account the broader and more intricate aspects of herding; they only looked at populations. Here, I show that the snake that eats reptile eggs, *Oligodon formosanus*, has a significant impact on the reproductive season of the lizard assemblage on Orchid Island. The complexity of ecosystems has long been acknowledged (Greene & Jaksic, 1983; Vitt & Zani, 1996), and the interactions and interactions of extremely diverse ecosystems, such as islands or lowland forests, can be used to describe the structure of natural communities (Greene & Jaksic, 1983; Raven, 1987; Vitt & de Carvalho, 1995; Grant, 1998). However, because the majority of ecological studies have only looked at one or two species, they have not addressed issues regarding how different species interact in complex systems. I provide ecological, morphological, and life history information on tropical reptiles found on Orchid Island in this chapter.

The dynamics and structure of communities can be significantly impacted by predators (Schmitz et al., 1997; Schmitz & Booth, 1997). Two significant studies that also looked at the effects of predation at various levels of organization in natural systems provide evidence for this generalization. Examining the lethal direct effects of predators on prey populations and figuring out how these direct effects cascade along food chains throughout the direct and indirect pathways of energy transfer for all species in the community is the first step in studying food web interactions (Sih, 1987; Schoener, 1989; Spiller & Schoener, 1990; Schmitz, 1994; Schmitz & Booth, 1997). Examining the mechanisms by which predators cause aversive changes in prey behavior or life history allocation is the second strategy (Sih, 1987; Dodson & Havel, 1988; Lima & Dill, 1990; Werner, 1991; Hik, 1995). Ecologists have recently realized how crucial it is to combine these two research projects in order to analyze the population and behavioral impacts of predators on community dynamics (Huang & Sih, 1990; Werner, 1992; Werner & Anholt, 1996). The majority of studies have previously been carried out under controlled and standardized laboratory conditions because it is challenging to conduct such studies in the field due to the complexity of most communities. Similar to this, islands' distinct geographic boundaries can act as a natural framework for researching patterns of relationships between species assemblages and the food web. I describe my field observations of Orchid Island's lizards, snakes, insects, plants, and some invertebrates in this chapter.

Traditional food web models ignore the egg stage and limit predator–prey relationships to adult individuals (Winemiller et al., 2001). However, when egg-eating animals are found in oviparous colonies of few species, such as the egg-eating snake *Oligodon formosanus*, which is found on Orchid Island where there are few reptile species, the egg stage is clearly important for mutualistic relationships. I have therefore used information on stomach contents and field observations to study the general food web structure as well as the egg-based food web of the reptiles on Orchid Island.

There has been a seasonal variation in mean clutch size in lizard populations, according to numerous authors (Williams, 1966b; Pianka, 1976; Nussbaum, 1981; DeMarco, 1989). According to their theories, selection factors such as resource accessibility, a competitive environment (MacArthur & Wilson, 1967; Pianka, 1972) or community environment (Murphy, 1968; Tinkle, 1969; Stearns, 1976) are selection factors that shape life history traits such as clutch size. The reproductive effort of a species must be optimized by various selection factors in order to reach maximum fitness. Each individual organism must properly allocate reproductive energy to each offspring in addition to optimizing reproductive effort (Williams, 1966a; Pianka, 1976; Wilbur, 1977; Congdon et al. 1983). The selection environment will optimize the quantity and size of the offspring, and there may be a trade-off between quantity and size. More energetic investment per offspring may increase parental fitness if offspring survival rises(DeMarco, 1989), but it may also reduce the number of offspring. In response to a more competitive environment later in the reproductive season, females produce larger offspring than they did earlier in the season (Ferguson & Bohlen, 1978). In order to deal with increased competition among themselves and the pressure from predators during the reproductive season, Ferguson and Bohlen (1978) hypothesized that lizards that are larger at hatching have a greater survival advantage. According to a different theory called "spreading the risk (bet-hedging)"(Nussbaum, 1981), females' access to food during late egg laying is unpredictable, so selection is typically more cautious in late clutches to ensure that each egg is at least minimally supplied. Few studies, however, have examined the effects of high predation pressure on clutch egg size, particularly in the presence of numerous snakes that consume eggs. In this study, I looked at Orchid Island lizard clutch characteristics to find out three things: (1) the seasonal clutch size of nine lizard species, (2) the impact of reptilian egg-eating snakes on this process, and (3) how females of *Eutropis longicaudata* produce large clutches early in the breeding season and small clutches later in the breeding season, hatch large-sized offspring.

In summary, the main objectives of this study were to (1) describe the reptile assemblage of tropical Orchid Island and its origin and evolution; (2) examine the timing of activity, habitat and microhabitat distribution, and resource use patterns (adult and egg food webs) among the nine species of lizards; (3) understand the reproductive patterns of the nine species, including differences in male and female body size, clutch size, and egg; (4) describe the ecology of eggs and how one reptile, the egg-eating snake, affects reproductive patterns, constant clutch size, and variation in nesting habitat in lizards; (5) describe how females produce large clutches of eggs in the late season; and (6) food web of the reptile on Orchid Island.

3.2 *Eutropis longicaudata* (EL)

3.2.1 Habitat Description

The majority of EL was first noticed in deciduous layers, in caves, or at the edges of forests. The distribution of EL's habitat and microhabitats shows that it is a thermoregulatory species, with most of its occurrences occurring in habitats with direct sunlight and microhabitats.

Body measurements include the length, weight, width, height, and diameter of the waist as well as the tail, lengths of the third and fourth toes, forelimbs, and hindlimbs. EL engaged in active reproduction from February to August. 153 samples in total were gathered. The mean body mass of adult males (46.8 g, $n = 51$; range:

Table 3.1 Morphology of 71 sexually mature male and 82 female *Eutropis longicaudata*

Trait	Male mean ± SE (min, max)	Females mean ± SE (min, max)	*p*-value
BM[a] (g)	46.8 ± 1.4 (26.1, 71.3)	36.9 ± 0.9 (26.6, 64.4)	<0.001
FLL (mm)	32.7 ± 0.2 (28.5, 36.4)	30.3 ± 0.2 (27.2, 36.1)	<0.001
FTL (mm)	8.1 ± 0.1 (6.8, 9.3)	7.6 ± 0.1 (6.3, 8.9)	<0.001
HD (mm)	11.7 ± 0.2 (9.6, 15.2)	10.3 ± 0.1 (8.1, 12.9)	<0.001
HL (mm)	26.5 ± 0.2 (22.8, 29.5)	23.7 ± 0.2 (21.2, 28.4)	<0.001
HLL (mm)	47.3 ± 0.3 (38.5, 52.0)	43.7 ± 0.3 (32.1, 48.3)	<0.001
HW (mm)	17.6 ± 0.2 (13.8, 20.6)	15.3 ± 0.1 (13.3, 18.3)	<0.001
SVL (mm)	118.7 ± 0.8 (100.9, 130.4)	113.5 ± 0.7 (98.1, 126.8)	0.01
TL[a] (mm)	208.8 ± 5.9 (112.5, 281.0)	194.7 ± 4.8 (116.0, 274.6)	0.07
TTL (mm)	8.0 ± 0.9 (6.1, 9.2)	7.6 ± 0.1 (6.0, 8.9)	<0.001
WD (mm)	17.7 ± 0.2 (14.5, 21.2)	16.6 ± 0.2 (13.1, 20.9)	<0.001

BM body mass, *FLL* forelimb length, *FTL* fourth toe length, *HD* head depth, *HL* head length, *HLL* hind limb length, *HW* head width, *SVL* snout-vent length, *TL* tail length, *TTL* third toe length, *WD* waist diameter

[a]Some individuals lost their tails while being caught; thus, only 51 males and 62 females were analyzed for TL and BM

26.10–71.31 g) was significantly higher than that of adult females (36.9 g, $n = 62$; range: 26.63–64.44 g), and the mean body length of adult males and females was 118.7 ($n = 70$) and 113.5 ($n = 78$) mm, respectively (Table 3.1). Males had longer, taller, and wider heads, longer limbs, and wider waist circumference than females when size effects were removed in addition to morphological differences between adult sexes based primarily on size (Table 3.1), while there were no sex differences in size between TLs of adult males (112.47–281.00 mm) and adult females (116.01–274.57 mm) at maturity (Table 3.1).

3.2.2 Food Habits

Sixty one of the 148 lizards (41.2%) with individual samples collected had food in their stomachs. On Orchid Island, Orthoptera (31.06%), Hymenoptera (20.45%), and Hemiptera (15.15%) made up the majority of the lizard's diet. The diet was primarily composed of four prey groups: Tettigoniidae, Cricket, Scarabaeidae, and Cicadellidae. Seven lizards' stomachs contained plant foods like seeds, leaves, and fruits (Table 3.2). Two of the four ELs consumed *Gekko hokouensis* and *Hemidactylus frenatus*, while the other two consumed the eggs of *Sphenomorphus incognitus* and *Eutropis multicarinatus*.

3.2.3 Female Reproductive Cycle

Only females with an SVL greater than or equal to 98.12 mm were regarded as adults because this was the smallest reproductively active female of EL. In February, all

Table 3.2 Diet summary for *Eutropis longicaudata* from Orchid Island. Frequency is the number of lizards containing each prey item

Prey type	Frequency	% frequency
Coleoptera	**27**	**20.50**
Cicindelidae,	4	3.03
Tenebrionidae	1	0.76
Elateridae	1	0.76
Curculionidae	3	2.27
Scarabaeidae	9	6.82
Unknown	9	6.82
Diptera	**1**	**0.80**
Hemiptera	**36**	**27.30**
Cicadellidae	9	6.82
Gymnocerata	12	9.09
Unknown	15	11.36
Hymenoptera	**3**	**2.27**
Formicidae	3	2.27
Lepidoptera	**4**	**3.03**
Sphingidae	4	3.03
Mantodea	**1**	**0.76**
Mantidae	1	0.76
Orthoptera	**41**	**31.10**
Grylidae	2	1.52
Acrididae	9	6.82
Tettigoniidae	25	18.90
Unknown	5	3.79
Pseudoscorpionida	**3**	**2.27**
Cheliferidae	3	2.27
Insect larvae	4	3.03
Spider	1	0.76
Vertebrate	**4**	**3.03**
Gekkonidae	2	1.52
Gekko hokounensis	1	0.76
Hemidactylus frenatus	1	0.76
Lizards eggs	**2**	**1.52**
Scincidae	2	1.52
Sphenomorphus incognitus	1	0.76
Eutropis multicarinata	1	0.76
Plant	7	3.79
Seeds	5	3.79
Leaves	1	0.76
Fruit	1	0.76
Total	132	100.00

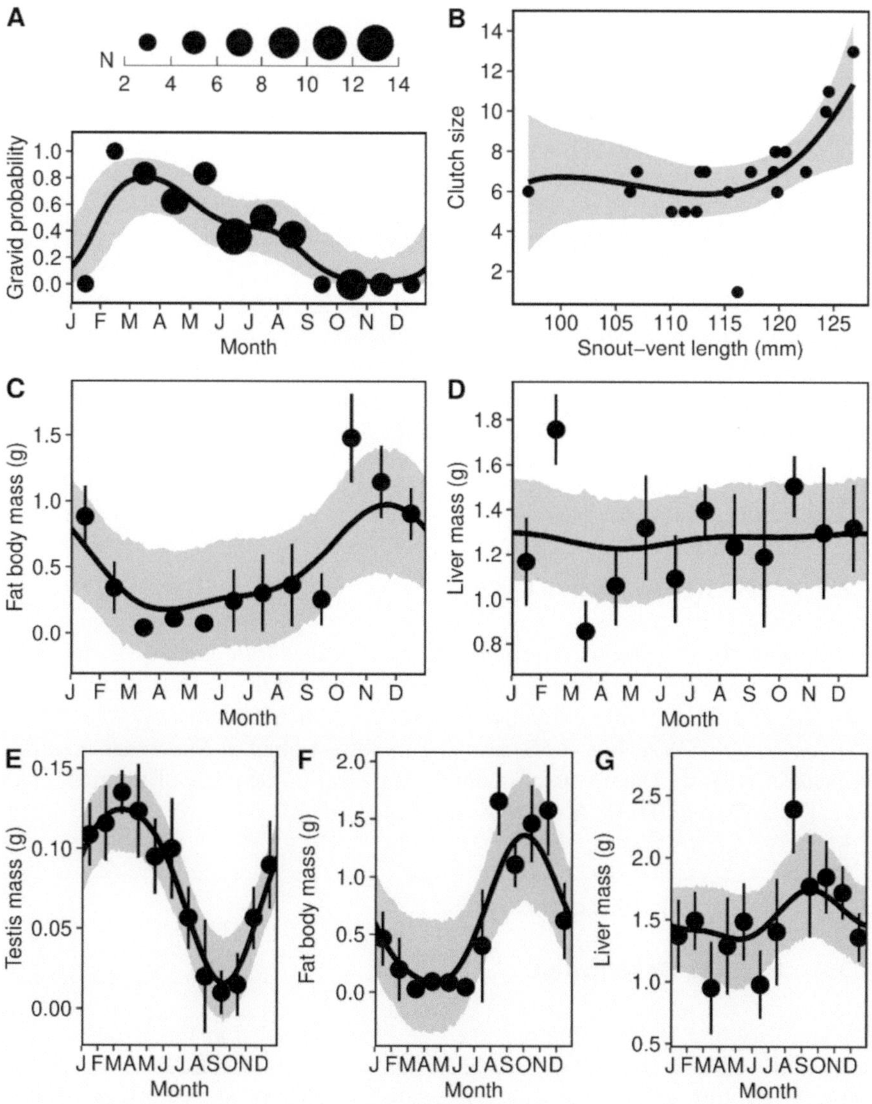

Fig. 3.1 Reproductive characters in *Eutropis longicaudata* on Orchid Island, Taiwan: monthly distribution of gravid probability (**a**). Adult females were reproductively active from February to August, and all adult females showed non-reproductive from September to January. Correlation between female body size (SVL) and the number of oviductal eggs (**b**). Monthly variation (mean ± SE) in mean fat body mass (**c**), liver mass (**d**) of females, and male mean left testis mass (**e**), fat body mass (**f**), and liver mass (**g**) of males during the annual reproductive cycle of *Eutropis longicaudata* on Orchid Island, Taiwan. (Redrawn from Huang (2006a), Copeia)

adult females were actively reproducing; in March and May, 83%; in April, 63%; in June, 36%; in July, 50%; and in August, 38%. (Fig. 3.1a). All adult females were not gravid from September to January (Fig. 3.1a).

On February 19 and March 22, respectively, the first females with oocytes and oviductal eggs were discovered. The first females with larger follicles and females with oviductal eggs were separated by 31 days. I take this to mean that this is roughly how long it takes for a female to lay a clutch of eggs. Two pieces of evidence support the hypothesis that females lay two clutches of eggs per reproductive season: (1) 5 of the 22 females who were captured in May and June 1998 still had enlarged yolk sacs after laying eggs; and (2) the reproductive season lasted roughly 6 months, which corresponds to the time period previously suggested for females to lay an egg clutch.

Clutch size was positively correlated with female SVL (Y = 0.227 X–19.414; $r = 0.68$; $n = 22$; $p < 0.001$) and averaged 6.55 ± 3.48 (X ± SD; range 1–13) (Fig. 3.1b). Female fat body and liver mass showed significant monthly variation ($F_{11,65} = 8.33$, $p = 0.000$ for fat body; $F_{11,65} = 2.82$, $p = 0.005$ for liver, ANOVA). Prior to the dormant season from October to January, fat body mass increased after the reproductive season from March to August (Fig. 3.1c). Although liver mass fluctuated between months, it did not do so as dramatically as fat body mass (Fig. 3.1d).

3.2.4 Male Reproductive Cycle

Left testicular mass increased from December and peaked from January to June, with a significant monthly variation ($F_{11,57} = 5.33$, $p < 0.0001$; ANCOVA, Fig. 3.1e). Male fat body mass varied significantly from month to month ($F_{11,57} = 11.25$, $p = 0.000$; ANCOVA, Fig. 3.1f), reaching its lowest point in March and reaching its peak in August. There was no monthly variation in the mass of the male liver ($F_{11,57} = 2.93$, $p = 0.004$; ANCOVA, Fig. 3.1g).

3.3 *Eutropis multicarinata* (EM)

3.3.1 Habitat Description

The majority of EM were first observed on concrete walls, in forest edges, or in deciduous layers. Nearly half of the population ($n = 10$) lived in the deciduous layer, followed by 40% ($n = 8$) on concrete walls and 10% ($n = 2$) on forest edges. Habitats on concrete walls were warmer than those on deciduous layers and forest edges. The habitat and microhabitat distribution of EM indicates that it is a thermoregulatory species, occurring most frequently in habitats with direct sunlight and microhabitats.

Body measurements included length, weight, waistline diameter, tail length, length of the third and fourth toes, forelimb and hindlimb length, and height of the head.

EM in this study was reproductively active from April to June. Twenty samples in all were gathered. While the mean body mass of adult males (18.37 g, $n = 10$; range: 10.72–21.15 g) was significantly higher than that of adult females (16.46 g, $n = 9$;

Table 3.3 Morphology of 11 sexually mature male and 9 female *Eutropis multicarinata*

	Males		Females		
	Mean ± SE	Range	Mean ± SE	Range	*p*-value
SVL (mm)	83.36 ± 1.72	73.19–89.20	84.01 ± 1.30	77.56–87.92	0.747
BM (g)	18.37 ± 1.19	10.72–21.15	16.46 ± 0.84	11.52–20.08	0.029
WL (mm)	13.72 ± 0.31	12.08–14.80	13.63 ± 0.21	12.75–14.40	0.228
TL (mm)	126.84 ± 1.19	66.18–161.49	127.84 ± 6.83	100.32–155.22	0.834
FLL (mm)	25.90 ± 0.56	22.35–28.03	25.01 ± 0.45	22.95–26.64	0.015
HLL (mm)	40.53 ± 0.76	35.78–43.40	38.18 ± 0.55	35.71–39.81	< 0.001
TTL (mm)	7.92 ± 0.17	7.26–8.81	7.35 ± 0.17	6.76–8.03	0.003
FTL (mm)	7.43 ± 0.18	6.27–8.27	6.77 ± 0.14	6.20–7.37	0.002
HL (mm)	17.66 ± 0.34	15.15–18.79	17.17 ± 0.20	16.01–17.90	0.011
HW (mm)	12.87 ± 0.38	10.43–14.18	11.78 ± 0.18	10.77–12.54	0.001
HD (mm)	7.46 ± 0.21	6.31–8.33	7.46 ± 0.10	7.05–7.96	0.797

BM body mass, *FLL* forelimb length, *FTL* fourth toe length, *HD* head depth, *HL* head length, *HLL* hind limb length, *HW* head width, *SVL* snout–vent length, *TL* tail length, *TTL* third toe length, *WL* waist length

range: 11.52–20.08 g), the mean body lengths of adult males and females were 83.36 ($n = 11$) and 84.01 ($n = 9$), respectively, and did not differ from each other ($p = 0.75$, ANOVA) (Table 3.3). Males had longer and wider heads, longer legs, and longer toes than females after the effect of size was taken into account, in addition to morphological differences between adult sexes based primarily on size (Table 3.3). Adult males (TL 66.18–161.49 mm) and adult females (TL 100.32–155.22 mm) did not differ in size at maturity based on gender ($p = 0.834$; Table 3.3).

3.3.2 Female Reproductive Cycle

Only females with SVL > 46.94 mm were classified as adults because the smallest reproductively active female of EM measured 46.94 mm in SVL. The average clutch size was 2.83 ± 0.75 (X ± SD; range: 2–4; $n = 6$).

3.4 *Shenomorphus incognitus* (SI)

3.4.1 Habitat Description

Most SIs were first discovered in caves, deciduous layers, or forest edges. In the deciduous layer, which was warmer than other habitats and where they foraged, nearly 79.5% ($n = 70$) of the individuals were found. The distribution of SI's habitat and microhabitats indicates that it is a thermoregulatory species, with most of its occurrences occurring in habitats with direct sunlight and microhabitats.

Table 3.4 Morphology of 45 sexually mature male and 43 female *Sphenomorphus incognitus*

	Males		Females		
	Mean ± SE	Range	Mean ± SE	Range	*p*-value
SVL (mm)	87.56 ± 0.74	82.65–93.92	79.78 ± 0.70	70.07–91.78	< 0.001
BM[a] (g)	16.54 ± 0.48	11.79–21.19	10.00 ± 0.87	6.61–15.18	< 0.001
WL (mm)	12.81 ± 0.20	10.40–16.64	10.54 ± 0.31	8.80–13.87	0.343
TL[a] (mm)	120.73 ± 3.25	85.42–147.04	90.38 ± 2.61	67.67–122.25	0.006
FLL (mm)	23.60 ± 0.18	20.91–26.19	20.43 ± 0.18	17.81–22.66	0.001
HLL (mm)	39.09 ± 0.27	33.73–42.80	34.04 ± 0.23	31.55–37.14	0.001
TTL (mm)	5.95 ± 0.06	4.87–7.21	4.50–6.26	4.50–6.26	0.027
FTL (mm)	5.47 ± 0.06	4.72–6.36	4.15–5.69	4.15–5.69	0.007
HL (mm)	17.67 ± 0.09	16.27–18.81	15.22 ± 0.12	14.15–17.08	< 0.001
HW (mm)	11.39 ± 0.08	10.43–12.63	9.61 ± 0.10	8.06–10.92	<0.001
HD (mm)	7.77 ± 0.08	6.53–8.82	6.55 ± 0.08	5.52–7.92	0.024

BM body mass, *FLL* forelimb length, *FTL* fourth toe length, *HD* head depth, *HL* head length, *HLL* hind limb length, *HW* head width, *SVL* snout–vent length, *TL* tail length, *TTL* third toe length, *WL* waist length

[a]Some individuals lost tails while caught; thus only 29 males and 30 females are analyzed in TL and BM

Body measurements included length, weight, waist circumference, tail length, length of the third and fourth toes, forelimb and hindlimb lengths, and height of the head.

SI in this study was reproductively active from March to July. A total of 88 specimens were collected. The mean body mass of adult males (16.54 g, $n = 45$; range: 11.79–21.19 g) showed significant difference from those of adult females (10.00 g, $n = 43$; range: 6.61–15.18 g; $p < 0.001$), and the mean snout-vent length of adult males (SVL: 87.56 mm; $n = 45$) are larger than that of females (79.78 mm; $n = 43$) ($p < 0.001$; Table 3.4). After removing the impact of SVL, males had longer, taller, wider heads and longer legs than females, with the exception of morphological differences between adult males and females based primarily on size (Table 3.4). In comparison to adult females (TL 67.67–122.25 mm), adult males (TL 85.42–147.04 mm) were significantly larger ($p = 0.006$; Table 3.4).

3.4.2 Food Habits

Twenty four of the 88 gathered lizards (27.3%) had food in their stomachs. On Orchid Island, hymenopterans (23.53%) and Orthoptera (35.29%) made up the majority of SI's diet. Crickets and ants were the main prey species. Two lizards consumed insect larvae, while two others had spiders in their stomachs (Table 3.5). *Ramphotyphlops braminus*, a blind snake, was eaten by two lizards.

Table 3.5 Diet summary for *Sphenomorphus incognitus* from Orchid Island. Frequency is the number of lizards containing each prey

Prey type	Vol. (mm^3)	% Vol.	Frequency	% frequency
Orthoptera				
Grylidae	2080	37.4	8	23.53
Acrididae	240	4.3	1	2.94
Tettigoniidae	660	11.9	3	8.82
Hemiptera				
Aphrophoridae	450	8.1	2	5.88
Cicadellidae	550	9.9	2	5.88
Gymnocerata	100	1.8	2	5.88
Hymenoptera				
Formicidae	110	2.0	8	23.53
Lepidoptera	220	4.0	1	2.94
Blataria				
Blaberidae	310	5.6	1	2.94
Spider	220	2.0	2	5.88
Insect larvae	350	6.3	2	5.88
Vertebrate				
Blind snake	270	4.9	2	
Total	5560	100.0	132	100.00

3.4.3 Female Reproductive Cycle

Only females with SVL > 70.07 mm were considered adults because these were the smallest reproductively active females of SI. Adult females were 50% of adult females were reproductively active between March and April, 33% between May and June, 60% between June and July, and 62.5% between June and July (Fig. 3.2a). From August to January, none of the adult females were gravid (Fig. 3.2a).

On March 22 and April 23, respectively, the first females with oocytes and oviductal eggs were discovered. The first females with larger follicles and females with oviductal eggs were separated by 32 days. I take this to mean that this is roughly how long it takes for a female to lay a clutch of eggs. There are two pieces of evidence that support the theory that females lay two clutches of eggs per reproductive season: (1) one of the 10 females that were captured in April 1998 still had enlarged yolk sacs after oviposition, and (2) the reproductive season lasted roughly 5 months, which is more than twice the estimated oviposition period mentioned above.

Clutch size averaged 4.00 ± 1.00 (X ± SD; range 3–5), independent of female body length (Y = 0.0384 X + 0.7627; $r = 0.278$; $n = 10$; $p = 0.437$) (Fig. 3.2b). Female liver mass showed significant monthly variation ($F_{9,32} = 4.05$; $p = 0.002$; ANCOVA, Fig. 3.2c).

Male reproductive cycle.

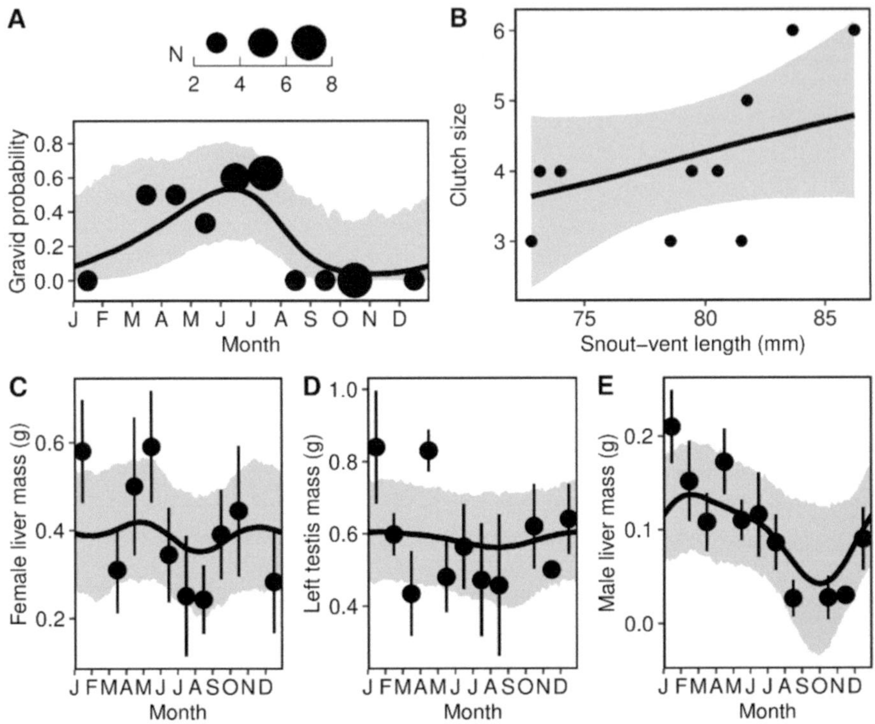

Fig. 3.2 Monthly distribution of gravid probability. (**a**) Adult females were reproductively active from March to July, and all adult females showed non-reproductive activity from August to February. Correlation between female body size (SVL) and the number of oviductal eggs (**b**). Variation (mean ± SE) in mean liver mass (**c**), male mean left testis mass (**d**), and liver mass (**e**) during the annual reproductive cycle of *Sphenomorphus incognitus* on Orchid Island, Taiwan. (Redrawn from Huang (2010), Zoological Studies)

Left testicular mass showed a significant monthly variation, rising from December and peaking from January to June ($F_{10,33} = 9.58, p < 0.0001$; ANCOVA, Fig. 3.2d). Male liver mass did not vary on a monthly basis (ANCOVA, Fig. 3.2e; $F_{10,33} = 1.17; p = 0.348$).

3.5 *Emoia atrocostata* (EA)

3.5.1 Habitat Description

All EA were found in reef areas near the sea. EA is a thermoregulatory species that is most frequently found in habitats with direct sunlight and microhabitats, according to the distribution of its habitat and microhabitats.

Table 3.6 Morphology of 84 sexually mature male and 55 female *Emoia atrocostata*

	Males		Females		
	Mean ± SE	Range	Mean ± SE	Range	*p*-value
SVL (mm)	76.60 ± 0.49	67.01–84.32	74.46 ± 0.51	68.59–82.69	0.005
BM[a] (g)	8.35 ± 0.21	5.11–12.66	7.06 ± 0.19	4.88–10.24	< 0.001
WL (mm)	9.82 ± 0.10	7.77–12.32	9.24 ± 0.11	7.76–11.59	0.008
TL[a] (mm)	105.64 ± 1.92	68.31–148.57	99.02 ± 2.18	57.64–122.26	0.049
FLL (mm)	24.77 ± 0.18	19.66–29.17	22.35 ± 0.14	19.50–24.68	< 0.001
HLL (mm)	37.12 ± 0.24	30.58–41.00	33.19 ± 0.25	32.09–48.27	< 0.001
TTL (mm)	6.52 ± 0.06	4.02–7.66	5.92 ± 0.06	4.62–6.62	< 0.001
FTL (mm)	7.12 ± 0.10	5.02–8.60	6.64 ± 0.06	5.63–7.87	< 0.001
HL (mm)	17.66 ± 0.12	15.31–19.76	16.21 ± 0.11	14.34–18.24	< 0.001
HW (mm)	10.03 ± 0.10	7.78–12.25	9.13 ± 0.08	7.78–10.87	< 0.001
HD (mm)	6.51 ± 0.09	5.20–11.38	5.84 ± 0.06	4.98–6.96	< 0.001

BM body mass, *FLL* forelimb length, *FTL* fourth toe length, *HD* head depth, *HL* head length, *HLL* hind limb length, *HW* head width, *SVL* snout–vent length, *TL* tail length, *TTL* third toe length, *WL* waist length

[a]Some individuals lost tails while caught; thus only 70 males and 47 females are analyzed in TL and BM

Body measurements include forelimb and hindlimb length, third and fourth toe length, head length, head width, and head height. Waist measurements are also taken.

EA was reproductively active from March to August. A total of 139 specimens were collected. The mean body length of adult males and females was 76.60 ($n = 84$) and 74.46 ($n = 55$) mm, respectively, with significant differences between them ($p = 0.005$, ANOVA), and the mean body mass of adult males (8.35 g, $n = 70$; range: 5.11–12.66 g) was significantly higher than that of adult females (7.06 g, $n = 47$; range: 4.88–10.24 g; $p < 0.001$) (Table 3.6). In addition to morphological differences between adult males and females based primarily on size, males had longer, taller, and wider heads, longer limbs, and wider waists than females when the effect of size was removed (Table 3.6). Adult males (TL 68.31–148.57 mm) were significantly larger than adult females (TL 57.64–122.26 mm) ($p = 0.049$; Table 3.6).

3.5.2 Food Habits

A mere 4 lizards out of 139 (2.9%) had food in their stomachs. EA consumed Hemiptera (Aphrophoridae; 2/4), Hymenoptera (Formicidae; 1/4), and crabs (1/4), all of which were present on Orchid Island. The lack of food in the reef area may be explained by this.

3.5.3 Female Reproductive Cycle

The smallest reproductively active females of EA measured 68.59 mm SVL; therefore, only females with SVL $\geqq$ 68.59 mm were considered adults. All adult females were reproductively active in March, 75% in April, 25% in May, 50% in June, 60% in July, and 57% in August (Fig. 3.3a). From September to February, all adult females were nonreproductive (Fig. 3.3a).

On March 22, the first females with oocytes were discovered, and on April 24, the first females with oviductal eggs were discovered. The first females with larger follicles and females with oviductal eggs were separated by 33 days. I take this to mean that this is roughly how long it takes for a female to lay a clutch of eggs. The reproductive season lasted about 6 months, which is five times longer than the time it takes a female to produce a clutch of eggs as described above. Additionally, three of the 20 females that were collected in May and June 1998 still had enlarged yolk sacs after egg production.

Clutch size averaged 2.00 ± 0.00 (X ± SD; range 1–3) and was not correlated with female body length (Y = 0.019 X + 0.5319; $r = 0.236$; $n = 20$; $p = 0.317$) (Fig. 3.3b). There was no monthly variation in fat body and liver quality in females ($F_{11,42} = 0.80$, $p = 0.638$ for fat body; Fig. 3.3c; $F_{11,42} = 5.96$, $p = 0.000$ for liver; ANCOVA, Fig. 3.3d). However, fat body mass was smaller during the reproductive season from March to August and increased before the dormant season from October to January (Fig. 3.3c).

3.5.4 Male Reproductive Cycle

Figure 3.3e illustrates the left testicular mass's significant monthly variation ($F_{11,71} = 2.43$, $p = 0.012$), which began in December and peaked from January to June. Male fat body mass exhibited monthly variation ($F_{11,71} = 2.94$, $p = 0.003$; ANOVA, Fig. 3.3f), that it did decrease to a minimum between February and June before rising to a peak in August. The mass of the male liver varied significantly from month to month ($F_{11,71} = 9.46$, $p < 0.001$; ANCOVA, Fig. 3.3g).

3.6 *Diploderma swinhonis* (DS)

3.6.1 Habitat Description

The majority of DS were first observed on tree trunks, in caves or at the edge of forests. Nearly 60% ($n = 74$) of the individuals were on tree trunks, 30% ($n = 37$) foraged for ants at forest edges, and the other 10% ($n = 11$) were found in caves or on slopes in concrete areas, where temperatures were higher than in other habitats. The habitat and microhabitat distribution of DS indicate that it is a thermoregulatory species, most often found in direct sunlight and microhabitats.

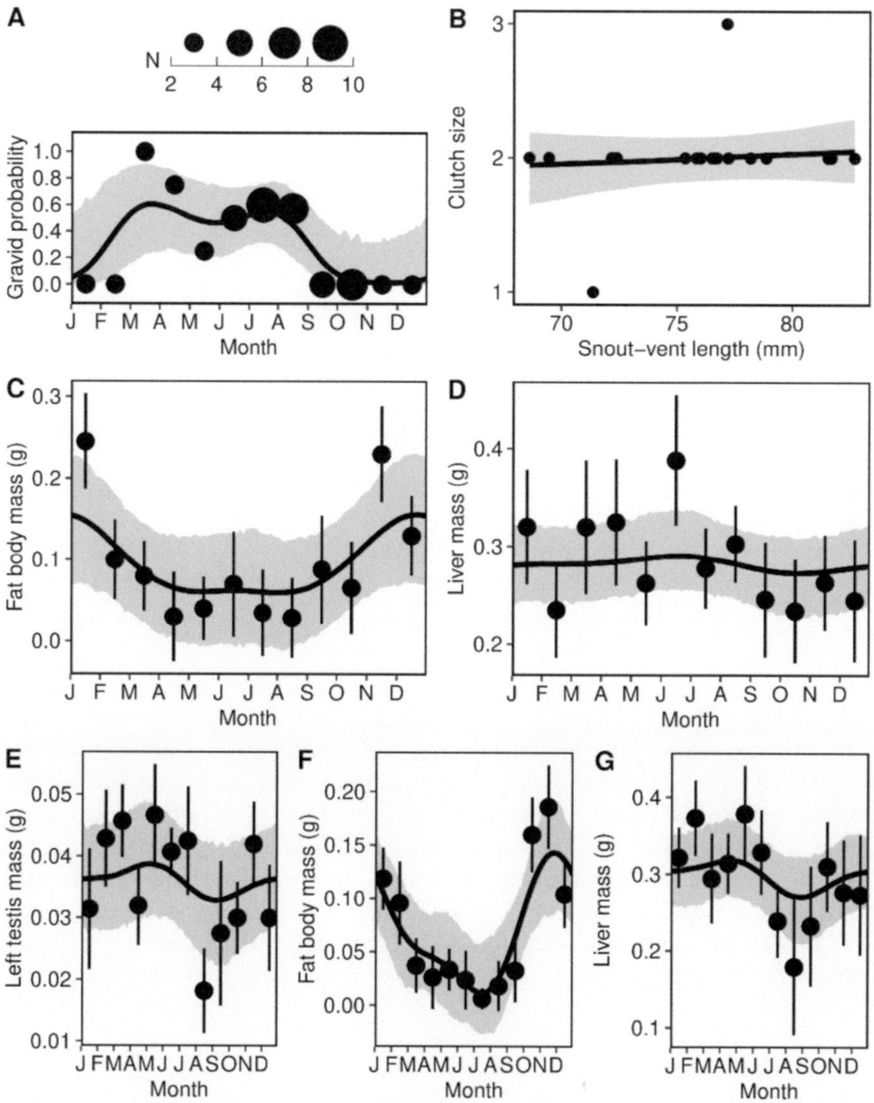

Fig. 3.3 Monthly distribution of gravid probability. (**a**) Adult females were reproductively active from March to August, and all adult females showed non-reproductive activity from September to February. Correlation between female body size (SVL) and the number of oviductal eggs (**b**). Variation (mean ± SE) in mean fat body mass (**c**), liver mass (**d**), and male mean left testis mass (**e**), fat body mass (**f**), and liver mass (**g**) during the annual reproductive cycle of *Emoia atrocostata* on Orchid Island, Taiwan. (Redrawn from Huang et al. (2011), Zoological Studies)

The following body measurements were taken: head length, head width, and head depth. Waistline diameter, tail length, forelimb and hindlimb lengths, and third and fourth toe lengths.

Table 3.7 Morphology of 89 sexually mature male and 37 female *Diploderma swinhonis*. Comparisons of SVLs between the sexes are based on the ANOVA test. Other morphological comparisons were made by ANCOVA with the LogSVL as covariate so that characters could be compared with the effect of size removed

	Males		Females		
	Mean ± SE	Range	Mean ± SE	Range	p-value
SVL (mm)	74.58 ± 0.40	62.08–81.9	69.31 ± 0.64	63.33–76.33	< 0.001
BM (g)	11.24 ± 0.21	6.23–15.25	9.32 ± 0.28	6.78–12.11	< 0.001
WL (mm)	8.44 ± 0.07	7.01–9.85	8.45 ± 0.09	7.43–9.43	< 0.001
TL (mm)	188.44 ± 2.0	57.40–216	172.73 ± 2.10	141.58–191.4	< 0.001
FLL (mm)	37.90 ± 0.20	32.80–41.8	34.85 ± 0.29	31.77–38.68	< 0.001
HLL (mm)	66.38 ± 0.31	57.70–72.1	60.55 ± 0.47	54.81–65.95	< 0.001
TTL (mm)	10.74 ± 0.07	8.08–12.05	10.00 ± 0.10	8.60–11.02	< 0.001
FTL (mm)	9.91 ± 0.07	7.63–11.40	9.15 ± 0. 09	8.18–10.05	< 0.001
HL (mm)	23.73 ± 0.17	19.30–27.1	21.08 ± 0.25	18.72–25.50	< 0.001
HW (mm)	13.39 ± 0.11	10.90–15.8	12.09 ± 0.15	10.69–14.10	< 0.001
HD (mm)	10.54 ± 0.09	8.23–12.80	9.70 ± 0.14	8.52–12.11	< 0.001

SVL snout–vent length, *BM* body mass, *WL* waist length, *TL* tail length, *FLL* forelimb length, *HLL* hind limb length, *TTL* third toe length, *FTL* fourth toe length, *HL* head length, *HW* head width, *HD* head depth

DS engaged in active reproduction from March to October. In total, 122 samples were gathered. The mean body mass of adult males (11.24 g, $n = 87$; range: 6.23–15.25 g) was significantly higher than that of adult females (9.32 g, $n = 32$; range: 6.78–12.11 g), and the mean body length of adult males and females was 74.58 ($n = 89$) and 69.31 ($n = 33$), respectively (Table 3.7). Males had longer heads, longer legs, and wider waistlines than females when the effect of size was taken away, in addition to morphological differences between adult males and females based primarily on size (Table 3.7). Size at maturity differed by sex in adult males (TL 57.40–216.7 mm) and adult females (TL 141.58–191.39 mm) ($p = 0.000$; Table 3.7).

3.6.2 Food Habits

One hundred and eleven lizards were examined individually, and 20 of them (18%) had food in their stomachs. On Orchid Island, Hymenoptera (53.33%) and Orthoptera (16.67%) made up the majority of the diet of DS. Ants (50%) and crickets (16.67%) made up the majority of the diet. Two lizards consumed spiders, while one had a bee in its stomach (Table 3.8).

Female reproductive cycle.

The smallest reproductively active females of DS measured 63.33 mm SVL; therefore, only females with SVL ≧ 63.33 mm were considered adults. The reproductive activity of adult females peaked in May and July, and peaked at 100% in

Table 3.8 Diet summary for *Diploderma swinhonis* from Orchid Island. Frequency is the number of lizards containing each prey

Prey type	Vol. (mm^3)	% Vol.	Frequency	% frequency
Orthoptera				
Grylidae	2802	56.2	5	16.67
Hemiptera				
Cicadellidae	603	12.1	2	6.67
Hymenoptera				
Apidae	192	3.8	1	3.33
Formicidae	180	3.6	15	50
Coleoptera				
Curculionidae	600	12	1	3.33
Spider	244	4.9	2	6.67
Scolopendromorpha	369	7.4	1	3.33
Insect larvae	100	2	3	10
Total	4990	100	30	100

March, 75% in April, 63% in August, 40% in September, and 33% in October. Adult females did not reproduce from November to February (Fig. 3.4a).

On March 22, the first females with oocytes were discovered, and on May 22, the first females with oviductal eggs were discovered. The first females with larger follicles and the first females with oviductal eggs were found 61 days apart. I take this to mean that this is roughly how long it takes for a female to lay a clutch of eggs. There are two pieces of evidence that show females lay two clutches of eggs per reproductive season: (1) 2 of the 15 females that were captured in May and June 2000 still had enlarged yolk sacs after egg production; and (2) the reproductive season lasted roughly 7 months, which is more than three times the time it would take females to lay one clutch of eggs according to the estimation made above.

Clutch size averaged 3.20 ± 0.86 (X ± SD; range 2–5) and was not correlated with female body length ($Y = 0.0261 X + 1.4368$; $r = 0.10$; $n = 15$; $p = 0.735$; Fig. 3.4b). Female fat body but not liver mass showed significant monthly variation ($F_{1,30} = 6.44$, $p < 0.05$ in fat body; ANCOVA; Fig. 3.4c; $F_{1,30} = 0.02$, $p > 0.05$ in livers; ANCOVA; Fig. 3.4d).

3.6.3 Male Reproductive Cycle

Significant monthly variation was observed in the left testicular mass ($F_{11,76} = 5.30$, $p < 0.001$; ANCOVA; Fig. 3.4e). Male liver mass varied significantly monthly, whereas fat body mass did not (fat body mass, $F_{11,76} = 0.46$, $p = 0.919$; ANCOVA, Fig. 3.4f; $F_{11,76} = 2.15$, liver mass, $p = 0.026$; Fig. 3.4g).

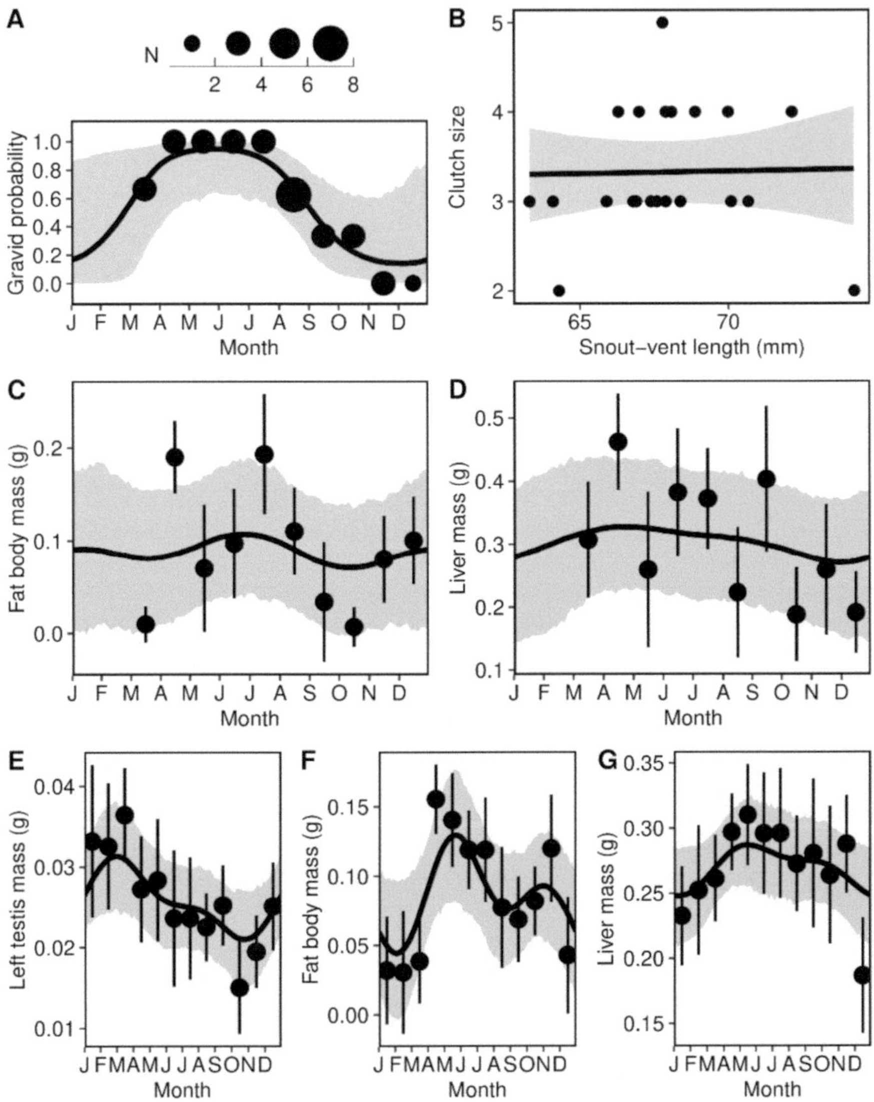

Fig. 3.4 Monthly distribution of gravid probability. (**a**) Adult females were reproductively active from March to October, and all adult females showed non-reproductive activity from November to February. Correlation between female body size (SVL) and the number of oviductal eggs (**b**). Variation (mean ± SE) in mean fat body mass (**c**), liver mass (**d**), and male mean left testis mass (**e**), fat body mass (**f**), and liver mass (**g**) during the annual reproductive cycle of *Diploderma swinhonis* on Orchid Island Taiwan. (Redrawn from Huang (2007), Zoological Science)

3.7 *Gekko hokouensis* (GH)

3.7.1 Habitat Description

The majority of the individuals—nearly 75.5% ($n = 105$) and 25% ($n = 34$)—were in abandoned homes and concrete holes, where temperatures were higher than in other habitats. The distribution of GH's habitat and microhabitats indicates that it is a thermoregulatory species that prefers habitats with direct sunlight and microhabitats.

Body measurements included body length, body weight, waistline diameter, tail length, forelimb and hindlimb length, third and fourth toe length, head length, head width, and head height.

GH in this study was in breeding season from January to September. A total of 139 specimens were collected. The mean snout-vent lengths of adult males and females were 60.69 ($n = 69$) and 61.54 ($n = 70$) mm, respectively, with no significant difference between them ($p = 0.122$, ANOVA), and the mean body mass of adult males (4.78 g, $n = 60$; range: 3.09–6.16 g) and adult females (4.88 g, $n = 64$; range: 2.86–7.12 g) did not differ significantly ($p = 0.245$; Table 3.9). In addition to morphological differences between adult males and females based primarily on size, males had longer, wider, and taller head dimensions and longer leg lengths than females when size effects were removed, but males did not differ in size at maturity (Table 3.9).

Table 3.9 Morphology of 69 sexually mature male and 70 female *Gekko hokouensis*

	Males		Females		
	Mean ± SE	Range	Mean ± SE	Range	*p*-value
SVL (mm)	60.69 ± 0.33	54.27–66.34	61.54 ± 0.40	54.09–68.04	0.122
BM[a] (g)	4.78 ± 0.44	3.09–6.16	4.88 ± 0.11	2.86–7.12	0.245
WL (mm)	7.20 ± 0.11	6.30–0.50	7.35 ± 0.07	6.30–8.90	0.428
TL[b] (mm)	56.07 ± 1.40	31.73–73.66	55.80 ± 1.32	28.28–76.32	0.918
FLL (mm)	18.42 ± 0.12	15.87–20.89	18.08 ± 0.12	15.24–20.65	0.001
HLL (mm)	25.29 ± 0.18	21.12–28.04	24.81 ± 0.20	20.30–27.69	0.003
TTL (mm)	4.71 ± 0.06	3.49–5.74	4.70 ± 0.05	3.71–5.73	0.488
FTL (mm)	5.00 ± 0.08	3.55–6.21	5.06 ± 0.05	3.98–6.20	0.501
HL (mm)	17.23 ± 0.08	15.59–18.70	16.96 ± 0.11	14.90–18.56	0.000
HW (mm)	11.89 ± 0.07	10.08–12.98	11.72 ± 0.19	10.02–16.52	0.008
HD (mm)	5.88 ± 0.05	4.87–7.04	5.79 ± 0.09	4.86–7.52	0.032

BM body mass, *FLL* forelimb length, *FTL* fourth toe length, *HD* head depth, *HL* head length, *HLL* hind limb length, *HW* head width, *SVL* snout–vent length, *TL* tail length, *TTL* third toe length, *WL* waist length

[a]Some individuals lost tails when caught; thus only 60 males and 64 females are analyzed in TTL and BM. Comparisons of SVLs between the sexes are based on the ANOVA test. Other morphological comparisons were made by ANCOVA with the SVL as covariate so that characters could be compared with the effect of size removed

3.7.2 Food Habits

Based on the individual samples collected, 16 of 139 lizards (11.5%) had food in their stomachs. On Orchid Island, the diet of GH consisted mainly of Orthoptera (35%) and Dermaptera (dermapterans) animals (15%). Two prey groups, crickets and earwigs animals, dominate the diet. The stomachs of two lizards contained spiders, and one ate a cockroach (Table 3.10).

3.7.3 Female Reproductive Cycle

The smallest reproductively active female measured 54.09 mm SVL; therefore, only females with an SVL of 54.09 mm were considered adults. All adult females were reproductively active in March, July, and August, with 75% in January, 27% in April, 50% in May, 33% in June, and 29% in September (Fig. 3.5a). All adult females were nonreproductive from October to December (Fig. 3.5a).

Clutch size averaged 2.0 ± 0.00 (X ± SD; range 2). Female liver mass showed monthly variation ($F_{10,58} = 3.33$, $p = 0.002$; ANCOVA; Fig. 3.5b).

3.7.4 Male Reproductive Cycle

The left testicular mass showed significant monthly variation ($F_{10,57} = 14.73$, $p = 0.000$; ANCOVA, Fig. 3.5c). Male liver quality showed significant monthly variation ($F_{10,57} = 2.63$, $p = 0.011$; ANCOVA, Fig. 3.5d).

Table 3.10 Diet summary for *Gekko hokouensis* from Orchid Island. Frequency is the number of lizards containing each prey

Prey type	Frequency	% frequency
Orthoptera		
Grylidae	7	35
Hemiptera		
Cicadellidae	2	10
Gymnocerata	2	10
Hymenoptera		
Formicidae	1	5
Coleoptera	1	5
Dermaptera	3	15
Lepidoptera	1	5
Blattaria	1	5
Spider	2	10
Total	20	100

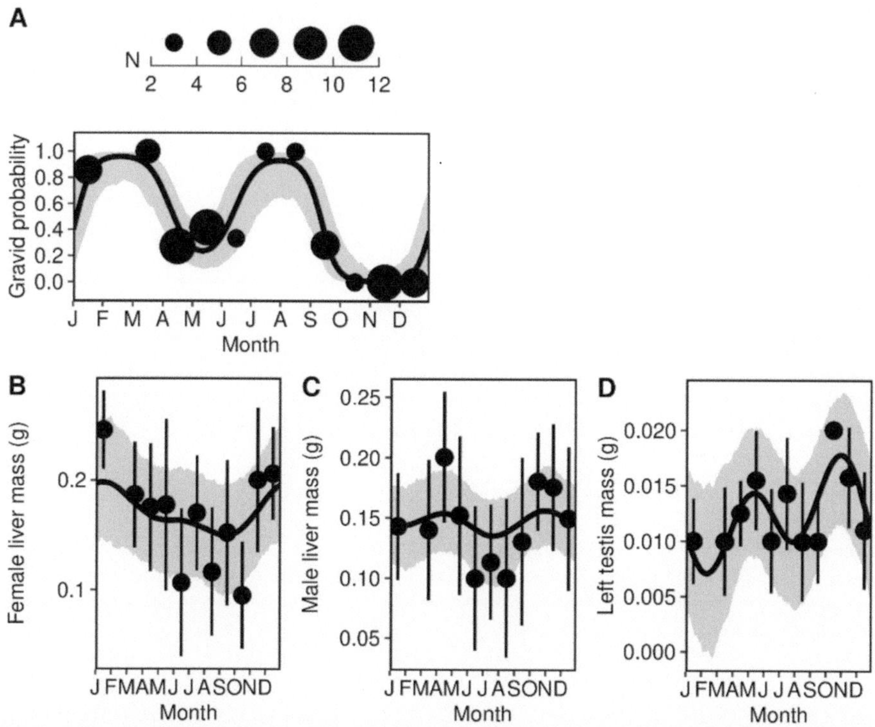

Fig. 3.5 Monthly distribution of gravid probability. (**a**) Adult females were reproductively active from January to September, and all adult females showed non-reproductive activity from October to December. Variation (mean ± SE) in female mean liver mass (**b**), male mean liver mass (**c**), and mean left testis mass (**d**) during the annual reproductive cycle of *Gekko hokouensis* on Orchid Island, Taiwan

3.8 *Gekko kikuchii* (GK)

3.8.1 Habitat Description

Most of GK was first observed in concrete holes or on abandoned houses. Nearly 68% ($n = 52$) of the observed individuals were in concrete holes, 25% ($n = 20$) were on abandoned houses, and 7% ($n = 5$) were in rock crevices where temperatures were higher than in other habitats. The habitat and microhabitat distribution of GK suggests that it is a thermoregulatory species that occurs most often in habitats with direct sunlight and microhabitats.

Body measurements include length, weight, girth, tail length, forelimb and hindlimb length, third and fourth toe length, head length, head width and head height.

GK was breeding actively from March to August. A total of 77 specimens were collected. The mean body lengths of adult males and females were 77.78 ($n = 48$)

Table 3.11 Morphology of 29 sexually mature male and 48 female *Gekko kikuchii*

	Males		Females		
	Mean ± SE	Range	Mean ± SE	Range	*p*-value
SVL (mm)	77.78 ± 0.85	62.79–90.40	72.73 ± 1.00	63.52–88.64	< 0.001
BM[a] (g)	10.06 ± 0.35	4.82–14.97	7.55 ± 0.37	4.39–13.49	0.067
WL (mm)	8.52 ± 0.13	6.30–10.05	7.96 ± 0.16	6.10–9.45	0.349
TL[a] (mm)	91.33 ± 2.18	46.27–112.65	82.71 ± 2.85	43.26–104.79	0.522
FLL (mm)	23.94 ± 0.25	19.50–27.70	21.99 ± 0.30	18.50–26.60	0.005
HLL (mm)	34.60 ± 0.33	28.19–39.60	32.05 ± 0.38	27.32–38.40	0.003
TTL (mm)	6.12 ± 0.08	5.08–7.45	5.66 ± 0.08	4.70–6.32	0.019
FTL (mm)	6.40 ± 0.15	2.50–7.98	6.01 ± 0.08	5.06–6.66	0.013
HL (mm)	21.36 ± 0.23	16.68–25.08	19.88 ± 0.25	17.33–23.91	0.118
HW (mm)	15.78 ± 0.30	11.36–26.60	14.02 ± 0.22	11.76–17.70	0.000
HD (mm)	7.05 ± 0.11	5.38–9.10	6.41 ± 0.10	5.17–7.20	0.193

BM body mass, *FLL* forelimb length, *FTL* fourth toe length, *HD* head depth, *HL* head length, *HLL* hind limb length, *HW* head width, *SVL* snout–vent length, *TL* tail length, *TTL* third toe length, *WL* waist length

[a]Some individuals lost tails when caught; thus only 25 males and 42 females are analyzed in TTL and BM. Comparisons of SVLs between the sexes are based on the ANOVA test. Other morphological comparisons were made by ANCOVA with the SVL as covariate so that characters could be compared with the effect of size removed

and 72.73 ($n = 29$) mm, respectively, which were significantly different from each other ($p < 0.001$, ANOVA), but the mean body mass of adult males (10.06 g, n = 42; range: 4.82–14.97 g) was not significantly greater than that of adult females (7.55 g, $n = 25$; range: 4.39–13.49 g) (Table 3.11). In addition to morphological differences between adult males and females based primarily on size, males had longer head widths, legs, and toes than females when the effect of size was removed (Table 3.11). There were no significant differences in size between adult males (TL 46.27–112.65 mm) and adult females (TL 43.26–104.79 mm) at maturity (Table 3.11).

3.8.2 Food Habits

Based on the individual samples collected, 5 of the 77 lizards (6.5%) had food in their stomachs. The diet of GK at Orchid Island consisted mainly of Orthoptera (4/5) and Hymenoptera (1/5).

3.8.3 Female Reproductive Cycle

The smallest reproductively active female of GK measured 63.52 mm SVL; therefore, only females with SVL≧63.52 mm were considered adults. All adult females were reproductively active in March, 67% in April, 50% in May, 33% in June, 25%

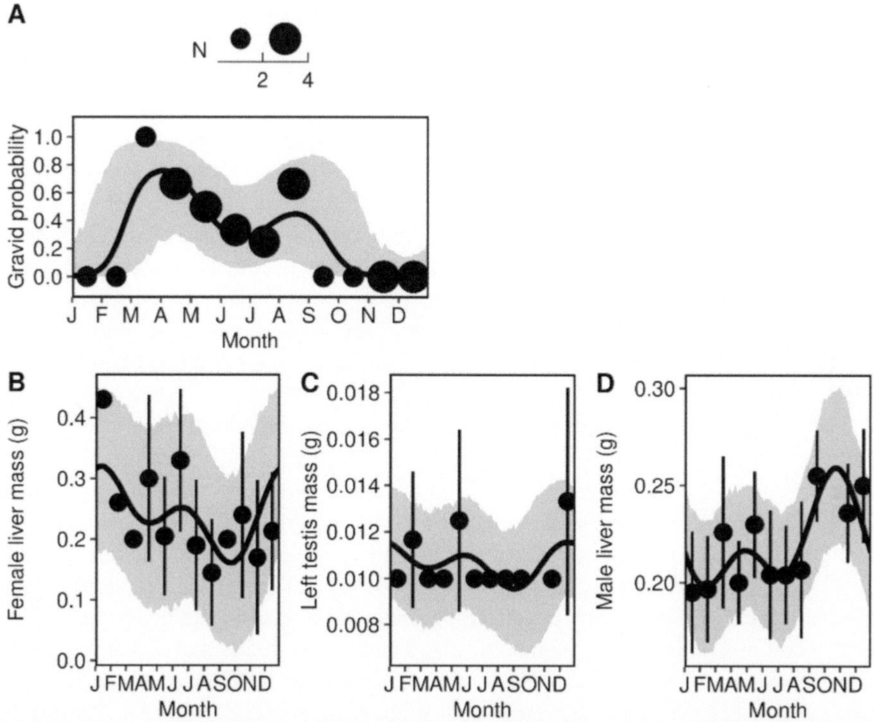

Fig. 3.6 Monthly distribution of gravid probability (**a**). Adult females were reproductively active from March to August, and all adult females showed non-reproductive activity from September to February. Variation (mean ± SE) in female mean liver mass (**b**), male mean liver mass (**c**), and testis mass (**d**) during the annual reproductive cycle of *Gekko kikuchii* on Orchid Island, Taiwan

in July, and 67% in August (Fig. 3.6a). From September to February, all adult females were non-gravid (Fig. 3.6a).

The mean clutch size was 2.0 ± 0.0 (X ± SD; range 2). The body cavity of both females and males was devoid of fat bodies. Liver mass showed significant monthly variation ($F_{11,16} = 4.06$, $p = 0.006$; ANCOVA, Fig. 3.6b).

3.8.4 Male Reproductive Cycle

Left testicular mass showed no monthly variation ($F_{10,36} = 0.92$, $p = 0.525$; ANCOVA, Fig. 3.6c). Male liver mass showed no significant monthly variation ($F_{10,36} = 0.69$, $p = 0.726$; ANCOVA, Fig. 3.6d).

3.9 *Hemidactylus frenatus* (HF)

3.9.1 Habitat Description

The majority of HF were first observed on houses, abandoned houses, or in concrete holes. Nearly 86.3% ($n = 88$) of the observed individuals were in houses, 7.8% ($n = 8$) in abandoned houses, and 5.9% ($n = 6$) in concrete holes, where temperatures were higher than in houses. The habitat and microhabitat distribution of HF indicates that it is a thermoregulatory species that occurs most often in habitats with direct sunlight and microhabitats.

Body measurements included body length, body weight, waistline diameter, tail length, forelimb and hindlimb length, third and fourth toe length, head length, head width, and head height.

HF in this study was reproductively active from March to July. A total of 102 specimens were collected. The mean body lengths of adult males and females were 53.29 ($n = 47$) and 49.78 ($n = 55$) mm, respectively, differed from each other significantly ($p < 0.001$, ANOVA), but the mean body mass of adult males (3.26 g, $n = 39$; range: 2.27–5.43 g) was not significantly greater than that of adult females (2.55 g, $n = 50$; range: 1.06–4.02 g; $p = 0.099$) (Table 3.12). In addition to morphological differences between adult males and females based primarily on size, when size effects were removed, males had wider heads and longer legs and toes than females (Table 3.12). There was no significant difference in size between adult males (TL 18.02–69.02 mm) and adult females (TL 23.21–57.19 mm) at maturity ($p = 0.068$; Table 3.12).

Table 3.12 Morphology of 47 sexually mature male and 55 female *Hemidactylus frenatus*

	Males		Females		
	Mean ± SE	Range	Mean ± SE	Range	*p*-value
SVL (mm)	53.29 ± 0.49	47.17–60.88	49.78 ± 0.70	46.94–54.71	< 0.001
BM[a] (g)	3.26 ± 0.11	2.27–5.43	2.55 ± 0.07	1.06–4.02	0.099
WL (mm)	6.09 ± 0.07	5.14–7.04	5.73 ± 0.06	4.93–7.10	0.587
TL[a] (mm)	50.78 ± 1.69	18.02–69.02	44.54 ± 1.21	23.21–57.19	0.068
FLL (mm)	15.75 ± 0.16	12.93–17.87	14.18 ± 0.16	12.46–16.35	< 0.001
HLL (mm)	21.71 ± 0.22	18.82–25.10	19.33 ± 0.13	16.88–21.70	0.000
TTL (mm)	3.96 ± 0.07	3.28–4.51	3.54 ± 0.04	2.77–4.28	0.001
FTL (mm)	4.33 ± 0.05	3.51–5.17	3.92 ± 0.06	3.01–4.78	0.005
HL (mm)	15.28 ± 0.14	13.06–17.45	14.29 ± 0.07	13.42–15.67	0.065
HW (mm)	10.05 ± 0.11	8.38–11.73	9.02 ± 0.08	8.04–10.41	0.001
HD (mm)	4.99 ± 0.06	4.02–6.21	4.64 ± 0.04	4.12–5.40	0.372

BM body mass, *FLL* forelimb length, *FTL* fourth toe length, *HD* head depth, *HL* head length, *HLL* hind limb length, *HW* head width, *SVL* snout–vent length, *TL* tail length, *TTL* third toe length, *WL* waist length

[a]Some individuals lost tails when caught; thus only 39 males and 50 females are analyzed in TTL and BM. Comparisons of SVLs between the sexes are based on the ANOVA test. Other morphological comparisons were made by ANCOVA with the SVL as covariate so that characters could be compared with the effect of size removed

3.9.2 Female Reproductive Cycle

The smallest reproductively active female of HF measured 46.94 mm SVL; therefore, only females with 46.94 mm SVL were considered adult. Adult females were 67% reproductively active in March, 60% in April, 57.1% in May, and 60% in June and July (Fig. 3.7a). All adult females were not reproductive between September and February (Fig. 3.7a).

Clutch size averaged 2.0 ± 0.00 (X ± SD; range 2). Females showed no significant monthly variation in liver mass ($F_{10,43} = 1.96$, $p = 0.062$; ANCOVA; Fig. 3.7b).

Male reproductive cycle.

The testicular mass did not vary and its mass was equal to 0.01 g. Male liver mass also showed no significant monthly variation ($F_{11,34} = 0.72$, $p = 0.716$; ANCOVA, Fig. 3.7c).

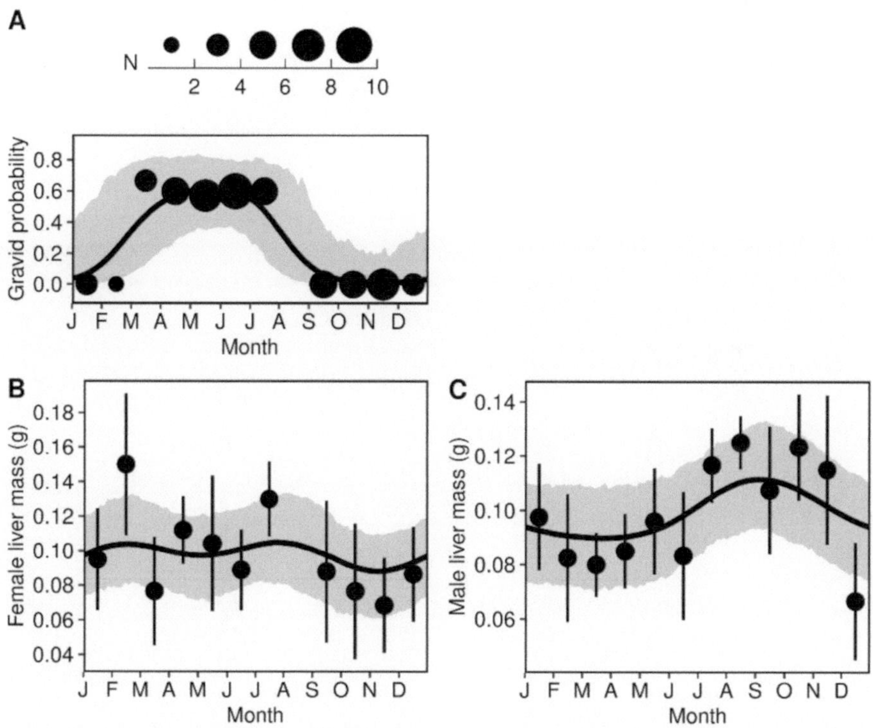

Fig. 3.7 Monthly distribution of gravid probability. (**a**) Adult females were reproductively active from March to July, and all adult females showed non-reproductive activity from August to February. Variation (mean ± SE) in female mean liver mass (**b**), and male mean liver mass (**c**) during the annual reproductive cycle of *Hemidactylus frenatus* on Orchid Island, Taiwan

3.10 *Takydromus sauteri* (TS)

3.10.1 Habitat Description

All TS were observed in grass (*Miscanthus sinensis*) microhabitats at the edge of the forest. Body measurements included body length, body weight, waistline diameter, tail length, forelimb and hindlimb length, third and fourth toe length, head length, head width, and head height.

In this study, TS was in the breeding season from February to August and in December. A total of 115 specimens were collected. The mean SVL of adult males and females was 64.2 ($n = 54$) and 67.08 ($n = 61$) mm, respectively, which were significantly different from each other ($p = 0.023$; ANOVA). While the mean body mass of adult males (3.87 g, $n = 40$; range: 1.47–5.77 g) was significantly different from that of adult females (4.50 g, $n = 61$; range: 1.93–7.11 g; $p < 0.001$) (Table 3.13). In addition to morphological differences between adult males and females based primarily on size, males had longer, taller, and wider heads, longer legs, and wider waist circumference than females after removing the effect of size (Table 3.13). There were sex differences in TL between adult males (134.1–299.3 mm) and adult females (131.90–283.67 mm; $p < 0.05$) at maturity (Table 3.13).

Table 3.13 Morphology of 59sexually mature male and 63 female *Takydromus sauteri*

	Males		Females		
	Mean ± SE	Range	Mean ± SE	Range	*p*-value
SVL (mm)	64.20 ± 0.80	50.31–73.70	67.08 ± 0.72	50.61–76.51	0.023
BM[a] (g)	3.87 ± 0.16	1.47–5.77	4.50 ± 0.16	1.93–7.11	< 0.001
WL (mm)	5.98 ± 0.07	4.68–7.27	6.63 ± 0.10	4.93–8.10	< 0.001
TL[a] (mm)	228.35 ± 6.92	134.1–299.3	211.68 ± 5.40	131.90–283.67	< 0.05
FLL (mm)	23.19 ± 0.26	18.20–27.07	22.67 ± 0.20	18.18–25.51	< 0.001
HLL (mm)	31.05 ± 0.37	24.45–35.71	29.78 ± 0.26	24.25–33.70	< 0.001
TTL (mm)	6.36 ± 0.10	5.02–7.88	6.12 ± 0.07	4.88–7.49	< 0.001
FTL (mm)	7.54 ± 0.10	6.03–9.09	7.28 ± 0.08	5.63–8.46	< 0.001
HL (mm)	15.60 ± 0.17	12.88–17.79	15.29 ± 0.13	12.71–17.58	< 0.001
HW (mm)	7.12 ± 0.08	5.76–8.40	7.07 ± 0.06	6.02–8.02	< 0.001
HD (mm)	5.07 ± 0.07	3.96–6.10	5.05 ± 0.05	3.98–5.94	< 0.001

BM body mass, *FLL* forelimb length, *FTL* fourth toe length, *HD* head depth, *HL* head length, *HLL* hind limb length, *HW* head width, *SVL* snout–vent length, *TL* tail length, *TTL* third toe length, *WL* waist length

[a]Some individuals lost tails while caught; thus only 40 males and 51 females are analyzed in TL and BM

3.10.2 Food Habits

Based on individual stomach samples, 5 of 115 (4.3%) lizards had food. On Orchid Island, the diet of TS consisted mainly of spiders (35.71%) and hemipterans (14.28%). Two prey categories, spiders and hemiptera, dominated the diet. One lizard had plant seeds in its stomach and one ate a cockroach (Table 3.14).

3.10.3 Female Reproductive Cycle

The smallest reproductively active female of TS had a body length of 66.82 mm; therefore, only females with a body length of 66.82 mm were considered to be adults. All adult females were reproductively active in July, 50% in February and April, 78% in March, 75% in May, 80% in June, 47% in August, and 20% in December (Fig. 3.8a). All adult females were not reproductive from September to November and after January (Fig. 3.8a).

The first females with oocytes were found on 20 March and the first females with oviductal eggs were found on 20 April. The interval between these dates is 31 days, so I interpret this as the approximate time when females produce a clutch of eggs. There are two pieces of evidence that females produce two clutches of eggs per reproductive season: (1) one of the four females collected on 30 March 2001 still had enlarged yolk sacs after egg production; and (2) the reproductive season lasted approximately 6 months, which is six times longer than the time it takes for a female to produce a clutch of eggs as described above.

Clutch size averaged 2.0 ± 1.0 (X ± SD; range 1–4) and was not correlated with female body length ($Y = 0.0379\ X - 0.3064$; $r = 0.156$; $n = 32$; $p = 0.395$; Fig. 3.8b). Both female fat body and liver mass showed significant monthly variation

Table 3.14 Diet summary for *Takydromus sauteri* from Orchid Island. Frequency is the number of lizards containing each prey

Prey type	Vol. (mm^3)	% Vol.	Frequency	% frequency
Orthoptera	20.4	6.32	2	7.14
Hemiptera				
Cicadellidae	18.8	5.83	2	7.14
Gymnocerata	16.6	5.15	4	14.28
Coleoptera				
Scarabaeidae	20.8	6.45	2	7.14
Lepidoptera	31.0	9.61	2	7.14
Spider	150.8	46.75	10	35.71
Plant				
Seeds	20.8	6.45	2	7.14
Cockroach	25.2	7.81	2	7.14
Snail	16.4	5.08	2	7.14
Total	320.8	100	28	100

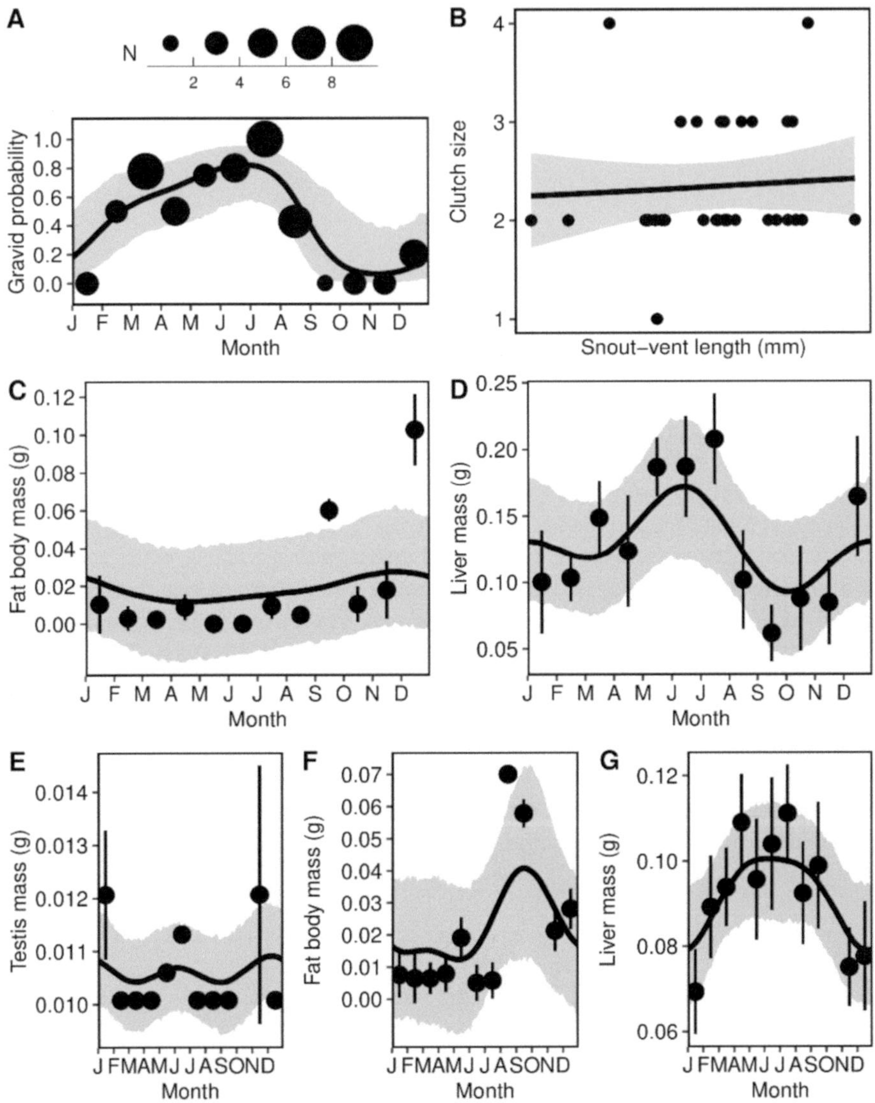

Fig. 3.8 Monthly distribution of gravid probability. (**a**) Adult females were reproductively active from February to August, and December, and all adult females showed non-reproductive activity from September to November and January. Correlation between female body size (SVL) and the number of oviductal eggs (**b**). Variation (mean ± SE) in mean fat body mass (**c**), liver mass (**d**), and male mean left testis mass (**e**), fat body mass (**f**), and liver mass (**g**) during the annual reproductive cycle of *Takydromus sauteri* on Orchid Island, Taiwan. (Redrawn from Huang (2006b), Zoological Science)

($F_{11,48} = 2.61, p = 0.011$ for fat body; ANCOVA; $F_{11,48} = 2.97, p = 0.004$ for liver; ANCOVA). Fat body mass was smaller during the reproductive season from February to August, but increased in September and December (Fig. 3.8c). In contrast, during the reproductive season, liver mass was higher than during the non-reproductive season from September to November (Fig. 3.8D).

3.10.4 Male Reproductive Cycle

Left testicular mass did not show significant monthly variation ($F_{10,42} = 4.87$, $p = 0.033$; ANCOVA, Fig. 3.8e). Male fat body mass showed significant monthly variation ($F_{10,42} = 4.82, p = 0.034$; ANCOVA, Fig. 3.8f), decreasing to a minimum value from January to August and increasing to a peak in September. Male liver mass showed no significant monthly variation ($F_{10,42} = 1.23, p = 0.302$; ANOVA, Fig. 3.8g).

3.11 Comprehensive Information on Nine Species of Lizards

3.11.1 Fauna

Orchid Island has a rich but less endemic tropical reptile fauna (Tables 3.15 and 3.16). Twenty one species of terrestrial reptiles have been recorded on Orchid Island (Tables 3.15 and 3.16). Green Island has a comparative count of 16 species (Tables 3.15 and 3.16). The total number of species shared by both species is 19 for the three anuran amphibians, *Hoplobatrachus rugulosa, Fejervarya limnocharis* and *Duttaphrynus melanostictus*, and 16 species of squamates: *Gekko hokouensis, Hemidactylus frenatus, Hemiphyllodactylus typus typus, Lepidodactylus lugubris, Diploderma swinhonis, Takydromus sauteri, Emoia atrocostata, Eutropis longicaudata, Sphenomorphus incognitus, Elaphe carinata, Oligodon formosanus, Psammodynastes pulverulentus, Amphiesma stolatum,* and *Calamaria pavimentata*, Only five species of lizards differ in their distribution between the two islands, and four species of lizards, *Gekko kikuchii, Lepidodactylus yami, Eutropis cumingi*, and *E. multicarinata*, are Philippine derivatives of Orchid Island, while one lizard, *Plestiodon leucostictus* is endemic to Green Island.

3.11.2 Habitat and Activities

There are 13 species of lizards on Orchid Island (Table 3.15). Two species, *Lepidodactylus yami* (Gekkoidae) and *Eutropis cumingi* (Scincidae) are restricted to a small microhabitat on the island. The former gecko is an endemic species and the latter *Eutropis* is dispersed from Luzon and Philippine Islands. The other two geckos, *L. lugubris* and *Hemiphyllodactylus typus typus*, are extremely rare. These four species will not be considered further in this book. The lizard assemblage

Table 3.15 Species of squamata and their origins on Orchid and Green Islands. Species with an asterisk (*) are endemic; (**) indicate not found in the present study

Family and species	Origin locality	Green Island
Gekkonidae		
Gekko hokouensis	Taiwan	Yes
G. kikuchii	Philippines	No
Hemidactylus frenatus	Taiwan	Yes
Hemiphyllodactylus typus	Taiwan	Yes
Lepidodactylus lugubris	Taiwan	Yes
*L. yami**	Philippine	No
Agamidae		
Diploderma swinhonis	Taiwan	Yes
Lacertidae		
Takydromus sauteri	Taiwan	Yes
Scincidae		
Emoia atrocostata	Taiwan	Yes
Eutropis longicaudata	Taiwan	Yes
Eutropis multicarinata	Philippine	No
Eutropis cumingi	Philippine	No
Sphenomorphus incognitos	Taiwan	Yes
Colubridae		
Elaphe carinata	Taiwan	Yes
Oligodon formosanus	Taiwan	Yes
Psammodynastes pulverulentus	Taiwan	Yes
*Amphiesma stolatum***	Taiwan	Yes
*Calamaria pavimentata***	Taiwan	Yes
Viperidae		
Trimeresurus stejneger	Taiwan	Yes
Cheloniidae		
Chelonia mydas	?	No
Typhlopidae		
Ramphotyphlops braminus	Taiwan	Yes

includes ten common species (Table 3.15) that feed primarily on invertebrates (see above). Six of these species are active foragers and four are sit-and-wait foragers (Table 3.17). Three of the gecko species (*G. kikuchii, G. hokouensis* and *H. frenatus*) are mainly nocturnal, sometimes appearing on buildings at the edge of the forest; the other seven species are diurnal. Some of the three species (*E. longicaudata, G. kikuchii* and *G. hokouensis*) use concrete holes as their nests; one species (*T. sauteri*) is strictly arboreal; and one species (*E. atrocostata*) occurs only on the rocky shorelines of coral reefs. *E. longicaudata* occurs in a variety of microhabitats, and the lizards in the Gekkonidae are more restricted in their microhabitat distribution (Table 3.17). The lizards *Diploderma swinhonis* are found on the ground only when foraging, and they spend most of their time in three locations (tree trunks, forest, and forest edge).

Table 3.16 Specific squamate richness and endemism in the tropical islands. Data for island groups outside Orchid Island and Taiwan were derived from Darevsky and Roitberg (1999) and sources cited therein

Island	Area (km^2)	Species	Endemic species	% endemic species
Orchid Island	46	30	2	3
Taiwan	34,000	86	26	30
Con Dao	58	15	2	13
Cat Ba	45	8	0	0
Cu Lao Ba Mun	21	14	0	0
Cu Lao Phon Vong	14	8	0	0
Cu Lao Pangjang	12	12	0	0
Hon Bai Can	6.6	8	1	13
Hon Thom	6.2	10	1	10
Hon Nor Way	5.8	2	1	50
Cu Lao Cham	5.5	9	0	0

Table 3.17 List of lizard species studied on Orchid Island, with time of activity (TA), foraging mode (FM) and general categorization of habitat use

Family and species	TA	FM	Habits	Habitat
Scincidae				
EL	Diurnal	Active	Terrestrial-arboreal	Low elevation: forest edges, leaf-litter, rock or concrete holes, cacti, human agricultural areas and shrubs
EM	Diurnal	Active	Terrestrial	Forest edges and leaf litter
EC	Diurnal	Active	Terrestrial	Near coral reef with vegetation
SI	Diurnal	Active	Terrestrial	Higher or lower elevation: Leaf litter, concrete holes and forest edges
EA	Diurnal	Active	Terrestrial	Coral reef near seawater
Agamidae				
DS	Diurnal	S & W	Terrestrial-arboreal	Tree trunks, forest and forest edges
Lacertidae				
TS	Diurnal	Active	Arboreal	Grass vegetation
Gekkonidae				
GH	Nocturnal	S & W	Terrestrial-arboreal	Human habitation, concrete holes (higher elevation) and tree trunk
GK	Nocturnal	S & W	Terrestrial-arboreal	Human habitation, concrete holes (higher elevation) and older coral in forest
HF	Nocturnal	S & W	Terrestrial-arboreal	Human habitation, concrete holes (higher elevation) and tree trunk

Species: *EL Eutropis longicaudata*, *EM E. multicarinata*, *EC E. cumingi*, *SI Sphenomorphus incognitos*, *EA Emoia atrocostata*, *DS Diploderma swinhonis*, *TS Takydromus sauteri*, *GH Gekko hokouensis*, *GK G. kikuchii*, *HF Hemidactylus frenatus*

3.11.3 Morphology

Based on the morphological characteristics of sexually mature animals, the smallest lizard was the nocturnal gecko (*H. frenatus*) (51.39 mm) and the largest was the long-tailed skink (*E. longicaudata*) (117.19 mm; Tables 3.1 and 3.12). The same results were obtained using data from individuals of all ages, with significant differences in body length between species ($F_{1,8} = 1744.64$, $p < 0.0001$; ANOVA). Examination of body size among species within the foraging line revealed no differences between sit-and-wait and active foragers (Mann Whitney test, $p = 0.11$).

Except for *G. hokouensis* and *E. multicarinata*, there was considerable variation in traits for each species (Table 3.18). In general, when sexual differences were observed, body mass and SVL were greater in males than in females. Four species of lizards (*E. longicaudata, D. swinhonis, S. incognitus*, and *E. atrocostata*) had larger body length and body mass in males than in females. For the three Gekkonidae species, *G. hokouensis, G. kikuchii, H. frenatus,* and the snake *O. formosanus*, male SVLs and body mass did not differ significantly from females; female SVLs of *T. sauteri* were larger than males, but body mass did not differ significantly between females and males. There was no significant difference in body mass between females and males. Similarly, SVLs of *E. multicarinata* did not differ significantly between males and females, but the body mass of males was heavier than that of females.

3.11.4 Reproductive Cycle

Figure 3.9 summarizes the reproductive cycle data obtained from the nine species. Keep in mind that in my sample, I have assembled the data by month for different years (due to rainy seasons or typhoons or government harvesting restrictions).

Table 3.18 Sexual size dimorphism (SSD) in ten species of reptile on Orchid Island, Taiwan

	Species									
	GH	GK	HF	EL	DS	SI	EA	EM	TS	OF
Observation										
M-M competition	+	+	+	+	+	+	?	—	—	—
Fecundity selection	—	—	—	+	—	—	—	—	—	?
SSD of SVL	2	1	1	1	1	1	1	2	3	2
SSD of body mass	2	2	2	1	1	1	1	1	2	2
Expectation										
SSD of body mass	1	1	1	1	1	1	1	1	2	1

Species: *GH Gekko hokouensis*, *GK G. kikuchii*, *HF Hemidactylus frenatus*, *EL Eutropis longicaudata*, *DS Diploderma swinhonis*, *SI Spenomorphus incognitos*, *EA Emoia atrocostata*, *EM E. multicarinta*, *TS Takydromus sauteri*, *OF Oligodon formosanus*

+: Yes; —: No;?: No data available; 1: Male > female; 2: Male = female; 3: Male < female; M-M: Male-male

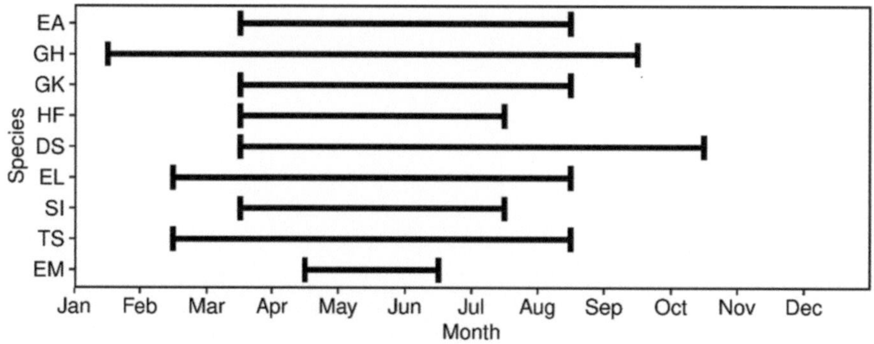

Fig. 3.9 Reproductive cycles of nine species of lizards on Orchid Island, Taiwan. *EA Emoia atrocostata, GH Gekko hokouensis, GK Gekko kikuchii, HF Hemidactylus frenatus, DS Diploderma swinhonis, EL Eutropis longicaudata, SI Sphenomorphus incognitus, TS Takydromus sauteri, EM Eutropis multicarinata*

Therefore, I have temporarily ignored any temporal effects, including the effect of food supply through rainfall in arid habitats. Due to the small population size and sample size of *E. multicarinata*, I have listed the available data here without further discussion. The longest reproductive cycle lasted 9 months, *G. hokouensis* from January to September, and the shortest was *H. frenatu* and *S. incognitus* from March to July. There was no difference in reproductive length between those lizards with rigid-shell eggs (all geckos) and those with flexible-shell eggs ($F_{1,6} = 0.01$, $p = 0.913$; ANOVA). With the exception of *S. incognitus* and *H. frenatus*, most lizards reproduced from March to August. This appears to coincide with the egg-laying period (May to August) of the Orchid Island green sea turtle *Chelonia mydas*.

3.11.5 Clutch Size and Egg Quality

Among Scincidae (four species, *E. lonicaudata, E. multicarinata, S. incognitus,* and *E. atrocostata*), clutch sizes, only EL was positively correlated with maternal SVL (Table 3.19). No significant correlations were found in other species. Egg quality was elastic in Scincidae on Orchid Island. Female *Eutropis* usually bury their eggs in soil, but some *E. longicaudata* place their eggs in concrete holes and exhibit parental care.

3.11.6 Agamidae (*D. swinhonis*) and Lacertidae (*T. sauteri*)

In both species, no significant correlation was found between maternal SVL and clutch size. The quality of eggs was elastic, and parental care was absent.

Table 3.19 Clutch size (CS), correlation between SVL and CS (CSC), egg quality (EQ), micro-habitat of egg nest (MEN), and parental care (PC) of lizard assemblage on Orchid Island, Taiwan

Family and species	CS (mean ± SD)	Range	CSC	EQ	MEN	PC
Scincidae						
EL	6.55 ± 3.48	1–13	Positive	Flexible	Expose (holes)	Yes
EM	2.83 ± 0.75	2–4	–	Flexible	Unknown	No
SI	4.00 ± 1.00	3–5	No	Flexible	Burry under soil	No
EA	2.00 ± 1.00	1–3	No	Flexible	Coral reef holes	No
Agamidae						
DS	3.20 ± 0.86	2–5	No	Flexible	Burry under soil	No
Lacertidae						
TS	2.00 ± 1.00	1–4	No	Flexible	Unknown	No
Gekkonidae						
GH	2.00 ± 0.00	2	No	Rigid	Attachable	No
GK	2.00 ± 0.00	2	No	Rigid	Attachable	No
HF	2.00 ± 0.00	2	No	Rigid	Un-attachable	No

Species: *EL Eutropis longicaudata, EM E. multicarinta, SI Spenomorphus incognitos, EA Emoia atrocostata, DS Diploderma swinhonis, TS Takydromus sauteri, GH Gekko hokouensis, GK G. kikuchii, HF Hemidactylus frenatus*

3.11.7 Gekkoidae (Three Species, *G. hokouensis, G. kikuchii,* and *H. frenatus*)

All geckos examined produced a clutch of two eggs (Table 3.19). Although the eggs of all geckos were hard-shelled, only those of *G. hokouensis* and *G. kikuchii* were attachable. They were often attached to tree trunks or concrete holes. Parental care was not observed in any of these geckos.

Figure 3.10 shows the relationship between mean adult female body size and mean SVL size for six lizard species (excluding three gecko species). Clutch size was highly correlated with female body size among lizards ($r = 0.82$, df $= 5$, $p = 0.013$). Among the eight species of lizards on Orchid Island, only *E. longicaudata* showed significant differences in mean egg production by month ($r = 0.82$, $p < 0.05$; Fig. 3.11). Examination of clutch size among species with different foraging patterns showed no difference between sit-and-wait and active foragers (Mann Whitney test, $p = 0.41$).

3.11.8 Food Webs

Adult food webs (Fig. 3.12) establish the basic structure and relationships among species; these relationships are derived from personal observations, previous studies

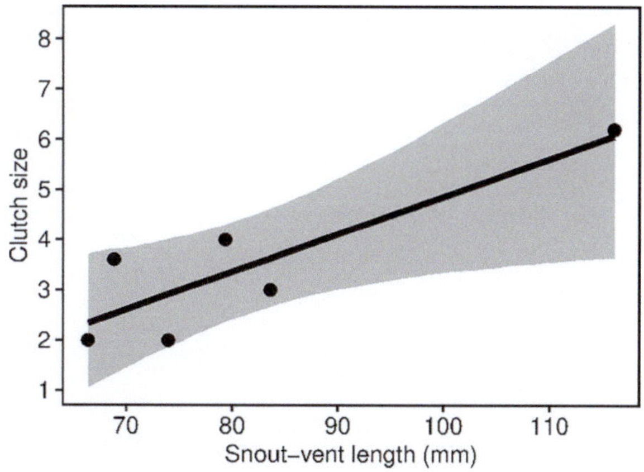

Fig. 3.10 Relationship between clutch sizes of six species of lizards and body sizes (SVL) on Orchid Island, Taiwan

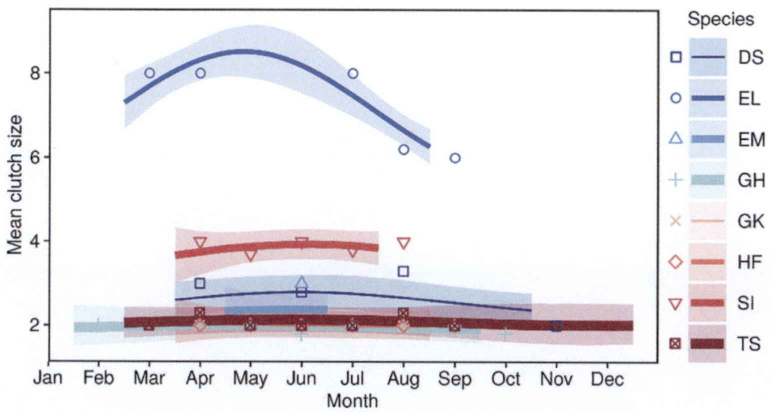

Fig. 3.11 Monthly variations of mean clutch sizes in eight species of lizards on Orchid Island, Taiwan. Only the *Eutropis longicaudata* showed a significant difference in the mean clutch size by month among the eight species of lizards on the Orchid Island (r-0.82, $p < 0.05$)

(see above for the diets of nine lizard species), and general biological knowledge. This form of network is entirely qualitative, with significantly higher resolution for species that are nutritionally close to reptiles.

As is often the case in food webs, reptiles and anurans are consistently identified as species, so there is no need to refine them further. For arthropods, I have used the insect order in most cases and distinguished between larval and adult Lepidoptera. These taxa are known to occur in the diet of *Eutropis longicaudata, Diploderma swinhonis*, and *Sphenomorphus incognitus*. Again, I use generic names for non-insect invertebrates (e.g., spiders) as well as for plants (e.g., leaves, seeds, or fruits).

I recorded two lizard eggs in the stomach of *E. longicaudata* and observed snakes that eat reptile eggs, *Oligodon formosanus* that eats turtle eggs and other lizards with

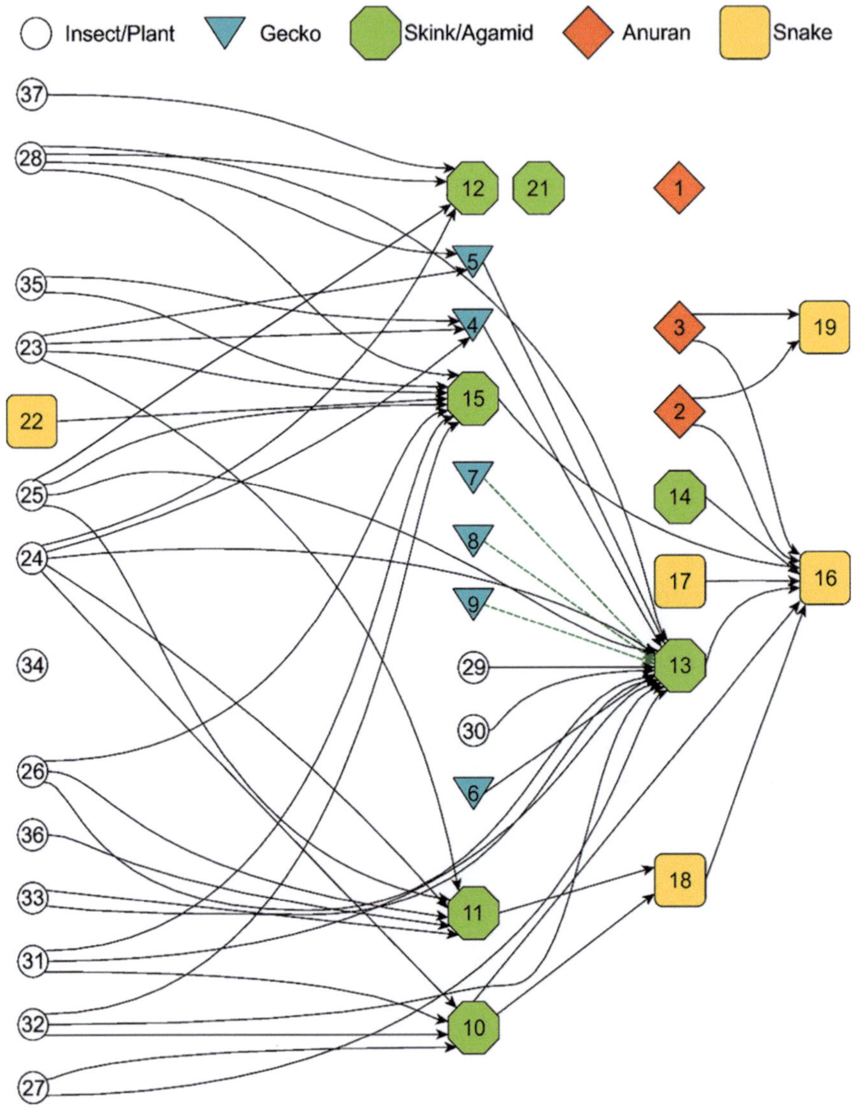

Fig. 3.12 Food web of anurans, reptiles and insects on Orchid Island, Taiwan. 1. *Duttaphrynus melanostictus*, 2. *Fejervarya limnocharis*, 3. *Hoplobatrachus rugulosa*, 4. *Gekko hokouensis*, 5. *G. kikuchii*, 6. *Hemidactylus frenatus*, 7. *Hemiphyllodactylus typus typus*, 8. *Lepidodactylus lugubris*, 9. *L. yami*, 10. *Diploderma swinhonis*, 11. *Takydromus sauteri*, 12. *Emoia atrocostata*, 13. *Eutropis longicaudata*, 14. *E. multicarinata*, 15. *Sphenomorphus incognitus*, 16. *Elaphe carinata*, 17. *Oligodon formosanus*, 18. *Psammodynastes pulverulentus*, 19. *Viridovipera stejnegeri stejnegeri*, 20. *Chelonia mydas*, 21. *Eutropis cumingi*, 22. *Ramphotyphlops braminus*, 23. Orthoptera, 24. Coleoptera, 25. Hemiptera, 26. Lepidotera, 27. Pseudoscorpionida, 28. Hymenoptera, 29. Mantodea, 30. Diptera, 31. Spider, 32. Insect larva, 33. Plant, 34. Dermaptera, 35. Blataria, 36. Snail, 37. Crab

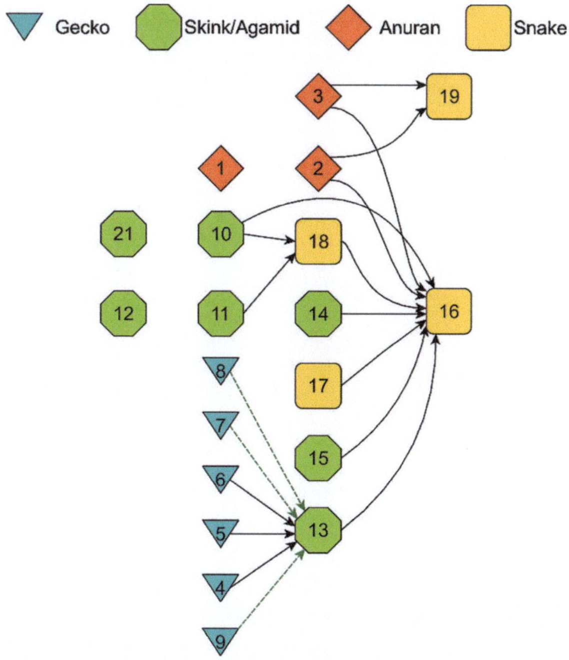

Fig. 3.13 Food web of herpetofauna on Orchid Island, Taiwan. Arabic numbers in the circle indicate the same species as Fig. 3.12

elastic eggs. However, these lizard adults were not foraging targets of the snake; therefore, it would be inappropriate to associate these lizards with the snake in a traditional food web. Therefore, two food webs were established on Orchid Island: an individual food web and an egg food web (Figs. 3.12 and 3.13).

In determining the trophic cascades between members, I included almost all documented linkages, even weak ones. For example, *G. hokouensis* and *H. frenatus* are the two food items of *E. longicaudata*. It is not surprising that it usually seems biologically more reasonable to expect all geckos to be associated with *E. longicaudata* than to expect them not to exist. However, I have ignored the association between anurans and insects because stomach data for these anurans, *Fejervarya limnocharis*, *Hoplobatrachus rugulosa*, and *Duttaphrynus melanostictus*, were not available in the current study.

3.11.9 The Individual Food Webs and Percentage Utilization of Prey Classes by Five Lizard Species

Among the 38 "species" in this network, 49 chains (different pathways from top predators to base species) were found (Fig. 3.12). The number of chains can be reduced by grouping the equivalent species together. The longest chain contains three links, the intermediate species is a gecko or lizard, and the top predator is a snake, *Elaphe carinata*. Using parentheses to indicate insects or plants or snails, and arrows for prey to predator, these chains can be shown as: (insect or plant or snail)-

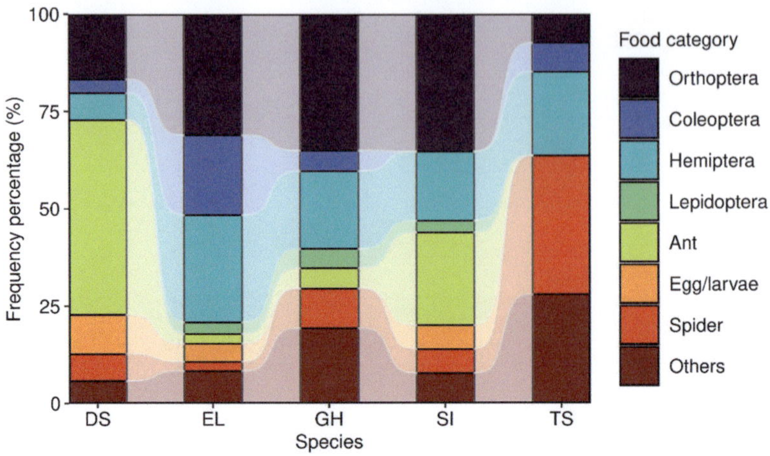

Fig. 3.14 Diets of five species lizards on Orchid Island, Taiwan. *GH Gekko hokouensis*, *DS Diploderma swinhonis*, *EL Eutropis longicaudata*, *SI Sphenomorphus incognitus*, *TS Takydromus sauteri*

> *Gekko hokouensis* or *G. kikuchii*- > *Eutropis longicaudata*- > *Elaphe carinata* or (insect or plant or snail)- > *Diploderma swinhonis*- > *Psammodynastes* pulverulentus- > *E. carinata*. *Eutropis longicaudata* feeds on 12 food types, with the most links, while the two food groups (Orthoptera and Hymenoptera) with the least amount of food in the Orchid Island diet are *G. kikuchii* (Fig. 3.12). Figure 3.13 is a simplified food web showing only reptiles and amphibians.

Orthoptera (grasshoppers) dominated the diet of *Eutropis longicaudata* (31.06%), *Shenomorphus incognitus* (35.29%), and *Gekko hokouensis* (35%). Ants dominated in *Diploderma swinhonis* (50%) and contributed significantly (23.53%) to the diet of *S. incognitos*. Spiders were important in the diet of *Takydromus sauteri* and *G. hokouensis*. Vertebrates were frequent contributors in the diet of *E. longicaudata* and *S. incognitus* (Fig. 3.14).

3.11.10 Egg Food Web

There are four relationships between reptile breeding on Orchid Island and the reptile egg-eating snake *Oligodon formosanus*: (1) eggs eaten by *O. formosanus*, including *Diploderma swinhonis, Takydromus sauteri, E. multicarinata, Sphenomorphus incognitus, Chelonia mydas,* and *Eutropis cumingi*; (2) Eggs are not available due to the viviparous reproductive mode of *Psammodynastes pulverulentus* and *Viridovipera stejnegeri* snakes; (3) eggs unavailable due to parental care in some individuals of *E. longicaudata*, and in *Elaphe carinata* due to predation on *O. formosanus* and parental care behavior; and (4) eggs unavailable due to rigid eggshells in *Gekko hokouensis, G. kikuchii, Hemidactylus frenatus, Hemiphyllodactylus typus typus, Lepidodactylus lugubris* and *L. yami* (Fig. 3.15).

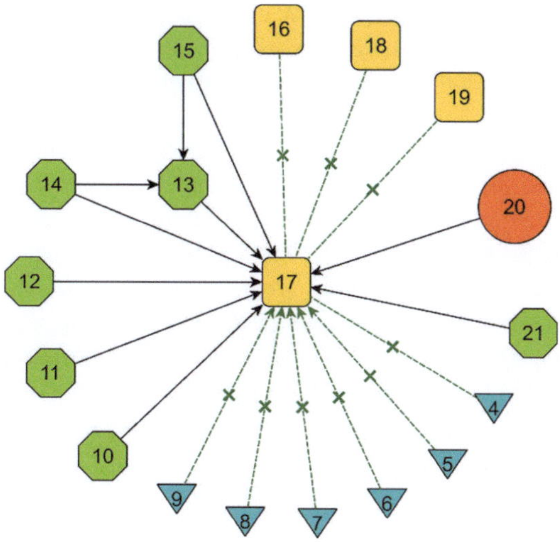

Fig. 3.15 Egg food web of reptile on Orchid Island, Taiwan. Arabic numbers and colors are the same species as Fig. 3.12

3.12 Discussion

3.12.1 Fauna

On Orchid Island, there are at least 21 different species of terrestrial squamates. Eighteen of these species can be found in Taiwan, one of which is endemic, and two others—*Eutropis cumingi* and *E. multicarinata*, from the Philippine island of Luzon. Sixteen species of presumably native reptiles can be found on the nearby Green Island (16 km2) (Lue et al., 1999) (Table 3.15). The only significant distinction between these two islands is that Green Island is home to the endemic *Plestiodon leucostictus* (Hikida, 1988), whereas the other four lizard species—*Gekko kikuchii, Lepidodactylus yami, Eutropis cumingi*, and *E. multicarinata* listed as Philippine elements in Orchid Island. Due to its proximity to Luzon, Orchid Island may be the reason why these four species exist there but not in Taiwan or Green Island.

According to the current taxonomy, Orchid Island and Green Island have nine species of lizards (*Gekko hokouensis, Hemidactylus frenatus, Hemiphyllodactylus typus typus, Lepidodactylus lugubris, Diploderma swinhonis, Takydromus sauteri, Emoia atrocostata, Eutropis longicaudata* (Table 3.15). They can be found in Taiwan as well. The paleogeography of these three islands, in particular the intricate geological history of the inundation of land bridges, which isolated them from Taiwan and later reconnected them during the Late Pleistocene sea-level lowering of Green Island, sheds light on the apparent similarity among the reptiles of these three islands.

The strait separating Taiwan from Orchid Island is more than 200 meters deep and 60 km wide. Heaney (1986) studied 120 m bathymetry to confirm that these two islands were isolated from each other during the Late Pleistocene when sea level dropped dramatically (Wang & Wang, 1990). Although there is no general consensus on the timing and exact extent of this dryland, there is no doubt that it existed and that faunal elements in the immediate area could have used the land as a dispersal route (Ota, 1999).

Two striking aspects of the Orchid Island biota are the high abundance of reptiles and the high diversity of gecko species. For example, the lizard *Takydromus sauteri* is a rare lizard in Taiwan, but is abundant on Orchid Island. A possible explanation is that Orchid Island (46 km^2) is too small to sustain populations of some predators (i.e., *Boiga kraepelini, Macropisthodon rudis*); therefore, lizard populations may be more dense than in Taiwan. As seen in Table 3.15, Orchid Island is inhabited by one endemic geckos, *L. yami*, which live in lowland rainforest, a habitat not found in either Green Island or Taiwan.

3.12.2 Habitat and Activity

From 1997 to 2001, I collected 950 individuals of nine species of lizards during fieldwork on Orchid Island. Two other species, *Lepidodactylus yami* (12 specimens, nocturnal) and *Eutropis cumingi* (14 specimens, diurnal) have been collected at the site. Both species were active around the patches of mixed false pineapple (*Pandanus odoratissmus*) around coral beaches. Both are rarely encountered and very difficult to collect (Ota & Huang, 2000). In addition, previous encyclopedias indicate that two snakes, *Amphiesma stolatum* and *Calamaria pavimentata*, may be present there (Lue et al., 1999), but I have not collected either species in my fieldwork since 1997.

Six species of lizards were active on Orchid Island for most of the day (Table 3.17). The few studies that have been done on the island's active lizards indicate that there is little variation in their activity patterns. In the Kalahari Desert, (Huey & Pianka, 1977) compared four sympatric species of *Eutropis* and discovered that, with one exception, their activity times were highly overlapping. The peak activity period and duration of the tropical lizards of northern Brazil (Vitt & de Carvalho, 1995) are most active from morning to evening, and are thus very similar to the lizards of Orchid Island in terms of peak activity and duration.

There may be significant differences in feeding habits among lizard groups depending on how each uses its microhabitat. For instance, of the nine species of lizards on Orchid Island, *Eutropis longicaudata* used the most microhabitats, using six (low elevation areas: forest edges, deciduous, rocky or concrete caves, cacti, human agricultural areas, and shrubs). Numerous investigations into other *Eutropis* have revealed that they inhabit a wide range of microhabitats, including bushes, grasses, trees, pineapples, fallen tree trunks, rocks, and even human homes (Vitt, 1991; Vitt & Blackburn, 1991; Vrcibradic & Rocha, 1996). *Eutropis longicaudata* has the most varied diet of any lizard on Orchid Island, consuming 24 species of food in 14 groups, primarily Orthoptera (Grylidae, Acrididae, and Tettigoniidae; 31%), Coleoptera (Scarabaeidae; 20.5%), and Hemiptera (Gymnocerata; 9.1%); it also consumes some plant material and some vertebrates. It is also true on Orchid Island that lizard diets are frequently closely related to their foraging microhabitats, activity times, and foraging strategies (Greene & Jaksic, 1983; Vrcibradic & Rocha, 1996). For instance, nocturnal species (like *G. hokouensis*) primarily consume crickets, whereas diurnal species (like *E. longicaudata, S. incognitus*, and *D. swinhonis*) primarily consume day-active arthropods (such as grasshoppers, bees, or ants). This suggests that some of the observed differences in feeding habits among them are caused by their various activity periods.

Active foragers tend to consume mostly low-mobility prey, while sit-and-wait foragers typically have more generalized diets and prefer high-mobility prey (Huey & Pianka, 1981). Except for *T. sauteri* and *E. atrocostata*, species in the Orchid Island active foraging guild share a great deal of microhabitats. Small lizard species, like *S. incognitus* and *D. swinhonis*, rely heavily on small prey like ants (Table 3.8), whereas larger species, like *E. longicaudata*, adopt a wide range of prey sizes, including large prey, especially other vertebrates (e.g., geckos; Table 3.2). Comparisons among the most common feeding species, however, reveal that both the active forager *Eutropis longicaudata* (stomach containing 14 categorizations; Table 3.2) and the sit-and-wait forager *Diploderma swinhonis* (7 categorizations; Table 3.8) eat a variety of prey, including both relatively stationary prey like insect larvae and highly mobile prey like Locustidae, Scarabaeidae, Formicidae, spiders , the consumption of both kinds of prey by *E. longicaudata*, as well as intermediate prey species like cicadas, Tettigoniidae, and moths, in addition to my own field observations, suggest that it straddles the line between sit-and-wait and active foragers, employing both foraging strategies concurrently.

3.12.3 Morphology

Species in the active foraging guild overlap considerably in microhabitat use (except *T. sauteri* and *E. atrocostata*; Table 3.17); smaller species of lizards (i.e., *S. incognitus*) rely to a large extent on small-bodied prey such as ants (Table 3.5), whereas large species (i.e., *E. longicaudata*) take a wide size range of prey, including large items, such as grasshopper or other vertebrates (i.e., geckos; Table 3.2). Large differences in body size among sympatric active foragers have been reported elsewhere (Vitt, 1991, 1995; Vitt & de Carvalho, 1995).

Differences in microhabitat use among the nine lizard species may be related to specific morphological characteristics such as differences in finger length, tail length, body shape, and body weight, or to their ability to tolerate environmental extremes such as seawater or higher temperatures. For example, the slender body shape and longer tail (three times their SVL) length of *T. sauteri* may facilitate movement in foliage compared to other species; the longer fingers of *E. longicaudata* enable them to climb tree trunks in addition to allowing them to move quickly on the ground; the higher tolerance to salt water of *E. acrostata* allows them can live close to rocky reefs. The relatively long fingers of lizards are also associated with arboreal nature, mainly because terrestrial species tend to have shorter fingers than those that regularly climb on trees, rocks, and other rough surfaces (Colli et al., 1992; Vrcibradic & Rocha, 1996).

I have established that sexual size dimorphism in body size exists in several species of Orchid lizards (Table 3.18). Sexual size dimorphism may be due to (1) male competition: males are larger than females because they thus gain more access to females (Stamps, 1977, 1983; Fitch, 1981; Zamudio, 1998); (2) female choice: larger males are selected for mates, possibly because of their size or traits associated with large size (Trivers, 1972); (3) fecundity selection: females are larger than males because of clutch size; and (4) reproductive investment: females may allocate more energy to reproduction at the expense of growth. Clearly, body size may be affected by a mixture of these factors (Grant, 1986; Anderson & Vitt, 1990), so care should be taken to determine the cause of any sex-specific body size, as well as the cause of overall sexual size dimorphism (Schoener, 1982).

Female reproduction is probably the most important factor affecting sexual size dimorphism (i.e., SVL and body mass) in lizards (Huang, 1996a, b, 1997b, 2006b, 2007). Because females devote most of their energy to producing embryos and, in some cases, to providing maternal care, this explanation applies particularly to species in which males do not exhibit competition for mates or territorial defenses. In this case, I expect that (1) males that gain intra-competition may have greater SVL and body mass than those females with or without reproductive fitness options because they allocate more energy to reproduction rather than growth. Among the 10 species of lizards on Orchid Island, the relationship between SVL and body mass in three species, *E. longicaudata, D. swinhonis*, and *S. incognitus*, is consistent with this hypothesis (Table 3.18). However, the body length of males of *G. hokouensis, G. kikuchi* and *H. frenatus* are either longer (*G. kikuchii* and *H. frenatus*) or equal to that of females (*G. hokouensis*); the body mass of males was not larger than that of females and therefore also inconsistent with expectations (Table 3.18); (2) the body masses of males might be heavier than those of females when the SVLs male larger and/or equal than females, between male without competition and females without fecundity selection. The *E. atrocostata* and *E. multicarinata* were coincident with expectation. However, the reptile-egg-eating snake, *O. formosanus* female body mass was heavier than that of males, which did not conform to this hypothesis; (3) the body mass would be no different between males and females if the SVLs of females were larger than those of males. The grass lizard, *T. sauteri* was consistent with this hypothesis.

I am not sure why the three Geckos (*G. hokouensis G. kikuchii* and *H. frenatus*) conflict with the first hypothesis. Larger males may be subject to higher predation pressure than smaller males on Orchid Island, where they overlap with their natural enemy *E. longicaudata* in shrubs and concrete holes (Table 3.17). Predation pressure may eliminate larger males because (1) larger males are territorial and therefore more vulnerable to *Eutropis* predation and (2) larger males are unable to escape from concrete holes when encountering predators, whereas smaller geckos can pass through gaps between rocks when *E. longicaudata* enters the cave. Despite their different routines, geckos (nocturnal) and *Eutropis* (diurnal) still encounter each other at dawn or dusk. Whether the small size selection comes from predation risk depends on the importance of body size to avoid predation, and I do not know either the frequency of predation attempts by predators of known size or the outcome (3) female geckos may not invest too much energy in reproduction; or (4) they may be influenced by phylogenetic history.

3.12.4 Food Web

3.12.4.1 Individuals

One of my purposes in explicitly documenting the food web structure of Orchid Island was to assess the strengths and weaknesses of the constructed food web. I believe that almost any food web construction is likely to be influenced by the weaknesses I have identified. The web appears to be more densely connected close to reptiles because my data focused on them and the predators or prey that they are. Additionally, the diversity of what I can find in the basal portion of the internet is severely constrained by the fact that I classify basal species by site rather than taxon. These biases could decrease the proportion of associations between various trophic levels and the accuracy of estimates of whole-web properties (Goldwasser & Roughgarden, 1993).

For example, in the food items of *E. longicaudata*, I combined Grylidae, Acrididae and its main prey Tettigoniidae (18.94%) in one Orthoptera (Fig. 3.12), which is the same number of links as *G. hokouensis* consuming only Grylidae (Fig. 3.12). *E. longicaudata* has a higher number of true links than shown (Fig. 3.12) and is more complex than the number of links in *G. hokouensis* (all three links in both are reduced to one). Thus, my method represents a compromise between different levels of resolution (Table 3.2). It improves the uniformity of resolution among insect taxa at the cost of giving up some data. Any attempt to integrate various sources may face the dilemma of finding a resolution that minimizes existing groups (discards information) while segmenting other groups (introduces guesswork).

After analyzing the dietary information of non-snake squamates, Vitt et al. (2003) came to the conclusion that historical variations among subgroups had an impact on the kinds of prey consumed. They discovered that during the evolution of squamates, there was a change in diet, with scleroglossans consuming primarily termites, grasshoppers, spiders, and insect larvae while iguanian diets contained a large proportion of ants, other hymenopterans, and beetles. They also claimed that because iguanians and scleroglossans have different diets, the latter have a competitive advantage during the day, forcing iguanians to change their microhabitats while gekkotans shift their activities to the night.

Summary data on the diet of the five lizard species (Fig. 3.14) are partially consistent with the predictions of (Vitt et al., 2003). For example, the main diet of *E. longicaudata* and *S. incognitus* (Scincidae) is orthopterans, including crickets and grasshoppers, while the lizard *D. swinhonis* mainly consumes ants (50%). However, insect larvae and eggs were the secondary prey of these two species (4.55% and 5.88%, respectively), which contrasts with the findings of Vitt et al. (2003).

3.12.4.2 Eggs

The egg-based food web focused specifically on the reptilian egg-eating snake *Oligodon formosanus*, as well as egg quality, daily activity, behavior, and other predators of the Orchid Island lizard group (Fig. 3.15). Although there have been numerous studies on adult food webs (Spiller & Schoener, 1990; Schmitz & Booth, 1997), our understanding of reptile egg predation in the context of community ecology remains limited. In this chapter, I tentatively divided Orchid Island reptile eggs into four groups with respect to their availability to various predators and the presence or absence of parental care (see subsequent chapter for more details).

First, the eggs of *Diploderma swinhonis, Takydromus sauter, E. multicarinata, Sphenomorphus incognitus, Chelonia mydas,* and *Eutropis cumingi* can be used by snakes *Oligodon formosanus*. Due to their soft outer shells. Some work has established that the eggs of most oviparous animals are subject to intense predation (Orians & Janzen, 1974). For example, 35% of eggs of temperate birds are eaten by predators (Lack, 1968), 50% of eggs of some tropical birds are predated (Ricklefs, 1969), and predation on the nests of many reptiles and amphibians is apparently severe (Tinkle et al., 1970). The prevalence of intense predation on eggs suggests that predators are an important selection factor for egg characteristics (Orians & Janzen, 1974). For example, some species have been reported to lay eggs in places that are not easily accessible to predators, such as nesting on sea cliffs, in trees or near ant and wasp colonies, or exhibiting cryptic coloration that precludes predator detection (Orians & Janzen, 1974). In this chapter, six species of reptiles either buried their eggs under soil or sand (*D. swinhonis, S. incognitus, E. multicarinata Chelonia mydas,* and *Eutropis cumingi*) or exposed their eggs in rock or concrete holes (*E. longicaudata*), which suggests that these eggs are easily found by snakes. I expect that these eggs are therefore more vulnerable than those of other lizards on Orchid Island.

Second, due to the viviparous mode of reproduction of both *Psammodynastes pulverulentus* and *Viridovipera stejnegeri,* the eggs are not available. Third, due to the parental care of the lizard (*E. longicaudata*, see following chapter) and the snake (*Elaphe carinata*), eggs are not available. Fourth, eggs could not be obtained because of the hard egg shells of the geckos *Gekko hokouensis, G. kikuchii, Hemidactylus frenatus, Hemiphyllodactylus typus typu, Lepidodactylus lugubris,* and *L. yami*. Therefore, the reproductive biology (reproductive cycle, clutch size, etc.) of the six species in the first group is most likely influenced by the snake *O. formosanus* on Orchid Island. *i* Reproductive cycle, clutch size, and egg quality.

Compared to mainland populations, many island populations start reproducing later (Stamps & Buechner, 1985). It is uncommon for both places to have the same onset time, and no island species has ever been discovered to begin breeding before mainland species. The defensive hypothesis is generally supported by these findings. The breeding season for *Eutropis longicaudata* began in the current study in February, 1 month earlier than it did in Taiwan (Huang, 2006c). Similarly, *Diploderma swinhonis* started breeding on Orchid Island and Taiwan at the same time, i.e., March (Huang, 2007). These findings don't seem to support the defense hypothesis, which could be explained by the fact that Orchid Island is home to numerous reptiles, including the egg-eating snake *Oligodon formosanus. Eutropis lonicaudata* exhibits parental care to prevent its eggs from being eaten by the snake, and thus can breed earlier than other Orchid Island lizards (Huang, 2006c). However, species that lack parental care run the risk of having their eggs eaten if they reproduce too soon. The best option for these lizards is to reproduce simultaneously in order to share the risk of predation and decrease mortality to the eggs of each species.

Thus, the reptilian egg-eating snake *O. formosanus* may have two effects on the breeding season: (1) species with hard eggshells, which can withstand snake predation, can breed earlier and longer than those with flexible eggshells; and (2) species with flexible eggshells may breed during the green sea turtles' breeding season, which runs from May to August, when turtle eggs predominate as a major food source for snakes. Snakes that eat reptile eggs don't affect the breeding season of this assemblage of lizards, according to studies of lizards with both flexible and hard eggshells. However, the breeding period of the lizard, *E. longicaudata*, a large lizard whose eggs could be a major resource for snakes on Orchid Island, lasted until August, and thus is coincident with the time that most snakes are aggregated at the beach where green sea turtles are nesting. This coincides with the time of year when most snakes gather on the beaches where green sea turtles are nesting. Additional lizards include *S. incognitus*, which ceased breeding in July, 1 month earlier than *E. longicaudata*. Similarly, *T. sauteri* breeds from March to August, which seems to coincide with the spawning period of Orchid Island green sea turtles from May to August.

If competition for territory on an island is intense, animals may produce fewer but more competitive juveniles rather than more but less competitive juveniles (Cody, 1966). According to this hypothesis, juveniles' larger bodies will help them find resources more successfully and have higher survival rates. Larger eggs or neonates may produce larger offspring in species that don't receive parental care. This type of life history change should result in fewer but larger juveniles being born on islands if the juveniles are fed at a higher rate or for a longer period of time, and parental care can have the same effect.

In this chapter, I describe that female *E. longicaudata* lizards can produce large clutches early in the breeding season and small clutches later in the season, but clutch size seasonality is not the same for other species (Fig. 3.11). Although there is no direct evidence for the size of those late and early hatchlings in the Orchid Island population, my data also indicate that small clutches (mean = 6.5) resulted in larger hatchlings than large clutches (mean = 7.5) in the *E. longicaudata* of mainland Taiwan (Table 3.20). It seems reasonable to expect that hatchlings from a clutch of 6 eggs may be larger than hatchlings from a clutch of 8 eggs, rather than being equal in body size to each other.

Late in the season, when competition for scarce resources is at its peak, parental investment models assume that females are subject to positive selection, resulting in larger hatchlings (Ferguson & Bohlen, 1978; Nussbaum, 1981). It is predicted that if larger hatchlings are available early in the reproductive season and there is no fitness advantage, females should produce more small-sized hatchlings. My results are

Table 3.20 Comparisons of body sizes of *Eutropis longicaudata* hatchlings with Taiwanese and Orchid Island populations. Snout-vent lengths are given only for new hatchling juveniles. Comparison of SVL is based on ANOVA, while other comparisons are based on ANCOVA (SVL as covariate)

	Taiwanese population ($n = 66$)		Orchid Island population ($n = 34$)			
Character	Mean	Range	Mean	Range	F	P
SVL (mm)	35.45	31.79–38.1	35.27	32.7–36.9	0.52	0.47
HL (mm)	10.11	9.48–10.58	10.38	8.42–11.5	9.32	0.003
HW (mm)	5.87	5.42–6.37	6.06	4.82–6.63	17.1	<0.001
HD (mm)	4.00	3.59–4.33	4.33	3.74–4.74	87.72	<0.001
TL (mm)	61.96	51.18–70.10	65.52	57.0–71.0	37.06	<0.001
BM (g)	0.96	0.66–1.28	1.07	0.77–1.20	37.64	<0.001

SVL snout-vent length, *HL* head length, *HW* head width, *HD* head depth, *TL* tail length, *BM* body mass

consistent with this model, with larger hatchling size lizards occurring later in the season when competition for resources is most intense. Although it is well established that larger hatching lizards have higher survival rates than smaller hatching lizards during periods of food shortage and/or increased predation pressure (DeMarco, 1989), there is little evidence that females choose to produce specific hatchling sizes. A demonstration that females can vary seasonal patterns of clutch size to correspond to variable timing during periods of intense competition would provide support for parental investment patterns. However, it is unlikely that females can predict in advance the type of competitive environment that hatchlings will experience and adjust hatchling size accordingly (DeMarco, 1989). However, females can predict food shortages and/or prey abundance or size from early and/or late daily food. For example, smaller Coleoptera (i.e., Cicindelidae, Tenebrionidae, and Curculionidae) are rarely found in the late season (from September to January), when the main diet of *E. longicaudata* is larger prey such as grasshoppers, Tettigoniidae. Thus, females that produce smaller young in larger clutches early in the season and larger young in smaller clutches late in the season may respond to changes in food size. In other words, there is a survival advantage for larger juvenile lizards in times of food shortage in the late season.

Questions remain as to why the other eight species of lizards have fewer hatchlings than long-tailed lizards throughout the season, and why hatchling numbers are highly correlated with female size in long-tailed lizards but not in other species. Shine and Greer (1991) proposed that invariable clutch sizes in lizards might arise from (1) adaptation: females with less variable clutch sizes have higher fitness; (2) as a by-product of selection for other traits, such as low mean clutch size, small body size and low variation in adult female size; or (3) phylogenetic history: lizards may have invariant clutch size and large body size because genetic variation in these traits has been lost in evolution; for example, all geckos (>90 genera, 850 species worldwide) apparently lay two eggs for clutch size (Shine & Greer, 1991).

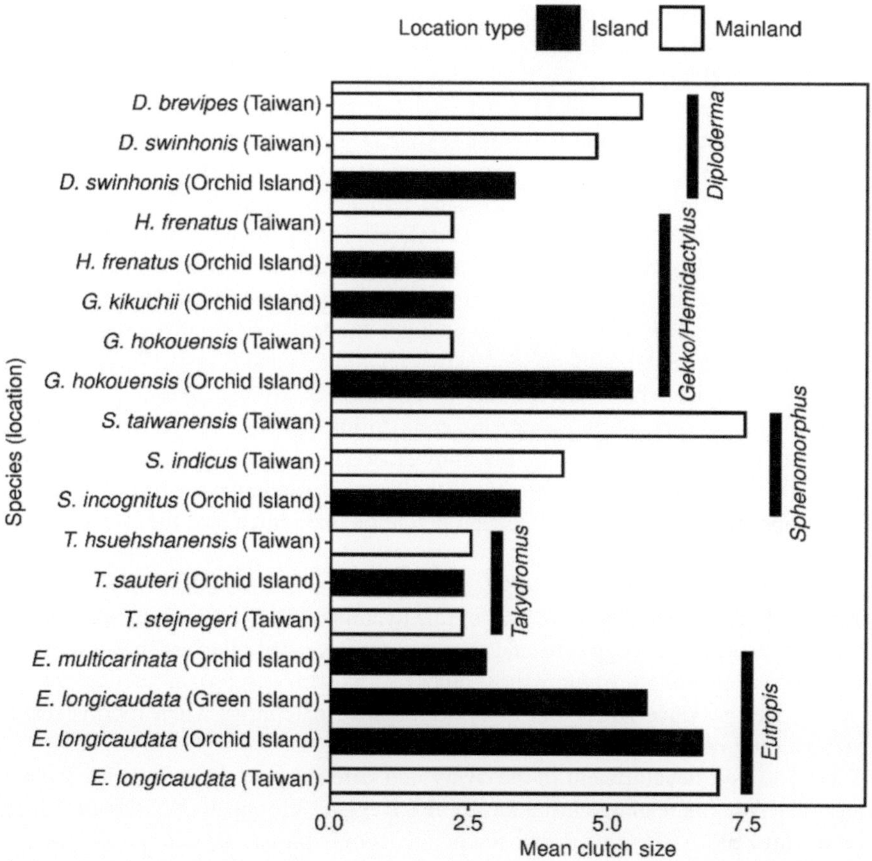

Fig. 3.16 Comparisons of the clutch sizes in conspecific and congeneric lizards among Taiwan, Green Island, and the Orchid Island. Black symbols indicate species distributed on offshore islands, open symbols indicate species distributed on mainland

The data on the two-egg clutches of the three gecko species appear to be consistent with their predicted phylogenetic history. However, another explanation may apply to other lizards on Orchid Island, namely that snake predation on eggs selects for lizards that have not evolved parental care, reducing their clutch size. I have shown that for lizards of the same species or genus, smaller clutch sizes occur more frequently in *Diploderma swinhonis* and *S. incognitus* on Orchid Island than in these taxa on Green Island and Taiwan (Fig. 3.16). Since larger clutches are more likely to be found by snakes, selection on Orchid Island clearly favors smaller clutches.

MacArthur and Wilson (1967) and Grant (1998) suggested that lizards on islands have smaller clutch sizes compared to mainland relatives. They attributed this change to persistent intra-specific competition at high densities in the absence of predators and intra-specific competition (Grant, 1998). My data suggest that the small clutch size is primarily the result of intra-specific competition, as the clutch size found in Taiwan is larger than that found on Orchid Island or Green Island (Fig. 3.16). For example, the average clutch size of *E. longicaudata* on Orchid Island was 6.8, larger than that on Green Island (mean = 5.5). Similarly, *D. swinhonis* laid 4.6 eggs in a clutch larger than the 3.1 or 3.2 clutches on Orchid Island and Green Island, respectively. In addition, unlike female *D. swinhonis* on Orchid Island, their second clutch in Taiwan produced more eggs than their first clutch(Cheng & Lin, 1977), and the clutch size of Green Island females was positively correlated with SVL ($n = 32$, $p = 0.000$, regression analysis; Huang unpublished data). These differences suggest that the reptilian egg-eating snake (*O. formosanus*) influences fecundity on Orchid Island, as females can adjust their investment to reduce snake predation (Fig. 3.15). My clutch size-predator model conceptualizes the trade-off between clutch size and predation intensity on Orchid Island (Fig. 3.17). Clutch size tends to be smaller when intra-specific competition is stronger, but *O. formosanus* reduces the intensity of intra-specific competition because it consumes a large number of eggs, thereby reducing the number of adult individuals; therefore, clutch size will tend to be consistent with that of the Taiwanese population due to predation. There appears to be a trade-off between predation and clutch size for Orchid Island reptiles.

The clutch size-predator model incorporates "intra-specific competition"in lizards, whereby when clutch size varies due to competition, lizards increase parental investment to increase egg size and produce similar-sized clutches throughout the breeding season, and they ignore most other factors such as seasonal fluctuations. As a result, they produce larger and more competitive hatchlings when food is scarce and hatchling densities are high.

In birds, egg development in the ovary and parental care are sequential, so they can adjust parental feeding during the period of egg development (Nussbaum, 1981). Nussbaum (1981) also noted that temperate ovoviviparous lizards usually do not have parental care, and they develop their eggs simultaneously in each clutch. They can cope with energy shortages at the onset of clutch development by absorbing one or more eggs, or they can cope with an abundance of energy by increasing the size of their eggs. Thus, unlike birds, lizards are likely to choose a smaller overall clutch

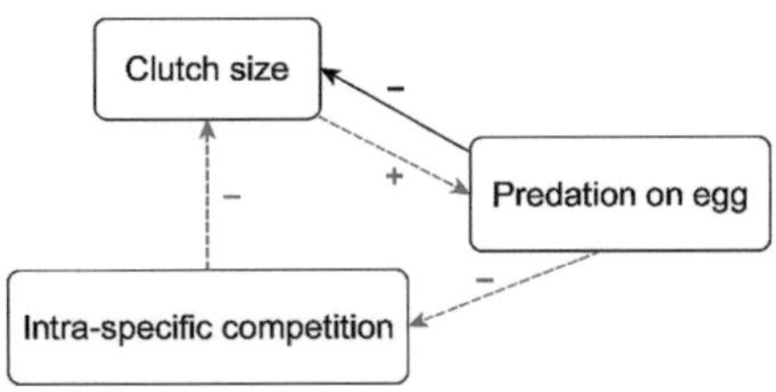

Fig. 3.17 The clutch size-predator model: relationships among clutch size, intra-specific competition, and predation on eggs. Clutch size tends to be smaller when intra-specific competition is stronger, but predators will decrease the intensity of intra-specific competition; thus, the clutch size will be larger because of predators (the causal relationships are represented by green dash arrows). However, larger clutch sizes are easily located by the egg-eating snake, *Oligodon formosanus*, hence larger clutch size will be eliminated by that predator (represented by the black arrow). – and + indicate the intensity of variables is decreasing and increasing, respectively

size and adjust their investment by increasing egg size because they do not have the opportunity to change their investment after the eggs are laid. However, *E. longicaudata* had parental care on Orchid Island and its clutch egg size (mean = 6.5) was larger than that of its Green Island counterpart without parental care (mean = 5.5), suggesting that the former invested more energy and time during egg-caring. Unlike birds, females of *E. longicaudata* neither incubate their eggs nor feed their young lizards. Therefore, if all factors were equal, they would probably lay more eggs. Although most of their eggs were exposed to higher predation pressure from the reptile egg-eating snakes on Orchid Island, *E. longicaudata* showed parental care and successfully defended its eggs from the snakes. Because these lizards invest more energy in caring for each clutch of eggs, they may lay more eggs than their Green Island counterparts.

The most important role in this book is the reptile-egg-eating snake, the kukri snake. The next chapter will introduce the kukri snake's territorial behavior and how the female snake defeats the male snake in the competition for the precious food source of sea turtle eggs.

References

Anderson, R. A., & Vitt, L. J. (1990). Sexual selection versus alternative causes of sexual dimorphism in teiid lizards. *Oecologia, 84*, 145–157.

Brown, J. L. (1975). *The evolution of behavior*. W. W. Norton and Company.

Cheng, H. Y. (1987). A review on annual reproductive and energetic patterns of five taxa of lizards in Taiwan for ten years. *Proceedings of the National Science Council, Republic of China, 11*, 313–321.

Cheng, H. Y., & Lin, J. I. (1977). Comparative reproductive biology of the lizards, *Japalura swinhonis formosensis*, *Takydromus septentrionalis* and *Hemidactylus frenatus* in Taiwan i. male reproductive cycle. *Bulletin of the Institute of Zoology, Academia Sinica, 16*, 107–120.

Cheng, H. Y., & Lin, J. I. (1978). Comparative reproductive biology of the lizards, *Japalura swinhonis formosensis*, *Takydromus septentrionalis* and *Hemidactylus frenatus* in Taiwan ii. Fat body and liver cycles of the males. *Bulletin of the Institute of Zoology, Academia Sinica, 17*, 67–74.

Cody, M. L. (1966). A general theory of clutch size. *Evolution, 20*, 174–184.

Colli, G. R., Araujo, A. F. B., Silveira, R., & Roma, F. (1992). Niche partitioning and morphology of two syntopic *Tropidurus* (Sauria: Tropiduridae) in Mato Grosso, Brazil. *Journal of Herpetology, 26*, 66–69.

Congdon, J. D., Gibbons, J. W., & Greene, J. L. (1983). Parental investment in the chicken turtle (*Deirochelys reticularia*). *Ecology, 64*, 419–425.

Darevsky, I. S., & Roitberg, E. S. (1999). A new subspecies of the rock lizard Lacerta caucasica (Sauria, Lacertidae) from the south-east of Chechen Republic of the Caucasus. *Russian Journal of Herpetology, 6*(3), 209–214.

DeMarco, V. G. (1989). Annual variation in the seasonal shift in egg size and clutch size in *Sceloporus woodi*. *Oecologia, 80*, 525–532.

Dodson, S. I., & Havel, J. E. (1988). Indirect prey effects: Some morphological and life history responses of *Daphnia pulex* exposed to *Notonecta undulata*. *Limnology and Oceanography, 33*, 1274–1285.

Duvall, D., et al. (1982). Environmental control of reptilian reproductive cycles. In C. Gans & F. H. Pough (Eds.), *Biology of the reptilia. Volume 13. Physiology D. Physiological ecology* (pp. 201–231). Academic Press.

Duvall, D., Schuett, G. W., & Arnold, S. J. (1993). Ecology and evolution of snake mating systems. In R. A. Seigel & J. T. Collins (Eds.), *Snakes: Ecology and behavior* (pp. 165–200). McGraw-Hill.

Ferguson, G. W., & Bohlen, C. H. (1978). Demographic analysis: A tool for the study of natural selection of behavioral traits. In N. Greenberg & P. D. MacLean (Eds.), *Behavior and neurology of lizards, an interdisciplinary colloquium* (pp. 227–243). Department of Health, Education, and Welfare, Public Health Service, Alcohol, Drug Abuse, and Mental Health Administration, National Institute of Mental Health.

Fitch, H. S. (1970). *Reproductive cycles in lizards and snakes*. University of Kansas.

Fitch, H. S. (1981). *Sexual size differences in reptiles*. University of Kansas.

Goldwasser, L., & Roughgarden, J. (1993). Construction and analysis of a large caribbean food web. *Ecology, 74*, 1216–1233.

Grant, P. R. (1986). *Ecology and evolution of Darwin's finches*. Princeton University Press.

Grant, P. R. (1998). *Evolution on islands*. Oxford University Press.

Greene, H. W., & Jaksic, F. M. (1983). Food-niche relationships among sympatric predators: Effects of level of prey identification. *Oikos, 40*, 151–154.

Heaney, L. R. (1986). Biogeography of mammals of SE Asia: Estimates of rates of colonization, extinction and speciation. *Biological Journal of the Linnean Society, 28*(1–2), 127–165.

Hik, D. S. (1995). Does risk of predation influence population dynamics? Evidence from the cycle decline of snowshoe hare. *Wildlife Research, 22*, 115–129.

Hikida, T. (1988). A new white-spotted subspecies of *Eumeces chinensis* (Scincidae: Lacertilia) from Lutao Island, Taiwan. *Japanese Journal of Herpetology, 12*, 119–123.

Huang, W. S. (1996a). Reproductive cycles and sexual dimorphism in the viviparous skink, *Sphenomorphus indicus* (Sauria: Scincidae), from Wushe, Central Taiwan. *Zoological Studies, 35*, 55–61.

Huang, W. S. (1996b). Sexual size dimorphism in the five-striped blue-tailed skink, *Eumeces elegans*, with notes on its life history in Taiwan. *Zoological Studies, 35*, 188–194.

Huang, W. S. (1997a). Reproductive cycle of the oviparious lizard *Japalura brevipes* (Agamidae: Reptilia) in Taiwan, Republic of China. *Journal of Herpetology, 31*, 22–29.

Huang, W. S. (1997b). Reproductive cycle of the skink, *Sphenomorphus taiwanensis*, in Central Taiwan. *Journal of Herpetology, 31*, 287–290.

Huang, W. S. (1998). Reproductive cycles of the grass lizard, *Takydromus hsuehshanensis*, with comments on reproductive patterns of lizards from the central high elevation area of Taiwan. *Copeia, 1998*, 866–873.

Huang, W. S. (2006a). Ecological characteristics of the skink, *Mabuya longicaudata*, on a tropical east Asian Island. *Copeia, 2006*, 293–300.

Huang, W. S. (2006b). Ecology and reproductive patterns of the grass lizard, *Takydromus sauteri*, in a tropical rain forest of an East Asian Island. *Journal of Herpetology, 40*, 267–273.

Huang, W. S. (2006c). Parental care in the long-tailed skink, *Mabuya longicaudata,* on a tropical Asian Island. *Animal Behaviour, 72*, 791–795.

Huang, W. S. (2007). Ecology and reproductive patterns of the agamid lizard, *Japalura swinhonis*, on an East Asian Island, with comments on the small clutch sizes of Island lizrds. *Zoological Science, 24*, 181–188.

Huang, W. S. (2010). Foraging behaviors of two sympatric ant species in response to lizard eggs. *Zoology, 113*, 85–90.

Huang, C., & Sih, A. (1990). Experimental studies on behaviorally mediated, indirect interactions through a shared predator. *Ecology, 71*, 1515–1522.

Huang, W. S., Greene, H. W., Chang, T. J., & Shine, R. (2011). Territorial behavior in Taiwanese kukrisnakes *(Oligodon formosanus*). *Proceedings of the National Academy of Science, 108*, 7455–7459.

Huey, R. B., & Pianka, E. R. (1977). Patterns of niche overlap among broadly sympatric versus narrowly sympatric Kalahari lizards (Scincidae: *Mabuya*). *Ecology, 58*(1), 119–128.

Huey, R. B., & Pianka, E. R. (1981). Ecological correlates of foraging mode. *Ecology, 62*, 991–999.

James, C. D., & Shine, R. (1985). The seasonal timing of reproduction: A tropical-temperate comparison in Australian lizards. *Oecologia, 67*, 464–474.
Kano, T. (1933). Zoogeography of Botel Tobago Islands (Kôtôsho) with a consideration on the northern portion of Wallace's Line. *Geographical Review of Japan, 9*(6), 475–491.
Kato, J., & Ota, H. (1994). Notes on the breeding of the five-lined skinks, *Eumeces elegans*. *Acta Zoologica Taiwanica, 5*, 77–81.
Kuroda, N. (1925). On the distribution of Japanese birds. *Journal of Geography (Chigaku Zasshi), 37*(7), 369–380.
Lack, D. (1968). *Ecological adaptations for breeding in birds*. Methuen and.
Lima, S., & Dill, L. M. (1990). Behavioral decisions made under the risk of predation: A review and prospectus. *Canadian Journal of Zoology, 68*, 619–640.
Lue, K. Y., Tu, M. C., & Shang, G. (1999). *A pictorial guide to the amphibians and reptiles of Taiwan*. Nature Magazine Publishing.
MacArthur, R. H., & Wilson, E. O. (1967). *The theory of Island biogeography*. Princeton University Press.
Murphy, G. I. (1968). Pattern in life history and the environment. *The American Naturalist, 102*, 291–303.
Nussbaum, R. A. (1981). Seasonal shifts in clutch size and egg size in the side-blotched lizard, *Uta stansburiana* Baird and Girard. *Oecologia, 49*, 8–13.
Orians, G. H., & Janzen, D. H. (1974). Why are embryos so tasty? *The American Naturalist, 108*, 581–592.
Ota, H. (1991). Classification of agamid lizards of the genus *Japalura* from the insular East Asia: A history of confusion. *Biological Magazine Okinawa, 29*, 21–28.
Ota, H. (1999). *Tropical islands herpetofauna: Origin, current diversity, and conservation*. Elsevier Science B. V.
Ota, H., & Huang, W. S. (2000). *Mabuya cumingi* (Reptilia: Scincidae): An addition to the Herpetofauna of Lanyu Island, Taiwan. *Current Herpetology, 19*, 57–61.
Ota, H., & Lin, J. T. (1994). Digestive tract contents of *Oligodon formosanus* and their implications on the adaptation for reptilian-eating snakes. *Journal of Taiwan Museum, 47*, 75–78.
Pianka, E. R. (1972). R- and K-selection of b- and d-selection? *The American Naturalist, 106*, 581–588.
Pianka, E. R. (1976). Natural selection of optimal reproductive tactics. *American Zoologist, 16*, 775–784.
Raven, P. H. (1987). The scope of the plant conservation program world-wide. In D. Bramwell, O. Hamann, V. H. Heywood, & H. Synge (Eds.), *Botanic gardens and the world conservation strategy* (pp. 19–29). Academic Press.
Ricklefs, R. (1969). An analysis of nesting mortality in birds. *Smithsonian Contributions to Zoology, 9*, 1–48.
Schmitz, O. J. (1994). Resource edibility and trophic exploitation in an old-field food web. *Proceedings of the National Academy of Sciences, 91*, 5364–5367.
Schmitz, O. J., & Booth, G. (1997). Modeling food web complexity: The consequence of individual-based, spatially explicit behavioral ecology on trophic interactions. *Evolutionary Ecology, 11*, 379–398.
Schmitz, O. J., Beckerman, A. P., & O'Brien, K. M. (1997). Behaviorally mediated trophic cascades: Effects of predation risk on food web interactions. *Ecology, 78*, 1388–1399.
Schoener, T. W. (1982). The controversy over interspecific competition. *American Scientist, 70*, 586–595.
Schoener, T. W. (1989). Food webs from the small to the large. *Ecology, 70*, 1559–1589.
Shine, R., & Greer, A. E. (1991). Why are clutch sizes more variable in some species than in others? *Evolution, 45*, 1696–1706.
Sih, A. (1987). Predations and prey lifestyles: An evolutionary and ecological overview. In W. C. Kerfoot & A. Sih (Eds.), *Predation: Direct and indirect impacts on aquatic communities* (pp. 203–224). University Press of New England.

Spiller, D. A., & Schoener, T. W. (1990). A terrestrial field experiment showing the impact of eliminating top predators on foloage damage. *Nature, 347*, 469–472.
Stamps, J. A. (1977). The relationship between resource competition, risk and aggression in a tropical territorial lizard. *Ecology, 58*, 349–358.
Stamps, J. A. (1983). Sexual selection, sexual dimorphism, and territoriality. In R. B. Huey, E. R. Pianka, & T. W. Schoener (Eds.), *Lizard ecology: Studies of a model organism* (pp. 169–204). Harvard University Press.
Stamps, J. A., & Buechner, M. (1985). The territorial defense hypothesis and the ecology of insular vertebrates. *The Quarterly Review of Biology, 60*, 155–181.
Stearns, S. C. (1976). Life history tactics: A review of the ideas. *The Quarterly Review of Biology, 51*, 3–47.
Tinkle, D. W. (1969). The concept of reproductive effort and its relation to the evolution of life histories of lizards. *The American Naturalist, 103*, 501–516.
Tinkle, D. W., Wilbur, H. M., & Tilley, S. G. (1970). Evolutionary strategies in lizard reproduction. *Evolution, 24*, 55–74.
Trivers, R. L. (1972). Parental investment and sexual selection. In B. Campbell (Ed.), *Sexual selection and the descent of man, 1871–1971* (pp. 136–179). Aldine Publishing Company.
Vitt, L. J. (1986). Reproductive tactics of sympatric gekkonid lizards with a comment on the evolutionary and ecological consequences of invariant clutch size. *Copeia, 1986*, 773–786.
Vitt, L. J. (1991). Ecology and life-history of the wide-foraging lizard *Kentropyx calcarata* (Teiidae) in Amazonian Brazil. *Canadian Journal of Zoology, 69*, 2791–2799.
Vitt, L. J. (1995). The ecology of tropical lizards in the caatinga of Northeast Brazil. *Occasional Papers of the Oklahoma Museum of Natural History, 1*, 1–29.
Vitt, L. J., & Blackburn, D. G. (1991). Ecology and life history of the viviparous lizard *Mabuya bistriata* (Scincidae) in the Brazilian Amazon. *Copeia, 1991*(4), 916–927.
Vitt, L. J., & de Carvalho, C. M. (1995). Niche partitioning in a tropical wet season lizards in the Lavrado area of Northern Brazil. *Copeia, 1995*, 305–329.
Vitt, L. J., & Zani, P. A. (1996). Ecology of the lizard *Ameiva festiva* (Teiidae) in southeastern Nicaragua. *Journal of Herpetology, 30*, 110–117.
Vitt, L. J., Pianka, E. R., Cooper, W. E. J., & Schwenk, K. (2003). History and the global ecology of squamate reptiles. *The American Naturalist, 162*, 44–60.
Vrcibradic, D., & Rocha, C. F. D. (1996). Ecological differences in tropical sympatric skinks (*Mabuya macrorhyncha* and *Mabuya agilis*) in southeastern Brazil. *Journal of Herpetology, 30*, 60–67.
Wang, L., & Wang, P. (1990). Late Quaternary paleoceanography of the South China Sea: Glacial–interglacial contrasts on an enclosed basin. *Paleoceanography, 5*(1), 77–90.
Werner, E. E. (1991). Nonlethal effects of a predator on competitive interaction between two anuran larvae. *Ecology, 72*, 1709–1720.
Werner, E. E. (1992). Individual behavior and higher order species interactions. *The American Naturalist, 140*, S5–S32.
Werner, E. E., & Anholt, B. R. (1996). Predator-induce behavioral indirect effects: Consequences to competitive interactions in anuran larvae. *Ecology, 77*, 157–169.
Wilbur, H. M. (1977). Propagule size, number and dispersion pattern in *Ambystoma* and *Asclepias*. *The American Naturalist, 111*, 43–68.
Williams, G. C. (1966a). Natural selection, the costs of reproduction, and a refinement of Lack's principle. *The American Naturalist, 100*, 687–690.
Williams, L. H. (1966b). Observations on the life history of the poplar hawk moth, *Laothoe populi* L. part I. Mating, egg laying and larval development and behaviour. *Proceedings of the Royal Entomological Society of London, 41*, 93–102.
Winemiller, K. O., Pianka, E. R., Vitt, L. J., & Joern, A. (2001). Food web laws or niche theory? Six independent empirical tests. *The American Naturalist, 158*, 193–199.
Zamudio, K. R. (1998). The evolution of female-biased sexual size dimorphism: A population-level comparative study in horned lizards (*Phrynosoma*). *Evolution, 52*, 1821–1833.

Territorial Behavior of Kukri Snake

4

Abstract

Island populations of Oligodon formosanus reveal rare territorial behavior in snakes. Females defend sea turtle nests for weeks, expelling other individuals until eggs hatch or are consumed. These nests provide large, persistent food resources, unlike typical snake prey. The species has specialized teeth for cutting eggs and uses tail display as defense, making combat risky—especially for males. These traits and resource conditions favor female territoriality, supporting the hypothesis that valuable, defensible resources drive the evolution of territorial behavior.

Keywords

Territory · Kukri snake · Sea turtle eggs · Tail injuries

An old Taiwanese saying, "Even if you mess with bears and tigers, don't mess with crazy women." It means that you should never try to upset women, because when they are angry, no man can bear it. Even if the female does not beat the male, she can still conquer him if she adopts the right strategy. For example, another saying: sacrifice what your opponent cannot sacrifice, and you will win. These examples are alive and well in human society. But how do animals work? Most of the vertebrates are dominated by males, who can fight better than females. But is this really the case? Are there any counter-examples?

Generally speaking, most vertebrates exhibit territorial behavior, such as early spring bird calls, frog chirps, or mammalian roars, often for courtship, protection of food resources, or their own territory. For many years, we observed the kukri snake, commonly known as the "reptile egg-eating snake," eating green sea turtle eggs in Orchid Island, and found that every year, green turtles that came ashore to lay eggs would lay more than 10–30 nests of eggs, scattered on a beach about 50 meters long

W.-S. Huang, *Reptiles in the Changing Climate*, Fascinating Life Sciences,
https://doi.org/10.1007/978-981-92-1887-5_4

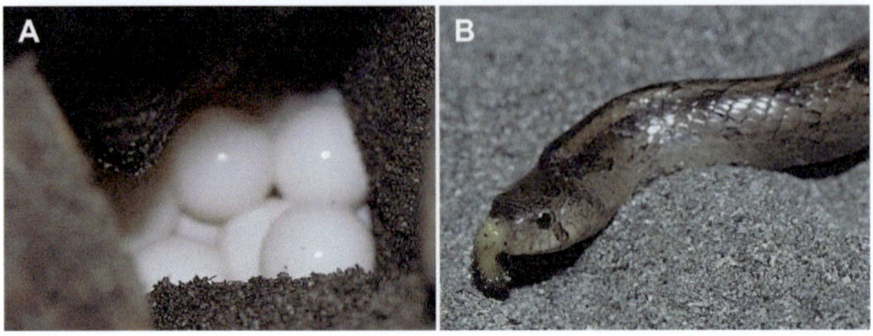

Fig. 4.1 (**a**) Female sea turtles come to Orchid Island beach every summer to lay eggs, and about 100 eggs are laid in the nest. (**b**) Kukri snake only eats the egg yolk but not the egg white. A snake that had just finished eating was startled by my presence and spat out the yolk through its nose

Fig. 4.2 The female snake uses her lower jaw teeth to hold the egg in place and her upper jaw teeth to cut open the shell of the egg, which is known as the can-eating method. A variety of ways for snakes to break eggs (**a**–**d**)

(Fig. 4.1a). About 0.5–8 hours after the mother buried the eggs, the kukri snake would appear near the nest and immediately burrow into the nest to consume on the eggs (Fig. 4.1b).

We observed that the kukri snake's eating the eggs was quite unique, first using its sharp upper teeth to hold the egg in place, then using its sharp mandibles to cut through the soft-shell eggs, and then sticking its head directly into the egg to suck the yolk instead of eating the egg white, thus neither opening its mouth wide to swallow the egg nor spitting out the egg shell, which is called a "novel open can eating method" (Fig. 4.2).

Another interesting finding is that the female snake will dig a hole above the green turtle's nest as a good place to eat the eggs, in addition to avoiding the hot sun, but also to occupy the territory to dominate the turtle's eggs and drive away other

Fig. 4.3 (**a**) When the turtle eggs are finished, the female digs a small hole next to the nest to avoid exposure to the high temperatures of the beach. (**b**) Only females or young snakes were allowed to enter the cave after the females had eaten the eggs, not the males. The picture shows four female snakes in the cave. (Photo B adapted from Huang et al. (2011), with permission)

Fig. 4.4 (**a**) Female snakes can sacrifice their entire tail to approach cloaca. (**b**) Because hemipenis hides near the male tail near the cloaca, the male snake can only allow part of the tail to be bitten off. (Photo A adapted from Huang et al. (2011), PNAS, with permission)

snakes that want to eat the eggs. What's more, after the entire clutch of eggs is eaten, the female can continue to hide in the hole and wait for the next green turtle to come ashore to lay its eggs first, saving her from the hassle of running around (Fig. 4.3). The cloaca of the kukri snake secretes an irritating smell that makes it easy to locate and attack, resulting in snakes that want to enter the nest often being attacked and even having their tails bitten off. So if people see a kukri snake with a tail-cut in Orchid Island, there is no need to make a fuss.

The kukri snake's battle for the green turtle's eggs led to the breaking of the tail, but we discovered an unprecedented area of behavior in snake research.

Since a female snake's tail is short and thin, even if it is bitten off, it has little effect, so if a female snake's tail has been bitten off, she will go into the hole and eat without fear of being bitten by another snake (Fig. 4.4a). However, the same tail-cut strategy does not apply to males, because the male snake's hemipenis are hidden in the front of the tail and is connected to some of the tail muscles (Fig. 4.4b). Despite this, male snakes often have difficulty resisting the fatal attraction of turtle eggs and often force their way across the mountain. If only a small section of the tail is lost and the muscles close to the hemi-penis are not injured, the male is lucky, but if it is too close to the genitals, the male cannot tolerate it and is often the one driven away from the nest. Under this mechanism of divergent feeding strategies by sex, females always have the upper hand in fights.

Females prefer to defend their nests by their tail-cuts, partly because they have the task of reproduction and have to make this sacrifice in order to obtain reproductive energy, and partly because their tail-cuts do not affect their movements or lives. Once a female snake has control of the nest, she can be assured of "food resources" and can lay more eggs as she matures and grows, allowing more of her genes to be passed on. The female snake defeats the male in order to protect the eggs and becomes the dominant sex, unlike the male dominance of the nest in other animals.

The behavior of the Orchid Island kukri snake in establishing its world's only survival territorial and its unique and interesting egg-eating technique, as well as the strong behavior of the female snake in repelling the male snake at the expense of its tail in order to take control of the nest, allow us to discover another unequal gender relationship and competition for survival, as well as to appreciate the mystery of the evolution of behavior in the biological world (Video 1: https://youtu.be/-jb_36-wuJE). This chapter provides the following interesting research findings.

4.1 Introduction

Identifying the selective pressures and environmental factors that promote the evolution of specific traits is a central goal of evolutionary biology. In contrast, widespread traits can be the most difficult to explain in terms of selective forces because their current distribution often provides little insight into the conditions under which they originally arose. In such cases, studying rare or lineage-specific instances of the trait can yield clearer information about its evolutionary origins than examining its common forms. For example, viviparity and parental care are nearly universal among living mammals and likely trace back to a single ancient origin in each case, making the selective advantages that drove their evolution difficult to discern (Blackburn, 1995). By comparison, both viviparity and parental care remain uncommon in reptiles and have arisen independently multiple times, so squamate reptiles provide a more favorable system for investigating the initial selective forces that favor these traits (Shine & Thompson, 2006; Huang & Wang, 2009).

When examining the evolutionary changes that shape social systems, a comparable pattern emerges . Territoriality, which entails the active defense (i.e., exclusion of conspecifics) of a specific area by one or more individuals, is one of the most common systems (Tarof et al., 2005). Critical resources like food supplies, shelters, or breeding opportunities are frequently the focus of the site of defense (Butchart et al., 1999). By granting more or less exclusive access to the resource in question, territoriality can increase personal fitness (Calsbeek & Sinervo, 2002). Mathematical models predict that territoriality develops when the advantages of having exclusive access to resources outweigh the costs (risk, effort, and time) of seeking out and avoiding competitors. This theoretical framework is frequently supported by empirical data on these costs and advantages in territorial species (Krebs & Davies, 1993; Sih & Mateo, 2001).

Territorial defense of locations with vital resources has independently evolved in many phylogenetic lineages, in animal from invertebrates to mammals (Stamps & Buechner, 1985; Crespi, 1986). Nevertheless, territoriality remains rare or absent in certain taxa. For instance, territorial social systems are common in many birds, mammals, lizards, frogs, and salamanders among the major lineages of terrestrial vertebrates, but they have never been documented in snakes(Greene, 1997). Although some snake species' males engage in combat to gain access to receptive females, there are about 2700 snake species left in existence that have no known long-term regional defense mechanisms (Gregory et al., 1987; Duvall et al., 1993; Greene, 1997). A single predator can often be supported within a potentially defensible-sized area, but most snakes consume relatively large prey and feed only sporadically. This difference between lizards and snakes may be due to their nutritional ecology, which likely requires snakes to range widely in search of sufficient food items. Theoretically, when only a portion of a population demonstrates territoriality, the group's identity will be determined by its capacity for territorial occupancy, such as its capacity to outperform rivals in combat (Davies, 1978).

On a field trip to a tropical Asian island, we learned about a species of snake's territorial behavior. This case is significant not only because it is the first documented instance of a territorial system in a snake species, but also because it offers an excellent chance to test hypotheses regarding the environmental factors (particularly defended food sources) and species traits (particularly intraspecific variation in resource holding potential) that are thought to have influenced the evolution of territoriality.

4.2 Arrivals of Snakes at Sea Turtle Nesting Locations in Order

We observed 105 green sea turtle nests (Fig. 4.5a) that were actively incubating eggs, and on 819 separate occasions, we successfully caught a total of 413 kukri snakes (274 adult males and 139 adult females). At sunrise and sunset and again after dark, marked snakes looked for nests. Snakes discovered the nests where we made focused observations between 0.5 and 8.5 hours ($n = 17$) after the females finished covering them (Fig. 4.5b). In contrast to the null hypothesis of equal numbers of each sex, male snakes were the first to arrive at all 17 nests (Fig. 4.6). The fact that there were more males than females in the entire beach population cannot account for this tendency for males to arrive first (see above, sex of first-arriving snakes is contrary to the null hypothesis of 66% of males.

4.2.1 Fate of Predator-Infested Turtle Nests

After 50–60 days of deposition, the remaining turtle eggs hatched ($n = 62$ nests). We always discovered a female snake and a number of unfertilized eggs when we dug up nests shortly after the majority of hatchlings had left. The females dug their burrows

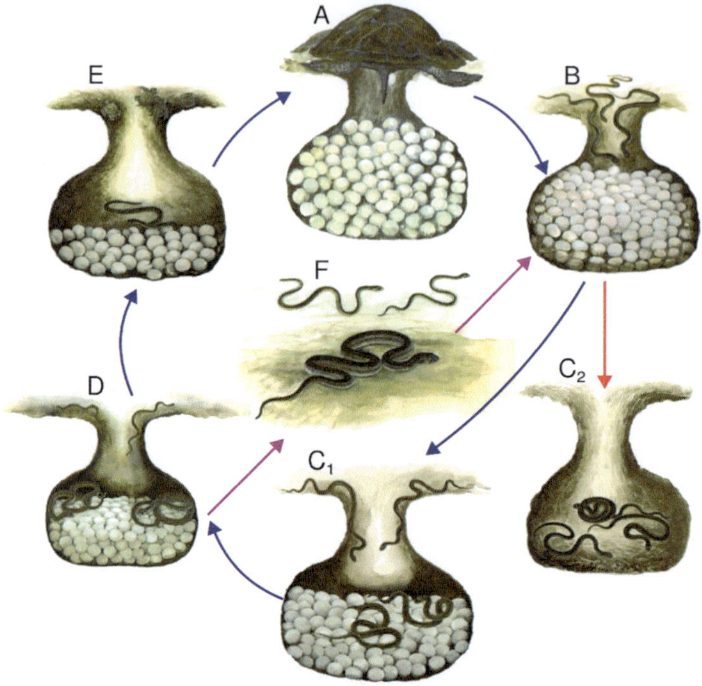

Fig. 4.5 The interactions between green sea turtle eggs and egg-eating snakes. (**a**) Most of green sea turtles laid eggs from June to September. (**b**) At sunrise and sunset and again after dark, marked snakes looked for nests. Snakes discovered the nests between 0.5 and 8.5 h after the female turtles finished covering them; male snakes were the first to arrive at all nests. (**c_1**) Later arriving females will occupy the nest and fight with the male snakes. (**c_2**) (Red arrow): if the female snake does not reach the turtle nest, the male will eat all the eggs. (**d**) A female snake took over the nest and drove all the males out of the nest. (**e**) A female snake could not finish all the eggs, and about 60 days later, the baby turtles hatched and left the nest to crawl away to the ocean. After about 20–30 years, the young turtles mature and return to the beach to lay their eggs. (**f**) (Purple arrow): the male snakes will wander outside the nest, waiting for a new turtle to lay eggs and will go to the nest first

to the surface after we covered these nests again, and we observed their heads poking out of the resulting holes (3.4 ± 0.4 cm in diameter; occupation time 8–22 days, $n = 30$). We verified that all unfertilized eggs had been consumed following the departure of these females.

4.2.2 Snakes' Actions When They First Enter Their Nests

The sequence in which snakes arrived at nests revealed distinct behavioral patterns by sex and depending on nest occupancy. When there were only male snakes at the nest, they all entered and stayed inside, likely eating the eggs (time from the first snake's arrival until all snakes left the nest: 5–24 h, mean $= 7$ h, $n = 35$ nests). In situations where males reached the nest ahead of females and maintained dominance

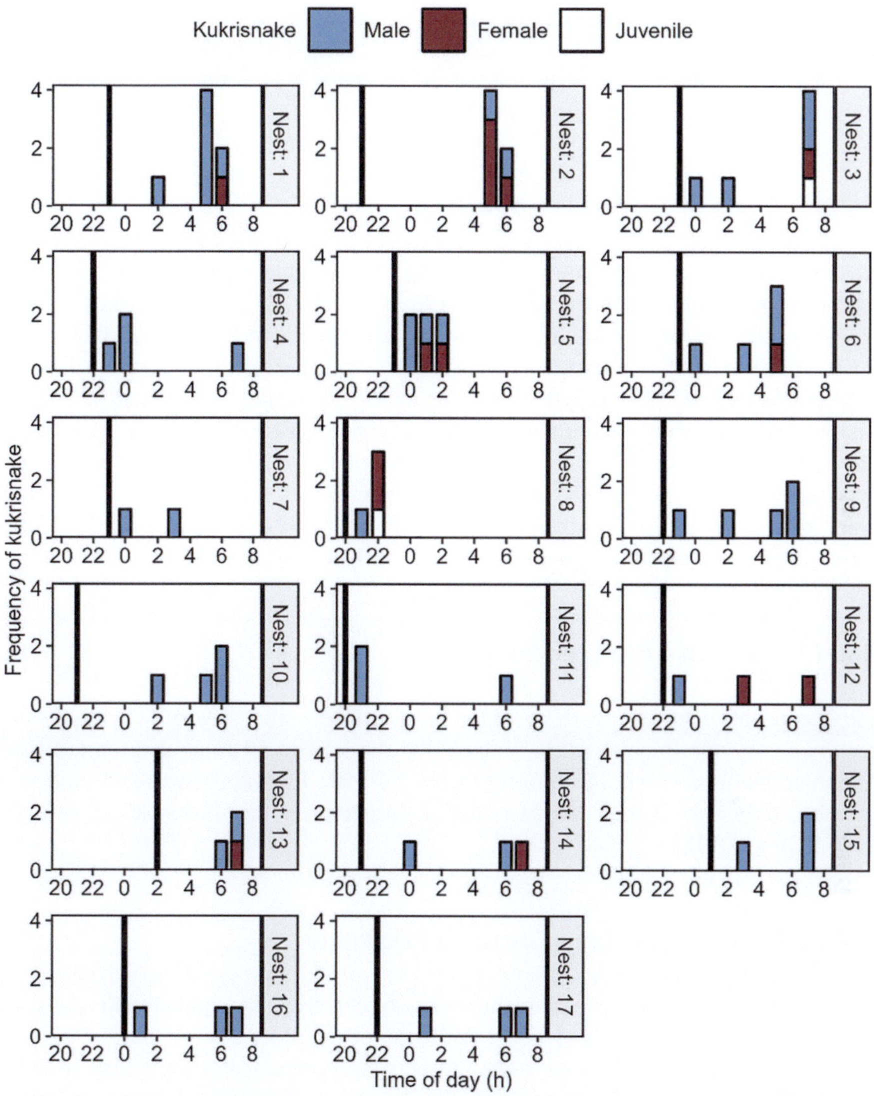

Fig. 4.6 The duration of delays between oviposition by sea turtles (*Chelonia mydas*) and the arrival of kukri snakes (*Oligodon formosanus*) at 17 turtle nests during a 12-h period after egg-laying by turtles. The vertical black line indicates the time when the sea turtle lays eggs

upon arrival, the entire clutch of turtle eggs was depleted rapidly. For the three nests with precise documentation of snake arrivals and departures, all snakes left within 5 days, no new snakes entered afterward, and later digging confirmed that every egg had been consumed (Fig. 4.5c1).

When females arrived at a nest, we observed other snakes being ejected forcibly (inferred from rapid retreats, and sometimes new wounds; Fig. 4.5c2). The ejected males frequently lingered nearby and made an attempt to re-enter the nest (82 were observed in 105 individuals). The likelihood of a late-arriving male staying in the nest was lower if a female was already there than it was if she wasn't (20 out of 25 = 80%, $\chi^2 = 23.6$, df = 1, $p < 0.0001$).

4.2.3 Trials of Forced Introduction

In our introduction trials, both juvenile and adult female snakes showed a strong tendency to remain inside the nest for at least 30 min (23 out of 26 trials in both cases), but it was only 0 out of 26 trials for adult males in forced introductions to nests occupied by 1–5 females. That is, despite the presence of a resident female, young snakes and other females, but not males, frequently succeeded in staying inside the nest, whereas adult males did not ($\chi^2 = 40.9$, df = 1, $p < 0.0001$). Additionally, these studies offer up-close views of snake interactions. During intra-specific encounters, males frequently had their tails bitten by their opponents as a warning sign of impending danger (Figs. 4.4b and 4.5d). As snakes leaving from turtle nests, we noticed that their tails had recently sustained injuries.

4.2.4 Trials in a Field Enclosure

Further proof that female deter male kukri snakes was provided by our field confinement experiments. All the males fled the nest in 14 trials that started with one male and one female. Contrarily, in the 14 trials involving two males and two females, only two snakes and five snakes, respectively, left the nest ($\chi^2 = 17.5$, df = 2, $p < 0.001$).

4.2.5 Factors That May Cause Tail Damage

Males were more susceptible to the negative effects of tail damage on their tails than females. Hemipenes are found in the tail (21% of the most anterior part of the tail, according to the dissection of eight museum specimens), but they may also be particularly vulnerable because male kukri snakes erect the hemipenes when displaying a defensive tail (Wüster & Cox, 1992). In accordance with this alleged sex difference in vulnerability, female kukri snakes had a higher incidence of tail loss greater than 45% than males (52 out of 139 = 38% vs. 69 out of 270 = 25%, Wald $\chi^2 = 15.06$, $p < 0.001$). However, males did not exhibit severe tail truncation, which was more prevalent in very large females. Large females lost nearly all of their tails in 14 instances; similarly, males who lost so many tails would also lose half of their tails (and thus their ability to reproduce).

Snakes from Orchid Island had a significantly higher frequency of tail damage (121 out of 413; 52 females and 69 males; 121 out of 413) f compared to those from Taiwan and mainland Asia (3 out of 48; Wald X2 = 9.09, 1 df, $p = 0.003$). Additionally, whereas Orchid Island snakes' tail damage was more severe than that of mainland snakes, it was typically only minor scarring.

4.3 Discussion

In the Taiwanese island population of kukri snakes, females actively defended sea turtle eggs against conspecifics for several weeks. This is the first time that territoriality (site defense) has been observed in detail in any snake species. Anecdotal reports on African anurans and long-tailed snakes (de Haan, 1984, 2003) suggest that other snakes may also exhibit territorial behavior, a possibility that warrants further investigation. Until other territorial behaviors of snakes are confirmed (even considering other phylogenetic hypotheses), the closest known living relative of *O. formosanus* is the scincomorph lizards, which diverged from the snake family 150 million years ago in the Jurassic (Hedges & Vidal, 2009) (Fig. 4.7).

First, we consider several limitations in our conclusions. Many critical behaviors took place underground, meaning several of our conclusions rely on inference rather than direct observation. For instance, we attribute the higher incidence of tail injuries in Orchid Island snakes to intraspecific fighting rather than attacks by other predator species. This conclusion is supported by the absence of any nocturnal snake predators within the study area, so individuals that exited turtle nests with fresh tail wounds almost certainly sustained those wounds that were caused by territorial defenders. The higher incidence of tail injuries in the Orchid Island snake (the only snake known to eat turtle eggs) is consistent with this explanation, as are the direct intraspecific conflicts we observed in experiments to lure snakes into nests that were already occupied by other individuals.

The apparent uniqueness of the territorial defense of female kukri snakes is consistent with the predictions of theoretical models based on (a) the economic defensibility of resources and (b) the ability to repel conspecific competitors. We consider these two questions in sequence below. For kukri snakes in Taiwan, the

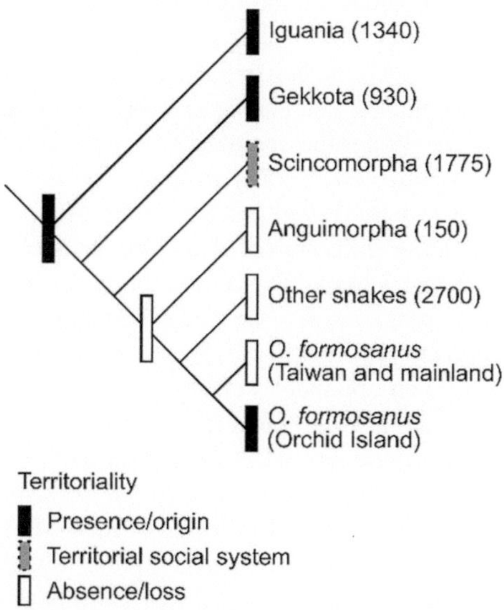

Fig. 4.7 The evolution of territoriality is mapped on the phylogeny of lizards and snakes [modified from (Coleman et al., 1993, Lee, 1997); number of species in parentheses]. According to MacClade (Martins, 1994), bars indicate the presence and origin (dark), the occurrence of territorial social systems (hatched), and the absence and loss (open) of territoriality. (Redrawn from Huang et al. (2011), PNAS)

turtle's nest includes a large prey resource that is available for several weeks after initial discovery. In one nest that we examined before and after snake predation, five snakes ate 102 eggs in 5 days, Fig. 4.5c1). Even taking into account 50% loss of eggshells and spillage, a nest with an average of 105 45-gram eggs (Huang et al., 2011) would supply enough biomass for five snakes (mean body mass 117 g) to ingest four times their own mass. The relatively small size of nests makes it easy to monitor and defend against intruders of the same species. To our knowledge, no other snake species has a feeding ecology that offers such a direct energetic reward for excluding conspecific competitors . Turtle eggs are (a) large for food, (b) highly aggregated, (c) sedentary, (d) without maternal protection, and (e) remain edible over an extended period. Plant resources frequently share these same characteristics (for example, we can imagine herbivorous iguana could plausibly defend a tree bearing edible foliage), but all snakes are carnivores (Greene, 1997). The sole form of prey that might provide a sustained food resource supply for snakes is a clutch of eggs; but while eggs are indeed often eaten by snakes(Huang et al., 2011; Lee et al., 2019), a clutch usually contains too little food to sustain a single snake for any prolonged duration.

The eggs and larvae of social insects satisfy many of the same conditions and serve as prey for blind snakes (Scolecophidians), a group of burrowing species characterized by shrunken eyes and highly modified mouthparts adapted for consuming these tiny food items (Greene, 1997; de Quieroz & Rodrigues-Robles, 2006). Despite sharing these nutritional traits, blind snakes lack the second prerequisite for resource defense evolution: the ability to drive out intraspecific competitors with weaponry. It is hard to envision interactions in which one snake causes injury to another. By contrast, the upper jaw teeth of *Oligodon formosanus* are very large and can cause deep cuts (H.W.G., pers. obs.). Enlarged maxillary teeth are a recurring adaptation in snakes that specialize in swallowing reptilian eggs and have arisen independently in multiple lineages (Scanlon & Shine, 1988). Swallowing and enlarged posterior teeth appear widely in the Cook's snake lineage and related genera, such as *Stegonotus* (Shine, 1991). Coincidentally, allowing the presence of economically defendable prey resources (large clutches of unattended reptile eggs) also facilitates the evolution of specialized dentition, promoting the ability of snakes to exclude conspecifics through serious injury during aggressive encounters.

Why is territorial defense performed by female kukri snakes rather than by larger and more numerous males? One possible explanation for this pattern is that sex differences in the fitness costs of intraspecific fighting. During intraspecific fighting, kukri snakes exhibit an unusual stereotypic display that includes tail waving at the opponent and half of the male's face upside down (Wüster & Cox, 1992). Thus, a tail bite by an opponent's enlarged teeth is far greater reproductive costs on males than on females. Even when the hemipenis is not directly injured, any shortening of the tail may seriously affect mating success (Shine, 1991; Shine et al., 1999). In contrast, damage to the short, slender tail of females does not significantly impair their movement or reproductive capability. Thus, unusual conditions favor the evolution of this surprising territorial behavior, so that only females can defend a persistent and abundant food resourcesat low additional cost. The apparent inability of males to stay in the nest with females in the territory may favor the active search of newly laid (and therefore unoccupied) nests by males, leading to a strong tendency for males to arrive at nests before females do.

From a broader perspective, the discovery of territoriality in a clade of approximately 2700 unclaimed snake species (Huang et al., 2011) highlights the fragility of current generalizations about social systems, even in well-studied taxa such as terrestrial vertebrates. Future studies, particularly in species-rich biota in tropical regions, will undoubtedly uncover additional instances of behavioral and social diversity. Species such as the kukri snake of Orchid Island, a single taxon within a diverse taxon, show independent evolution of a complex character, providing a strong opportunity to test the accuracy of existing explanatory frameworks for these characters.

A female snake cannot finish all the eggs, and after about 60 days, the eggs hatch into baby turtles and head for the sea (Fig. 4.8), and after about 20–30 years, the turtles mature and go back to Orchid Island to lay eggs (Fig. 4.5e). In order to protect its food resources, a female snake will naturally have many battles (Fig. 4.9), and during the battles it will be injured in one way or another.

Generally speaking, male vertebrates tend to have shorter lifespans than females. The main reason is that high concentrations of male androgen damage the immune system, or because mitochondrial mutations inherited from the mother have a greater impact on sons than on daughters. There is little discussion of how the frequency of injuries from male–female fights affects lifespan, or how these injuries impact the lifespan of female snakes. The next chapter will use the lifespan of female snakes to refute the male hormone and mitochondrial mutation hypotheses.

Fig. 4.8 About 60 days later, the baby turtles hatched, and they ran wildly toward the ocean. They run wildly toward the ocean

Fig. 4.9 There are seven male snakes in total in this picture, and they are the males that were driven out of egg nests by the females occupying the nests

References

Blackburn, D. G. (1995). Saltationist and puncturated equilbrium models for the evolution of viviparity and placentation. *Journal of Theoretical Biology, 174*, 199–216.

Butchart, S. H. M., Seddon, N., & Ekstrom, J. M. M. (1999). Polyandry and competition for territories in bronze-winged jacanas. *Journal of Animal Ecology, 68*, 928–939.

Calsbeek, R., & Sinervo, B. (2002). The ontogeny of territoriality during maturation. *Oecologia, 132*, 468–477.

Coleman, K., Rothfuss, L. A., Ota, H., & Kardong, K. V. (1993). Kinematics of egg-eating in the specialized Taiwan snake, *Oligodon formosanus* (Colubridae). *Journal of Herpetology, 27*, 320–327.

Crespi, B. (1986). Size assessment and alternative fighting tactics in *Elaphrothrips tuberculatus* (Insecta: Thysanoptera). *Animal Behaviour, 34*, 1324–1335.

Davies, N. B. (1978). *Behavioural ecology: An evolutionary approach* (1st ed.). Blackwell Scientific.

de Haan, C. C. (1984). Dimorphism et comportement sexuel chez *Malpolon monspessulanus*: considerations sur la denomination subspecifique insignitus. *Bulletin de la Société herpétologique de France, 30*, 19–26.

de Haan, C. C. (2003). Extrabuccal infralabial secretion outlets in *Dromophis*, *Mimophis* and *Psammophis* species (Serpentes, Colubridae, Psammophiini). A probable substitute for 'self-rubbing' and cloacal scent gland functions, and a cue for a taxonomic account. *Comptes Rendus Biology, 326*, 275–286.

de Quieroz, A., & Rodrigues-Robles, J. (2006). Historical contingency and animal diets: The origins of egg-eating snakes. *The American Naturalist, 167*, 682–692.

Duvall, D., Schuett, G. W., & Arnold, S. J. (1993). Ecology and evolution of snake mating systems. In R. A. Seigel & J. T. Collins (Eds.), *Snakes: Ecology and behavior* (pp. 165–200). McGraw-Hill.

Greene, H. W. (1997). *Snakes: The evolution of mystery in nature*. University of California Press.

Gregory, P. T., Macartney, J. M., & Larsen, K. W. (1987). Spatial patterns and movements. In R. A. Seigel, J. T. Collins, & S. S. Novak (Eds.), *Snakes: Ecology and evolutionary biology* (pp. 366–395). Macmillan.

Hedges, S. B., & Vidal, N. (2009). Lizards, snakes, and amphisbaenians (Squamata). In S. B. Hedges & S. Kumar (Eds.), *The timetree of life* (pp. 383–389). Oxford University Press.

Huang, W. S., & Wang, H. Y. (2009). Predation risks and anti-predation parental care behavior: An experimental study in a tropical skink. *Ethology, 115*, 273–279.

Huang, W. S., Greene, H. W., Chang, T. J., & Shine, R. (2011). Territorial behavior in Taiwanese kukrisnakes *(Oligodon formosanus). Proceedings of the National Academy of Science, 108*, 7455–7459.

Krebs, J. R., & Davies, N. B. (1993). *An introduction to behavioural ecology* (3rd ed.). Wiley-Blackwell.

Lee, M. S. Y. (1997). The phylogeny of varanoid lizards and the affinities of snakes. *Philosophical Transactions of the Royal Society of London Series B-Biological Sciences, 352*, 53–91.

Lee, C.-Y., Pike, D. A., Tseng, H.-Y., Hsu, J.-Y., Huang, S.-L., Shaner, L. P.-J., Liao, C.-P., Manica, A., & Huang, W.-S. (2019). When males live longer: Resource-driven territorial behavior drives sex-specific survival in snakes. *Science Advances, 5*, eaar5478.

Martins, E. P. (1994). Phylogenetic perspectives on the evolution of lizard territoriality. In L. J. Vitt & E. R. Pianka (Eds.), *Lizard ecology: Historical and experimental perspectives* (pp. 117–144). Princeton University Press.

Scanlon, J., & Shine, R. (1988). Dentition and diet in snakes: Adaptations to oophagy in the Australian elapid genus *Simoselaps*. *Journal of Zoology, 216*, 519–528.

Shine, R. (1991). Strangers in a strange land: Ecology of the Australian colubrid snakes. *Copeia, 1991*, 120–131.

Shine, R., & Thompson, M. B. (2006). Did embryonic responses to incubation conditions drive the evolution of reproductive modes in squamate reptiles? *Herpetological Monographs, 20*, 159–171.

Shine, R., Olsson, M. M., Moore, I. T., Lemaster, M. P., & Mason, R. T. (1999). Why do male snakes have longer tails than females? *Proceedings of the Royal Society B: Biological Sciences, 266*, 2147–2151.

Sih, A., & Mateo, J. (2001). Punishment and persistence pay: A new model of territory establishment and space use. *Trends in Ecology and Evolution, 16*, 477–479.

Stamps, J. A., & Buechner, M. (1985). The territorial defense hypothesis and the ecology of insular vertebrates. *The Quarterly Review of Biology, 60*, 155–181.

Tarof, S. A., Ratcliffe, L. M., Kasumovic, M. M., & Boag, P. T. (2005). Are least flycatcher (*Epidonax minimus*) clusters hidden leks? *Behavioral Ecology, 16*, 207–217.

Wüster, W., & Cox, M. J. (1992). Defensive hemipenis display in the kukri snake, *Oligodon cyclurus*. *Journal of Herpetology, 26*, 238–241.

When Males Live Longer: A Case from Territorial Female Snakes

5

The life span of a person who is not a pine tree, who can get the immortal.

Abstract

Studies show that female Oligodon formosanus defending green turtle nests attack males, often targeting tails where reproductive organs are located. Although this leads to higher injury rates and shorter lifespans in females, the nests provide crucial, abundant food. This energy boost allows females to produce more offspring, increasing population success. In areas without nests, lifespan differences disappear. Thus, despite high costs, aggressive defense is favored because it enhances reproductive output—making it a high-risk, high-reward "big fight" strategy.

Keywords

Sexual conflict · Territory · Life span · Female-defeat-male

From "Hibiscus Pond" by Cao Pi of the Wei Jin Dynasty. Cao Pi's father, Cao Cao, also had a poem with the first line: "Although the divine tortoise lives long, there is still time for it to end; The snake rides on the fog, but finally becomes ashes," which mean even though the divine tortoise has a very long life span, there is still time for it to end, and even though the dragon snake can fly in the fog, it will eventually die and turn into ashes. Both of the above poems illustrate that there is an end to life.

The Hungarian poet Petofi Sandor wrote in 1823: "Life is precious, but love is more valuable. If for the sake of freedom, both can be thrown away." It means that human life is very valuable, and if we lose it, everything is gone. However, if people do not have love, life becomes dull and boring, even if the meaning of living.

W.-S. Huang, *Reptiles in the Changing Climate*, Fascinating Life Sciences,
https://doi.org/10.1007/978-981-92-1887-5_5

However, the most important thing is freedom; even with love and life, it is still not happy. It is clear that freedom is the most important thing, and for the sake of freedom, love and life can be sacrificed.

Living animals have two major goals: survival and reproduction. In this case, survival is life and reproduction is love. Although the life span of each species may vary, the main goal is to pass on its genes. Generally speaking, life expectancy refers to the length of an individual's life, but perhaps life is immortal if we look at it from the genetic level. For example, your parents' genes are passed on to you, and although they have passed away, their genes remain in you. Even if you don't get married, your genes are not passed on, but your siblings still have some of your genes. That is to say, genes are immortal.

Basically, in vertebrates in general, males have shorter lifespans than females (including humans). Many theories have attempted to explain this common phenomenon; the best known of these include the "mother's curse hypothesis"and the "hormone hypothesis." The mother's curse hypothesis suggests that all offspring inherit their mitochondrial DNA from the mother, and that if there are defects in the mitochondria DNA, the female individual is largely unaffected; instead, males slowly accumulate these defects to the point where they die. Therefore, the average life expectancy of surviving individuals is greater in females than in males. The hormone hypothesis suggests that males have a much higher concentration of androgens than females, and that high concentrations of androgens can greatly damage our immune system, making us vulnerable to outside germs and reducing our lifespan. On the other hand, the high concentration of androgens will also increase the aggressiveness of animals; for example, most of the animals in the field are males, and males fight fiercely with each other for the field, food, and mating rights, and are easily injured and die. Females are gentler, so they live longer.

These theories all revolve around males, and it seems that it is not so easy to test these hypotheses [that males have shorter life spans than females] from males. If we can find a female animal that dominates and defeats males and has a lower lifespan than males, then it is clear that competition is a major factor in lifespan. In this case, both the "mother's curse hypothesis"and the "hormone hypothesis"could be disproved. Fortunately, the reptile egg-eating snake female is always the dominant in Orchid Island, attacking the intruder and defeating the male, while the male fled as soon as he encountered the female. This unique phenomenon provides a good example of how we can verify that males have shorter lifespans than females.

5.1 Introduction

Males typically live shorter lives than conspecific females in most animal groups, including mammals (e.g., mammals (Owen-Smith, 1993, Moore & Wilson, 2002, Lemaître & Gaillard, 2012), birds (Liker & Székely, 2005), amphibians (Quiroga et al., 2015), fish (Candolin, 1998), and insects (Keller & Genoud, 1997) (but see (Parker & Plummer, 1987, Réale et al., 1996, Orell & Belda, 2002, Hyslop et al., 2009, Low et al., 2010, Tatarnic et al., 2014) for some exceptions)). Males of many species grow faster than females do, necessitating more resources during development as well as injury and stress from aggressive interactions, which is thought to be the main driver of this survival asymmetry (Candolin, 1998, Lemaître & Gaillard, 2012). The highest ranking males, for instance, had higher testosterone and glucocorticoid levels (stress hormones) in a long-term study of social rank-related stress in a natural population of prairie baboons (*Papio cynocephalus*), suggesting that males at the top of the hierarchy may be more expensive than males in lower ranks (Gesquiere et al., 2011). Six out of seven nonhuman primates studied in the study displayed intense male competition for mates, and the males of these species had shorter life spans than the females of the same species (Bronikowski et al., 2011). The only species in which both sexes lived equally long lives was the muriquis (*Cebus capucinus*), possibly because there was less male competition in this species (Bronikowski et al., 2011).

Because species-specific strategies are frequently fairly rigid, it is a limitation of most studies that try to test expensive competition as a cause of increased mortality that they must rely on comparative data. It is possible to correct for phylogenetic non-independence, but it can be difficult to quantify lifestyle differences among species that might have an impact on survival.

Here, we present a thorough investigation of the Kukri snake (*Oligodon formosanus*), a species of territorial snake (Huang et al., 2011), which offers an ideal system to study the relationship between competition and survival. The species is a specialist predator of small reptile eggs, mostly those of lizards and other snakes, but at the location of our study, Kukri snakes also eat sea turtle eggs (Huang et al., 2011). Because it was the females at both study sites who protected and competed for eggs as a resource, male survival rates should have been higher than they were. A severe storm that eroded the beach where turtles were nesting at one of the study sites caused a loss of territorial behavior, allowing us to compare the two pre-storm sites (when females were defending eggs at both sites) with the post-storm site. This allowed us to analyze the role of territoriality (when eggs and associated territorial behavior were present at only one site). We verified that at both study sites, eggs were a significant resource that influenced population density. Due to their territorial behavior, females grow more quickly than males in the presence of eggs, but this behavior also increases injury rates, which reduces survival rates and shortens lifespans.

5.2 Turtle Eggs' Significance in Determining Kukri Snake Populations

At Little Paiday (LP, 1997–2007), we caught and marked 476 kukri snakes (274 adult males, 139 adult females, and 63 juveniles; Fig. 5.1a), with 919 recaptures. At Tungching, we caught and marked 509 kukri snakes (302 adult males, 172 adult females, and 35 juveniles (T, 1997–2012) with 632 recaptures (Fig. 5.1b). Snake populations decreased significantly for both adults; Fig. 5.1b) and juveniles (Fig. 5.1b). The snake population in Tungching significantly declined after a storm

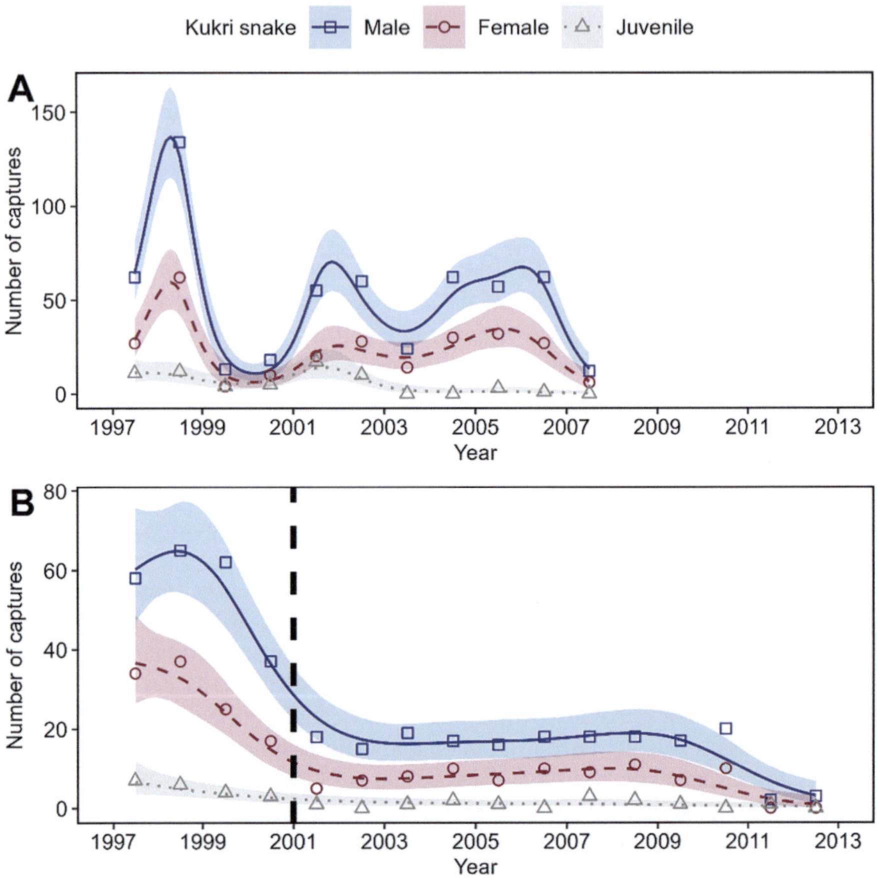

Fig. 5.1 Number of individual kukri snakes (*Oligodon formosanus*) captured annually at our study sites in Little Paiday Bay (where sea turtles nested throughout our study (**a**) and Tungching (where sea turtles nested from 1997 to 2000, and nests were absent from 2001 to 2012 (**b**), shown separately for adults and juveniles. The vertical dashed line indicates that a severe storm eroded the beach. (Redrawn from Lee et al. (2019), Science Advances)

destroyed the turtle nesting beach. Because Little Paiday, another study site, which was not subject to beach erosion despite being close to Tungching, did not experience similar changes, we are confident that this decline is a direct result of resource changes. Consequently, there was no difference in snake abundance between the two sites before the storm. This suggests that turtle eggs are a major resource driving population density in kukri snakes, at least at our study site.

5.2.1 Sex-Specific Differences in Size

In spite of being more aggressively territorial, females were smaller than males at both sites: at Little Paiday, males had a mean SVL length of 51.78 mm, whereas females measured only 48.86 mm; at Tunching, males measured 50.94 mm compared to 48 mm in females.

5.2.2 Growth Rate

The next test was to see if faster growth rates helped females acquire eggs. By utilizing an ANOVA and its associated multiple comparisons, we were able to compare growth rates (daily increase in body mass) between sexes and study locations. The body mass were increased in females [(0.894 g/day; $n = 23$) vs. 0.446 g/day (males, $n = 76$), $p < 0.001$], and 1.069 g/day [(females, $n = 10$) vs. 0.518 (males, $n = 24$), $p = 0.011$], in Little Paiday and Tungching, respectively (Fig. 5.2). When snakes of the same sex were compared between study sites, growth was statistically similar for males or females (males: $p = 0.175$; females: $p = 0.469$). As a result, females grew more quickly than males at both sites, which is consistent with their greater access to turtle eggs.

5.2.3 Injury Rates Based on Sex

Then, we investigated whether females incurred a territorial cost in defending this valuable resource by doing more harm when eggs were available. When turtle eggs were present at both study sites, gender-specific injury rate patterns were seen (Figs. 5.3 and 5.4). Prior to 2001, injury rates between the two sites were statistically similar for both sexes (odds ratio = 1.237, $p = 0.689$; Fig. 5.3), but overall, females

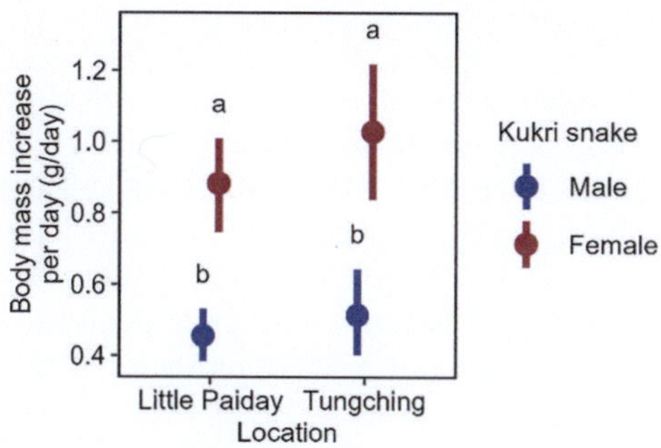

Fig. 5.2 The mean daily increase in body mass (± 95% CI) Mean ± 95% CI of Kukri snakes (*Oligodon formosanus*) body mass increase per day at our study sites in Little Paiday Bay (1997–2007) and Tungching (1997–2000) study sites on Orchid Island, Taiwan. Lowercase alphabets letters summarize indicate the results of pairwise multiple comparisons. (Redrawn from Lee et al. (2019), Science Advances)

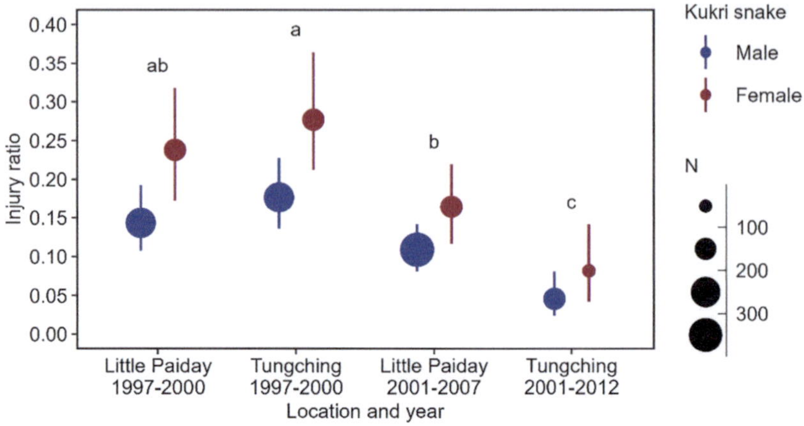

Fig. 5.3 Ratios of injuries incurred by male and female Kukri snakes (*Oligodon formosanus*) at Little Paiday Bay (where sea turtles nested throughout our study) and Tungching (where sea turtles nested from 1997 to 2000, and nests were absent from 2001 to 2012), shown separately by sex. Lowercase alphabets summarize the results of pairwise multiple comparisons among location-year groups. (Redrawn from Lee et al. (2019), Science Advances)

Fig. 5.4 Injury on a female's dorsal back (**a**), and a scar near tail in a male on Orchid Island (**b**), related to intraspecific agonistic encounters in turtle nests

sustained significantly more physical injuries than males (odds ratio = 1.848, $p < 0.001$; Fig. 5.3). Although female injury rates were still higher than male injury rates, this pattern changed after 2001 when turtle eggs at Tungching ceased to be available (Fig. 5.3). Injury odds at Tungching decreased and were statistically different from those at Little Paiday (odds ratio = 0.331, $p = 0.003$). Thus, even in the absence of turtle eggs, females may fight more frequently than males, but the intensity of fighting increases noticeably in the presence of this important resource.

5.2.4 Apparent Survival Rates of Snakes of Different Sexes

Male survival was higher than female survival at both sites when turtle eggs were present (Little Paiday, 0.661 vs. 0.357, pairwise $t_2 = 7.352$, $p = 0.018$; Tungching,

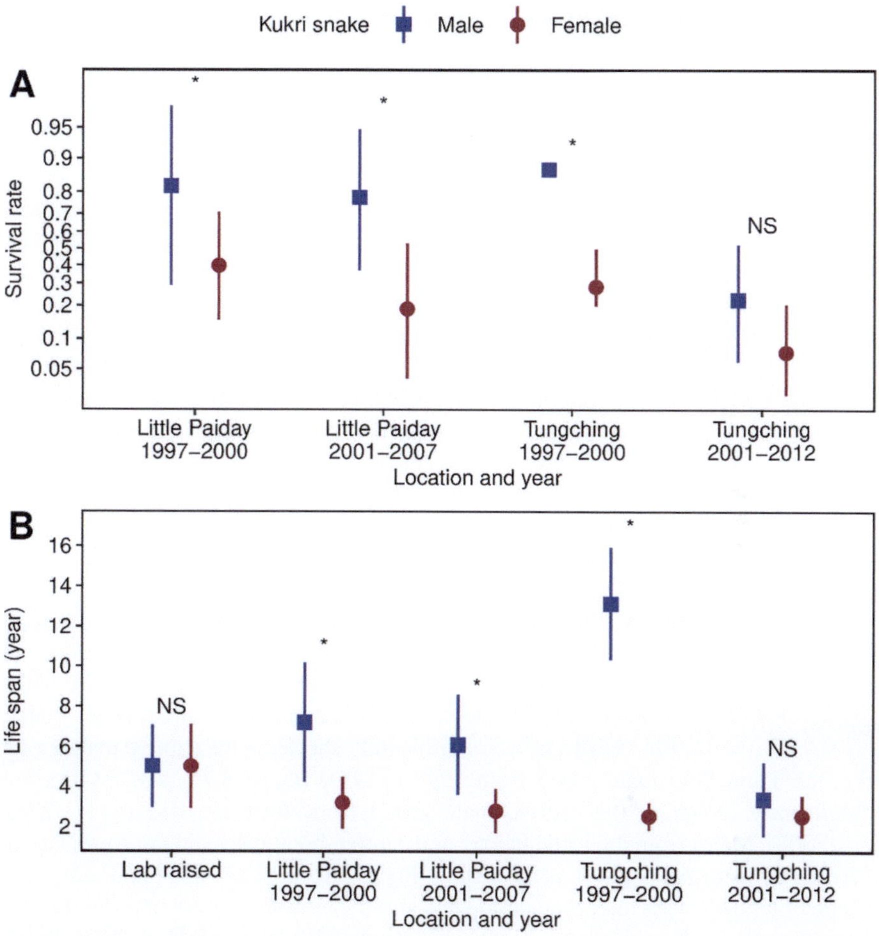

Fig. 5.5 Survival rates (**a**) and estimated longevity (years; **b**) of male and female kukri snakes (*Oligodon formosanus*) at Little Paiday Bay (where sea turtles nested throughout our study) and Tungching (where sea turtles nested from 1997 to 2000, and nests were absent from 2001 to 2012). Annual apparent survival rates were calculated using mark-recapture analysis, and lifespan estimates are based on these survival rates. (Redrawn from Lee et al. (2019), Science Advances)

0.877 vs. 0.30, pairwise $t_2 = 5.284$, $p = 0.034$; Fig. 5.5a). However, when there were no more eggs in Tungching between 2001 and 2012, there was no significant difference in sex survival (0.409 vs. 0.201, pair $t_{11} = 1.696$, $p = 0.118$; Fig. 5.5a). In contrast, the difference in sex survival persisted at the Little Paiday, where they also had eggs (0.645 vs. 0.33, pair $t_6 = 2.525$, $p = 0.045$; Fig. 5.5a).

5.2.5 Snake Sex Life Span in the Wild and in Captivity

The life expectancy (Ax) of males in Little Paiday was calculated using our mark-release-recapture data to be more than twice as long as that of females both before and after the storm ($t_2 = 5.29$, $p = 0.034$ in 1997–2000; $t_6 = 3.13$, $p = 0.02$ in 2001–2007, Fig. 5.5b). Similar to this, male longevity in Tungching was significantly higher than female longevity before the storm ($t_2 = 7.33$, $p = 0.018$; Fig. 5.5b). However, after the storm had passed and the nest had vanished, there was no difference in longevity between the sexes ($t_{11} = 1.14$, $p = 0.276$; Fig. 5.5b). As a result, we can draw the conclusion that the existence of valuable resources and the territorial behavior that goes along with them may be the cause of the difference in longevity between males and females. Contrarily, disparities in survival and life expectancy between the sexes vanished when turtle eggs ceased to be present in Tunching (after 2000), though the apparent decline in male longevity may have been caused by a decrease in overall food availability over a wider area. Both sexes sustained injuries less frequently after the loss of eggs, and the difference in injury frequency between males and females was no longer significant [see also (Huang et al. 2011)], suggesting that the absence of sea turtle eggs as a food source may have resulted in decreased aggressiveness.

By contrasting males and females raised in the laboratory, where access to food was not limited, we tested whether we could identify intrinsic sex differences in longevity. Males were 5.02 years old, with a range of 1.8–9 years old ($n = 12$), while females were 5.01 years old, with a range of 3–10, according to statistics ($t_{24} = 0.01$, $p = 0.991$; $n = 14$; Fig. 5.5b). Ax was significantly shorter in the captive male group than the Tungching male group from 1997 to 2000 ($t_{14} = 8.75$, $p < 0.001$), but significantly longer in the captive male group from 2001 to 2012 ($t_{22} = 2.92$, $p = 0.008$). Males from the Little Paiday and captive-bred males did not significantly differ in terms of Ax ($t_{20} = 1.91$, $p = 0.07$; Fig. 5.5b). In contrast, Little Paiday and Tungching females at either time period had significantly shorter Ax than laboratory-reared females ($t_{22} = 4.80$, $p = 0.004$, LP; $t_{15} = 4.39$, $p = 0.003$, T 1997–2000; $t_{24} = 5.13$, $p = 0.0012$, T 2001–2012; Fig. 5.5b).

5.3 Discussion

Our research on Taiwanese kukri snakes shows how sex-specific survival and lifespan can be highly plastic and significantly influenced by environmental factors, particularly resource availability and related territorial defenses. The field studies we present demonstrate that sex-specific mortality differences in this species are primarily caused by aggressive interactions. Data from animals kept in captivity, where there are no discernible differences in life expectancy and no evidence of physical conflict, expensive exercise, or energy expenditure for reproduction, further support this conclusion.

In a variety of phylogenetic species, including invertebrates and vertebrates, lifespan differences between the sexes have been linked to reproductive behaviors

(Candolin, 1998; Liker & Székely, 2005; Lemaître & Gaillard, 2012; Tatarnic et al., 2014). Studies on primate mortality have revealed that males in species with relatively more intense male-to-male competition for mate acquisition experience higher rates of aging and shorter lifespans (compared to their conspecific females), whereas species with less intense female-specific internal conflicts do not show intersex differences in aging patterns and lifespans (Bronikowski et al., 2011). This pattern indicates that male aging in polygynous species may be accelerated in part by intermale competition (Bronikowski et al., 2011). By utilizing natural experiments that enable us to examine such differences between populations of the same species, our study supports this conclusion. Additionally, our research independently verified this theory in species with opposing sex roles: females engage in territorial aggression, supporting the idea that aggression (or an environment that supports aggression) functions as a key driver of sex-specific lifespan patterns rather than sex itself.

Some of the theories put forth that link shorter lifespans in males to intrinsic factors are not supported by our findings. For instance, the "curse of motherhood hypothesis" predicts a higher mutational load that is passed down from females and is harmful to male-specific traits in the mitochondrial genome (Camus et al., 2012). Additionally, it has been hypothesized that having higher levels of the stress hormones and testosterone in men can have detrimental effects (Gesquiere et al., 2011). Contrarily, we think that territorial aggression is a major predictor of longevity and that this behavior is frequently influenced by the environment or predetermined by background circumstances. Female snakes, which no longer exhibit territorial aggressive behavior when eggs are no longer available, are an easy way to show the malleability of territorial aggressive behavior. Humans are one of several primate species whose mortality rates have been demonstrated to be phenotypically plastic. Mortality rates differ significantly between populations of a given species, with populations living in more challenging environments having higher mortality rates (Bronikowski et al., 2011).

What advantages do female snakes who defend turtle nest sites have? Although it appears that this territorial behavior increases the ability of female snakes to produce calories, they do not live longer than is typically seen after storms, when eggs are not readily available (and resources are actually more scarce). When turtle eggs are present, population density and offspring numbers are higher than when turtle eggs are not present (Little Paiday 1997–2007), suggesting that the main benefit of territorial defense and egg consumption is higher reproductive output (Tungching 2001–2012). Female snakes must have an adequate supply of energy and nutrients to invest in reproduction, which can be provided by a high-quality food source like sea turtle eggs. Females may maximize their reproductive output at the expense of their lifespan because the advantages of egg consumption may outweigh the drawbacks of territorial aggression (i.e., shorter lifespan).

In addition to feeding primarily on sea turtle eggs during the summer, the Kukri snake also consumes other lizard eggs with elastic eggshells during other times of the year. The eggs of the largest species, the long-tailed skink, are the largest and are laid in open concrete holes, where their scent easily disperses, making them easy for snakes to detect and locate. What impact does this have on the long-tailed skink, and how does it adapt? The next chapter focuses on this topic with detailed research and surprising findings.

References

Bronikowski, A. M., Altmann, J., Brockman, D. K., Cords, M., Fedigan, L. M., Pusey, A., Stoinski, T., Morris, W. F., Strier, K., & Alberts, S. C. (2011). Aging in the natural world: Comparative data reveal similar mortality patterns across primates. *Science, 331*, 1325–1328.

Camus, M. F., Clancy, D. J., & Dowling, D. K. (2012). Mitochondria, maternal inheritance, and male aging. *Current Biology, 22*, 1717–1721.

Candolin, U. (1998). Reproduction under predation risk and the trade-off between current and future reproduction in the threespine stickleback. *Proceedings of the Royal Society B: Biological Sciences, 265*, 1171–1175.

Gesquiere, L. R., Learn, N. H., Simao, M. C. M., Onyango, P. O., Alberts, S. C., & Altmann, J. (2011). Life at the top: Rank and stress in wild male baboons. *Science, 333*, 357–360.

Huang, W. S., Greene, H. W., Chang, T. J., & Shine, R. (2011). Territorial behavior in Taiwanese kukrisnakes *(Oligodon formosanus)*. *Proceedings of the National Academy of Science, 108*, 7455–7459.

Hyslop, N. L., Meyers, J. M., Coopers, R. J., & Norton, T. M. (2009). Survival of radio-implanted eastern indigo snakes (*Drymarchon couperi*) in relation to body size and sex. *Herpetologica, 65*, 199–206.

Keller, L., & Genoud, M. (1997). Extraordinary lifespans in ants: A test of evolutionary theories of ageing. *Nature, 389*, 958–960.

Lee, C.-Y., Pike, D. A., Tseng, H.-Y., Hsu, J.-Y., Huang, S.-L., Shaner, L. P.-J., Liao, C.-P., Manica, A., & Huang, W.-S. (2019). When males live longer: Resource-driven territorial behavior drives sex-specific survival in snakes. *Science Advances, 5*, eaar5478.

Lemaître, J. F., & Gaillard, J. M. (2012). Male survival patterns do not depend on male allocation to sexual competition in large herbivores. *Behavioral Ecology, 24*, 421–428.

Liker, A., & Székely, T. (2005). Mortality costs of sexual selection and parental care in natural populations of birds. *Evolution, 59*, 890–897.

Low, M., Arlt, D., Eggers, S., & Pärt, T. (2010). Habitat-specific differences in adult survival rates and its links to parental workload and on-nest predation. *Journal of Animal Ecology, 79*, 214–224.

Moore, S. L., & Wilson, K. (2002). Parasites as a viability cost of sexual selection in natural populations of mammals. *Science, 297*, 2015–2018.

Orell, M., & Belda, E. J. (2002). Delayed cost of reproduction and senescence in the willow tit *Parus montanus*. *Journal of Animal Ecology, 71*, 55–64.

Owen-Smith, N. (1993). Comparative mortality rates of male and female kudus: The costs of sexual size dimorphism. *Journal of Animal Ecology, 62*, 428–440.

Parker, W. S., & Plummer, M. V. (1987). Population ecology. In R. A. Seigel, J. T. Collins, & S. S. Novak (Eds.), *Snakes: Ecology and evolutionary biology* (pp. 253–301). Macmillan.

Quiroga, L. B., Sanabria, E. A., & Marangoni, F. (2015). Sexual size dimorphism and age in *Odontophrynus* cf. *barrioi* (Anura: Odontophrynidae) from the Monte Desert, Argentina. *Journal of Herpetology, 49*, 627–632.

Réale, D., Boussés, P., & Chapuis, J. L. (1996). Female-biased mortality induced by male sexual harassment in a feral sheep population. *Canadian Journal of Zoology, 74*, 1812–1818.

Tatarnic, N. J., Cassis, G., & Siva-Jothy, M. T. (2014). Traumatic insemination in terrestrial arthropods. *Annual Review of Entomology, 59*, 245–261.

Parental Care in Long-Tailed Skinks

6

Abstract

Unexpectedly, long-tailed skinks only attacked genuinely dangerous snakes. When non-threatening sympatric lizards—such as Swinhoe's tree lizard, *Sphenomorphus incognitus*, and *Eutropis multicarinata*—were introduced to nests, no aggression occurred, suggesting the ability to recognize familiar neighbors. Moreover, skinks engaged snakes only when eggs were present; in the absence of eggs, they either fled or ignored potential threats. These findings indicate that parental defense is highly context-dependent. Overall, this study provides insight into the evolution and dynamics of parental care, showing that the behavior of long-tailed skinks on Orchid Island is primarily aimed at protecting the nesting environment and enhancing offspring survival.

Keywords

Maternal care · Long-tailed skink · Egg survival · Kukri snake · Cost-and-benefit model

6.1 Adaptation of Parental Care

Mother's love is the greatest of all, especially at the highest point of socialization of human beings. For example, the poet Meng Jiao of the Tang Dynasty wrote a poem called "The Wandering Son's Song"

"*The thread from my dear mother's hand*
Was sewn in the clothes of her wandering son. For fear of my belated return,
Before my leave they were closely woven. Who says my heart like a blade of grass
Could repay her love's gentle beams of spring sun?"

W.-S. Huang, *Reptiles in the Changing Climate*, Fascinating Life Sciences,
https://doi.org/10.1007/978-981-92-1887-5_6

Fig. 6.1 Long-tailed skinks choose higher temperature concrete holes (**a**) for their nests when laying eggs (**b**)

Of course, it is difficult for a socialized animal to behave like a human, let alone one that has not been socialized at all. But even animals without any socialization can sometimes show care for their offspring. For example, poisonous frogs carry tadpoles on their backs, etc.

After the birth of eggs or hatchlings, the majority of amphibians and reptiles tend to abandon their young to fend for themselves; for example, 98% of lizards and 95% of snakes do this. The evolutionary mystery of animal care behaviors has been a crucial area of study for scientists because reptiles are the ancestors of birds and mammals that exhibit parental care behaviors.

The population of long-tailed skinks (*Eutropis longicaudata*) in Taiwan does not exhibit any parental care behavior, but the long-tailed skinks on Orchid Island exhibit specific parental care behavior that influences the rate at which well-tended eggs hatch. For instance, after the eggs are laid, long-tailed skinks will continue to care for them in the nest (Fig. 6.1a, b). The mother will bury or conceal the eggs, turn them over to aid in hatching, and even remove eggs that have been parasitized by fungi to prevent the entire clutch from becoming infected in order to ensure a smooth hatch. Even in the face of a large snake, the mother lizard will attack the snake and push it out of the nest to protect the eggs when the kukri snake, which is known for eating eggs, invades the nest (Fig. 6.2; Video 2: https://youtu.be/uT7pPlUtDu8).

6.2 Introduction

Some studies suggest that it is crucial to illustrate the costs and advantages of parental care because the majority of animals experience the greatest risk of mortality during the first stages of their life cycles (Olsson & Shine, 1998; Roff, 2002). The level of risk to future reproductive potential, the value of current offspring in comparison to the expected future contribution of parents, and the anticipated improvement in offspring welfare as a result of such parental expenditures should all be taken into consideration when determining the level of parental effort in any activity(Pressley, 1981). Natural selection should therefore favor the best level of

Fig. 6.2 *O. formosanus* preparing to enter skink nests to eat eggs

defense based on the risks that the parents take and the expected benefits that the offspring will receive (Shine, 1988).

There are four types of parental care for reptiles: (1) egg companying, in which parents stay nearby their eggs without ever defending their nests; (2) when incubating eggs, parents raise nest temperatures above ambient levels to encourage hatching; (3) active parental protection of the nest against potential predators; and (4) crocodiles have a unique case of complex active defense mechanisms, which include guarding eggs and hatchlings, breaking into nests, and transferring eggs and hatchlings by mouth (Shine, 1988). Despite the rarity of reports of parental care in reptiles, a recent investigation has revealed that pit viper females stay in the nest until the young are shed, and that the dominant. When compared to their behavior prior to birth or following neonates shed their skin, females can sometimes be more aggressive toward natural predators (Greene et al., 2002). Over 140 species of reptiles (3%) have been reported as having parental care, but many of these records are insufficient or lacking. For instance, few reports of actual defense against predators in the natural environment have been made; the majority of records are based on associations between females and eggs, and there is no concrete evidence of parental care (Somma, 1990).

Maternal aggression, which can range from early gestation to offspring independence, is an essential component of parental care and includes increased aggression toward conspecifics or heterospecifics (Figler et al., 2001), and can occur anywhere from their raising time (Archer, 1988; Maestripieri, 1992). However, only crocodiles have reports of maternal aggression in reptiles (Rand, 1968; Shine, 1988). After spawning, female crocodiles hang around the nest's area and frequent it on a regular basis. While non-predatory species are ignored, potential predators close to the eggs may be attacked (Dietz & Jackson, 1979). Females assist with opening the nest and transporting the newly hatched hatchlings to the water. For several days or weeks after hatching, the young remain with their parents, who actively protect them from any would-be predators.

6.2.1 Parental Care: *E. longicaudata* Females and Males in Relation to the Snake *O. formosanus*

6.2.1.1 *E. longicaudata* Female

I introduced snakes (20 in total) into holes containing female skinks 177 times over the course of 3 years (using 20 snakes, 107 skinks without eggs, and 70 skinks with eggs). In nests with eggs, females typically attacked snakes aggressively (53 attacks, 11 escapes, and 6 ignores; Fig. 6.3a); however, in nests without eggs, they typically ignore or escaped (9 ignores; Fig. 6.3b and 12 attacks, 86 escapes; Fig. 6.3c). Even when they initially escaped, the majority of lizards went back to their burrows (55 returned and 15 abandoned their nests, $Z = 4.66$, $p < 0.05$, Sign test).

Fig. 6.3 When the snake enters the nest, the female skink reacts in three ways: by attacking the snake (**a**), by ignoring the snake (**b**), and by escaping (**c**)

6.2.1.2 *E. longicaudata* Male

In 58 holes with male *E. longicaudata*, I put snakes in and discovered that none of the skinks attacked the invaders; all of the males left the holes right away.

6.2.1.3 Snake Size and Parental Care Are Related

In order to compare the differences in clutch size between the two groups, I divided all responses into two categories: females attacking and females not attacking (both ignoring and fleeing). In comparison to females who were not attacked, attacking females had a larger mean clutch size (mean = 6.7; range: 3–11; $n = 30$) than non-attacking females (mean = 4.2; range: 2–7; $n = 12$; U = 775.3, $p = 0.0003$, Mann-Whitney test).

6.2.1.4 Experiments on Other Intruders

Conspecific and other lizards were typically ignored by egg-laying females (males *S. incognitus* had 5 attacks, while *D. swinhonis* had 0 attacks and 56 were ignored. *E. longicaudata* had 2 attacks, but 54 of them were ignored. The females always fled when I put the lizard-eating snake *E. carinata* in the hole (2 initially attacked before fleeing, 54 fled right away; compare total test for all types of invaders; Chi-square = 221.65, df = 6, $p < 0.0001$), and the males never did either. The lizards never came back when the invader was *E. carinata*.

6.2.1.5 Long-Tailed Lizards' Parental Care of *O. formosanus* and Egg Survival

The majority of egg losses were brought on by ant and snake attacks or fungal infections. If fungus or ants attack the eggs, the eggshell is still visible at the nest site; if snakes eat the eggs, the eggs vanish. Snake predation is the main cause of egg mortality in unprotected nests.

On Orchid Island, more eggs than unprotected eggs hatched (median = 81%, range: 50–100% for the protected group; median: 18%, range: 0–100% for the unprotected group; U = 3638.19, $p < 0.0001$; Mann-Whitney U test). *O. formosanus* was the culprit behind all egg losses in unprotected nests.

6.2.2 Discussion

6.2.2.1 Parental Care

One of the most fundamental and primitive forms of parental care is egg guarding, and by understanding how it evolved, we may be able to learn more about the beginnings of parental care in other animals (Reynolds et al., 2002). In this case, the experimental removal of females was followed by a sharp decline in hatching success (only 70%), illuminating the adaptive importance of long-tailed skink egg guarding. The idea that egg guarding fundamentally enhances parental fitness is supported by the fact that few females abandon their eggs in the wild. This outcome is in line with that of female treefrogs, where early egg abandonment by females was also connected to a reduction in hatching success of almost 50%. (Zink, 2003).

Clearly, maternal aggression in long-tailed skinks is primarily brought on by the need to protect eggs from *O. formosanus* attack. Skinks, however, display a distinct form and degree of maternal defensive behavior that is responsive to regional factors. The size of the nest affects how aggressive the mother is. Females who abandon their eggs have smaller nests, and empty nests rarely or never cause maternal aggression. Similar phenomena have been observed in fish (Oliveira & Almada, 1998), lizards (Chapple, 2003; O'Connor & Shine, 2003; 2004), and mice (Giordano et al., 1984). When a fish's chick is taken out of its mother's mouth, the aggressiveness level drops quickly. A day later, the aggressive characteristics and color pattern are no longer distinguishable from those of non-chick females (Oliveira & Almada, 1998). Similar to this, when female rodents are separated from their young, their level of aggression significantly decreases (Giordano et al., 1984).

In the empty nest test, 12 out of 107 skink females exhibited aggressive behavior, but 10 of these 12 females were pregnant. The neurohormonal mechanisms that support the defense of eggs in the nest may have had an impact on them.

Parents of multiple-nesting birds may experience severe time and/or energy restrictions as well as conflicts between reproduction, incubation, and territory establishment (Sanz & Tinbergen, 1999). In birds, the size of the nest is typically thought to be constrained by either (1) the ability of the parents to raise their young under their physiological conditions (Lack, 1947) in conjunction with territory quality or (2) the cost to the parents of raising an excessive number of young (subsequent fecundity and/or survival) (Daan et al., 1996). In the current study, *E. longicaudata* did not raise or protect its young, in contrast to other animals like birds. As demonstrated in this study, a larger clutch size may be preferable in a lizard that displays maternal care because the cost of guarding is the same for larger or smaller clutches. Another theory, however, contends that social aggression experiences, such as subsequent encounters with the same species, can cause aggression or escape (Kudryavtseva, 2000). For instance, mice that are repeatedly given the chance to fight show more aggression than individuals who have never had this opportunity (Parmigiani & Brain, 1983). According to Huang (2004), in *E. longicaudata*, smaller females are more likely to engage in nesting and escape behavior because they may be unfamiliar with snake encounters when they are in charge of protecting eggs. To test this learning hypothesis using marker individuals with known histories, more investigation will be required.

6.2.2.2 Experimental Invaders

Prey employ a variety of tactics to lessen their chances of being captured by predators, such as avoiding high-risk microhabitats or reducing activity when predators are most active (Lima & Dill, 1990; Mathis & Smith, 1993). The success of anti-predator behavior patterns frequently determines the survival of prey, with potential predators only engaging in anti-predator behavior after coming into contact with prey (Greene, 1988). Furthermore, it is crucial for prey to be able to distinguish between predators and other animals. According to Hirsch and Bolles (1980), failing to recognize predators may increase the likelihood of being captured during an encounter, and defending against non-predators may be a waste of time and energy

(Hirsch & Bolles, 1980), and defensive responses to non-predators may be a waste of time and energy (Hansen & Slagsvold, 2003).

After egg laying, female skinks stay in the nest and spend a lot of time watching over their eggs while they are being incubated. *O. formosanus*, a potential egg predator, is attacked, but other species that aren't egg predators are ignored (i.e., conspecific males *E. longicaudata*, *Diploderma swinhonis*, and *S. incognitus*). However, the participating long-tailed skinks immediately fled and never made it back to their nests after coming into contact with the predatory snake *E. carinata*. In exchange for the eggs, female skinks sometimes put them in greater danger from rapacious snakes by taking greater risks. These findings imply that the advantages of abandonment may frequently depend on the female's capacity to protect the nest or the possibility that the abandoned female will produce another clutch during the same breeding season. Aside from non-predator lizards, my observations suggest that long-tailed skinks have distinct reactions to the egg-eating snake *Oligodon formosanus* and the lizard predator *E. carinata*. To find out which cues cause lizards to attack, flee, or ignore intruders, more research is required.

6.3 The Cost and Benefit Model of Parental Care in Long-Tailed Skink

As mentioned earlier, only Orchid Island has mother lizard care behavior, while Taiwan does not. Female lizards in the Taiwan population lay eggs in exactly the same environment as on Orchid Island, but leave the concrete hole immediately after laying and do not provide protection. The parental care behavior of the Orchid Island long-tailed skink is a recent development and an evolutionary behavior in progress. Compared to other animals with parental care behaviors, such as birds or mammals, the parental care behavior of the long-tailed skink is not well developed. Perhaps this is a manifestation of ongoing evolutionary behavior that deserves more in-depth investigation. This study may provide a glimpse into the origin of animal parental care and the cause and effect of the interaction. Therefore, the time and cost of parental care in long-tailed skinks and the increase of offspring survival rate by nest site selection in long-tailed skinks should be investigated in more depth in future studies.

The time and cost of parental care determines the willingness of the parent to continue caring for the offspring, and except in a few cases, few parents are willing to sacrifice themselves to protect their offspring because if they die, the offspring will probably not survive either. Therefore, the parents must balance their own contribution with the best interests of the offspring. If the cost exceeds the benefit, the parental care will stop. The cost of parental care for reptiles can be divided into the following: reduced food intake, reduced maternal survival rate, and the chance of having a second clutch. For example, the production cost—the mother loses the opportunity to give birth due to prolonged maternal care, or even if she has a second clutch, the quantity or quality of the second clutch is reduced due to less foraging and less energy storage due to prolonged parental care; and the survival cost—the mother loses her life due to prolonged parental care, or is more likely to die from bacterial infection if she does not eat for a long time. The best reward is that the more offspring are hatched, the better, so that the genes of the offspring can be passed on in large numbers. How can all these costs and rewards (benefits) be balanced between long-tailed skink parents and offspring? The author's article in 2007 also showed that parental care of long-tailed skinks does not waste too much of the mother's energy, i.e., the cost to the mother does not sacrifice her survival rate or the number or quality of her second litter because of parental care (Huang, 2007a, b).

When the benefits of providing care outweigh the cost of the act, parents are expected to develop strategies for raising eggs or young. Protecting the energy requirements of offspring comes at a price, as does delaying future fertility and raising predation risk. Using lizard populations that varied in their expression of maternal care, we tested ecological conditions favorable to parental care evolution using a cost-benefit model. We discovered that predators are crucial in the evolution of maternal care because (1) it is impossible for maternal care to evolve when parental protection puts more pressure on predators, (2) it is impossible for maternal care to evolve when both parents and offspring are under little threat from predators, and (3) maternal care can only advance if parents can successfully defend their young against predators without putting themselves at increased risk of being attacked. Our investigation of one of the few vertebrate species known to show interpopulation variation in maternal care expression clearly supports some hypothesized scenarios in which maternal care should evolve (such as when nests are located in exposed areas, parents can protect their young from predators, and eggs have a short incubation period), but not others (e.g., when nests are scarce, life history strategies are "dangerous," low breeding frequency, and poor environmental conditions). We draw the conclusion that numerous pathways can result in the evolution of parental care from a non-caring state, even in a single population of a wide range of species.

When the benefits of providing care result in higher offspring survival than adult survival and the costs of future reproduction, parental care develops (Shine, 1988; Clutton-Brock, 1991; Greene et al., 2002). Parental care benefits offspring by increasing their chances of survival, which is relatively simple to measure, but the costs to parents are complex and thus harder to identify (Reguera & Gomendio, 1999). Parenting has two costs: (1) reproductive costs, such as missed mating opportunities when providing long-term parental care, which could reduce future fecundity or the number of offspring produced in the next reproductive event due to lower energy intake; and (2) survival costs, where long-term parental care decreases parental survival. According to evolutionary theory, parents adjust their parenting costs to maximize fitness based on changes in their own costs and the advantages to their offspring (Winkler, 1987).

There have been a number of mathematical models of parental care put forth, and the majority of them involve a cost-benefit trade-off between the parent and the offspring (Winkler, 1987; Shine, 1988; Reynolds & Gross, 1990; Clutton-Brock, 1991; Rosenheim, 1999; Besnard et al., 2002). The ecological form of the cost and benefit functions, however, is unknown, which makes it challenging to empirically test the models, and this poses a significant limitation in predicting the evolution of parental care (Clutton-Brock, 1991). Since all populations express care to some extent, it is common to test models using focal species that differ in the type or intensity of care given or to investigate interspecific variation in care strategies. Our understanding of the direct ecological effects of parental care evolution at the population level is limited because the vast majority of species either have no parental care at all or have parental care.

Despite being widely dispersed throughout Southeast Asia, the long-tailed skink (*Eutropis longicaudata*) is only known to have maternal care in one island population (Huang, 2006c; Huang et al., 2013). This presents a rare opportunity to investigate how the ecological environment has supported the evolution of parental care. While nesting female long-tailed skinks on Orchid Island protect their nests during incubation, females from at least 13 other populations of the species immediately abandon their nests after laying eggs (Huang et al., 2013). This pattern appears to be a result of the apparent increased predation pressure on Orchid Island, which necessitates maternal care to ensure successful reproduction (Huang et al., 2013). Using ecological data from a population that provides maternal care and two other populations lacking maternal care, we developed a cost-benefit model of parental care and tested it. We concentrated on the negative effects of maternal care, where mothers gain from higher rates of egg hatching but pay a price in terms of survival and fewer opportunities for future reproduction for female lizards who build nests. This goes beyond other models where parental care simply changes in response to a trade-off between growth rate and future reproduction, or the connection between parental care and fecundity (Clutton-Brock, 1991). Parental care should eventually develop from non-caring ancestors as a result of the trade-off between costs and benefits, and should be continued if the long-term benefits outweigh the costs. Our research suggests that the evolution of maternal care has been significantly influenced by natural selection, which directly affects females and their capacity to procreate. We incorporate our research into theoretical models that suggest mechanisms for the development of parental care (Shine, 1988), and we find some hypotheses to be supported but not others.

6.3.1 Study Population and Empirical Data

We gathered information on long-tailed skink nest size from Orchid Island (22°02′N, 121°34′E), 60 km southeast of Taiwan, Green Island (22°40′N, 121°28′E), 33 km southeast of Taiwan, and mainland Taiwan (SanDiMen, Pingtung County; 22°42′N, 120°38′E) between 2001 and 2010. Long-tailed skinks nested at each location in drainage holes in a concrete retaining wall that was situated next to a mountain road. See the following references for detailed information on the study locations and procedures (Huang, 2006c; Huang & Pike, 2011; Huang et al., 2013). Egg incubation failed for a number of reasons, such as fungus infection or ant or predation by egg-eating snake (*Oligodon formosanus*) (Huang et al., 2013). The maternal concern of the Orchid Island long-tailed skink, however, was to deter snakes from eating

Table 6.1 Predation, competition, and parental care attributes of three long-tailed skink populations. Individual predation pressure was assessed using the number of lizard predator species present in each location; intensity of egg predation was estimated using numbers of egg-eating snakes (*Oligodon formosanus*) (Huang et al. 2011) encountered during our study; and intra-specific competition was estimated using skink density. Data for hatching success with and without parental care are from (Huang et al., 2013)

Variable	Location Mainland Taiwan	Green Island	Orchid Island
Individual predation	Strong	Weak	Weak
Egg predation	Weak	Weak	Strong
Parental care	No	No	Yes
Clutch size (eggs)	6.7	6.0	6.5
Hatching success with parental care	59%	55%	81%
Hatching success without parental care	62%	51%	18%
Relative change in hatching success with care	−3%	4%	63%
Cost-benefit model without maternal care	1.15	0.06	−1.83
Cost-benefit model evolving maternal care	0.95	0.30	2.27

lizard eggs, which were more plentiful on this island than at our other two study sites. Snakes are the main egg predators (Huang et al., 2013). In order to ascertain the protective impact of keeping snake predators out (such as female lizards on Orchid Island), we carried out a manipulative experiment (Huang et al., 2013). In order to estimate the protective effect that maternal care would have in our two study populations that did not exhibit maternal care, we glued netting to the ducts that housed the lizards' nests (Table 6.1) (Huang et al., 2013). As a control, we also carried out these tests on Orchid Island (Table 6.1) (Huang et al., 2013).

6.3.2 Model Overview

We employed a game-theoretic evolutionary stabilization strategy (ESS) model (Maynard-Smith, 1977). The game theoretic model predicts that the other parent's investment will be impacted by one parent's abandonment as well as the selection pressure to continue providing parental care for both males and females (Table 6.2) (Maynard-Smith, 1977). The game-theoretic model can be used to define four distinct stabilization strategies (ESS) for the evolution of parental care: (1) both parents express parental care and either parent may give up care if the costs outweigh the benefits; (2) females abandon offspring while males give care; (3) females give care while males abandon; or (4) both sexes leave the nest, leaving no parental care. Here, we concentrate on the circumstances in which maternal care may develop because parental care for male reptiles is essentially nonexistent (Shine, 1988). Here, we focus on the conditions under which maternal care may evolve (ESS1 and ESS3). The evolutionary stabilization strategy ESS1 describes the long-tailed skink

Table 6.2 Payoff matrix and four evolutionary stable strategies (ESS) from the game theory model (Maynard-Smith, 1977). W: egg without care; w: egg with care; P0: egg survival without parental care; P1: egg survival with uni-parental care; P2: egg survival with bi-parental care; P2 > P1 > P0; p: male deserts and then obtains another mating opportunity. ESS1: both sexes show care; wP2 > WP1, otherwise female deserts; wP2 > wP1 (1 + p), otherwise male deserts. ESS2: female deserts and male cares; WP1 > wP2, otherwise female cares; WP1 > WP0 (1 + P), otherwise male deserts. ESS3: female cares and male deserts; wP1 > WP0, otherwise female deserts; wP1 (1 + p) > wP2 otherwise male cares. ESS4: both sexes desert; WP0 > wP1, otherwise the female cares; WP0 (1 + p) > WP1, otherwise the male cares

			Female	
			Care	Desert
Male	Care	♀	wP2	WP1
		♂	wP2	WP1
	Desert	♀	wP1	WP0
		♂	wP1 (1 + p)	WP0 (1 + p)

populations on Green Island and mainland Taiwan, where both parents abandon the nest, whereas ESS3 describes the population on Orchid Island, where females tend to incubate eggs and males abandon (Table 6.2). We can calculate the benefits in terms of the number of eggs laid and the percentage of these eggs that hatch into baby lizards because the returns to these two strategies are correlated with egg production and survival.

6.3.3 Benefits of Parental Care

The successful birth of offspring is a benefit of parental care. According to Huang and Pike (2011), in long-tailed skinks, the proportion of eggs laid that hatch depends on both the incubation temperature and the likelihood that predators will find and eat the eggs (Huang, 2006c). Eggs exposed to relatively moderate temperatures have higher hatching success (Huang & Pike, 2011), which increases the benefit of providing parental care. Similar to how successful attacks by females on egg-eating snake predators maintain the advantages of parental care (Huang, 2006c; Hunag et al., 2013).

6.3.4 Costs of Maternal Care

There are four costs associated with maternal care: (1) the energy required to produce an egg clutch; (2) the loss of future mate opportunities (when the female stays at the eggs, she loses future mate opportunities and gains energy for the next clutch of eggs); (3) predator resistance (the lizard's capacity to defend the nest from obligate egg predators); and (4) the potential for predation (maternal care can be achieved by increasing the exposure of predators during the care period by increasing predator exposure and reducing survival rates). The total cost cannot be greater than the value of the females themselves when these costs are viewed as point values in the model.

6.3.5 Model Development

The model devoid of maternal care is what represents the populations of long-tailed skinks found in Taiwan and Green Island.

$$\text{Clay} + \text{Begg}^{*}\ (1 - \text{X}) = \text{constant} \qquad (6.1)$$

Equation 6.1 states that Clay represents the cost of a clutch of eggs laid by a female lizard, Begg represents the increase in egg hatching rate, and X represents the percentage of eggs that do not hatch.

We assume that a female lizard equals ten points and that producing an egg clutch requires 30% of the female lizard's body mass, or three points (Huang, 2006c). The female lizard is awarded ten points if every egg in a clutch hatches; the value can vary depending on how quickly the eggs hatch. A female lizard is assumed to lay eggs twice during her lifetime, with at least one of the young from each clutch maturing into an adult. This means that each female will replace herself and one mate during her lifetime (20 lifetime points), maintaining a constant population size. By way of illustration, if a female lays four clutches of eggs over the course of her lifetime, each clutch will represent five points. These presumptions can be modified for life history characteristics.

The population of long-tailed skinks on Orchid Island is represented by an evolved model of parental care.

$$\text{Clay} + \text{Cdelay} + \text{Cpredation} + \text{Cantipredator} + \text{Begg}^{*}\ (1 - \text{X}) = \text{Constant} \qquad (6.2)$$

In Eq. 6.2, Clay equals the cost to females of postponing reproduction while protecting eggs, Cpredation equals the risk to females of caring for predation, and Cantipredator equals the cost incurred when females engage in anti-predatory behavior.

The cost of future reproduction, the risk of predation, and the cost of deterring predators are costs associated with not providing care if females develop maternal care. Delaying future reproduction was likely the most significant cost to females in our study system (Huang, 2007a). The result is that Cdelay $= -3$ (in Eq. 6.1, we assume that a clutch represents three points and that the greatest benefit to females is that a clutch is not predated by snakes, which results in all eggs hatching). Here, $X = 0$ and $-3 + 6*(1–0) = 3$. In this case, $X = 1$, $-3 + 6*(1–1) = -3$; the female only loses a maximum of three points for delaying the second clutch. The female's capacity to evade lizard predators and deter egg predators, respectively, is necessary for both Cpredation and Cantipredator. Due to the female investing three points in the nest, these expenses can total up to seven points. The reptile-egg-eating snake *Oligodon formosanus*, which does not eat lizards, is easily repelled by female lizards in our study system, so we assumed that Cantipredator was negligible (i.e., 0).

6.3.6 Examining the Model Without Parental Care

Using empirical data, we solved the model without parental supervision (Eq. 6.1) (Table 6.1). For the populations of mainland Taiwan and Green Island, the advantages of not providing care outweighed the costs, but not for the population of Orchid Island (Table 6.1). For the Orchid Island population to survive without maternal care (i.e., to get a positive value for the equation), either Clay must be decreased or Begg must be increased, X must be higher than 0.46, or, given the current hatching success of eggs without maternal care (18%), females must increase their clutch size to >17 eggs [known the maximum clutch size is 13 eggs] (Huang, 2006a)]. If eggs are not shielded from predators during incubation, long-tailed skinks will be exterminated from Orchid Island due to the small clutch size and high rate of egg predation in the absence of maternal care.

6.3.7 Examining the Evolving Maternal Care Model

Using empirical data, we solved the evolved maternal care model (Eq. 6.2) (Table 6.1). When eggs were protected from predation, Orchid Island's B_{egg} increased from 18% to 81%, but in the other two populations, the change was insignificant [<4% change; Table 6.1 (Huang et al., 2013)]. This was because egg predators were not included in the experiment, and maternal defense of the nest was simulated. Therefore, only the population of Orchid Island offered care benefits that outweighed the costs. The values for the models with and without maternal care were equal when empirically derived hatching success data were used as inputs (rather than data from predator exclusion experiments; Table 6.1) for the Green Island population. The absolute value of the model with evolved maternal care was higher than that of the model without evolved maternal care (i.e., 0.06). As a result, lizards on Green Island and the mainland of Taiwan should continue to bury their eggs after oviposition (which is common in most areas of the species).

6.3.8 Discussion

Understanding how animals make decisions in complex environments and the tactics they use under realistic circumstances is a major goal of behavioral and evolutionary ecology (Schmitz et al., 1998). Our cost-benefit model analyzes the development of maternal care in complex environments using an optimized framework (Maynard-Smith, 1977; Stearns, 1992; Schmitz et al., 1998) and yields four key conclusions: (1) Maternal care can evolve from a non-maternal care ancestor if the benefits outweigh the costs; (2) In species lacking predator-resistant behavior, the benefits of maternal care can increase with ecological or environmental factors; (3) In individuals actively defending their nests from predation, the benefits and costs of maternal care increase; and (4) Predators play a significant role in the evolution of maternal care. The evolution of parental care poses a significant evolutionary

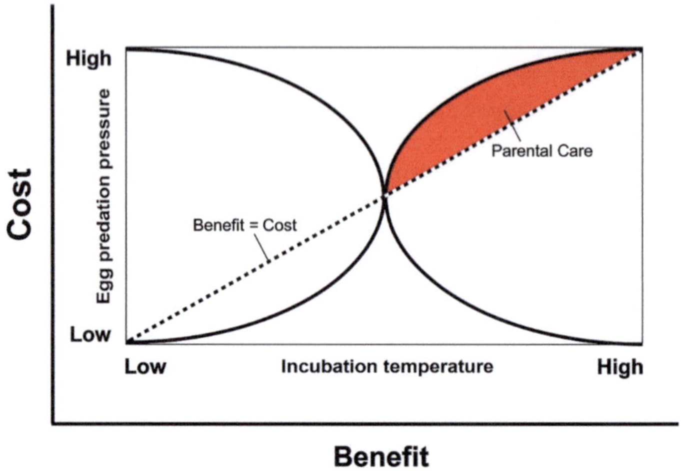

Fig. 6.4 A hypothetical cost-benefit model describing the evolution of parental care in long-tailed skinks (*Eutropis longicaudata*), showing how parental care can evolve based on the relationship between incubation temperature (the benefit is increased hatching success, x-axis) and egg predation pressure from snakes (the cost of laying eggs in populations with varying snake population size, y-axis). Note that these relationships are hypothetical, and provide only one example of how these two variables could influence the evolution of maternal care (similar relationships can be developed between other variables of interest). Parental care has a high likelihood of evolving in instances where female lizards select nest sites that are relatively high (i.e., inside of a retaining wall as compared to in natural habitats) but in which there are large numbers of egg-eating snakes (and thus high risk of any eggs left unattended becoming depredated). This matches the situation on Orchid Island, Taiwan. Maternal care is unlikely to evolve when predation pressure is relatively low, which matches the situation on Green Island and mainland Taiwan. (Originally published from Huang and Pike (2013), PLOS ONE, licensed under CC BY 4.0.)

challenge because it requires behavior to adapt in a way that benefits offspring. This is the path through which parental care develops, according to systematic developmental studies from a wide range of taxa [reviewed by (Clutton-Brock, 1991)].

We currently know very little about the specifics of the relationship between parental investment and offspring fitness, despite the numerous costs and benefits of parental care. This relationship probably involves one or more threshold effects (e.g., see Fig. 6.4 for a hypothesized relationship between selection of a nest site with appropriate incubation temperatures and predation pressure). Measurement issues with parents' reproductive costs are numerous, though. The impact of parental expenditure on offspring fitness may be obscured by individual differences in phenotype or genetic quality. Additionally, manipulating parental expenditure experimentally is necessary, even though it is challenging to measure reproductive costs and offspring gains using lab or field experiments (Clutton-Brock, 1991). Assumptions about the structure of cost and benefit functions are a necessary component of theoretical models of parental investment. Only a small number of in-practice predictions about the nature or degree of parental investment are based on ecology because it is impossible to adequately address all of these functions. For instance, it is currently unknown whether parents should spend more energy on raising good-quality offspring because they are more likely to survive or less energy on raising unfit offspring because of the greater fitness impact of parental spending. These uncertainties make us wary of predictions based solely on theoretical principles because we do not know whether environmental factors affect parental decisions. Our aim is to use biologically significant parameters to understand how these parameters may relate to the evolution of parental care, which may be a problem in our model as well.

Why maternal care in Taiwanese skink population has not evolved?

There are two reasons why maternal care hasn't evolved in the Taiwanese long-tailed skink population. First of all, female lizards in Taiwan, where there are numerous lizard predators [for example, at least four species of snakes; Table 6.1 (Huang et al., 2013)], are in great danger of being eaten while protecting their nests. Risk may rise with the diversity of lizard predator species. The X-value will rise concurrently with the mother lizard being attacked because the eggs are more vulnerable to attack without the mother's protection. According to Table 6.1 (Huang et al., 2013), the experimental exclusion of vertebrate predators in the Taiwanese population revealed that care and abandonment had nearly equal benefits in terms of hatching success. In contrast to Orchid Island populations, which showed a 63% increase in hatching success, Taiwan and Green Island populations did not experience a significant increase in hatching success when vertebrate predators were excluded [Table 6.1 (Huang et al. 2013)]. When there is no positive net benefit from maternal care, natural selection will favor nest abandonment.

No matter whether their species exhibits aggression toward egg predators, animals will seek additional benefits if the environment permits, according to our cost-benefit model. For instance, higher temperatures have an impact on the long-tailed skink's egg mortality and incubation time on Orchid Island (Huang & Pike, 2011). Incubated eggs in warmer concrete walls had an 88% hatching success rate compared to natural nests' 67%, and they hatched 7 days earlier (Huang & Pike, 2011). When the mother stays in the nest throughout the incubation period, a shorter incubation period in a warmer environment reduces maternal care and enables newly hatched young to have a longer season of activity (and thus grow in size) before winter. Therefore, regardless of the female lizard's capacity to ward off predators or even to provide maternal care at all, these concrete walls are likely to draw female lizards due to the advantages of hatching eggs. This prediction is supported by the populations of Taiwan and Green Island, which have not evolved maternal care but do lay eggs inside concrete walls.

Predation is a strong selective force that affects the behavior, morphology, and life history of prey species (Lima & Dill, 1990; Chivers et al., 2001). Life history and morphology studies have received a disproportionate amount of attention in studies on how predation affects prey reactions. However, it has been demonstrated that exposure to predators has an impact on prey evolution and behavioral changes (e.g., Crowl & Covich, 1990; Ball & Baker, 1996; Chivers et al., 2001). According to the current cost-benefit analysis, both individual predators and reptilian egg predators have played a significant role in the evolution of maternal care in long-tailed skinks.

Smaller species on the island are unlikely to exhibit maternal care as a result of direct contact with potential predators because they are unable to successfully fend off snakes that eat eggs. However, it might be challenging for maternal care to evolve when predators are scarce (or when several predators are present at once), as the costs of doing so might not outweigh the advantages. For instance, numerous predators, many of which prey on lizards, are present in Taiwan's skink populations. Similar to Green Island, where there aren't any predators at all, maternal care would only have a small positive impact (Table 6.1).

6.3.9 Cost and Benefit Analysis in the Framework for Hypothesis Testing

Our research led to eight hypotheses covering (1) the characteristics of the external environment, including nesting sites and (2) the reproductive characteristics of the species, both of which must result in increased fitness of parental actions. These eight hypotheses can be integrated with existing hypotheses about the mechanisms underlying the evolution of parental care (Shine, 1988). We discuss each of these hypotheses in turn.

6.3.10 Characteristics of the External Environment

1. Harsh and unpredictable environment. According to this hypothesis, parental care is more prevalent in settings with scarce adult resources during egg care (Tinkle & Gibbons, 1977). For example, if resources are limited before or during parental care, parents might increase the rate of food search during nest guarding, which might increase the risk of predation on the parent or offspring and lessen the advantages of providing maternal care (Tinkle & Gibbons, 1977). However, if food supplies are plentiful when the nest is being built, another hypothesis is that parents might not necessarily consider the trade-off between egg protection and food hunting. As a result, the female lizard would spend less time away from the nest and have more energy to protect it. In this instance, parental care may greatly increase the likelihood that eggs will survive. For instance, long-tailed skinks reproduced during the months of February through September, when food resources are plentiful (Huang 2006a, b, c), defying the harsh and unpredictable environment hypothesis while supporting the hypothesis of abundant food resources. Despite the fact that skinks occasionally depart from their nests to forage in the forest close to concrete walls (Huang, 2006c), contrary to the harsh and unpredictable environment hypothesis, but supporting the abundant food resource hypothesis. Despite the fact that skinks occasionally depart from their nests to forage in the forest close to concrete walls (Huang et al., 2013), these trips may be infrequent or so brief as to not increase the risk of predation on mothers or eggs. This hypothesis could be addressed using more thorough behavioral observations, which would help to make this point clearer.

Fig. 6.5 A female lizard, *Sphenomorphus incognitus*, laid her eggs, and when she buried the nest, the egg-eating snake, *Oligodon formosanus* immediately came to the nest to eat the eggs. Even though the female lizard tried to bite the snake, the whole nest was eventually eaten by the snake

2. Selection on a "risky" life history strategies. According to our model, parental care will become more prevalent when reproductive output is lower. This suggests that compared to long-lived species, short-lived species are more likely to have evolved parental care. Due to their lower reproductive output compared to long-lived species, short-lived species may be more likely to use "risky" reproductive techniques (Williams, 1966; Shine, 1988). Because it could make the parent vulnerable to predators, providing for offspring is a "risky" behavior. In lizards, late-maturing or large species are more likely to provide parental care (Williams, 1966; Tinkle, 1969). Tinkle (1969) argued that long-lived repetitive-operating species, particularly those with brief annual breeding periods, should typically provide parental care. This hypothesis is not supported by the long-tailed skink, a species with an iteroparous reproductive life and a large body size in comparison to other sympatric lizards. This discrepancy could be explained by lizards' capacity for self-defense against predators. For example, *Sphenomorphus incognitus*, a smaller lizard, was unable to defend its nest from *Oligodon formosanus* [(Tseng & Huang, 2012); Fig. 6.5] because its head width [8.1–10.9 mm; (Huang, 2010)] was smaller than the chest diameter of *O. formosanus* [11.5–14.5 mm; (Huang, 2004)], but *Eutropis longicaudata* females have head widths [13.3–18.3 mm; (Huang, 2006a)] larger than the chest diameter of the snakes; thus they are able to defend nests from invading snakes.
3. Exposed versus concealed nests. In his review of parental care in lizards, Tinkle (1969) hypothesized that female lizards who don't accompany eggs may be more at risk than lizards who lay eggs in secret nests. Likewise, predators may be more likely to find eggs that hatched in an exposed location as opposed to one that was hidden or buried. Therefore, species that do not bury their eggs should experience more frequent evolution of parental care (Noble & Mason, 1933; Shine, 1988). Despite the fact that long-tailed skinks in their natural habitat burrow their eggs beneath rocks, eggs laid by females in retaining wall nests are left exposed while they are incubating. These retaining walls are where females nest in populations

that either express maternal care or do not, indicating that maternal care is not always the result of egg exposure per se.

4. Nesting habitat suitability. When adults live in the same habitat where eggs are laid, parental care is anticipated to be more prevalent. This is due to the fact that parents' home environments, which also act as a valuable resource, may make them less susceptible to predators. In lizards, where males are more likely to be territorial and females are more likely to exhibit parental care, this hypothesis is challenging to test (Shine, 1988). Long-tailed skinks do not support this hypothesis because females typically spend time in their natural habitat outside of the nesting season but move to retaining walls during that time (Huang, 2006c). A related hypothesis addresses the abundance of potential nesting locations. Parental care may change as new females arrive and destroy existing nests when nest sites are few in comparison to the number of females that are capable of laying eggs. This circumstance has been discussed in relation to island populations sea turtles (Girondor et al., 2002) and iguanas (Rand, 1968). Although there were almost 1200 identical potential nests in the study area, we never saw more than 50 nesting females at once. Our data do not support this hypothesis because the number of suitable nesting locations was significantly higher than the number of breeding females.
5. The ability of parents to protect eggs from predators. Parental care should evolve more frequently in species where parents have the capacity to deter egg predators if one of the main benefits of parental care is to ward off potential egg predators (Shine, 1988; Fraipont et al., 1996; Greene et al., 2002). For instance, parental care is typical of venomous snakes (Viperidae, over 50% of species, and Serpentidae, 41% of species) as well as the largest snakes (Pythoninae). Similar to this, only 7.5% of Viperidae species have been reported to have parental relationships. These Viperidae species tend to be large or very venomous [e.g., *Psammophylax, Rhabdophis*; (Shine, 1988)]. This prediction is supported by the fact that parental care is common among intimidatingly large crocodiles. On Orchid Island, the long-tailed skink seems to be the only lizard species large enough to successfully fend off egg-eating snakes due to its relatively large size (Tseng & Huang, 2012).
6. Reproductive attributes. In species with low reproductive frequency, parental care is more likely to have evolved. Parental care may have a higher chance of evolving in species that only lay one large clutch of eggs per reproductive season as opposed to species that lay many small, irregular clutches of eggs. This is so because it is probably less expensive to care for one clutch of eggs than it is to care for several clutches over a longer period of time. This prediction seems to be supported by data from several studies (Tinkle, 1969; Bull & Shine, 1979). Parental care should have evolved more frequently in temperate species than in tropical ones, though multiple clutches are more common in tropical species than in temperate species. In contrast, parental care is more typical in tropical species of reptiles, such as Iguanidae, Pythonidae, Crocodylidae, and *Crocodylus* (Shine, 1988). Egg protection by long-tailed skinks is inexpensive and the same for females with large and small clutches (Huang, 2007a). This prediction is

unfounded because some long-tailed skinks can produce two clutches of eggs during a breeding season.

7. Short incubation periods. Parental care should be developed when the incubation period is short because females spend less time and costs providing for their offspring. A comparison of the incubation period of species with eggs versus those without eggs supports this hypothesis (Tinkle & Gibbons, 1977). However, if large species, which frequently lay large eggs (which need longer incubation times), are overrepresented in species that do not guard at eggs, this interspecific comparison may be tainted. This is supported by our field data, which show that long-tailed skink eggs hatch more quickly on Orchid Island than in Taiwan (Huang, 2007a). However, given that females in populations with and without maternal care nest in two habitat types—some bury their eggs in the ground (invisible to natural enemies) and some lay their eggs in retaining walls [visible to natural enemies]—incubation period may not have played a significant role in the evolution of maternal carein this species (Huang, 1994). Additionally, the length of maternal care varies depending on how frequently snakes try to eat the eggs, not always lasting the entire incubation period (Huang et al., 2013). As a result, this prediction might be overly straightforward.
8. Parental involvement improves fitness. When parental care improves egg hatching rates, it is more likely to evolve. Our model predicts that the likelihood that parental care will develop increases with hatch rate (benefit, Begg). Compared to eggs in natural nests, long-tailed skink eggs laid on concrete walls have a higher hatching success rate (Huang & Pike, 2011). Higher incubation temperatures for eggs may have a significant impact on how parental care has evolved. This hypothesis could shed light on why tropical species display parentage more frequently than temperate species (Shine, 1988). The benefits would be consistent with our model because eggs in tropical environments are exposed to higher temperatures, which in many species results in higher hatching success. Furthermore, because the majority of embryonic temperatures are higher in endotherms than in ectotherms, our model may suggest that endotherms will demonstrate parental care more frequently than ectotherms. This may help to explain why parent-child relationships are present in almost all mammals and birds.

6.3.11 Conclusion

Because we know so little about the processes that shape the development of parent-offspring relationships, it is challenging to make predictions about how parents should support their offspring. Our general cost-benefit framework can be used to investigate the simultaneous effects of a sizable number of biologically significant variables on quantitative estimates of parental care. According to this method, at least five different factors contributed to the evolution of maternal care in long-tailed skinks, including the energetic investment in a clutch, the delay in future fecundity, the risk of parental predation, the capacity to protect eggs from predation, and the improved hatching success brought on by maternal care. The development of parental care depends on the relative magnitude of the costs and benefits of these variables. Therefore, different pathways can result in the evolution of parental care from a non-care state even in a single population of a wide variety of species.

6.4 Green Sea Turtles Trigger Behavioral Evolution on Land Lizards

It is surprising that sea creatures in their natural state have such a significant impact on the behavioral evolution of land animals! This chapter is the result of long-term observations and studies that demonstrate that green turtles do indeed trigger the behavioral evolution of a terrestrial lizard, the long-tailed skink, the most important agent of which is the egg-eating snake, *O. formosanus*.

A look at animal behavior and evolutionary ecology reveals a major focus: understanding how animals best adapt to their environment by adopting behaviors or life histories in a complex environment. Because it is difficult to intuitively determine which strategies an animal should adopt to adapt to its complex environment, much of what is known about the evolutionary direction of parental care or life history is derived from theoretical models, and real field observations are rare.

Although many animal species have been found to exhibit parental care behaviors, the mechanism of parental care evolution in reptiles is unclear. Most studies of parental care behavior have shown that it most likely evolved when offspring or eggs were exposed to harsh environments or high levels of predation or intraspecific competition. Among these factors, environmental conditions in particular play the most important role, as they can contribute to an animal's willingness to develop parental care behaviors. For example, seasonal changes and the length of reproductive seasons, food availability, and predator feeding may influence the evolution of parental care behaviors. When these many influences are taken into account, animals make the best decisions in response to their environment; for example, animals in the most difficult environments will take the least amount of parental care time to obtain the most benefit.

In many animals, the amount of time parents are willing to spend caring for their offspring may be related to their knowledge and interaction with predators. Although there are many reports of animal responses to predator encounters, there is no information on how a large population of reptile egg-eating predators, *O. formosanus*, may have influenced the evolutionary causes of offspring care in the long-tailed skink (*Eutropis longicaudata*). In contrast to other parental care animals (e.g., mammals, birds, and crocodiles), most reptiles do not raise their hatched young, but only care for their eggs; egg care is the most basic and primitive form of parental care, so understanding its evolution is biologically significant for the beginnings of other parental care animals.

In the past few decades, I have devoted myself to studying the parental care of long-tailed skinks on Orchid Island, and a series of studies published by me from 2006 to 2010 on parental care of long-tailed skinks on Orchid Island in Taiwan. I found that long-tailed skinks stayed in the nest to care for the eggs after they were laid; when the kukri snake, which specializes in eating reptile eggs, crawled into the nest, the mother lizard immediately attacked the snake and pushed it out of the nest, and this egg care behavior continued until the eggs hatched. The study clearly demonstrates the dramatic effect of the mother lizard's egg-guarding behavior on reproductive success. The study also shows that this aggressive behavior only occurs during the reproductive season and that female lizards ignore snake entry and exit during the rest of the year. Interestingly, long-tailed skinks in Taitung do not exhibit the same protective behavior as Orchid Island long-tailed skink, one possibility being that there are fewer reptilian egg-eating snakes, *O. formosanus*, in Taitung. In other words, Orchid Island long-tailed skinks have developed strong egg protection behaviors to ensure the safety of their offspring in the face of reproductive challenges such as egg-eating snakes. So how did Orchid Island's large population of kukri snakes come about? Is it related to the eggs of green sea turtles? How do sea turtles and snakes evolve long-tailed skinks through the food chain to develop parental care behaviors?

The main content of this chapter is briefly described as follows: In summer, the green sea turtle beach on Orchid Island lays eggs to attract *O. formosanus* to come and eat the eggs to support a large population of *O. formosanus*. Although there are also long-tailed skinks in Taiwan that lay their eggs in cement holes, they do not have egg protection behavior. The main reason is that there are no large groups of *O. formosanus* eating eggs here; therefore, female long-tailed skinks leave immediately after laying eggs.

In addition, we had to find an island similar to Orchid Island, and the species on the island had to be similar to Orchid Island as a control group, but only green sea turtles came ashore to lay eggs; fortunately, Green Island had this condition. Therefore, in order to answer the behavioral evolutionary process, we must answer the following four questions.

1. Are long-tailed skink eggs really eaten by snakes?
   Our field experiments revealed that the egg-eating rates of lizards in Pingtung and Green Island, Taiwan (without large population of *O. formosanus*) and Orchid Island (with large population of *O. formosanus*) with and without protection (with plastic nets to seal the holes) were not very different (about 2–4% difference); however, at Orchid Island, 81% of eggs were eaten by snakes without parental protection (difference of 62%).
2. Density of snakes?
   In Taiwan and Green Island long-tailed skink egg-laying sites, the density of kukri snake is quite low (1 to 2 snakes per year); but Orchid Island is quite high (15 or more snakes per year; and 45 or more in Orchid Island sea turtle egg-laying sites).
3. What is the relative relationship between the number of sea turtles laying eggs ashore on Orchid Island and the number of snakes?
   The larger the number of turtles laying eggs on shore, the greater the number of snakes on the beach; the more snakes on the laying-egg sites of long-tailed skink, the lower the number of snakes on the beach; i.e., when the number of turtles

laying eggs on shore is high, the chance of snakes appearing on long-tailed skink is low.
4. When snakes moved to long-tailed skink egg-laying sites in large numbers, the rate of lizard eggs being eaten increased.

Parental care is a crucial characteristic and is defined as any type of behavior that improves the fitness of offspring by providing protection or food for most endotherms (birds and mammals) and some ectotherms (e.g., reptiles, amphibians, fish, and insects). Many taxa have developed a wide range of parental care techniques, from relatively simple nest-keeping habits to more intricate interdependent relationships between parents and offspring, like actively feeding offspring (Clutton-Brock, 1991; Farmer, 2000; Klug & Bonsall, 2009). It is known that parental care occurs frequently and independently across animal lineages (Tallamy, 1984; Shine, 1988; Wesolowski, 1994; Gross, 2005; Klug & Bonsall, 2009; Gilbert & Manica, 2015), but it is difficult to empirically study how parental care evolved from carefree ancestors because expressions of care often remain consistent within each species. As a result, rather than focusing on the effects of not providing any care at all, most research on this subject has relied on theoretical models (e.g., Gross & Sargent, 1985; Burley & Johnson, 2002; Reynolds et al., 2002; Roff, 2002; Manica & Johnstone, 2004; Klug et al., 2013) or investigations of predictors of how much care was provided by parents.

Protecting young or eggs from predators is thought to have been the primary motivation for the evolution of parental care in many taxa (Wilson, 1975). Predators have been shown to drive the evolution of many traits, including avoidance (e.g., reduced feeding and activity, and use of refuges) [e.g., (Gordon & Kulig, 1996), crypsis (Stevens & Merilaita, 2009), aposematism (Mappes et al., 2005; Tseng et al., 2014), mimicry (Speed, 1993), and masquerading (visual similarity to inedible objects) (Skelhorn et al., 2011)]. Although predation pressure (i.e., high offspring mortality in the absence of care) is often cited as the most likely factor driving the evolution of caregiving behavior from a no-care state (Clutton-Brock, 1991; Klug & Bonsall, 2009), there is little direct evidence of this relationship in natural systems [see also (Brown et al., 2010)]. This is due to the fact that comparing populations with different parental care expressions is necessary to elucidate the evolutionary origins of care. By contrast, for instance, the ecological characteristics and selection pressures affecting several populations of a single species that displays parental care, but not in others. Nearly all species that exhibit parental care always express this behavior in some way, even though some species show geographic variation in the type and intensity of care given (Clutton-Brock, 1991; Webb et al., 1999).

The long-tailed skink (*Eutropis longicaudata*) is one species that does display intraspecific variation in parental care. One population of Orchid Island, off the coast of Taiwan, has mothers who actively guard their nests from egg predators after they lay eggs, in contrast to lizards from other populations, which always leave their nests right away after doing so (Huang et al., 2013). The evolution of parental care on Orchid Island is most likely a result of predation by the egg-eating snake *Oligodon formosanus*, as this predator is unusually abundant there. Although anecdotal evidence suggests that nearby green sea turtle eggs (*Chelonia mydas*) may be an important food source for this snake, the reason for such high snake abundance is unknown (Huang et al., 2011). Thus, Long-tailed skinks offer the chance to empirically study the transition from an unconcerned ancestor to parental concern.

In this chapter, we demonstrate the genetic basis for parental care on Orchid Island, quantify the role of predation in promoting this behavior there, and investigate the reasons why predation pressure is unusually high at this location. To accomplish this, we first performed a common garden experiment to test the effectiveness of anti-snake behavior in captive lizards on Orchid Island and in two populations of no care lizards (Green Island and mainland Taiwan) that are genetically undifferentiated from one another (Huang et al., 2013). When comparing the number of lizards removed from guarded mothers on Orchid Island with the number of lizards removed from their nests on Orchid Island, we were able to determine the causes of egg mortality in unguarded lizard nests in these three populations. The cause of egg mortality in unguarded nests in these three populations was subsequently investigated by contrasting nests on Orchid Island, where mothers' guards had been removed, with nests at two other sites, as well as nests at all three sites that had been artificially shielded from snake predation. To understand the role of this trophic link between terrestrial and marine habitats in the evolution of lizard maternal care, we also looked at the annual and seasonal relationships between *Chelonia mydas* nesting, egg-eating snake population dynamics and density, and predation rates at long-tailed skink nests on Orchid Island.

6.4.1 Genetic Basis of Parental Care on Orchid Island

Ten captive female lizards with eggs, 12 from Taiwan's mainland and 14 from Green Island, all without any prior experience with nesting or snake encounters, were used in our study on common garden maternal care. Despite the low overall genetic differentiation between these populations, all ten of the females from Orchid Island violently attacked egg-eating snakes when they came into contact with them, while none of the females from the other two sites did so. This suggests that egg-protective behavior has a genetic basis (Huang et al., 2013) .

6.4.2 Hatching Success and Predation Sources of Lizard Eggs

The lowest rate of lizard eggs hatching in the absence of maternal care was found on Orchid Island (18%), while rates were significantly higher in the other two populations (62% on Green Island and 52% on Taiwan; $\chi_{2,2} = 106.74$, $p < 0.0001$; Fig. 6.6a). Key environmental factors, such as nest temperature and relative humidity, did not differ significantly between sites (mean ± SE: Orchid Island: 30.0 ± 0.30 °C[$n = 125$], Green Island: 30.3 0.30 °C[$n = 45$], Mainland: 30.8 0.21 °C [$n = 248$]; $F_{2,\,415} = 1.\,57$, $p = 0.21$) and therefore cannot account for variations in hatching success.

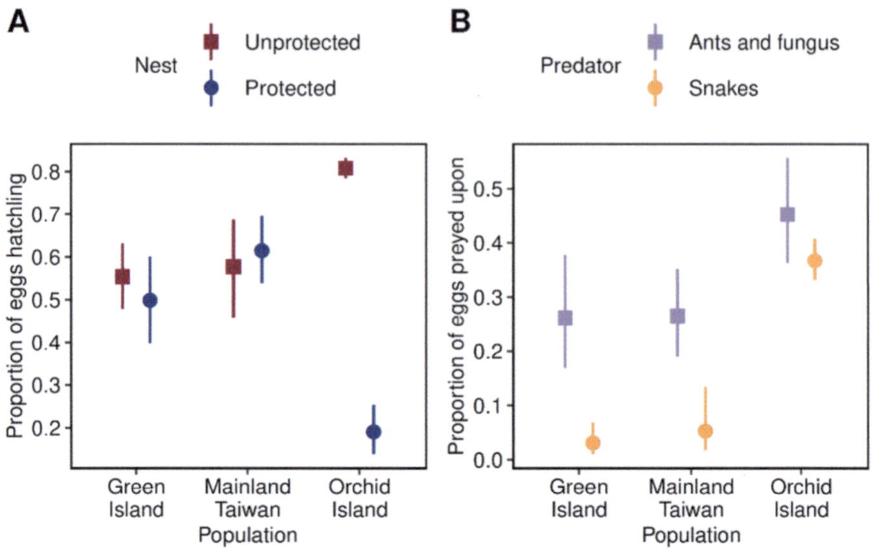

Fig. 6.6 Predation rates of lizard nests and associated predators. Proportion of long-tailed sun skink eggs hatching from three populations without any form of nest protection (**a**, for the mainland, Green Island, and Orchid Island, $n = 108$, 60, and 450 eggs) and when nests were protected by predator-excluding mesh ($n = 1000$, 70, and 156 eggs). Known causes of egg mortality, expressed as the numbers of eggs preyed on by snakes or ants and fungi (**b**). (Redrawn from Pike et al. (2016), Scientific Reports, licensed under CC BY 4.0)

The cause of lizard egg mortality varied significantly among populations ($\chi_{2,2} = 143.07$, $p < 0.0001$), with snakes predating most eggs at Orchid Island, but ants and fungal infections causing most deaths at the other two sites (Fig. 6.6b). Even when data on eggs that died from unknown causes were recorded as snake predation, Orchid Island had the highest rate of snake predation (compare all three sites: $\chi_{2,2} = 89.60$, $p < 0.0001$; compare only the two sites with the highest assumed rate of snake predation, Orchid Island vs. Taiwan, $\chi_{2,1} = 86.30$, $p < 0.0001$).

At Orchid Island ($n = 1000$ eggs, $\chi_{2,1} = 200.89$, $p < 0.0001$), experimental exclusion of vertebrate predators from exposure to lizard eggs significantly increased hatching success, but it had no significant impact at the other two sites (Taiwan: $n = 70$ eggs, $\chi_{2,1} = 0.31$, $p = 0.58$; Green Island: $n = 156$ eggs, $\chi_{2,1} = 0.17$, $p = 0.68$). This proves that the only population in which maternal nest guarding has a protective effect in Orchid Island. Thus, the initial emergence of maternal guardian nests on Orchid Island may have been influenced by the high predation pressure of egg-eating snakes.

6.4.3 Density of Snakes Nearby Lizard Nests

Given the primary function of snake predation on Orchid Island, between 1997 and 2008, we examined the prevalence of this predator on concrete walls where lizard nests were found at three sites. Orchid Island had a higher density of snakes (number of snakes per year) than the other two sites (mean $\pm$ SE $= 0.91 \pm 0.71$ for Green Island and 0.36 ± 0.45 for Taiwan, $\chi_{2,2} = 6.56$, $p = 0.0001$; Fig. 6.7).

6.4.4 Turtle Nesting as a Factor in Lizard Abundance

In Taiwan, Orchid Island is a significant green sea turtle nesting location and may serve as a seasonal food source for snakes that consume eggs. We conducted monthly surveys on Orchid Island to gauge the abundance of egg-eating snakes and made a distinction between snakes caught in habitats where lizards nest in the inland and habitats where sea turtles nest on the beach. Our goal was to ascertain whether this food source is a significant driver of snake density and, consequently, the predation pressure experienced by lizards. Snake densities peaked in the months when sea turtles nested (May to October) and were highest in years with high sea turtle nesting effort across the island (pooling estimates for inland and beach sites; Fig. 6.8a; $R^2 = 0.79$, $F_{1,9} = 33.54$, $p < 0.0001$). The relationship between turtle nesting and snake abundance was caused by different processes than those that caused lizard nesting, as there was no correlation between annual turtle and lizard nest abundance ($n = 8$ years, Pearson's correlation $r = 0.06$, $p = 0.88$), nor between snake and lizard nest abundance ($r = 0.11$, $p = 0.64$).

The strong negative correlation between snake abundance in the two habitats (annual: $r = -0.66$, $F_{1,12} = 8.99$, $p = 0.011$; monthly: $r = -0.99$, $F_{1,3} = 176.34$, $p < 0.001$; Fig. 6.8b) suggests that snake populations are likely to migrate between beach and inland sites. Snakes were more numerous on the beach than inland. Individual snakes that were markedly recaptured helped to support this finding:

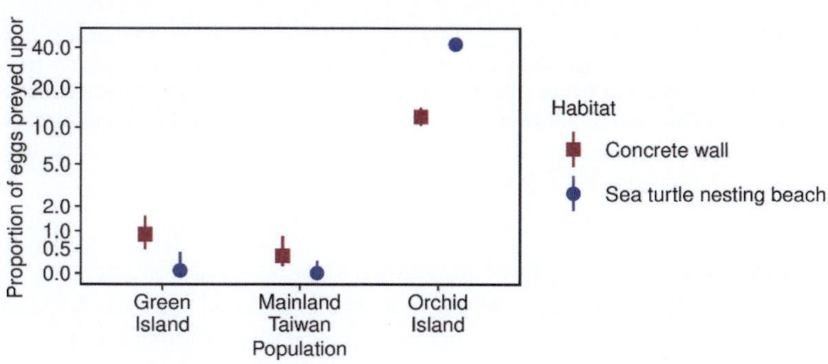

Fig. 6.7 Snakes observed on concrete walls on Green Island and Orchid Island when turtle eggs on Orchid Island are available (May–October) and not available (November–April) from 1997 to 2008. (Redrawn from Pike et al. (2016), Scientific Reports, licensed under CC BY 4.0)

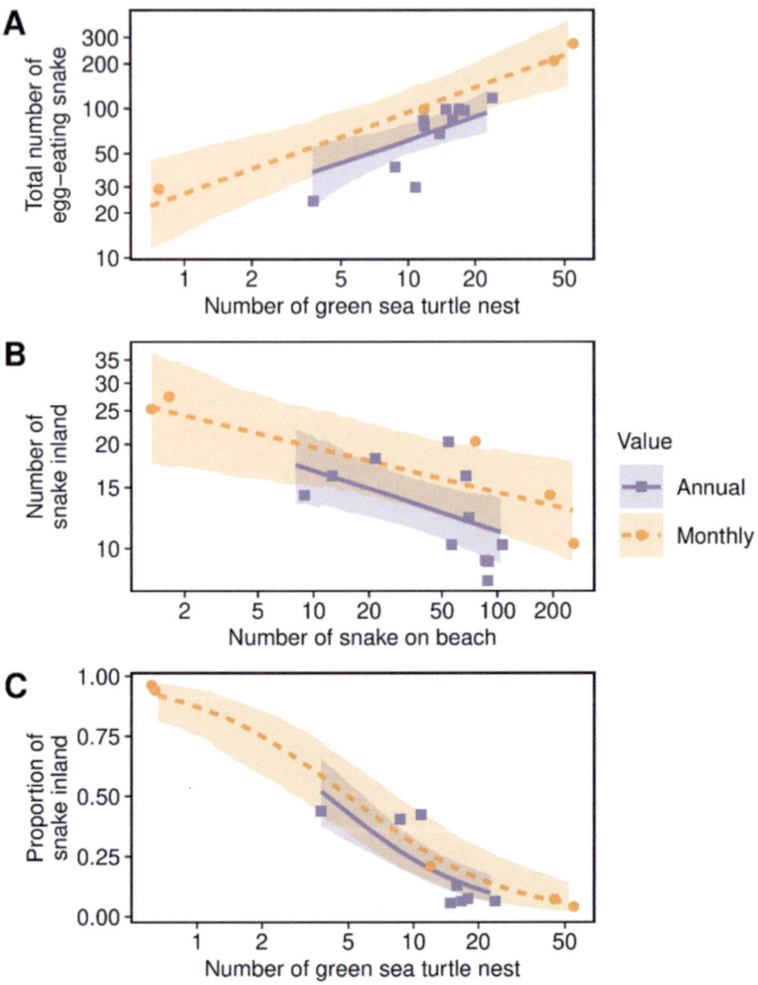

Fig. 6.8 Relationships between sea turtle nesting and egg-eating snake abundance on Orchid Island, Taiwan. Egg-eating snake abundance depends upon sea turtle nest availability (**a**). When more egg-eating snakes are at the sea turtle nesting beach, fewer are found inland (**b**). Ultimately, there is a negative relationship between sea turtle nest availability and the proportion of snakes inland (**c**). Circles show annual values ($n = 11$) and squares show monthly values ($n = 5$). (Redrawn from Pike et al. (2016), Scientific Reports, licensed under CC BY 4.0)

72% of snakes that were recaptured moved between beach and inland sites (beach: $n = 463$ marked snakes, 109 recaptured snakes, 78 relocated snakes). Additionally, we discovered a negative correlation between the number of inland snakes and the presence of sea turtle nests (year: $r = -0.83$, $F_{1,9} = 20.77$, $p = 0.001$; month: $r = -0.86$, $F_{1,3} = 8.43$, $p = 0.06$; Fig. 6.8c), suggesting that sea turtle eggs serve as a significant source of food for snakes.

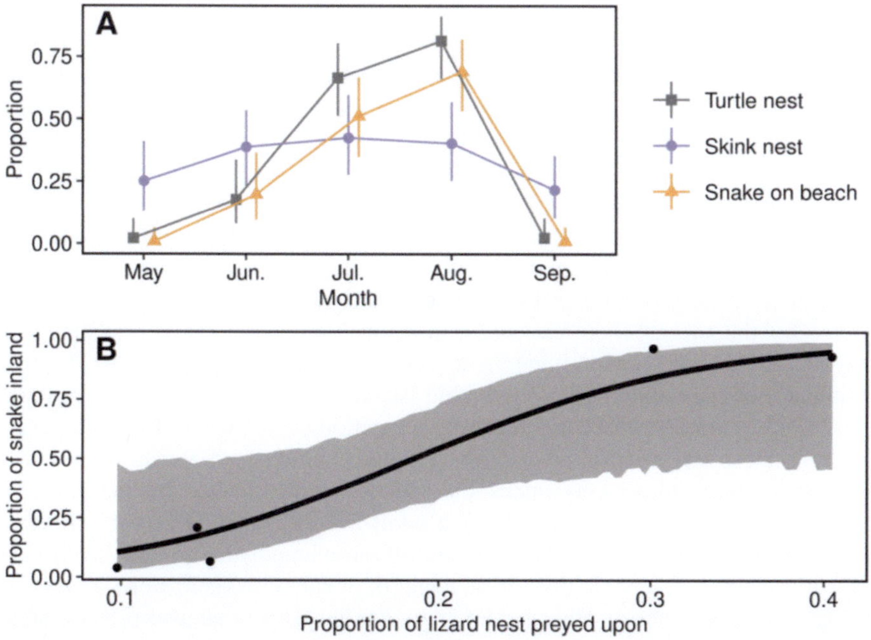

Fig. 6.9 Seasonal patterns of nesting activity and nest predation on Orchid Island, Taiwan. Sea turtle nesting is highly seasonal and coincides with egg-eating snake abundances at the beach, whereas lizard nesting occurred in each month (**a**) $n = 5$. Proportions represent the number of monthly observations as a function of the total number. In May and September, $> 99\%$ of snakes are inland, at the lizard nesting site. Predation rates of lizard nests are highest when proportionately more snakes are inland (**b**). (Redrawn from Pike et al. (2016), Scientific Reports, licensed under CC BY 4.0)

With the exception of May and September, lizard nesting was fairly consistent month to month, unlike the highly seasonal supply of sea turtle eggs (Fig. 6.9a). More snakes moved inland when there were fewer lizard nests because snake migration prioritized turtle nests ($r = -0.81$, $F_{1,3} = 12.68$, $p = 0.038$; Fig. 6.9a). Although monthly predation rates were highest when there were more snakes in the interior ($r = 0.95$, $F_{1,3} = 28.13$, $p = 0.013$; Fig. 6.9b), annual snake predation rates on Orchid Island lizard nests remained stable ($n = 390$ nests; annual predation rates 10.2–15.6% from 2000 to 2008). Turtle nest availability decreased at the same time that there was a positive correlation between snake abundance in the inland and lizard egg predation rates (Fig. 6.8c). In the event that turtle nests aren't available, snakes switch between lizard nests inland and turtle nests on the beach, establishing a distinct foraging pattern.

6.4.5 Discussion

There is a genetic foundation for the egg-protective behavior of female lizards on Orchid Island, despite the low level of genetic differentiation among lizard populations and the lack of maternal care in other populations (Huang et al., 2013). This characteristic appears to have developed in response to the unusual levels of predation present at the location, where a large number of egg-eating snakes are responsible for the majority of the egg mortality. However, egg-eating vertebrate predators have only had a very minor effect on reproductive success in other lizard populations (Huang & Pike, 2013). Although turtle eggs appear to be a major food source for the snakes, their seasonal availability means that snake predation is a major threat to lizard eggs, driving the evolution of care in this population. The high abundance of egg-eating snakes on Orchid Island is a result of green turtle nesting.

We offer thorough proof that maternal protection (in the form of egg guarding) evolved in response to snake predation on eggs on Orchid Island. Because the lizards are large enough to actively attack the snakes but the snakes are devoted egg predators and do not physically threaten adult lizards, guarding is very effective (Coleman et al., 1993). Given the outcomes of our manipulation to keep snakes out of artificial nests, it appears that the advantages of care elsewhere were simply not available. The absence of other predators on Orchid Island that also prey on adult lizards may have contributed to the evolution of this behavior, but given the results of our manipulation (or negligible). Therefore, the evolution of care at Orchid Island is a combination of extremely high benefits (acquired only because of the extremely high density of egg-eating snakes) and relatively low costs (at least in terms of survival, changes in body condition, fecundity, or time in subsequent nests relative to egg-laying time) (Huang, 2007a).

There is evidence that marine animals, like sea turtles, can occasionally bring a seasonal influx of nutrients to terrestrial communities, significantly influencing their structure and evolutionary dynamics despite spending most of their lives outside of these communities (Lundberg & Moberg, 2003; Helfield & Naiman, 2006). The most well-studied example of this phenomenon is the annual migration of wild salmon (*Oncorhynchus* spp.) from the North Pacific Ocean to freshwater streams and lakes, where they spawn and die. Adult salmon support populations of numerous animal species, such as grizzly bears(Levi, 2012) and provide nutrients to freshwater and terrestrial ecosystems, which in turn affect riparian structure and dynamics (Helfield & Naiman, 2006; Hocking & Reynolds, 2011). These results highlight the complexity of interspecific interactions across ecosystems as well as potential cascading effects on the ecology and life histories of different species.

The concrete barriers that the majority of the lizards on Orchid Island use as nesting sites have only been there since 1995, so we cannot say with certainty when skinks nesting on Orchid Island first developed egg-guarding behavior. It's possible that egg-eating snakes will find nests more alluring because of the concentration of nests on these artificial substrates, but it seems unlikely that this trait has become fixed in the population in such a short amount of time. It seems more likely that the lizards evolved to guard their nests due to the seasonal nature of turtle nesting, and that this pressure increased with the creation of concrete barriers. The species is a prime target for the search for genetic mutations that may underlie guarding behavior because of the low genetic differentiation between Orchid Island and nearby populations. We demonstrate here that gene-based parental guarding can occur and be fixed in a population in response to sufficient predation pressure on eggs. Whether there are "parental genes" or not remains to be seen.

6.5 The Cost of Egg Caring in a Skink

Why don't the majority of animal species care for their eggs or young? According to the cost hypothesis, parental care would be unavoidably expensive unless a sizable fitness trade-off is made, which would reduce food intake, the likelihood of survival, and/or the subsequent fecundity of reproductive females. Therefore, it is expected that parental care will evolve most frequently in species where costs are minimal or unimportant. Although parental care has been reported in more than 140 species of reptiles, only a small number of records have conclusively shown the value of this care. The "cost" of maternal activity and survival in long-tailed skinks (*Eutropis longicaudata*), as well as egg attendance rates and body size changes in females during and between seasons after providing for their young, are all topics covered in this chapter. I used this information to investigate whether female lizards must necessarily pay a significant cost for parental care. According to my research, a mother's length, body weight, recapture rate, or fecundity may not necessarily suffer significant energetic losses as a result of prolonged, intense parental care.

It is anticipated that parental care will evolve as a result of a trade-off between the advantages of increasing offspring survival and the costs of reducing adult survival and future reproduction (Shine, 1988; Clutton-Brock, 1991; Greene et al., 2002; Huang, 2006c, 2007a; Huang & Wang, 2009; Huang et al., 2013; Huang & Pike, 2013). The costs to parents have been shown to be more challenging to identify and quantify than the benefits to offspring, which are frequently fairly obvious and easily measured (Reguera & Gomendio, 1999). The costs of parental care can be divided into two categories: reproductive costs, which include missed opportunities for mating as a result of prolonged parental care, reducing future fecundity or reducing the number of offspring at the next reproduction due to lower energy intake; and survival costs, which include prolonged parental care as a result of food requirements and may increase parental susceptibility to predators or pathogens (Shine, 1988; Huang & Pike, 2013).

Why do ectotherms, such as amphibians and non-avian reptiles, less frequently display parental care than endotherms like birds and mammals do? Parental care appears to be closely linked to endothermy in the evolution of vertebrates, and this relationship is supported by morphological, physiological, and behavioral characteristics (Farmer, 2000; O'Connor & Shine, 2004). We need to learn more about the costs and benefits of parental care in ectothermic terrestrial vertebrates in order to better understand this hypothesis. Many fishes and amphibians exhibit complex parental behavior (Annett et al., 1999, Burrowes, 2000, Cook et al.,

2001; O'Connor & Shine, 2004), but the closet phylogenetic and ecological comparison to terrestrial endotherms come from those of reptiles. Parental care is ubiquitous in crocodiles (Shine, 1988), unknown in turtles(Iverson, 1990), and sporadically distributed in squamates (Shine, 1988; Somma, 1990). Several lizard taxa display long-lasting egg/offspring-maternal associations, despite the fact that females in the majority of squamate species depart soon after laying eggs (O'Connor & Shine, 2004; Huang, 2006c). Thus, lizards provide a good opportunity to examine the relatively simple forms of cost of parental care in a robust comparison with the more complex conditions found in endothermic vertebrates.

The primary costs of parental care may result in a decrease in food intake, survival, and subsequent fecundity. Depending on the location chosen and whether the eggs have an odor that can be detected by predators, lizards with eggs may be more vulnerable to them. Also, lizards that spend a lot of time with their eggs might not have as many opportunities to eat. As a result, the accumulation of reserves required to produce a second clutch of eggs may be slowed down or prevented (Shine, 1988). Information on reptile body mass (in terms of food intake), survival rates, and subsequent fecundity, however, is still lacking and poorly understood (Clutton-Brock, 1991).

As a result of changes in organ size and energy reserves, adult vertebrates' physical characteristics frequently change over time (Cichon, 2001). Several studies have shown how crucial it is to preserve parental reserves in order to successfully complete incubations without jeopardizing parental survival (Kellett & Alisauskas, 2000; Bolduc & Guillemette, 2003; Gorman & Nager, 2003). Animals' quality varies significantly throughout their life cycle in relation to migration, overwintering, and reproduction (Cichon, 2001), and females are known to suffer significant quality loss during incubation. Similarly, patterns of quality variation in reptiles that exhibit parental care have rarely been reported.

The measurement of reproductive costs has limitations, and almost all attempts to do so have ignored many aspects of parental health that might be impacted by prior attempts at reproduction, such as the viability of subsequent offspring (Shine, 1988; Clutton-Brock, 1991). Instead of illuminating the nature of the relationship between reproductive output and subsequent parental survival or reproductive success, almost all studies have concentrated on determining whether reproductive costs are present. These challenges emphasize the need for more research into parental care's unknown facets, such as how much it affects variables that affect reproductive costs, like survival, reproduction rates, and the timing of second clutch.

During the majority of the active season, long-tailed skinks are dispersed over a wide area, but during the seven-month breeding season from February to August, they congregate in high densities at hole entrances to lay eggs (Huang, 2006b). In addition, on one occasion in 1997, I discovered a female lizard with her jaws clamped to the head of an *Oligodon* while the two lizards were motionless at the base of a cliff. I have seen female long-tailed skinks attack egg-eating snakes *O. formosanus* in rocky areas or holes in concrete walls at least 15 times. My research on the maternal reactions and experimental oviposition of long-tailed skink females was made easier because many of them use man-made concrete holes as their nesting locations.

In the low elevation rainforest of central Orchid Island, at a study site with concrete walls, all field data were collected. These concrete retaining walls have plastic pipes embedded within them that are at least 10 cm in diameter to allow runoff to pass through. They were constructed to help prevent landslides. For the long-tailed skink, these holes make excellent egg-laying locations because some of the holes are currently blocked by rocks and other debris. Over long time scales, the temperature at the nesting locations is comparatively constant. Since the difference in temperature between day and night is very small—never exceeding 10 °C—warmer temperatures are produced during periods of low sunlight. Typically, the temperature rises at sunrise, peaks at noon, and then slowly drops after sunset, reaching its lowest point in the early morning. In comparison to the mean temperature of natural nests, which was 29.7 °C ($n = 1355$, $t = 24.47$, $p < 0.0001$, paired *t*-test; Huang, 2004), concrete-hole nests had a mean temperature of 30.6 °C.

6.5.1 Duration of Egg Guarding

The proportion of nest guards decreased over time after egg laying; that is, most females stayed in the nest for at least 1 week, but some started to leave after 1 week. By the time of hatching, less than 50% of nests were guarded by females (Fig. 6.10).

6.5.2 Loss of Body Mass and Length in Egg-Guarding Females

6.5.2.1 Intra-seasonal Variations

Females gained weight on days 5 and 8 compared to day 1 as they foraged, as measured by the mean body weight loss ($n = 64$) on days 4, 5, and 8 relative to day 1 of 2.1, –0.28, and –0.57 g, respectively (Fig. 6.11). Thus, the loss in body mass during the first week of egg guarding was balanced, on average, by an increase in

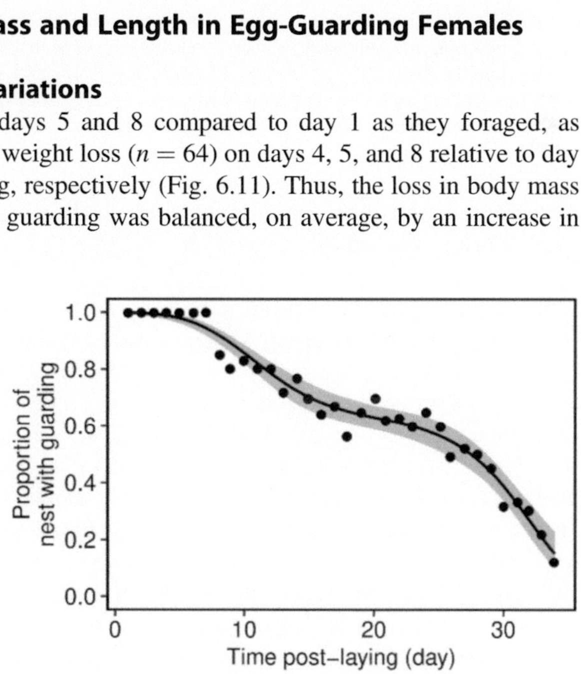

Fig. 6.10 Relationship between the proportion of nests with guarding *Eutropis longicaudata* females and the time after eggs were laid. The ratio of egg-guarding time spent on nests decreased with the duration of incubation. (Redrawn from Huang (2007a, b), Ecological Research)

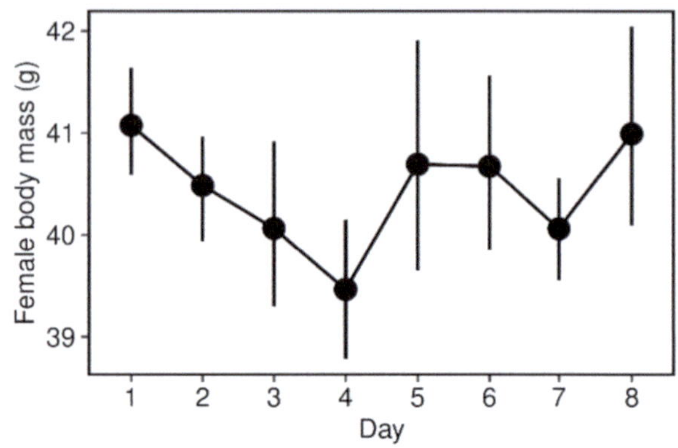

Fig. 6.11 Variation of *Eutropis longicaudata* female body mass after they engaged in egg-guarding behavior on Orchid Island. (Redrawn from Huang (2007a, b), Ecological Research)

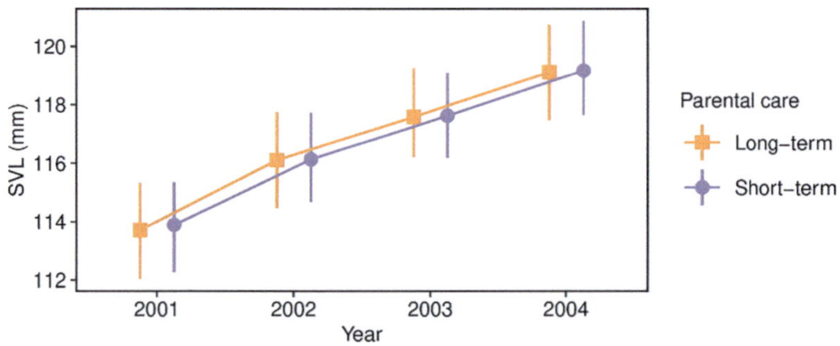

Fig. 6.12 Comparison of the changes in snout-vent length (SVL) of *Eutropis longicaudata* females between those engaging in long- and short-term parental care from 2001 to 2004 on Orchid Island. Sample sizes, 27, 20, 15, and 15, and 27, 24, 18, and 17 individuals were captured and re-captured in the long-term and short-term parental care groups, respectively, from 2001 to 2004. (Redrawn from Huang (2007a, b), Ecological Research)

mass during subsequent foraging; however, body mass did not return to pre-laying values during incubation.

6.5.2.2 Seasonal Variations

From the individuals recaptured each year, the average growth rates of body length (mm/year) and body weight (g/year) were estimated (4 years). The data from the intermediate group (17 to 27 days) were not used because they might have tainted the comparison results because all skinks displayed parental care in the first week and some withdrew after that. I arbitrarily divided them into two groups: the long-term parental care group (28 to 35 days) and the short-term parental care group (9 to 16 days). There was a significant difference in the number of days of in parental care between the two groups ($t = 42.3$, $p < 0.0001$; t-test). Figure 6.12 shows the mean SVL growth rate (mm/year) estimated from 4 years of mark-recapture data. There were no significant differences between the two groups. Similarly, there were no significant differences in mean weight gain rates (g/year) between the two groups.

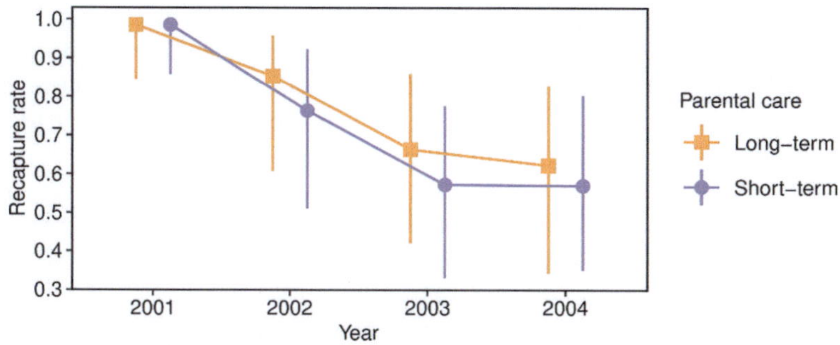

Fig. 6.13 Comparisons of recapture (survival) rates of *Eutropis longicaudata* females between those engaging in long- and short-term parental care from 2001 to 2004 on Orchid Island. (Redrawn from Huang (2007a, b), Ecological Research)

Differences between long-term and short-term parental care groups in recapture rates, fecundity, and the time it takes for the second clutch to be produced.

Total recapture rates for female lizards given long- and short-term maternal care were 74%, 56%, and 56%, and 89%, 67%, and 63%, respectively, from 2002 to 2004. These rates did not significantly differ (Fig. 6.13). Recapture (survival) rates were not significantly different between the long-term parental care group and the short-term parental care group, according to a log-likelihood ratio test. Between the long-term and short-term parental care groups, there were no significant differences in fecundity over the course of 4 years. Similar to this, there were no significant differences in the length of time it took for recaptured lizards to lay a second clutch of eggs.

6.5.3 Discussion

Relationship between maternal care and body quality and size.

In this study, I estimated the costs associated with reproduction using short-term (8 days) and long-term (4 years) changes in body mass, recapture rate (survival) , and fecundity variables (clutch size and time to produce a second clutch). I discovered that females did not incur significant costs associated with caring for eggs either within-season or between-season, regardless of whether they displayed long-term or short-term parental care. No matter their age, the majority of skink body size measurements taken over 4 years showed an increase in size. This finding rejects the hypothesis that parental care of living lizards reduces their chances of fecundity or survival in the future (Williams, 1966). Furthermore, my data indicate that, for both long-term and short-term maternal care, the magnitude of small increases in maternal quality and size was unaffected by the caring period. Most importantly, all of the female participants in the 4-year observation period survived and appeared healthy at the end of the study. Additionally, the two groups' body conditions appeared to be comparable and stayed within the typical range seen in earlier studies of wild-caught, non-breeding females [mean SVL of 115.9 mm;(Huang, 2006a)]. This finding is unequivocal proof that females maintain significant body reserves while providing for their offspring, possibly by engaging in subsequent active foraging activities away from the nest. As a result, they did not appear to pay a price for nest construction in terms of their survival.

My finding is that there was no significant decline in quality during parental care, which may be due to better access to food or a rise in concrete pore temperature. The findings of this study imply that body condition may affect parental investment choices in long-tailed skink females who guard eggs. The quality of people first decreased during guarding and then improved during the following foraging activities, indicating that their physical condition may have an impact on decisions regarding parental work.

According to some authors' hypotheses (Carrillo & Aparicio, 2001; Dearborn, 2001), there is a trade-off between feeding and defensive behavior, so nest defenses rise when food resources are more plentiful (Martin, 1992). If so, an increase in the abundance of specific prey animals, particularly Orthoptera, Coleoptera, and Hemiptera, could account for the seasonal increase in nest defense (Huang, 2006b). Despite losing mass at the start of the defensive behavior, individuals gained it back during subsequent foraging excursions, indicating that food availability may not be a major factor in this species' decision to abandon its nest. Long-tailed skink females may be able to devote more energy to raising their young if the environment is more hospitable and predictable (i.e., stable temperature and humidity, and/or plenty of food available). Accordingly, it is unlikely that the observed body quality dynamics can be explained by resource depletion brought on by the costs of providing for one's parents. Similar to this, a recent study found that pythons spent 2 months coiled around their eggs, going without food. Less than 6% of their initial body mass was lost from them. These unexpected findings were attributed by the authors to the pythons' low energy expenditure because they shiver to generate heat to aid in egg incubation (Aubret et al., 2005).

Species that nest in cooler climates may rely more on endogenous reserves during incubation than species that nest in tropical climates because they must endure lower temperatures and possibly less food availability (Kellett & Alisauskas, 2000). Low ambient temperatures may increase egg cooling rates, which may make high incubation stability essential for embryo development and female energy conservation. High incubation stability is needed to provide a more stable thermal environment for developing embryos. In addition, if food is scarce during incubation, lizards living in colder climates may heavily rely on their stored reserves. However, the tropical lizard *E. longicaudata* that lives on Orchid Island and lays eggs in concrete holes might experience more consistent and warmer temperatures than other lizards living in colder environments. As a result, they might not experience a significant loss of body mass during incubation compared to other species because less heat is needed to maintain egg temperature in warmer weather, and also because endogenous reserves are less depleted than in other lizards.

6.5.3.1 Recapture Rates, the Timing of the Production of the Second Clutch, and the Relationship Between Maternal Care and Fecundity

The assumption that caring for eggs or offspring reduces survival or fecundity or delays second clutch production is a major component of most functional arguments for the cost of parental care (Clutton-Brock, 1991). Surprisingly, there is little reliable data on how much it costs to reproduce in order to track changes in parental survival or reproductive success, and the validity of the data used to measure these costs is a contentious issue (Reznick, 1985; Partridge & Harvey, 1988; Clutton-Brock, 1991). If there is any proof, it comes from either mammals or invertebrates. For instance, numerous studies have demonstrated a negative correlation between gamete production and offspring rearing and parental longevity or litter size (Eisenberg, 1981; Bell, 1984). Due to the rarity of parental care in such ectotherms, few attempts have been made to calculate the fecundity or delayed cost of a second reptile litter. Long-term parental care and short-term parental care did not significantly affect the reproductive output or the time it took to produce second clutches in the current study of long-tailed skinks, and these findings did not seem to follow the same pattern as in the other animals mentioned above. This pattern contrasts with a variety of egg-carrying reptiles, which have low survival rates due to poor mobility (Shine, 1988) and are therefore more vulnerable to predators. Pregnant females lose about 25% of their non-pregnant running speed. When compared to the non-breeding season, parental mortality is higher in many birds (Lack, 1966).

Because parents who provide long-term care devote too much time and energy to the current clutch, which in turn may cause them to consume less energy and delay the development of the next clutch, differences in parental care periods are occasionally linked to delayed production of a second clutch of eggs. A slower growth rate, on the other hand, might prevent future fertility. The animal will have less reproductive success in either scenario (Shine, 1988). Several egg-laying species have reported reduced feeding, or even complete cessation of feeding (Fitch, 1970; Neill, 1971; Harlow & Grigg, 1984). According to the above observations, fasting while providing for hatchlings causes specific anorexia rather than a lack of encounters with suitable prey. Many of these reports on nest attendance discuss the eggs' and females' persistence.

However, after the first week of nonstop egg guarding, long-tailed skink females frequently leave the eggs to go foraging. In general, the gain in mass during the subsequent foraging period more than makes up for the loss of body mass during the initial week of egg guarding.

6.5.3.2 Egg-Guarding Time

In environments where they are exposed to eggs, long-tailed skinks have an important job to do: guard against potential natural enemies (*O. formosanus*). In order to protect their eggs from snakes that eat reptiles, females guard their eggs in concrete holes for at least a week; if they leave early, their eggs may be eaten by snakes quickly, as the odor of exposed eggs quickly attracts snakes (Huang, 2006c). Additionally, odors can spread much faster and farther in a concrete hole than they can in a nest that is buried. For this reason, female long-tailed skinks remain with them for at least a week following egg laying. However, some long-tailed skinks begin to leave the nest after this point, possibly due to body condition or because the eggs' odor is no longer as potent an attractant as it was in the first few days after laying. Despite the advantages of egg guarding for successful hatching, there may also be high costs due to a decline in future fecundity. I previously stated that at least some long-tailed skink females lay two clutches of eggs during the reproductive season (Huang, 2004). As a result, females who abandon their eggs may be more likely to lay another clutch. Because of the high predation pressure during the first week of hatching and possibly because the stronger odor of fresh eggs makes them easier to find by egg-eating reptile snakes, it appears that only skink females that abandon their eggs a few days before hatching have the benefit of doing so (in terms of future fecundity).

This companion behavior is crucial in terrestrial systems because it indirectly induces the proximity of predator and prey (and, consequently, potential egg odor transmission). Previous studies of predator-prey interactions have demonstrated that prey animals may display a variety of reactions after being exposed to predators, including changes in their life history, behavior, or morphology (Lima & Dill, 1990; Sih & Moore, 1994; Murray & Jenkinst, 1999). Anti-predator behavior is often initiated by direct contact between prey animals and predators, through less direct contact with predator chemical cues, or through prey avoidance (Lima & Dill, 1990; Murray & Jenkinst, 1999; Lopez & Martin, 2001), and detection of chemical cues by snake predators has been extensively studied in snakes and lizards (Cooper, 1994). The detection of chemical cues by snake predators has been extensively studied in snakes and lizards (Greene, 1988; Cooper, 1994), but little research has been conducted on indirect contacts, such as the lizard egg odor and snakes hypothesized in this study.

Since long-tailed skinks exhibit parental care, do they recognize the eggs themselves or the nest? If lizards from other regions are swapped, will they exhibit the same behavior? If snakes frequently visit the nest, how will the lizard mothers behave? The next chapter will discuss this in detail.

References

Annett, C. A., Pierotti, R., & Baylis, J. R. (1999). Male and female parental roles in the monogamous cichlid, *Tilapia mariae*, introduced in Florida. *Environmental Biology of Fishes, 54*, 283–293.

Archer, J. (1988). *The behavioural biology of aggression*. Cambridge University Press.

Aubret, F., Bonnet, X., Shine, R., & Maumelat, S. (2005). Energy expenditure for parental care may be trivial for brooding pythons, *Python regius*. *Animal Behaviour, 69*, 1043–1053.

Ball, S. L., & Baker, R. L. (1996). Predator-induced life history changes: Antopredator behavior costs or facultative life history shifts. *Ecology, 77*, 1116–1124.

Bell, G. (1984). Measuring the cost of reproduction. II. The correlation structure of the life tables of five freshwater invertebrates. *Evolution, 38*, 314–326.

Besnard, A., Gimenez, O., & Jean-dominique, L. (2002). A model for the evolution of creching behaviour in gulls. *Evolutionary Ecology, 16*, 489–503.

Bolduc, F., & Guillemette, M. (2003). Incubation constancy and mass loss in the common eider *Somateria mollissima*. *Ibis, 145*, 329–332.

Brown, J. L., Morales, V., & Summers, K. (2010). A key ecological factor drove the evolution of biparental care and monogamy in an amphibian. *The American Naturalist, 175*, 436–446.

Bull, J. J., & Shine, R. (1979). Iteroparous animals that skip opportunities for reproduction. *The American Naturalist, 114*, 296–316.

Burley, N. T., & Johnson, K. (2002). The evolution of avian parental care. *Philosophical Transactions of the Royal Society of London Series B: Biological Sciences, 357*, 241–250.

Burrowes, P. A. (2000). Parental care and sexual selection in the Puerto Rican cave-dwelling frog, *Eleutherodactylus cooki*. *Herpetologica, 56*, 375–386.

Carrillo, J., & Aparicio, J. M. (2001). Nest defence behaviour of the eurasian kestrel (*Falco tinnunculus*) against human predators. *Ethology, 107*, 865–875.

Chapple, D. G. (2003). Ecology, life history, and behavior in the Australian scincid genus *Egernia*, with comments on the evolution of complex sociality in lizards. *Herpetologica Monographs, 17*, 145–180.

Chivers, D. P., Kiesecker, J. M., Marco, A., DeVito, J., Anderson, M. T., & Blaustein, A. R. (2001). Predator-induced life history changes in amphibians: Egg predation induces hatching. *Oikos, 92*, 135–142.

Cichon, M. (2001). Body-mass changes in female collared fly catchers: State-dependent strategy. *The Auk, 118*, 550–552.

Clutton-Brock, T. H. (1991). *The evolution of parental care*. Princeton University Press.

Coleman, K., Rothfuss, L. A., Ota, H., & Kardong, K. V. (1993). Kinematics of egg-eating in the specialized Taiwan snake, *Oligodon formosanus* (Colubridae). *Journal of Herpetology, 27*, 320–327.

Cook, C. L., Ferguson, J. W. H., & Telford, S. R. (2001). Adaptive male parental care in the giant bullfrog, *Pyxicephalus adspersus*. *Journal of Herpetology, 35*, 310–315.

Cooper, J. W. E. (1994). Chemical discrimination by tongue-flicking in lizards: A review with hypotheses on its origin and its ecological and phylogenetic relationships. *Journal of Chemical Ecology, 20*, 439–487.

Crowl, T. A., & Covich, A. P. (1990). Predator-induced life-history shifts in a freshwater snail. *Science, 247*, 949–951.

Daan, S., Deerenberg, C., & Dijkstra, C. (1996). Increased daily work precipitates natural death in the kestrel. *Journal of Animal Ecology, 65*, 539–544.

Dearborn, D. C. (2001). Body condition and retaliation in the parental effort decisions of incubating great frigatebirds (Fregata minor). *Behavioral Ecology, 12*, 200–206.

Dietz, D. C., & Jackson, D. R. (1979). Use of American alligator nests by nesting turtles. *Journal of Herpetology, 23*, 510–512.

Eisenberg, J. F. (1981). *The mammalian radiations : An analysis of trends in evolution, adaptation, and behavior*. University of Chicago Press.

Farmer, C. G. (2000). Parental care: The key to understanding endothermy and other convergent features in birds and mammals. *The American Naturalist, 155*, 326–334.

Figler, M. H., Blank, G. S., & Peeke, H. V. S. (2001). Maternal territoriality as an offspring defense strategy in red swamp crayfish (*Procambarus clarkii*, Girard). *Aggressive Behavior, 27*, 391–403.

Fitch, H. S. (1970). *Reproductive cycles in lizards and snakes*. University of Kansas.

Fraipont, M. D., Clobert, J., & Barbault, R. (1996). The evolution of oviparity with egg guarding and viviparity in lizards and snakes: A phylogenetic analysis. *Evolution, 50*, 391–400.

Gilbert, J. D. J., & Manica, A. (2015). The evolution of parental care in insects: A test of current hypotheses. *Evolution, 69*, 1255–1270.

Giordano, A. L., Siegel, H. I., & Rosenblatt, J. S. (1984). Effects of mother-litter separation and Reunion on maternal aggression and pup mortality. *Physiology and Behavior, 33*, 903–906.

Girondor, M., Tucker, A. D., Rivalan, P., Godfrey, M. H., & Chevalier, J. (2002). Density-dependent nest destruction and population fluctuations of Guianan leatherback turtles. *Animal Conservation, 5*, 75–84.

Gordon, D. M., & Kulig, A. W. (1996). Founding, foraging, and fighting: Colony size and the spatial distribution of harvester and nests. *Ecology, 77*, 2393–2409.

Gorman, H. E., & Nager, R. G. (2003). State-dependent incubation behaviour in the zebra finch. *Animal Behaviour, 65*, 745–754.

Greene, H. W. (1988). Antipredator mechanisms in reptiles. In C. Gans & R. B. Huey (Eds.), *Biology of the reptilia. Volume 16. Ecology B. Defense and life history* (pp. 1–152). Alan R. Liss.

Greene, H. W., May, P. G., Hardy, D. L., Sr., Sciturro, J. M., & Farrell, T. M. (2002). Parental behavior by vipers. In G. W. Schuett, M. Hoggren, M. E. Douglas, & H. W. Greene (Eds.), *Biology of the vipers* (pp. 179–205). Eagle Mountain Publications.

Gross, M. R. (2005). The evolution of parental care. *The Quarterly Review of Biology, 80*, 37–46.

Gross, M. R., & Sargent, R. C. (1985). The evolution of male and female parental care in fishes. *American Zoologist, 25*, 807–822.

Hansen, B. T., & Slagsvold, T. (2003). Rival imprinting: Interspecifically cross-fostered tits defend their territories against heterospecific intruders. *Animal Behaviour, 65*, 1117–1123.

Harlow, P., & Grigg, G. C. (1984). Shivering thermogenesis in a brooding diamond python, *Python spilotes spilotes*. *Copeia, 1984*, 959–965.

Helfield, J. M., & Naiman, R. J. (2006). Keystone interactions: Salmon and bear in riparian forest of Alaska. *Ecosytems, 9*, 167–180.

Hirsch, S. M., & Bolles, R. C. (1980). On the ability of prey to recognize predators. *Zeitschrift für Tierpsychologie, 54*, 71–84.

Hocking, M. D., & Reynolds, J. D. (2011). Impacts of salmon on riparian plant diversity. *Science, 331*, 1609–1612.

Huang, W. S. (1994). Report on egg clutch size of the long-tailed skink, *Mabuya longicaudata* from Taiwan. *Journal of taiwan museum, 47*, 45–47.

Huang, W. S. (2004). *Reptile ecology and the evolution of parental care on a tropical Asian Island*. PH. D. Cornell University.

Huang, W. S. (2006a). Ecological characteristics of the skink, *Mabuya longicaudata*, on a tropical east Asian Island. *Copeia, 2006*, 293–300.

Huang, W. S. (2006b). Ecology and reproductive patterns of the grass lizard, *Takydromus sauteri*, in a tropical rain forest of an East Asian Island. *Journal of Herpetology, 40*, 267–273.

Huang, W. S. (2006c). Parental care in the long-tailed skink, *Mabuya longicaudata*, on a tropical Asian Island. *Animal Behaviour, 72*, 791–795.

Huang, W. S. (2007a). Costs of egg caring in the skink, *Mabuya longicaudata*. *Ecological Research, 22*, 659–664.

Huang, W. S. (2007b). Ecology and reproductive patterns of the agamid lizard, *Japalura swinhonis*, on an East Asian Island, with comments on the small clutch sizes of Island lizrds. *Zoological Science, 24*, 181–188.

Huang, W. S. (2010). Foraging behaviors of two sympatric ant species in response to lizard eggs. *Zoology, 113*, 85–90.

Huang, W. S., & Pike, D. A. (2011). Climate change impacts on fitness depend on nesting habitat. *Functional Ecology, 25*, 1125–1136.

Huang, W. S., & Pike, D. A. (2013). Testing cost-benefit models of parental care evolution using lizard populations differing in the expression of care. *PLoS One, 8*, e54065.

Huang, W. S., & Wang, H. Y. (2009). Predation risks and anti-predation parental care behavior: An experimental study in a tropical skink. *Ethology, 115*, 273–279.

Huang, W. S., Greene, H. W., Chang, T. J., & Shine, R. (2011). Territorial behavior in Taiwanese kukrisnakes *(Oligodon formosanus)*. *Proceedings of the National Academy of Science, 108*, 7455–7459.

Huang, W. S., Lin, S. M., Dubey, S., & Pike, D. A. (2013). Predation drives interpopulation differences in parental care expression. *Journal of Animal Ecology, 82*, 429–437.
Iverson, J. (1990). Nesting and parental care in the mud turtle, *Kinosternon flavescens*. *Canadian Journal of Zoology, 68*, 230–233.
Kellett, D. K., & Alisauskas, R. T. (2000). Body-mass dynamics of king eiders during incubation. *The Auk, 117*, 812–817.
Klug, H., & Bonsall, M. B. (2009). Life history and the evolution of parental care. *Evolution, 64*, 823–835.
Klug, H., Bonsall, M. B., & Alonzo, S. H. (2013). The origin of parental care in relation to male and female life history. *Ecology and Evolution, 3*, 779–791.
Kudryavtseva, N. N. (2000). An experimental approach to the study of learned aggression. *Aggressive Behavior, 26*, 241–256.
Lack, D. (1947). The significance of clutch size. *Ibis, 89*, 302–352.
Lack, D. (1966). *Population studies of birds*. Clarendon Press.
Levi, T. (2012). Using grizzly bears to assess harvest-ecosystem tradeoffs in salmon fisheries. *PLoS Biology, 10*, e1001303.
Lima, S., & Dill, L. M. (1990). Behavioral decisions made under the risk of predation: A review and prospectus. *Canadian Journal of Zoology, 68*, 619–640.
Lopez, P., & Martin, J. (2001). Chemosensory predator recognition induces specific defensive behaviours in a fossorial amphisbaenian. *Animal Behaviour, 62*, 259–264.
Lundberg, J., & Moberg, F. (2003). Mobile link organisms and ecosystem functioning: Implications for ecosystem resilience and management. *Ecosystems, 6*, 0087–0089.
Maestripieri, D. (1992). Functional aspects of maternal aggression in mammals. *Canadian Journal of Zoology, 70*, 1069–1077.
Manica, A., & Johnstone, R. A. (2004). The evolution of paternal care with overlapping broods. *The American Natualist, 164*, 517–530.
Mappes, J., Marples, N., & Endler, J. A. (2005). The complex business of survival by aposematism. *Trends in Ecology & Evolution, 20*, 598–603.
Martin, T. E. (1992). Interaction of nest predation and food limitation in reproductive strategies. *Current Ornithology, 9*, 163–167.
Mathis, A., & Smith, J. F. (1993). Fathead minnows, *Pimephales promelas*, learn to recognize northern pike, *Esox lucius*, as predators on the basis of chemical stimuli from monnows in the pike's diet. *Animal Behaviour, 46*, 645–656.
Maynard-Smith, J. (1977). Parental investment: A prospective analysis. *Animal Behaviour, 25*, 1–9.
Murray, D. L., & Jenkinst, C. L. (1999). Perceived predation risk as a function of predator dietary cues in terrestrial salamanders. *Animal Behaviour, 57*, 33–39.
Neill, W. T. (1971). *The last of the ruling reptiles: Alligators, crocodiles and their kin*. Columbia University Press.
Noble, G. K., & Mason, E. R. (1933). Experiments on the brooding habits of the lizards *Eumeces* and *Ophisaurus*. *American Museum Novitates, 619*, 1–29.
O'Connor, D., & Shine, R. (2003). Lizards in 'nuclear families': A novel reptilian social system in *Egernia saxatilis* (Scincidae). *Molecular Ecology, 12*, 743–752.
O'Connor, D., & Shine, R. (2004). Parental care protects against infanticide in the lizard *Egernia saxatilis* (Scincidae). *Animal Behaviour, 68*, 1361–1369.
Oliveira, R. F., & Almada, V. C. (1998). Maternal aggression during the mouthbrooding cycle in the cichlid fish, *Orechromis mossambicus*. *Aggressive Behavior, 24*, 187–196.
Olsson, M., & Shine, R. (1998). Timing of parturition as a maternal care tactic in an alpine lizard species. *Evolution, 52*, 1861–1864.
Parmigiani, S., & Brain, P. F. (1983). Effects of residence, aggressive experience and intruder familiarity on attack shown by male mice. *Behavioral Processes, 8*, 45–47.
Partridge, L., & Harvey, P. H. (1988). The ecological context of life history evolution. *Science, 241*, 1449–1455.

Pike, D. A., Clark, R. W., Manica, A., Tseng, H. Y., Hsu, J. Y., & Huang, W. S. (2016). Surf and turf: Predation by egg-eating snakes has led to the evolution of parental care in a terrestrial lizard. *Scientific Reports, 6*, 22207.

Pressley, P. H. (1981). Parental effort and the evolution of nest-guarding tactics in the threespine stickleback, *Gasterosteus aculeatus* L. *Evolution, 35*, 282–295.

Rand, A. S. (1968). A nesting aggregation of iguanas. *Copeia, 1968*, 552–561.

Reguera, P., & Gomendio, M. (1999). Predation costs associated with parental care in the golden egg bug *Phyllomorpha laciniata* (Heteroptera: Coreidae). *Behavioral Ecology, 10*, 541–544.

Reynolds, J. D., & Gross, M. R. (1990). Costs and benefits of female mate choice: Is there a lek paradox? *The American Naturalist, 136*, 230–243.

Reynolds, J. D., Goodwin, N. B., & Freckleton, R. P. (2002). Evolutionary transitions in parental care and live bearing in vertebrates. *Philosophical Transactions of the Royal Society of London, Series B, 357*, 269–281.

Reznick, D. (1985). Costs of reproduction: An evaluation of the critical evidence. *Oikos, 44*, 257–267.

Roff, D. (2002). *Life history evolution*. Sinauer Associates, Inc.

Rosenheim, J. A. (1999). Characterizing the cost of oviposition in insects: A dynamic model. *Evolutionary Ecology, 13*, 141–165.

Sanz, J. J., & Tinbergen, J. M. (1999). Energy expenditure, nestling age, and brood size: An experimental study of parental behavior in the great tit *Parus major*. *Behavioral Ecology, 10*, 598–606.

Schmitz, O. J., Cohon, J. L., Rothley, K. D., & Beckerman, A. P. (1998). Reconciling variability and optimal behaviour using multiple criteria optimization models. *Evolutionary Ecology, 12*, 73–94.

Shine, R. (1988). Parental care in reptiles. In C. Gans & R. B. Huey (Eds.), *Biology of the Reptilia Vol. 16 Ecology B. Defense and life history* (pp. 275–330). Alan R. Liss.

Sih, A., & Moore, R. D. (1994). Delayed hatching of salamander eggs in response to enhanced larval predation risk. *The American Naturalist, 142*, 947–960.

Skelhorn, J., Rowland, H. M., Delf, J., Speed, M. P., & Ruxton, G. D. (2011). Density-dependent predation influences the evolution and behavior of masquerading prey. *Proceedings of the National Academy of Sciences, 108*, 6532–6536.

Somma, L. A. (1990). A categorization and bibliographic survey of parental behavior in lepidosaurian reptiles. *Smithsonian Herpetological Information Service, 81*, 1–53.

Speed, M. P. (1993). When is mimicry good for predators? *Animal Behaviour, 46*, 1246–1248.

Stearns, S. C. (1992). *The evolution of life histories*. Oxford University Press, Oxford, UK.

Stevens, M., & Merilaita, S. (2009). Animal camouflage: Current issues and new perspectives. *Philosophical Transactions of the Royal Society B: Biological Sciences, 364*, 423–427.

Tallamy, D. W. (1984). Insect parental care. *Bioscience, 34*, 20–24.

Tinkle, D. W. (1969). The concept of reproductive effort and its relation to the evolution of life histories of lizards. *The American Naturalist, 103*, 501–516.

Tinkle, D. W., & Gibbons, J. W. (1977). The distribution and evolution of viviparity in reptiles. *Miscellaneous Publications Museum of Zoology University of Michigan, 154*, 1–54.

Tseng, H. Y., & Huang, W. S. (2012). *Sphenomorphus incognitus* (Brown forest skink) parental care. *Herpetological Review, 43*, 141.

Tseng, H.-Y., Lin, C.-P., Hsu, J.-Y., Pike, D. A., & Huang, W.-S. (2014). The functional significance of aposematic signals: Geographic variation in the responses of widespread lizard predators to colourful invertebrate prey. *PLoS One, 9*, e91777.

Webb, J. N., Houston, A. I., McNamara, J. M., & Szekely, T. (1999). Multiple patterns of parental care. *Animal Behaviour, 58*, 983–993.

Wesolowski, T. (1994). On the origin of parental care and the early evolution of male and female parental-roles in birds. *The American Naturalist, 143*, 39–58.

Williams, G. C. (1966). Natural selection, the costs of reproduction, and a refinement of Lack's principle. *The American Naturalist, 100*, 687–690.
Wilson, E. O. (1975). *Sociobiology: The new synthesis*. Belknap Press of Harvard University Press.
Winkler, D. W. (1987). A general model for parental care. *The American Naturalist, 130*, 526–543.
Zink, A. G. (2003). Quantifying the costs and benefits of parental care in female treehoppers. *Behavioral Ecology, 14*, 687–693.

Parental Care and Relative Issues

7

Abstract

The evolution of maternal care in egg-laying terrestrial vertebrates remains unclear, particularly whether it arises from egg recognition or territorial behavior. Most species either lack care or show it consistently, limiting inference. However, in long-tailed skinks (*Eutropis longicaudata*) on Orchid Island, females uniquely exhibit nest guarding, unlike other populations. They remain with eggs during incubation and actively repel egg-eating snakes (*Oligodon formosanus*). Remarkably, females defend the original nesting site even when eggs are removed or replaced, guard any conspecific eggs, and show this behavior before laying. These findings suggest maternal care may originate from territorial instincts rather than egg recognition, with nest guarding representing an early stage in the evolution of parental care.

Keywords

Mechanism of parental care · Egg guarding · Nest defense · Filial cannibalism

7.1 Does Maternal Care Evolve Through Egg Recognition or Through Territorial Orientation?

In light of the fact that long-tailed skinks modify their parental care routine and cut the length of their homecoming due to predators, does a female protect her nest or her eggs? The long-tailed skink is typically thought to be guarding her eggs, but is this really the case?

Are the eggs or nest of a long-tailed skink are they known? In order to determine whether the long-tailed skink would protect the eggs in place or return to the original nest to protect the nest, the mother and some or all of the eggs were moved to the next nest in this chapter and tested with a kukri snake. If the long-tailed skink

W.-S. Huang, *Reptiles in the Changing Climate*, Fascinating Life Sciences,
https://doi.org/10.1007/978-981-92-1887-5_7

defends the eggs already laid, it can be assumed that it will recognize its own eggs and attack the kukri snake. However, if the long-tailed skink always returns to the original nest, whether or not the eggs are present, it can be demonstrated that the long-tailed skink does not recognize its own eggs and is instead defending the nest rather than the eggs. This chapter may be able to shed light on the causes of the evolution of parental care by examining whether the long-tailed skink's parental care behavior is a byproduct of nest protection.

7.1.1 Introduction

Early life mortality risk can be significantly decreased by parental care (Clutton-Brock, 1991; Roff, 2002). Therefore, when eggs or young are exposed to unfavorable environments, high rates of predation, parasitism, or fierce competition from conspecifics, parental care is most developed (Clutton-Brock, 1991; Klug & Bonsall, 2009). The risk of parental predation, the value of the current offspring in comparison to the parent's expected future contribution, and the anticipated improvement in offspring welfare as a result of such parental expenditures should all be taken into consideration when determining the level of parental investment in the protection of offspring (Pressley, 1981; Clutton-Brock, 1991). Thus, based on parental risk and expected benefits to offspring, natural selection should favor the best level of care (Shine, 1988; Clutton-Brock, 1991).

Different vertebrate lineages have very different parenting styles. The majority of research has concentrated on birds, mammals, and fish, which exhibit the most advanced types of parental care (Clutton-Brock, 1991). Nest guarding, which has been demonstrated for a variety of taxa, including amphibians (Salthe & Mechan, 1974), birds (Lack, 1968), fishes (Perrone & Zaret, 1979; Gross, 2005), invertebrates (Hinton, 1981), and Reptilia (Shine, 1988), is thought to be one of the most fundamental forms of parental care. To associate parents with their eggs and thereby facilitate the evolution of parental care in oviparous species, it may be necessary to be able to distinguish one's own eggs from those of conspecifics. However, the fact that many species—even those that display more sophisticated parental care—cannot identify their own eggs suggests that there are alternate pathways. For instance, some species will protect a specific nest regardless of the presence of eggs (Rothstein, 1975; Vitt & Cooper Jr., 1989; Peterson, 2000). This happens frequently in bird species, many of which continue to raise heterospecific eggs that parasitic birds have laid in their nests; these species are likely unable to tell their own eggs apart from those of other species (Rothstein, 1975, 1982b). Similarly, female scorpion lizards (*Plestiodon* [formerly *Eumeces*] *laticeps* and *fasciata*) and red-backed salamanders (*Plethodon cinereus*) guard their eggs during incubation, but they are unable to distinguish them from those of other females (Vitt & Cooper Jr., 1989; Peterson, 2000). Contrarily, many species can reliably tell their own eggs apart from hetero- or even conspecific eggs (Forester, 1979; Rothstein, 1982a; Mehlis et al., 2010). Placental (living) species of reptiles frequently use chemical (odor) cues to identify their offspring (e.g., Main & Bull, 1996; Lena & de Fraipont, 1998; Head et al., 2008), but oviparous species may not have this capability (Vitt & Cooper Jr., 1989). These two behavioral tactics—egg defense, which is based on direct egg recognition, and nest site defense, which is based on territorial behavior—are obviously different from one another, but they both support the evolution of more sophisticated forms of care. Birds (Rothstein, 1975; Wesolowski, 1994), fish (Perrone & Zaret, 1979; Gross, 2005), and salamanders (Forester et al., 1983; Peterson, 2000) have been the subjects of studies that distinguish between these two pathways, and such studies are uncommon in reptiles. Collectively, these studies show that both species are significantly impacted. Together, these studies show evidence in favor of both mechanisms, and even closely related species have different capacities for identifying individual eggs or even for merely defending a nest. Because parental care is a feature that has been highly conserved throughout evolution, this taxonomic focus restricts the generalizability of conclusions about its origin and evolution (Clutton-Brock, 1991). This makes it difficult to determine how parental care evolved from ancestral non-caring states.

Because most species of Squamata reptiles (lizards and snakes) lack parental care, they make excellent study subjects for this issue. For instance, only about 1.3% of lizard species and 2.8% of snake species stay in their nests during incubation (Shine, 1988). Even less frequently seen is nest guarding behavior; *Eutropis longicaudata* is the only other species known to engage in this behavior, despite the fact that most lizards protect their nests from conspecifics (Wiewandt, 1982). As a result, parental reptiles are the exception rather than the rule. As of now, the majority of research on offspring recognition in reptiles has focused on viviparous species and assessed whether mothers and offspring can be distinguished from one another (e.g., Main & Bull, 1996; Lena & de Fraipont, 1998; Head et al., 2008). However, studies on parental egg recognition have only been conducted on two closely related species of scorpion lizards that can distinguish conspecific eggs from eggs of other genera but not from their own eggs (Vitt & Cooper Jr., 1989).

The cause of parental care evolution in the majority of egg-laying terrestrial vertebrates is still unknown. Does it occur when egg recognition makes it easier to identify by females, or does it develop as an indirect result of parental upkeep of a nest's immediate surroundings? In both situations, the expenses of parental care must outweigh the advantages; however, in this instance, our attention is on the immediate mechanisms that encourage close contact between parents and offspring. Mothers should only care for their own eggs if egg recognition makes parental care behavior more likely; however, if female territorial behavior makes parental care behavior more likely, then females should only protect the site of initial egg production.

To find out which of these factors influences maternal care, we conducted field experiments in a particular population of long-tailed skinks (*Eutropis longicaudata*). Long-tailed skinks lay their eggs under rocks or buried in soil throughout the majority of their geographic range, and then the females leave the nest without continuing to take care of the young [(Huang, 1994); Fig. 7.1]. However, females congregate in large numbers to lay eggs in drainage holes that run through concrete retaining walls on Orchid Island, 60 km off the southeast coast of Taiwan. The

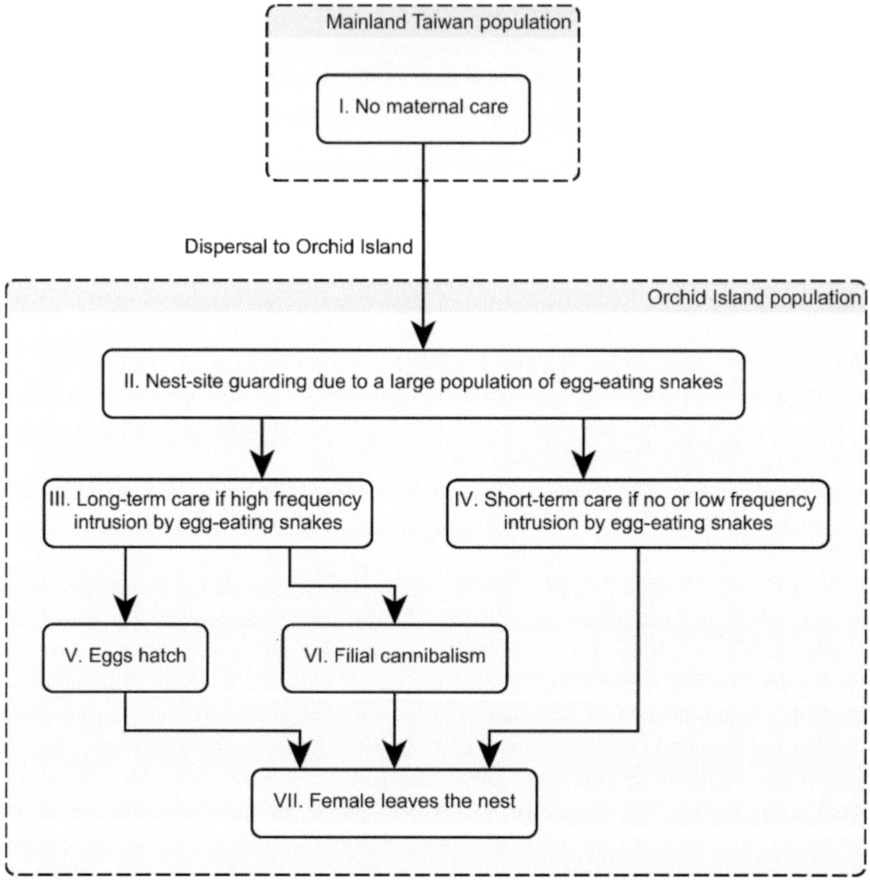

Fig. 7.1 Probable evolutionary pathway of parental care in long-tailed skinks, *Eutropis longicaudata*, showing differences in the behavior of nesting females on mainland Taiwan vs those on Orchid Island, off the southeastern coast of mainland Taiwan (after Huang, 2006a, b, c, 2007a, b, 2008a, b; Huang & Wang, 2009). (Redrawn from Huang and Pike (2011a, b), Journal of Evolutionary Biology)

distinctive feature of females that nest in these altered habitats is that they protect their eggs from egg-eating snakes (*Oligodon formosanus*) while they are incubating (Fig. 7.1). On other islands and on Taiwan's mainland, female nesters use these drainage holes (built as early as 1989), but lizards that nest there immediately abandon their eggs after they are laid. This stands in stark contrast to other parental care species, where a species may express care either constantly or infrequently [although the nature and length of care may vary(Clutton-Brock, 1991)]. The number of snakes that prey on eggs is decreased by nest-guarding females, who are resistant to these snakes' predatory instincts (Huang, 2006c, 2007a). Females will prolong nest-guarding when snake predation attempts rise (Huang & Wang, 2009) and in extreme circumstances, they may even consume their own clutch of eggs ["filial cannibalism"(Huang, 2008a); see other chapter]. Filial cannibalism illustrates a trade-off between preserving high-risk eggs and consuming them to gain energy for upcoming reproductive endeavors (Huang, 2008a, b). Females leave the nest to go short-term foraging while providing maternal care in order to regain qualities lost during incubation, which lowers the cost of this behavior (Huang, 2007a). The high nest density and physical similarity of the nests (a string of nearly identical holes through a concrete wall) imply that females should be able to recognize eggs and/or nests in some way. Otherwise, after foraging, females might unintentionally return to the wrong nest, wasting time and energy without the benefit of offspring survival (Clutton-Brock, 1991; Peterson, 2000; Huang, 2006a, b, c). Direct egg identification is more likely than nest identification, according to the nearly identically high density of potential nests (Peterson, 2000). However, some species are unable to recognize their own eggs, and territoriality near nest sites may be a factor in these species' ability to evolve parental care. In order to identify the causal mechanisms that facilitate the initial evolution of parental care in this distinct population of a widespread species, we looked into which of these mechanisms the long-tailed skink possesses.

7.1.2 Does the Location of the Nest Affect Nest Guarding?

No matter how many eggs were moved to a nearby nest with the females, they all managed to flee the invading snakes. Within 2 h of the displacement test, all escaped females made their way back to the original nest site. No matter how many eggs were present, however, when the females returned to the original nest site, they attacked the invading snake preferentially. We compare female responses to invading snakes in both new and original nests because responses from females in both new and original nests were independent of the number of eggs in the nest. Female behavior was strongly influenced by nest site; at the original nest site, females attacked snakes 83% of the time, but never at the new nest site.

Females attacked invasive snakes preferentially in our disturbance control trials. These attack rates were significantly higher than those of undisturbed females before testing. We predicted that when we relocated females to a new nest site, which involved disturbing females, they would be more likely to defend themselves against invading snakes because this result suggests that disturbed females are more likely to attack invading snakes. Contrarily, we discovered that females who were relocated to a new nest site had significantly lower attack rates than those in our disturbed control group. Additionally, after returning to their original nests, displaced females showed attack rates comparable to those of females in the disturbance control group. Therefore, the subsequent response of females to egg-eating snakes in the original nest was not significantly affected by our displacement test ($P > 0.05$).

7.1.3 Do Females Guard Conspecific Eggs?

The likelihood of females attacking snakes in the presence of their own eggs versus conspecific eggs was not significantly different (Fig. 7.2a); females attacked snakes in 13 of 14 trials when their own eggs were present, whereas females attacked snakes in 12 of 14 trials when conspecific eggs were present (Fig. 7.2a). As a result, once a female selects a nest, she guards any eggs there against snakes, whether they are her own or conspecific.

7.1.4 Do Females Defend Their Nests from Egg Predators Before Laying Eggs?

According to Huang (Huang, 2006a, b, c), both gravid and non-gravid females had similar body measurements. Despite this, gravid females were noticeably more likely than non-gravid females to attack intruders (Fisher's exact test, $p < 0.00001$; Fig. 7.2b).

7.1.5 Discussion

Orchid Island is the only place where long-tailed skinks display nest-guarding behavior (Fig. 7.1), possibly as a result of the relatively high density of egg predators there (Huang, 2004). Female long-tailed skinks defend incubating eggs from predators by attacking egg-eating snakes, which raises the rate of egg hatching (Huang, 2006a, b, c). The goal of the current study was to comprehend the processes

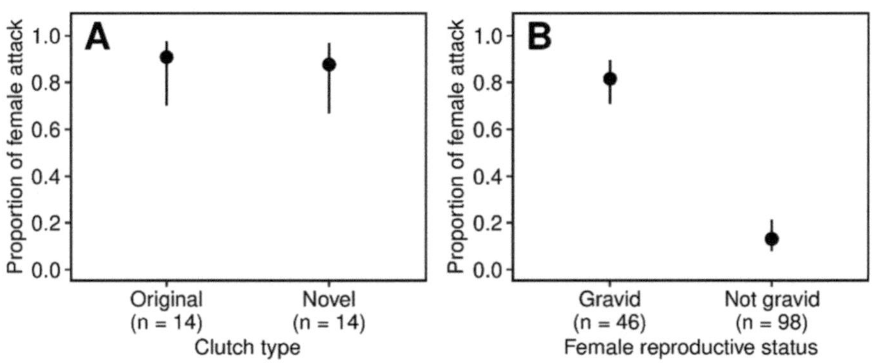

Fig. 7.2 Proportion of attacks by maternal-caring long-tailed skinks, *Eutropis longicaudata*, toward intruding egg-eating snakes (*Oligodon formosanus*) in (**a**) original nests containing the mothers' own eggs vs those of a conspecific; and (**b**) gravid vs non-gravid females. Females generally attacked snakes when the nest-site contained eggs, or when gravid, but escaped from predators when not gravid. Note that in a separate experiment some females continued to defend the original nest-site in which they laid eggs, even after we removed all of the eggs (see Results for more details). (Redrawn from Huang and Pike (2011a, b), Journal of Evolutionary Biology)

that led to the evolution of this nest protection behavior. We discovered that nest-guarding females (1) exhibited anti-predator behavior only in the nest where the eggs were first laid, regardless of how many eggs were still in the nest (or even if there were none), (2) did not defend their own eggs in nests other than the one where they were laid, (3) protected any eggs in the original nest (even if they were conspecific), and (4) displayed anti-predator behavior toward egg-eating snakes before the eggs were a hatchling.

These findings collectively imply that long-tailed skinks are unable to recognize their own eggs (as evidenced by an inability to protect their own eggs in an unfamiliar nest and a tendency to protect conspecific eggs placed at their original nest site). Our earlier research, in which we successfully removed chemical traces from nest sites but left the eggs and associated chemical traces in the nest, showed that long-tailed skink females were unable to successfully return home (Huang & Pike, 2011b). When chemical cues were taken away from the eggs but left at the nest site, however, the females quickly went back to the nest (Huang & Pike, 2011b). An alternative explanation for our findings is that, regardless of the type of eggs present, in an unfamiliar environment (such as when moving to a new nest site), the costs of anti-predator behavior outweigh the advantages of protecting the eggs. The low risk that lizards face when attacking snakes that only consumed eggs, and the lack of egg discrimination within the female's own nest, however, contradict this hypothesis. Under a variety of circumstances, long-tailed skinks protect their original nest sites from invading predators (e.g., after being disturbed, when their own eggs or eggs of the same species are present, when no eggs are present, and during gestation). Therefore, in this species, maternal care appears to serve a purpose other than protecting the eggs directly—namely, direct protection of the nest. Finally, rather than facilitating direct recognition of eggs through more complex mechanisms, the new evolution of parental care in Orchid Island long-tailed skinks may simply be a byproduct of nest site guarding (a form of territoriality) (Clutton-Brock, 1991). This discovery is consistent with those of other living lizard species that guard their young after birth (O'Connor and Shine 2004; Sinn et al., 2008; While et al., 2009) as well as oviparous species that are unable to tell their own eggs from those of conspecifics (Vitt & Cooper Jr., 1989).

Parental care animals ought to have ways to locate their eggs if they become separated from them (Peterson, 2000). Nest-guarding long-tailed skinks cannot recognize their eggs despite foraging away from the nest during prolonged parental care (up to 35 days; Huang, 2007a, b). According to Peterson (2000), direct egg recognition mechanisms are preferred if nests are clustered or challenging to identify using environmental cues, whereas indirect nest recognition mechanisms are preferred if there is little chance of confusing a neighboring nest with one's own. For instance, the red-backed salamander (*Plethodon cinereus*) can relocate experiment-placed nests but is unable to distinguish its own eggs from those of unfamiliar females (Peterson, 2000). Although it might be uncommon in nature, this provides proof that nest sites can be identified indirectly rather than directly by looking at the eggs (Vitt & Cooper Jr., 1989). Our research disproves this theory; nesting females gather near concrete walls to build their nests, and multiple nests are frequently found close together (50–300 cm away). In this case, direct nest identification should prevail over indirect nest identification (Peterson, 2000). Therefore, it is unlikely that more sophisticated forms of parental care developed solely as a result of direct egg recognition; however, in some instances, nest site guarding behavior (without direct egg recognition) may have served as a crucial stepping stone for the development of more sophisticated egg recognition. This theory is supported by studies of birds; many species can be categorized as accepting or rejecting new eggs that are placed in the nest, such as when a parasitic bird lays eggs in the nest of another species [reviewed in (Rothstein, 1982b)]. Although some species that accept parasitic eggs will occasionally remove them from the nest, this appears to be based on other cues, such as when the eggs are damaged or broken (Rothstein, 1982b). Species that reject parasitic eggs have a true egg recognition system.

One of the most unexpected results of our study was that both females with eggs to protect and females without any eggs to protect (because we had moved them to a new nest site) would attack invading egg predators. Only females engage in this territorial behavior, and it only happens when they are feeding or caring for eggs (Huang, 2006a, b, c; this study). These findings suggest that nest-guarding behavior develops through physiological changes during pregnancy and provide strong evidence that a form of territoriality, rather than direct egg recognition, drives this behavior [for additional examples of changes in anti-predator behavior during egg production see (Cooper & Burghardt, 1990, Downes & Bauwens, 2002)]. Anti-predator behavior can be a general reaction to any potential predator, even those who do not actually consume their prey and do not pose a threat (Blumstein, 2006; Ferrari et al., 2008; Webb et al. 2009, 2010). Long-tailed skinks, however, only protect their nests from snakes that eat eggs, while other reptile species go unnoticed (females do, however, escape from lizard-eating snakes, which are a direct threat to mothers; Huang, 2006a, b, c). As a result, this behavior is specific to females during the reproductive cycle as well as the main egg predator (Huang, 2006a, b, c).

Given this, why do some long-tailed skinks manage to flee from snake predators even though the snakes don't actually threaten the females? One explanation is that females may recognize which predators pose a threat to them due to previous encounters with them (Ferrari et al., 2005). Because of the potential consequences of attacking the wrong predator (for an extreme case-female die), it may be advantageous to be more cautious of unfamiliar predators in this situation. The possibility that individual differences exist in the relative costs and benefits of providing maternal care is an equally tenable hypothesis. For instance, because they have less time to forage while protecting their eggs, females who are less physically fit may have higher costs. For some females who routinely abandon their nests after predator invasions, the cost of long-term maternal care may be higher in this situation.

The Orchid Island long-tailed skink's evolutionary pathways are adaptable and specifically tuned to the threat posed by single-egg predators (Huang, 2006c, 2007a). This conclusion is not necessarily in conflict with the plasticity of parental care time during predator encounters (Huang & Wang, 2009). It's possible that in this particular species group, maternal care first evolved because of territoriality around the nest, and that the length of care is based on a plastic response to predator abundance (Huang & Wang, 2009). Nest guarding is a simple, early stage of parental care that paves the way for the development of more advanced parental care (Clutton-Brock, 1991). Once parental care has developed, additional ecological factors may have an impact on the type or length of care, which is consistent with closely related species displaying various parental care patterns (Clutton-Brock, 1991). Our study offers novel insights into the origin and evolution of parental care in terrestrial vertebrates because nest-guarding is one of the earliest types of parental care, is uncommon in squamate reptiles, and has only recently evolved in one population of long-tailed skinks.

7.2 Predation Drives Interpopulation Differences in Parental Care Expression

Understanding why the biological world is so diverse, including the evolution of behavioral ecology, is one of the central issues in evolutionary biology. The majority of research, however, has been done primarily in the lab to look into the effects and immediate causes of behavioral ecology because the ecological world is so complex. The best place to study the evolution of behavioral ecology in nature, on the other hand, is on the small islands, which have a relatively simple ecological system. This chapter compares the clustering relationships between populations from Pingtung, Taiwan (without parental care behavior) and populations from Orchid Island (with parental care behavior) using proximate mechanisms to find out why Taiwanese populations lack parental care behavior and why populations from Orchid Island do. Are parental care behaviors influenced by adult-based predation, egg-based predation, or both in the evolution of these behaviors? Regarding the remote mechanism, we would like to compare the DNA variations between the populations of long-tailed skinks from Orchid Island and those from Kaohsiung, Pingtung, Green Island (the three above mentioned locations do not exhibit parental care behavior). The kukri snake may consume the eggs even if they are laid by individuals without parental care if they are dispersed to Orchid Island, and both populations will not pass on their genes. If there is a significant DNA difference, it can be predicted that it is related to things like island isolation mechanisms, behavior, etc. If there are no differences, it is evident that the Orchid Island long-tailed skink population's parental care behavior evolved adaptively through vigorous natural selection.

The expression of parental care within a species is frequently a mandatory (evolutionary) trait, even though evolutionary theory predicts that widespread species should vary in their expression of parental care according to local selection pressures. Within its vast geographic range, the lizard *Eutropis longicaudata* only exhibits maternal care in one population, but it is not clear why this pattern exists. We tested whether this intraspecific variation is a fixed trait within populations and whether predator abundance explains this perplexing pattern using interchange and predator exclusion experiments. According to their native populations, female wild-caught lizards that underwent translocation always guarded or laid eggs. Contrarily, most lizards that are raised in communal gardens and then released as adults adopt a maternal care strategy that accepts the colony, even if their parents come from a colony that doesn't practice maternal care. Egg predation represented a significant fitness cost in populations where female lizards displayed egg-care behaviors, but egg care outweighed this potential cost by increasing hatching success. These findings imply that predators can act as a motivating factor for parental care expression in circumstances where it is not typically present and that local selection pressures are sufficient to cause behavioral divergence regarding parental care expression.

Animals have developed a wide range of parental care strategies to increase the survival of their offspring, from guarding the nest site. to engaging in more complex feeding behaviors (Clutton-Brock, 1991; Wesolowski, 1994; Klug & Bonsall, 2009). These techniques have independently evolved numerous times in various animal lineages, offering compelling evidence of their effectiveness as evolutionary techniques to improve fitness (Tallamy, 1984; Shine, 1988; Clutton-Brock, 1991; Gross, 2005). When the fitness benefits of protecting offspring outweigh the costs of providing care, parental care should evolve from the ancestral state of nonparental care, according to evolutionary theory (Tallamy, 1984; Clutton-Brock, 1991; Wesolowski, 1994). Accordingly, general life history environments favoring parental care consist of low egg or juvenile survival without care, relatively long egg incubation periods, and relatively short juvenile stages (Clutton-Brock, 1991; Klug & Bonsall, 2009), whereas general ecological environments favoring parental care consist of food availability and predation risk (Lack, 1968; Tallamy & Wood, 1986). In general, when there is a high risk of predation on eggs or offspring, species are more likely to evolve parental care (Lack, 1968; Tallamy & Wood, 1986).

Any action parents take to improve their offspring's chances of survival after spawning or delivery is referred to as parental care (Shine, 1988; Greene et al., 2002). For example, in a single species group of cichlids (*Sarotherodon galilaeus* Linnaeus 1758) and birds (*Remiz pendulinus* Linnaeus 1758), parental care can be provided by both parents or by either sex alone (Persson & Öhrström, 1989 (Persson & Öhrström, 1989, Balshine-Earn & Earn, 1998). Additionally, the type and degree of parental care behaviors can change depending on the development of succeeding generations [reviewed in (Magnhagen, 1992)] or on the environment in a given area (Dunn & Robertson, 1992; Wheelwright et al., 1992; Székely & Cuthill, 1999). Consequently, many widely distributed species exhibit variation in the timeframe, nature, or level of parental care within and between populations (e.g., Dunn & Robertson, 1992; Wheelwright et al., 1992; Székely & Cuthill, 1999; Webb et al., 1999); however, nearly all species that exhibit parental care are known to do so within their geographic range. In contrast, examples of expressing parental care in some populations but not others are extremely rare; known examples include frogs (Martins et al., 1999) and, if maternal care is thought to continue throughout pregnancy, lizard species with both oviparous and viviparous populations (e.g., Heulin et al., 1999). Other uncommon situations include species that produce young abandoning their nests, which almost always results in the death of the young (e.g., Burley & Johnson, 2002; Reynolds et al., 2002). As a result, parental care is almost always present throughout the range of the majority of species. Our understanding of the selection pressures that lead to the evolution of parental care is constrained by this evolutionary pattern (Clutton-Brock, 1991; Wesolowski, 1994; Gross, 2005). This also seems to go against the notion that widespread species should express parental care based on predation risk in order to maximize offspring survival because they are subject to various selection pressures across their geographic range. Then why don't the majority of widely dispersed species practice progressive parental care, where adults modify their level of caregiving in response to the risk of predation on their offspring? Only species where offspring can survive without either parent are affected by this situation (e.g., many reptiles, amphibians, and invertebrates; Shine, 1988).

Long-tailed skinks (*Eutropis longicaudata* Hallowell 1857) are widely distributed throughout Southeast Asia, with only one population on Orchid Island, located on the southeastern coast of Taiwan, expressing parental care (Huang, 2006c). In this population, females that lay their eggs inside concrete retaining walls stay in the nest throughout incubation and actively fend off egg-eating snakes (*Oligodon formosanus* Günther 1872) from coming inside and consuming the eggs (Huang, 2006c). The only time anti-snake behavior occurs is before egg laying (Huang & Wang, 2009). However, females that construct nests nearby in natural habitats (under rocks and logs) never protect their eggs while they are being incubated (Huang & Pike, 2011a). By exposing their eggs to higher incubation temperatures, which shorten incubation time, increase the proportion of eggs that hatch, and increase the growth and survival rates of offspring, females that nest on retaining walls have an advantage in terms of fitness (Huang & Pike, 2011a). However, this behavior has a cost because, in contrast to eggs buried under cover in their natural habitat, eggs laid within the retaining wall are exposed to predators throughout the incubation process (Huang & Pike, 2011a). Females from 12 other populations of long-tailed skinks (11 from Taiwan and one from Green Island, all nesting within concrete retaining walls) abandoned their eggs right away after oviposition despite the increased risk of predators discovering eggs laid within these retaining walls (Huang, unpublished data). Given that only females nest in retaining walls on Orchid Island, parental care expression is not a fixed interspecific trait (i.e., mandatory; as it appears to be in all other species). Not all females in many other taxa stay with their eggs during the entire incubation period (but they still express initially and thus provide maternal care; Huang & Wang, 2009). When there are few egg-eating snakes, females leave the nest early in the incubation period, but when there are many snakes, females continue to defend the nest until the eggs hatch. This difference is due to the risk of egg predation (Huang & Wang, 2009). This suggests that the development of maternal care has been significantly influenced by the pressure of local predation. In order to directly test whether exposing female lizards to local ecological conditions that they do not typically encounter changes the expression of parental care behaviors with the source population, we can test this hypothesis using a swap experiment.

An alternative explanation for the parental care behavior in long-tailed skinks is that geographic isolation has resulted in significant genetic divergence, with divergent populations now providing maternal care. This suggests that other populations would not express maternal care because they lack the necessary skills to do so, even in an ecological setting where other (genetically distinct) populations do so. Therefore, we would anticipate that the population of long-tailed skinks on Orchid Island represents a very divergent lineage (perhaps even a separate, but morphologically similar species). Even during the low sea level of the last glacial period, Orchid Island was never connected to other islands because the straight line dividing Taiwan and Orchid Island is deep enough (>1 km). As a result, it is unlikely that Orchid Island and other populations share any haplotypes, and it is possible that different genetic lineages exist there. There may be a genetic component to parental care; for instance, mice lacking the fosB gene do not raise their offspring (Brown et al., 1996). Although our knowledge of the genetic mechanisms underlying parental care is limited to mammals (a class that includes almost exclusively species that consistently display parental care), other genes and experiential learning also have a significant impact (Rosenblatt, 1994; Brown et al., 1996). Since parental care originally evolved from ancestors that did not provide care (Tallamy, 1984; Shine, 1988; Clutton-Brock, 1991), all populations of long-tailed skink may have genes that encode expression of parental care, but there may be some ecological triggers for this behavior.

Both ecological factors and genetic history may contribute to the indirect expression of parental care, so these two pathways—local adaptation and genetic differentiation through behavioral plasticity—are not mutually exclusive. We predict that (1) females moving between populations should adapt their behavior to fit that of the receiving population because behavioral plasticity is a feature of parental care and depends on the local environment, and (2) because nest guarding serves to protect eggs from predators (Huang, 2006c), Orchid Island should have significantly more predators than other populations. (3) Populations expressing care should have higher nest survival rates as a result of maternal care. (4) There shouldn't be much of a genetic difference between populations expressing parental care and other populations that don't. If our predictions come true, it will be clear that local selection pressures can affect this behavioral innovation and that predation plays a significant role in the plasticity of parental care expression.

7.2.1 Interchange Experiment

After laying eggs, each wild adult female released from SanDiMen to Orchid Island immediately abandoned her nest (two female lizards each abandoned their nests in 2008, 2009, and 2010; Table 7.1). These were all eaten by snakes that ate eggs. Only two females known to exhibit maternal care were discovered in SanDiMen in 2008, and both of them were protecting their eggs (Table 7.1). We observed a similar pattern when we repeated this interchange with a larger sample size: all encountered nesting female lizards expressed or did not express maternal care, consistent with their native population, rather than changing their behavior to match the receiving population (Table 7.1). Between release sites, there were comparable rates of

Table 7.1 Results from reciprocal translocation experiments, moving lizards from the population expressing maternal care (Orchid Island) to a population not expressing maternal care (SanDiMen, located on the main island of Taiwan) and vice versa. We performed two experiments, the first of which moved gravid wild-caught adult females from one population to the other, and the second of which captured a different set of gravid females from the wild and raised the progeny to maturity in captivity. We ran the first experiment twice (in both 2007 and 2010) due to low recapture rates. The last column summarizes the percentage of females known to have expressed maternal care, given as the percentage of females which laid eggs and expressed care, and then as the percentage of all females observed expressing care (i.e., including females who were not observed to be gravid and thus were not known to lay eggs). We could not assess the reproductive condition (gravid, not gravid) of females that were not recaptured

Population		Number of lizards				
Origin	Recipient	Expressing maternal care	Not expressing maternal care (not gravid)	Not recaptured	Total	Rate of parental care expression (including females which did not lay eggs)
Wild-caught females translocated in 2007 and recaptured from 2007 to 2011						
Orchid Island	SanDiMen	2	0	18	20	100%
SanDiMen	Orchid Island	0	14 (10)	6	20	0%
Orchid Island	SanDiMen	18	10 (10)	12	40	100%
SanDiMen	Orchid Island	0	25 (15)	15	40	0%
Wild-caught females translocated in 2010 and recaptured in 2011						
Orchid Island	Orchid Island	28	5 (5)	12	45	100% (87.5%)
SanDiMen	SanDiMen	0	15 (12)	17	32	0%
Progeny raised in captivity from wild-caught females, released in 2010						
Orchid Island	SanDiMen	6	10 (10)	30	46	100% (37.5%)
SanDiMen	Orchid Island	13	10 (10)	9	32	100% (56.5%)

encountering nesting females, non-nesting females (females who did not lay eggs), or no females at all Table 7.1). Even in a strange environment, adult females from Orchid Island were more likely to demonstrate parental care than adult females from SanDiMen Table 7.1). For 1 to 9 days (a mean of 3.5 days), female lizards from Orchid Island remained with their eggs at SanDiMen. This strongly implies that, regardless of the environmental circumstances present at the time of nest construction, adult females caught in the wild who have had prior reproductive experience consistently decide whether to express maternal care.

Both experimental groups significantly changed the expression of maternal care in lizards that were born and raised to maturity in captivity in comparison to the control group. All female lizards bred on Orchid Island and released there as adults showed maternal care, as did those relocated to SanDiMen (Table 7.1). Female lizards that were transported to SanDiMen and released there, however, were never seen to exhibit maternal care; in contrast, 100% of those that moved to Orchid Island did. This provides strong evidence that the ability to express parental care can be a behaviorally plastic trait since it shows that females from both populations are capable of doing so when growing up in a garden environment.

The length of maternal care varied significantly among recipient populations for lizards that were born and raised to maturity in captivity. Females taken captive originating from Orchid Island and released there, as well as females taken captive originating from SanDiMen and released to Orchid Island, differed from females taken captive originating from SanDiMen and released on Orchid Island (mean durations of 17.4 and 20.3 days, range 8–31 and 8–33 days, respectively) stayed in the nest for a considerable amount longer than those from Orchid Island and released to SanDiMen (mean duration = 3.2 days, range: 1–5 days).

7.2.2 Abundance of Egg-Eating Snakes and Hatching Success

The threat of predation causes females on Orchid Island to guard their nests while the eggs are being laid. If so, populations of long-tailed skinks without maternal care would have lower snake populations and, as a result, would have higher egg hatching success than those on Orchid Island. This is exactly what we discovered; the two Orchid Island sites had significantly more egg-eating snakes than the two sites with no maternal care for lizards (Chi-square = 156.73, df = 3, $p < 0.0001$; Fig. 7.3). There were 9–20 times more snakes present on Orchid Island than at the other two sites. Lizard nests at the Orchid Island site were significantly more preyed upon by snakes than those at the other two sites (Chi-square = 69.15, df = 3, $p < 0.0001$; Fig. 7.3). Snake abundance correlated with predation rates of egg-eating snakes on lizard eggs. Although snakes at these sites were 9–20 times more numerous than at other sites, predation rates were only twice as high as those observed for the two populations that did not express relatedness (Chi-square = 1.09, df = 1, $p = 0.296$; Fig. 7.3). Lizard eggs at the two Orchid Island sites were preyed upon by snakes at similar rates (Chi-square = 1.09, df = 1, $p = 0.296$; Fig. 7.3). Snake density was unrelated to the presence of lizard nests; Little Paiday Bay ($n = 173$) and SanDiMen ($n = 156$) on Orchid Island had a significant number of lizard nests, whereas Green Island ($n = 54$) and Tungching ($n = 37$) on Orchid Island had significantly fewer lizard nests.

7.2.3 Effectiveness of Parental Care Among Different Populations

Orchid Island had the lowest hatching success when maternal care was absent, whereas the other two populations (Fig. 7.4) had significantly higher hatching success. For the two populations that lacked maternal care, has similar hatching

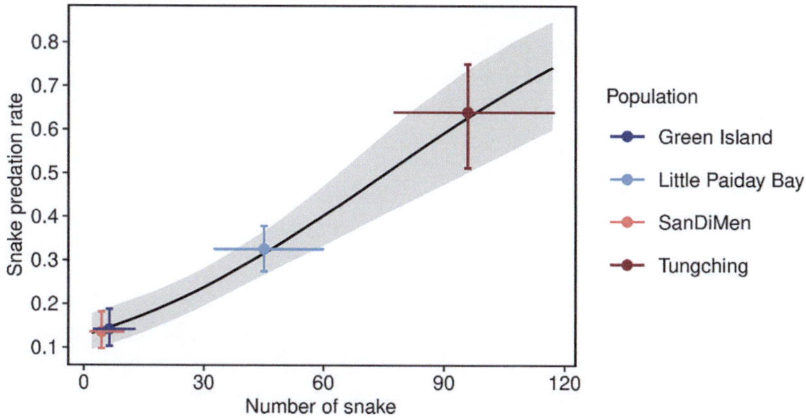

Fig. 7.3 Egg-eating snake (*Oligodon formosanus*) numbers and the associated percentage of lizard nests preyed upon by snakes from two insular lizard populations not expressing parental care (SanDiMen on the main island of Taiwan and Green Island) and two insular lizard populations from Orchid Island expressing parental care (Little Paiday Bay and Tungching). Snake numbers are the total number of snake encounters along a 2 km stretch of concrete retaining wall from 2001 to 2004, and the total number of lizard nests is as follows: Sandimen $n = 156$, Green Island $n = 54$, Little Paiday Bay $n = 173$, Tungching $n = 37$. (Redrawn from Huang et al. (2013), Journal of Animal Ecology)

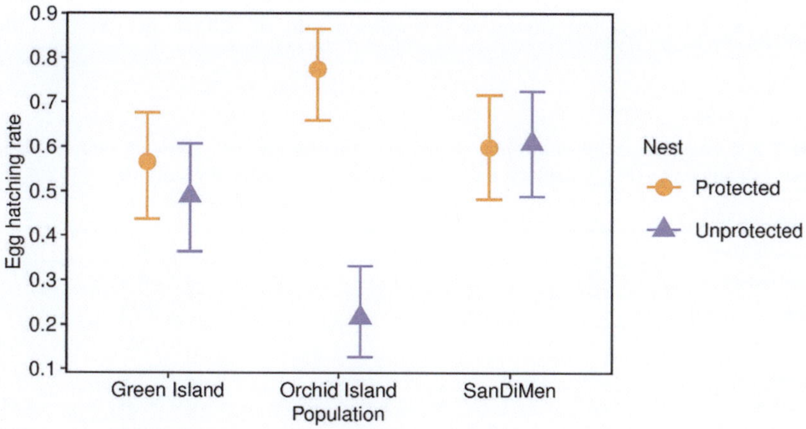

Fig. 7.4 Experimental demonstration of the fitness benefits derived from maternal care in three different populations. Predation rates of lizard nests are expressed as the proportion of long-tailed skink eggs hatching from the three populations without any form of nest protection (i.e., without a nest-guarding female present) for populations on Taiwan (SanDiMen), Green Island, and Little Paiday Bay on Orchid Island, and when vertebrate predators were excluded from accessing eggs using mesh netting. The only population in which maternal care would result in a fitness benefit is from Orchid Island, where females naturally express maternal care. (Redrawn from Huang et al. (2013), Journal of Animal Ecology)

success (Fig. 7.4). With regard to lizard nests without maternal care, the experimental exclusion of vertebrate predators from access to lizard eggs significantly increased hatching success at Orchid Island (Fig. 7.4), but there were no significant effects at the other two sites (Fig. 7.4). This demonstrates that Orchid Island is the only population where maternal nest guarding directly improves fitness by raising the likelihood that eggs will hatch.

7.2.4 Genetic Diversity

Four segregation sites and five different haplotypes totaling 1137 bp were visible in the sequences of 76 long-tailed skinks (Genbank accession numbers: JF689928-JF689932; Table 7.2; Fig. 7.5). The majority of the haplotypes (63/76 or 83%) belonged to the most common H1 haplotype. This haplotype was present at all sampling sites, including Green Island and Orchid Island (Table 7.2; Fig. 7.5). The other haplotypes (H2-H5; all haplotypes have a mutation separating them from H1) are rare and private, and have a very limited distribution (Fig. 7.5). FST and corrected ST values for pairs between Taiwan and Green Island, Taiwan and Orchid Island, and Green Island and Orchid Island were 0.02142, –0.00092, 0.00000 and 0.02139, –0.00085, 0.00000, respectively (p-values >0.05 for all pairs). AMOVA indicated that most of the genetic variation was within populations (73.91%), while inter-island variation was much lower (Table 7.3). We estimated tMRCA at 49,750–18,020 years ago (using a divergence rate of 2.85% per MY1; Gübitz

Table 7.2 Long-tailed skink (*Eutropis longicaudata*) molecular data, showing sample sizes (n), number of haplotypes (N_{hap}), haplotype diversity (h), and nucleotide diversity (π)

Sample locality	Population	n	N_{hap}	h	π
Orchid Island	1	10	1	0.0000	0.0000
Green Island	2	10	2	0.2000	0.0002
Island of Taiwan		56	4	0.3695	0.0004
BeiNan	3	5	1	0.0000	0.0000
TaiMaLi	4	3	1	0.0000	0.0000
DaWu	5	1	1	0.0000	0.0000
LaiYi	6	10	2	0.5556	0.0005
TaiWu	7	10	2	0.5333	0.0005
WanAn	8	2	1	0.0000	0.0000
SanDiMen	9	3	1	0.0000	0.0000
TianLiao	10	10	2	0.3333	0.0003
ShanLin	11	1	1	0.0000	0.0000
HsinHua	12	5	1	0.0000	0.0000
DaNei	13	4	1	0.0000	0.0000
SanJiePu	14	6	1	0.0000	0.0000
Total		76	5	0.3060	0.0003

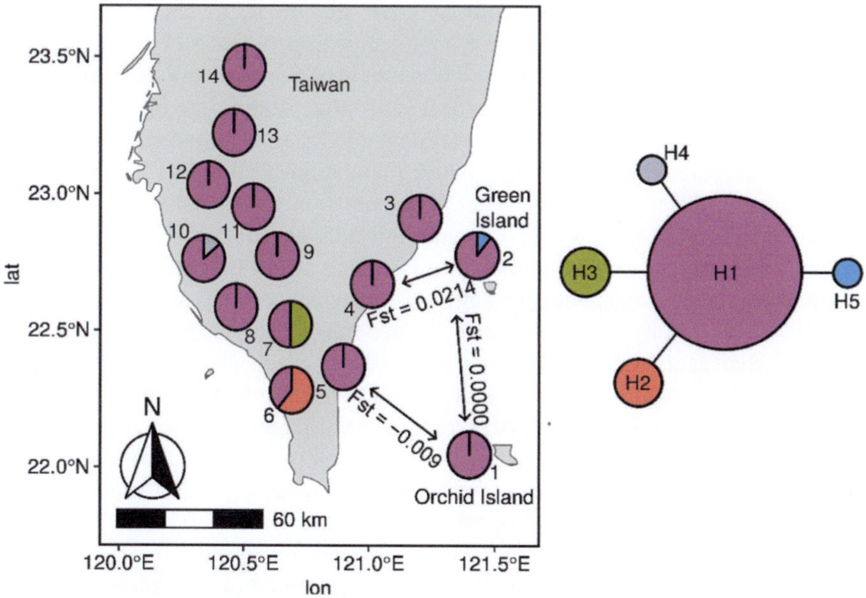

Fig. 7.5 Sampled populations of *Eutropis longicaudata*, showing F_{ST} values between the island of Taiwan and Green and Orchid Islands. Also shown are haplotype frequencies within populations and a statistical parsimony network of haplotypes (H1-H5). Populations are numbered as in Table 7.1: 1 = Orchid Island, 2 = Green Island, 3–14 = island of Taiwan. (Redrawn from Huang et al. (2013), Journal of Animal Ecology)

Table 7.3 Analysis of molecular variance results

Genetic variation	Sum of squares	% of variation	Fixation indices	Significance tests
Among groups	0.378	−29.46	$\Phi_{CT} = -0.29458$	$p < 0.001$, df = 2
Among populations within groups	5.160	55.54	$\Phi_{SC} = 0.42905$	$p < 0.001$, df = 11
Within populations	6.633	73.91	$\Phi_{ST} = 0.26086$	$p = 0.49$, df = 62

et al., 2000; Paulo et al., 2001), suggesting that these island populations diverged approximately after the last glacial period.

7.2.5 Discussion

We were curious as to why long-tailed skinks display indirect maternal care throughout their geographic range, while the majority of other species are forced to express parental care (for exceptions see Martins et al., 1999). We emphasize that even though many animal species differ in the kind of parental care they express or in how long they express it, almost all of them still do so to some extent. Two fascinating patterns emerged from our interchange experiments: (1) Adult females in the wild exhibited the same behaviors as their native populations consistently, even reproducing in new populations for 3 years after interchange ($n = 2$ females), but (2) individuals born and raised in a communal garden environment were able to show parental care regardless of their native populations (Table 7.1). Therefore, the lack of genetic differentiation between populations on Orchid Island and those in other populations (Table 7.3, Fig. 7.5) does not fully account for potential regional variations in maternal care expression (see Heulin et al., 1999 for similar results for viviparous and oviparous populations of *Lacerta vivipara* Lichtenstein 1823). Given the lack of genetic differences between populations (Table 7.3; Fig. 7.5) We might anticipate that ecological factors on Orchid Island, such as increased predation pressure on lizard eggs, would cause otherwise noncaring females to remain at the eggs during incubation given the lack of genetic differences between populations (Table 7.3; Fig. 7.5). According to their species or place of origin, females who had previously nested consistently expressed or did not express parental care, whereas captive females from both native populations who had never previously nested did so on Orchid Island. Therefore, regardless of changes in the selective pressure to do so, a female lizard is likely to continue expressing parental care in future reproductive activities once she has done so initially.

How do lizards that are nesting "decide" whether to protect their eggs while they are being incubated? Since our captive lizards never came into contact with any potential predators before we released them as adults, interaction with predators during ontogeny is unlikely to play a role. Because some of these naive creatures show maternal care, it is possible that the individual's encounters with predators before nest-building (along with the individual's propensity to learn) played a role. We cannot, however, rule out the fascinating possibility that lizards are eavesdropping on the nesting habits of their own kind and using those habits as a guide to their own behavior. To prove these theories, more research is required.

Females from Orchid Island spent longer at the nest site (8–35 days) than SanDiMen females expressing maternal care (1–5 days). We believe that this (as well as our reduced ability to find these females; Table 7.1) is a result of these females' incapacity to recognize the numerous new predators they may come across there (e.g., Downes & Adams, 2001). For example, the only lizard predator on Orchid Island is the snake *Elaphe carinana*, a long-tailed skink that is easily identified and escapes (Huang & Wang, 2009). SanDiMen, however, supports five additional species of predatory snake species from three different families (Colubridae, Elapidae, Viperidae). As a result, we believe that females at SanDiMen who attempt to stay at nest sites to provide maternal care either abandon their nests after coming into contact with these predators (see Huang & Wang, 2009) or become prey to them.

We discovered over a four-year period that Orchid Island had significantly more egg-eating snakes than other places, and that these snakes were responsible of higher egg predation rates (Fig. 7.3). The low rate of egg survival in the absence of parental care is one of the key ecological characteristics associated with parental care

(Clutton-Brock, 1991; Klug & Bonsall, 2009), which is consistent with the interpopulation variation we noticed. When female long-tailed skinks guarded their nests on Orchid Island, they were very effective at deterring egg-eating snakes (these snakes posed no threat to females; Huang, 2006a, b, c). This appears to be at odds with the theory that egg-eating snakes are to blame for the island's apparent high rate of predation, as females should be able to defend their eggs. However, the snake population is 9–20 times greater on Orchid Island than it is elsewhere, but egg predation is only twice as prevalent (Fig. 7.3). The snakes rely on additional food sources, such as sea turtle eggs, so nest-guarding females are quite effective in keeping them away (Huang et al., 2011)]. The vast majority of egg-eating snakes on Orchid Island can be narrowed down to (1) snakes preying on eggs when females leave the nest to forage (Huang, 2007a); (2) lizard predators (such as the snake *Elaphe carinata*) scaring off females (Huang, 2006c); or (3) females abandoning the nest before the eggs hatch (Huang, 2007a, b).

There was little genetic diversity, among populations and no clear structure between Orchid Island and other populations, according to our findings (Fig. 7.5). It is obvious from this that there are no different genetic strains present on the various islands, and the phylogeographic pattern of long-tailed skinks observed on these islands is very recent. Because of the recent colonization of these islands following the Last Glacial Maximum (LGM), which significantly changed the distribution of the region's fauna and flowers, there may not be much phylogeographic structure (Dynesius & Jansson, 2000; Harrison et al., 2001). The global cooling during the LGM may have eliminated long-tailed skinks from the region, and subsequently long-tailed skinks recolonized the region during the post-LGM warming. The ability of reptiles to drift across oceans for long distances may help with recolonization (Calsbeek & Smith, 2003; Vidal et al., 2008). This has been reported in other Taiwanese reptiles (Lin et al., 2008), and is consistent with long-tailed skinks reaching their northernmost distribution limit in southern Taiwan (Shang, 2001). We cannot rule out the possibility of a genetic component to the variations in parentage expression between populations, given the rapid, adaptive genetic changes that have been observed in the field (Hairston et al., 2005). For population genetic studies to infer migration and colonization, more research is required. According to what is currently known about lizards and the genetic data that is available, this behavior must have developed quickly. Our findings regarding the interchangeability of captive female lizards and the prevalence of predators support the idea that predators play a significant role in how parents express their parental care.

Our research shows that predation plays a significant role in the evolution of parental care expression in a single population, as shown by the lack of genetic differences between populations, the consistency of parental care expressed by wild-caught female lizards with their native populations, the adoption of parental care expressions of recipient populations by some captive lizards, and the higher rates of predators and predation in populations with maternal care. In this large monophyletic population, predation has been crucial to the evolution of parental care expression. Further ecological research may help to shed light on why there are so many more egg-eating snakes on Orchid Island than in other populations. This is one of the few species that contradicts the conventional paradigm that parental care is an evolutionary response that manifests uniformly throughout a species' range (despite the fact that the nature and length of parental care can vary geographically), and it suggests that parental care can be a sign of behavioral plasticity between individuals and populations. Since almost all endotherms provide parental care (Clutton-Brock, 1991; Klug & Bonsall, 2009), ectotherms are most likely to exhibit this pattern, even though it may be more prevalent than depicted in the literature (e.g., our study; Martins et al., 1999). The lack of new reports is not surprising given the general dearth of research on parental care in ectotherms, particularly terrestrial ectotherms (Tallamy, 1984; Shine, 1988; Clutton-Brock, 1991; Stahlschmidt, 2011).

7.3 The Filial Cannibalism in the Long-Tailed Skink

When I was a child, I was warned by my parents not to climb on top of the barn to watch the mother cat take care of her newborn kittens, otherwise she would move or bite them to death.

The long-tailed skink is not endemic to Taiwan, but it is the largest native lizard in Taiwan. The advantage of its size is that it often wins fights with others, but head-to-head fights between animals are not unprovoked, and the purpose of the fights is either to fight for food, habitat or to protect its young (eggs). This chapter is the story of a long-tailed skink protecting its eggs, which in turn leads to a fascinating case of filial cannibalism.

The story begins with the other main character, a reptile egg-eating kukri snake. In the past, the beaches of Orchid Island were dotted with green turtles that came ashore in the summer to lay eggs, and the large number of eggs provided ample food for the kukri snakes. Of course, the turtle eggs are not the only food they eat on Orchid Island; they also eat the eggs of other reptiles. There are also three species of snake on Orchid Island: the Green bamboo viper (*Viridovipera stejnegeri*), the Tea Spotted Snake (*Psammodynastes pulverulentus*), and the Stinky snake (*Elaphe carinata*). However, because the former two species are viviparous and do not lay eggs, and because the latter species is fierce, egg-protective and natural enemies of the kukri snakes, they will be eaten or attacked before they can eat its eggs. Therefore, the kukri snakes will seek out smaller lizard eggs.

The order of egg size for reptiles on Orchid Island is: Green sea turtle (about 45–60 cc), Long-tailed skink (1.7–1.9 cc), Swinhoe's tree Lizard (egg size about the same as the long-tailed skink), Boulenger Lizard (0.8–1.0 cc), Rocky shore island Lizard (about the same as Boulenger Lizard) and some smaller lizard eggs such as the Yami's Scaled-toed gecko or the rare Half-leaf-toed gecko, Sauter's Grass Lizard and other lizards. The hard-shelled egg of the gecko is not the food of kukri snakes, and the Rocky Shore Island Lizard's habitat is on a coastal reef where the snake simply won't climb up the sharp rocks. It is therefore quite difficult for a kukri snake to eat the eggs of a rocky shore island lizard. So, apart from green turtle eggs, the eggs of the Long-tailed skink are the largest source of food for the kukri snakes.

The long-tailed skink usually chooses a warm habitat to lay its eggs in, and the temperature of the concrete drainage hole is indeed much warmer than natural habitats, according to years of egg-laying temperature measurements. This completely eliminates the intrusion of the egg-eating kukri snake, as the lizard can simply defend one side of the opening to prevent the snakes from entering and feeding on the eggs; fights between the lizard and the kukri snakes are often seen, and all are won by the lizard, so the eggs are protected. Although there are also lizards in Taiwan that choose concrete drainage holes to lay their eggs, they leave the nest immediately after laying and do not have parental care. The main reason for this is that, apart from the absence of large populations of kukri snakes in Taiwan, it may also be related to the natural enemies of many other long-tailed skinks, for example, we have found snakes that eat long-tailed skinks in our sample area in Pingtung, such as the stinky snakes, Umbrella snake (*Bungarus multicinctus multicinctus*), Glass cobra (*Naja naja*), tortoise shell snake(*Protobothrops mucrosquamatus*), and other snake predators.

Between 2001 and 2003, I observed that female long-tailed skinks were found to eat their own eggs (filial cannibalism) due to the invasion of their nests by kukri snakes; however, the answer to this egg-eating behavior is still pending. But why would a long-tailed skink eat its own eggs? The mechanism of filial cannibalism may be related to three factors: (1) an increase in the frequency of filial cannibalism when the frequency of nest invasion by the predator kukri snake increases; (2) an energy adaptation—newly laid females are in great need of energy, so some individuals will eat their own eggs to replace the energy lost through parental care or egg production; (3) filial cannibalism is most prevalent in pregnant females, as eggs are exposed to the elements. As the eggs are exposed to being eaten by kukri snakes, pregnant females are most in need of replenishment, and instead of having their eggs eaten by snakes, they will eat them first to more or less replenish their energy. This adaptive energy feedback mechanism has been demonstrated in this chapter.

In the natural environment, animals adopt behavioral strategies that are often surprising and puzzling. The relationships that have evolved over time are dizzying, yet so intricate and logical. Although the evolution of the behavior of the long-tailed skink has been confirmed, the implications of this behavior are still worthy of further investigation, for example, does it affect the mechanism of speciation? More interesting stories will be uncovered.

In birds and other mammals, infanticide—killing other conspecific's offspring—is quite common; for example, a new lion king kills the cubs of another lion in his pride and gives birth to his own cubs. However, it is quite rare to kill (eat) one's own cubs (eggs). During 2001–2003, I observed that female lizards were found to eat their own eggs due to the invasion of nests by the kukri snake; however, the answer to this filial cannibalism behavior is still unresolved. After years of persistent research, the answer has finally been hard to come by: [Although filial cannibalism has been rarely reported in insects and fishes, this chapter is the first major finding of filial cannibalism in vertebrates].

Filial cannibalism, or eating one's own offspring, has been extensively researched in fish and is relatively common in some animal species. However, there have only been a few reports of whole-nest infant cannibalism by terrestrial vertebrate parents. I discuss the existence of whole-clutch filial cannibalism in long-tailed skinks (*Eutropis longicaudata*) on Orchid Island in this chapter. Most long-tailed skink females will attack or ignore invaders they can defend themselves against, such as the snake that eats reptile eggs *Oligodon formosanus* and the tree lizard *Diploderma swinhonis*, but as the frequency of invasion by *O. formosanus* rises, the long-tailed skink begins to engage in whole-clutch filial cannibalism. The best course of action (especially for pregnant females) when feeling threatened by *O. formosanus* may be to consume their entire clutch of eggs. This skink's behavior has developed over time, making this a unique example of whole-clout filial cannibalism in a female reptile. Reduced egg production and age of spawning, two existing hypotheses for whole clutch filial cannibalism behavior., imply that cannibalistic parents may consume entire clutches of eggs when the care is more expensive than the anticipated benefits. According to the data in this chapter, the presence of predators is primarily what causes whole clutch filial cannibalism As a result, I put forth a predation risk hypothesis for whole-nest filial cannibalism, according to which long-tailed skinks are more likely to engage in filial cannibalism as predation risk increases in succeeding generations.

Over the past few decades, the study of filial cannibalism (eating one's offspring) has gained popularity in evolutionary biology, and empirical evidence supports a number of its applicability hypotheses. Partial filial cannibalism (consumed a portion of the clutch) (Rohwer, 1978; Sargent, 1992) and whole nest filial cannibalism (consumed the entire clutch) (Coleman et al., 1993; Lindström & Sargent, 1997; Manica, 2002a) have both been well documented in some systems in the laboratory and in the field. According to the energy-based hypothesis, filial cannibalism represents a compromise between the success of current and future reproduction (Rohwer, 1978; Sargent, 1992; Manica, 2002b). This hypothesis predicts that animals should eat some eggs to meet their future energy needs for survival and reproduction. If these needs are sufficiently high, they should eat the entire clutch (Rohwer, 1978).

After reviewing pelagic fish, Manica (Manica, 2002a, b) summarized information on whole clutch filial cannibalism and discovered that the majority of predictions were interpreted as follows. (1) Fish population size, i.e., parents are more likely to engage in filial cannibalism when fish population size decreases, making the benefits of care in the experiment less likely (Manica, 2002b); (2) timing of filial cannibalism, i.e., the incidence of filial cannibalism decreases with increasing egg age (Lavery & Keenleyside, 1990; Manica, 2002a); (3) the physical condition of the parents, i.e., the frequency of filial cannibalism increases when the male condition decreases (Petersen, 1990; Okuda & Yanagisawa, 1996a); (4) availability of mating, i.e., males with cannibalism always re-mate within a shorter period of time than males that eat their young on the first day of care (Okuda & Yanagisawa, 1996b); (5) filial cannibalism of sex, i.e., males exhibit more cannibalistic behavior than females, i.e., males are more likely to filial cannibalism than females in two-parent care (Schwank, 1986; Lavery & Keenleyside, 1990); and (6) paternity certainty, i.e., cannibalistic behavior should be more common in nests with lower parental certainty.

Some animal species have been observed to engage in female filial cannibalism, in which females eat live eggs, possibly to quickly improve their physical condition and increase the likelihood that their offspring will survive and reproduce successfully in the future as well as their foraging success (Somma, 2003; Lourdais et al., 2005). However, females should only engage in cannibalism when it is absolutely necessary to escape their situation. They might eat a clutch of eggs if they have to give up a clutch, and the chances of any eggs surviving unattended are slim. Although the circumstances that lead to females eating their entire clutch are largely unknown, it appears that factors like expected future reproduction will affect filial cannibalism behavior (Williams, 1966).

Although the practice of filial cannibalism is common among some taxa (such as insects and fish), research on it in terrestrial vertebrates like reptiles has been sparse. For example, an assassin worm has been described (*Rhinocoris tristis*) in which males typically consume the eggs surrounding the worm's eggs, which are almost certainly parasitized by wasps, and proposed that males use the eggs as an alternative food source (Thomas & Manica, 2003). However, most reports on reptiles focus on female cannibalism (destroying eggs or killing the young of other colony members), possibly to increase the chances that their own unkilled offspring will thrive due to reduced competition with other young (O'Riain et al., 2000; Macedo et al., 2001; Sherman, 2003).

I have previously written about the parental care behaviors of the long-tailed skink (*Eutropis longicaudata*), which included active defense against snakes that eat eggs (*Oligodon formosanus*). The parental care period lasted between 7 and 35 days, and the females guarded the eggs for at least a week to keep them safe from snakes that eat eggs (mean 26 days). I have previously shown that the removal of *E. longicaudata*'s maternal egg guarding behavior at the spawning site as a result of egg predation by the main predator of the reptilian egg-eating snake *O. formosanus*, led to a significant decrease in hatching success (Huang, 2006c). For this species, this anti-predator behavior is specific to Orchid Island and is absent from the rest of mainland Taiwan (Huang, 2006c). Long-tailed skink also identifies the predatory snake *Elaphe carinata*, and when the species enters the nest of *E. longicaudata*, the skink escapes and does not return (Huang, 2006c).

Although it has been predicted in the fish literature that filial cannibalism behavior should be influenced by predation risk, this prediction is not supported by any empirical data. According to (Lindström, 1998), no additional sand goby (*Pomatoschistus minutus*) eggs disappeared when egg-predatory crabs (*Carcinus maenas*) were present, and no additional eggs disappeared when egg-predatory crabs weren't present. This finding suggests that the behavior of the species' filial cannibals is unaffected by the presence of predators. In this paper, I examine a new example of the effect of predation risk on the whole-nest filial cannibalism behavior of maternally cared-for long-tailed skinks. The predation risk hypothesis predicts that when egg-eating snakes enter the nest, maternally caring skinks will consume their own eggs at a high frequency while ignoring non-predator sympatric lizards; and that skinks with future nests should consume their own eggs at a higher frequency than those without nests. Therefore, I recorded an experimental manipulation in which the parental female *E. longicaudata* would consume its entire clutch of eggs when the frequency of nest invasion by the egg predator *O. formosanus* increased to a certain level.

7.3.1 Immediate Response of *E. longicaudata* Females to Intruders

Females in nests with eggs typically responded negatively to *O. formosanus*; in the once-daily treatment, 17 attacked, 2 escaped, and 1 ignored the intruder, compared to 53 attacks (17, 18 and 18 for the first, second, and third invasions, respectively), 3 brief abandonments of the nest (1, 1 and 1 for the first, second, and third invasions, respectively), and 4 ignores (2, 1 and 1 for the first, second, and third invasions, respectively). while most skinks ignored the invading *D. swinhonis* (1 attacked, 1 abandoned the nest for a short period of time, 18 ignored the invaders in one treatment/day, while 5 (2, 2, and 1 for the first, second, and third invasions, respectively) attacked, 4 (1, 2. and 1) abandoned the nest for a short period of time, and 51 (17, 17, and 17 for the first, second, and third invasions, respectively) ignored the invaders in 3 treatments/day ($\chi2 = 0.83$ for three invasions with df = 2; by Chi-squared test). Skinks encountering *O. formosanus* were very significantly different compared to skinks encountering *D. swinhonis* in both once-daily ($\chi^2 = 29.8$, $p < 0.0001$) and thrice-daily ($\chi^2 = 23.7$, 29.8 and 29.8, $p < 0.0001$; df = 2) treatments. All escaped females returned to their nests within 2 h of the invasion test. Skinks did not differ in their response to the once-a-day and thrice-a-day treatments.

7.3.2 Response of *E. longicaudata* 24 h After Encountering an Invader

I classified all female responses into four categories: no change, egg hiding, filial cannibalism, and abandonment. *D. swinhonis* was largely ignored by female long-tailed skinks that guarded eggs, and after the invasion of *D. swinhonis*, female long-tailed skinks at 40 test nests (including once and three treatments per day) showed no change. The phenomenon of filial cannibalism could be seen when the invader *O. formosanus* was taken out of the test nests. In the three daily treatments, the filial cannibalism behavior in response to *O. formosanus* was more prevalent ($p = 0.01$, df = 3; by Fisher exact examination; Fig. 7.6). Similarly, filial cannibalism behaviors were more frequent when the invader was *O. formosanus* than when the invader was *D. swinhonis* ($p < 0.0001$, df = 3; examined exactly by Fisher), in natural nests ($p < 0.0001$, df = 3; examined exactly by Fisher) with three treatments/day, but not with one treatment/day ($p = 0.11$ and 0.82; df = 3 when *O. formosanus* was compared with *D. swinhonis* and with natural nests, respectively;

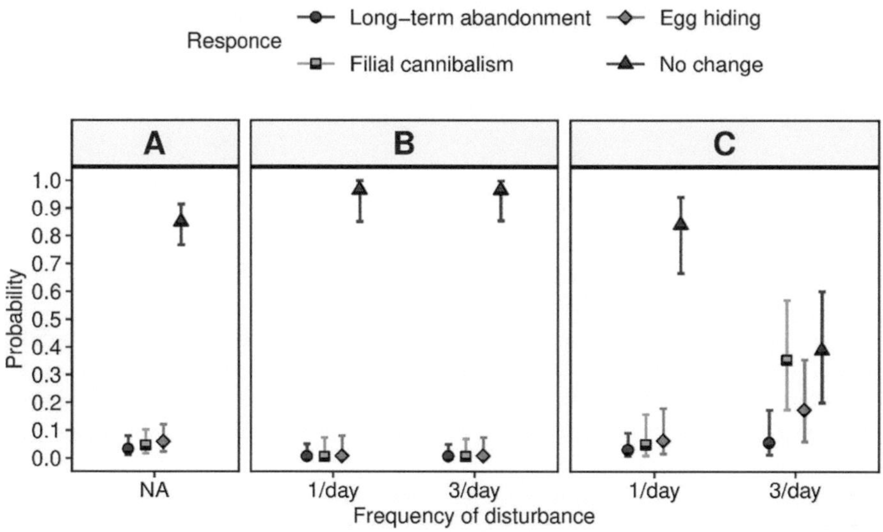

Fig. 7.6 Natural nests ($n = 80$) (**a**) and four 24-h responses by egg-guarding female skinks (*Eutropis longicaudata*) (no change, egg hiding, filial cannibalism, and escape) when faced with two intruders: a sympatric lizard, *Diploderma swinhonis* (DS) ($n = 20$ for each treatment) (**b**), and the reptile egg-eating snake, *Oligodon formosanus* (OF) ($n = 20$ for each treatment) (**c**). Intruders were experimentally released into nests once/day and three times/day. Egg-guarding females generally ignored *D. swinhonis*. Filial cannibalism increased when the frequency of *O. formosanus* being placed into guarded nests increased from once/day to three times/day. $n =$ the sample size. (Redrawn from Huang (2008a, b), Behavioral Ecology)

Fisher exact examination). There was no significant difference between the natural nest and *D. swinhonis* treatments ($p = 0.10$, df $= 3$; Fisher exact test).

7.3.3 Relationship Between Filial Cannibalism Behavior and Gravid Females

The long-tailed skink consumed every egg in each of the 12 cases of egg loss incidents mentioned above. Three of the nests in each of the three treatments per day were consumed by unidentified animals, so these nests were not used for further investigation. Both once-daily and three times-daily treatments found two natural invasions of *O. formosanus* in different nests, and no significant differences in natural invasions were found between the two treatments ($p = 1$, df $= 1$; by Fisher exact examination). Overall (including once-a-day and three-a-day treatments), nine cases of filial cannibalism (by eight gravid females and one non-gravid female) and 31 cases of non-filial cannibalism (by one gravid female and 30 non-gravid females) were recorded in response to *O. formosanus* invasion. Non-filial cannibalism included no change, egg hiding and escape. These results were significantly different ($p < 0.0001$, df $= 1$; by Fisher exact test). The frequency of filial cannibalism was significantly different between *E. longicaudata* (8 instances of filial cannibalism out of 9 gravid females) and *D. swinhonis* (6 gravid female lizards; no examples of filial cannibalism among the 6 gravid female lizards; $p = 0.04$, df $= 1$; examined by Fisher's precision).

7.3.4 Discussion

7.3.4.1 Predation Risk Hypothesis of Whole-Clutch Filial Cannibalism in Long-Tailed Skink

My research indicates that female long-tailed skinks may start eating their own eggs shortly after being startled by a kukri snake. Therefore, the current data support my hypothesis that long-tailed skinks will eat the female's eggs when egg-eating snakes *O. fomosanus* invade the nest with a high frequency of invasion. When the egg predator *O. fomosanus* invades frequently, the long-tailed skink will consume her eggs but ignore the non-predator sympatric lizard *D. swinhonis*. This response of the long-tailed skink may reflect a trade-off between current reproductive value (ignoring the non-threatening invader *D. swinhonis*) and future reproduction (in the case of a high frequency of invasion by the egg predator *O. fomosanus*, the long-tailed skink will consume her eggs but ignore the non-predator sympatric lizard *D. swinhonis*) which suggests that the long-tailed skinks engage in whole-clutch filial cannibalism behavior to maintain the energy requirements of fed females for future reproduction.

In an aquarium study, it was discovered that male sand goby (*Pomatoschistus minutus*) are sensitive to the costs and benefits of parental care, that care may be expensive if the nest site is small, and that when battling egg-predatory crabs (*Carcinus maenas*), the interaction with the crab is riskier because the male may be injured. Males may become more noticeable to predators after interacting with crabs. Because the males did not consume the entire brood, the study also demonstrated that the presence of egg predators had no impact on filial cannibalism behavior (Lindström, 1998). Similarly, introduced conspecifics and egg predators into the nests of scissor-tailed snakes (*Abudefduf sexfasciatus*), but neither invasion type resulted in whole-clutch filial cannibalistic behavior, even though the invasion of conspecifics increased partial cannibalism due to a decrease in the certainty of parentage(Manica, 2004).

Increased clutch size and clutch age are two basic predictions supported by studies of whole-clutch (brood) filial cannibalism in teleost, but parental reproductive status is not discussed (Manica, 2002b). However, the current study suggests that whole-clutch filial cannibalism behavior in long-tailed skinks may be more influenced by the parents' reproductive status (whether they are pregnant or not) than by the size of the clutch. Although pregnant females were more likely to eat their clutches than non-pregnant females, there was no difference in clutch size between skinks that engaged in filial cannibalism and those that did not. I did not investigate the behavior of filial cannibalism in relation to clutch age. Because pregnant females have more urgent energy needs than non-pregnant females, we would anticipate that filial cannibalism would happen more frequently in pregnant females. This is in line with pregnant female's increased energy requirements (Huang, 2006a). Alternately, it's possible that pregnant females have better chances of having offspring in the future, which might affect how they decide whether to engage in filial cannibalism in eggs. It is clear that females maintain significant body reserves throughout parental care, possibly by engaging in subsequent active foraging outside the nest, as previous research suggests that the loss in female quality during subsequent foraging is quickly balanced by an increase in quality and that there is no significant difference in SVL and body mass between females that exhibit parental care and those that do not (Huang, 2007a). Therefore, it doesn't seem like the energy requirement hypothesis fully accounts for the recent findings of filial cannibalism in long-tailed skinks. Furthermore, the frequency of invasions by the predator *O. formosanus* is directly correlated with the behavior of long-tailed skinks engaging in filial cannibalism, but not with other non-threatened lizard species. Therefore, it would seem that parental reproductive status and the risk of predation both play a role in skinks engaging in filial cannibalism.

Despite the fact that the reason for egg loss in *E. longicaudata* has not been identified, the phenomenon of filial cannibalism as a result of more frequent invasions has been more thoroughly studied. Third causalities may also be involved in this phenomenon of maternal filial cannibalism. First, compared to other nesting animals, *E. longicaudata* requires very little energy to build a nest because the nest holes are already there. However, the high frequency of invasion may lead to filial cannibalism. In contrast, some animals that engage in filial cannibalism make a significant investment in their offspring [see, however, (Manica & Johnstone, 2004)]. For instance, the significant costs associated with pregnancy in captive rainbow trout (*Epicrates cenchria maurus*) include not only the depletion of energy stores but also the reduction of female dorsal muscle due to the mobilization of protein reserves during pregnancy (Lourdais et al., 2004). Consequently, the post-partum female is weak and takes several weeks to recover from this condition (the quality of a clutch of eggs is roughly half that of the mother's body mass). In this sense, eating non-viable eggs might make it easier for the mother to regain her energy and quickly rebuild her body's muscle mass.

Second, with the invasion of *O. formosanus* between the two treatments, the frequency of filial cannibalism increased. It is reasonable to assume that the female long-tailed skink was able to repel *O. formosanus* during the first invasion, but as invasions became more frequent, the long-tailed skink began to see *O. formosanus* as a threat to her eggs, and she began to consume her own eggs during the three treatments per day. An alternative hypothesis is that the predation risk associated with filial cannibalism may arise from the energy expenditure required by female skinks to frequently defend against predators in dense nests as opposed to nests in their natural environment. As a result, this may be an ideal location for predators to congregate, leading to more energy-intensive work for female skinks.

Accordingly, the available data imply that increased predation may be a cue for female long-tailed skinks that their current nest is less valuable than potential future nests. As a result, female long-tailed skinks start eating their own eggs to conserve energy for potential future nests and to recover some energy for the current nest.

Thirdly, if the costs of providing care outweigh the benefits, whole-clutch filial cannibalism behavior may reflect the termination of that care and the recruitment of resources from the nest for future reproductive success. Egg-bearing skinks use whole-clutch filial cannibalism as the best way out of their bad situation because a permanently abandoned nest site will be quickly discovered and eaten by *O. formosanus*.

7.3.4.2 Relationship Between Long-Tailed Skink Filial Cannibalism Behavior and Invasion Frequency

I discovered that the frequency of *O. formosanus* invasions and the reproductive status of females were directly correlated with egg loss. After higher invasion rates, there was an increase in filial cannibalism behaviors. According to my research, females of the *E. longicaudata* species consume entire clutches and defend eggs regardless of the size of the clutch. These findings differ from those of the sand goby and scissor-tailed snake, possibly because long-tailed skinks can always defeat predators without endangering themselves or their eggs. As a result, they are not in danger of suffering harm (Huang, 2006c). But when filial cannibalism takes place, *E. longicaudata* will consume the entire nest, leaving no eggs behind.

7.3.5 Conclusion

Invasion frequency and the occurrence of filial cannibalism in female long-tailed skinks are closely correlated. It has been demonstrated that filial cannibalism behavior frequently manifests in pregnant females as invasion frequency increases. According to my research, pregnant long-tailed skink females are more likely to engage in filial cannibalism than non-pregnant ones. Long-tailed skink females in this population may engage in filial cannibalism to conserve as much energy as they can for use in the population later on when confronted with potential egg-eating snakes (*O. formosanus*). Even though there is a good chance of defeating a defendable invader like *O. formosanus*, it is likely that filial cannibalism is the best option when invasion frequency is high. Females can invest in their own eggs by shielding them from predators or attacks. The current study of filial cannibalism caused by predation risk in long-tailed skinks appears to be consistent with the notion that filial cannibalism is a trade-off between present and future reproductive success (Lindström, 1998; Manica, 2002b).

7.4 Examining Parental Care Behavior in Relation to Predator Risk: An Experimental Study on a Tropical Skink

Parental care can be difficult to understand and is hardly ever demonstrated in reptiles. It is still unknown how predators affect the timing of parental care in terrestrial vertebrates. In this chapter, we show that female long-tailed skinks (*Eutropis longicaudata*) can adapt to interspecific invaders that are placed in skink nests for experimental manipulation. Long-tailed skinks significantly increased their maternal care time when egg-eating reptilian snakes (*Oligodon formosanus*) invaded more frequently. In contrast, female long-tailed skinks were able to recognize two sympatric lizards (*Sphenomorphus incognitus* and *Diploderma swinhonis*) that did not prey on long-tailed skink eggs or adults and did not attack these species; thus, the time spent on egg care was similar to that of *S. incognitus* or *D. swinhonis* when compared to the control treatment. The long-tailed skink also identified the predatory snake *Elaphe carinata*, which escaped when the species entered the nest and never returned. These findings highlight the possibility of predators having an impact on the amount of time parents spend caring for their eggs, and that the length of this behavior depends on the mother's ability to defend herself. The current study offers evidence that may provide insights into the origins of parental care from reptiles to birds and mammals, as egg care is the most fundamental and primitive form of parental care.

Despite the fact that parental care has been found to be crucial for many animal species, little is known about how parental care behaviors in vertebrates and the specific behaviors used by reptiles developed (Shine, 1988; Huang, 2006c). Numerous studies on parental care have revealed that when eggs or young are exposed to unfavorable environments, high predation rates, parasitism, or intense competition with conspecifics, this behavior is most highly developed (Clutton-Brock, 1991). Environmental factors may be a significant factor in controlling parental willingness to participate in parental care. For instance, the environment's seasonality, the length of the breeding season, the availability of food, and predation may all have an impact on how parental care evolved. It is anticipated that the animal will spend the least amount of time possible in its riskiest developmental stages when all of these variables are in balance.

Despite attempts by (Reynolds et al., 2002) to address phylogenetic transitions in a variety of vertebrates, precise timing of parental care is not available in many cases, and the point of evolutionary transition from unattended to parental care or from part-time to full-time care remains unclear. It is expected that part-time (short-term) care may be an ancestral transition to full-time (long-term) care at the beginning of the evolution from no care to parental care, but the causal mechanism remains unclear.

The length of time parents spend caring for their offspring in many animal species may depend on intricate interactions between predator encounters and parental reactions to various predator types. In contrast to other animals like birds, mammals, and crocodiles, most reptiles only provide parental care while they are in the egg stage (Shine, 1988). Understanding the immediate factors that affect parental care decisions in the context of egg guarding, one of the most fundamental and primitive forms of parental care, may provide insights into the evolution of parental care in other animals (Reynolds et al., 2002).

The largest lizard in Taiwan, *Eutropis longicaudata*, has a wide geographic distribution (Huang, 2006a). This species is abundant in the lower elevations of the southern part of the mainland Taiwan and on nearby offshore islands, including Orchid Island. In the majority of its geographic range, *E. longicaudata* females bury their eggs five centimeters deep in the ground or under rocks (natural nests). Following this, the females leave the nest and do not provide parental care (Huang, 2004), with an incubation period of 26 to 31 days (Huang, 2006c). On the mainland Taiwan, females do not exhibit parental care for *E. longicaudata*. However, some female lizards congregate in large numbers to lay eggs in the holes of concrete retaining walls on Orchid Island, which is off the southeast coast of Taiwan. The plastic pipes (mostly 10 cm in diameter) were installed into the concrete retaining wall to allow runoff to drain. The wall was constructed in 1997 to aid in the prevention of landslides. Long-tailed skinks can nest successfully in holes because some of the hole entrances have been blocked by rocks and other debris since 2001 and because the air temperature inside the hole rises quickly on sunny days and stays warmer at night than in other natural nests (Huang, 2006b). Since 2001, long-tailed skink parental care has been noted on this island. Long-tailed skinks' ability to protect their eggs from being eaten by the reptilian egg-eating snake *Oligodon formosanus* was previously shown to have a significant negative impact on hatching success at this egg-laying site (Huang, 2006c). In contrast to natural nests, long-tailed skink females remain in the nest after laying eggs in holes in concrete walls, protecting the nest from this squamate predator (Huang, 2006c).

The "cost" of maternal activity and survival after long-tailed skink females provided parental care, as well as egg attendance time, and intra- and inter-seasonal body size changes, were discussed in an earlier chapter (Huang, 2007a). The percentage of females who guard the nest after egg-laying gradually declines over time, with most staying for at least a week before starting to leave. The average decline in female quality during the first week of egg guarding was countered by an improvement in quality during the following foraging period, indicating that food availability was not a major factor dictating the length of parental care in this Orchid Island population. Long-term parental care (>28 days; mean SD, 31.6 2.2 days; $n = 27$) and short-term parental care (9–16 days; 12.5 2.3 days; $n = 27$) were also identified as two patterns of care duration in this population.

Long-tailed skinks have different patterns of parental care duration (long- and short-term patterns), but the causes of these differences are still unknown. Long-tailed skink mothers attacked *O. formosanus*, a possible egg predator, while ignoring other species that weren't egg predators (i.e. conspecific males *E. longicaudata*, *Diploderma swinhonis* and *Sphenomorphus incognitus*). But when the accompanying *E. longicaudata* came into contact with a Stinky snake (*Elaphe carinata*), which preys on adult long-tailed skinks, it fled right away and didn't come back to the nest (Huang, 2006c, 2007a). We observed the timing and behavioral response (the first day an egg appeared in the nest) used for parental care after releasing four sympatric squamate reptiles (invaders) into the nests and compared it to the timing of parental care in control nests without invaders in order to determine whether predators (on eggs or individuals) induce or influence the timing of parental care in *long-tailed* skinks.

Table 7.4 shows the mean and standard deviation of snout-vent length (SVL), body mass (BM) and clutch size (CS) for the different experimental treatments of long-tailed skink. The results of one-way ANOVA for SVL, BM and CS are not

Table 7.4 The mean and standard deviation (in parentheses) of snout-vent length (SVL), body mass (BM), and clutch size (CS) for different experimental treatments of *Eutropis longicaudata* (EL) and sympatric intruders *Oligodon formosanus* (OF), *Sphenomorphus incognitus* (SI), *Diploderma swinhonis* (DS), and *Elaphe carinata* (EC)

Intruder	Frequency of intruding	Intruder property		ML host/nest property		
		SVL (mm)	BM (g)	SVL (mm)	BM (g)	CS
OF	High	377.3 (11.4)	40.1 (6.5)	120.4 (1.1)	10.1 (0.7)	6.0 (0.7)
OF	Low	380.6 (10.7)	40.4 (6.2)	120.5 (1.2)	10.0 (0.4)	6.3 (0.8)
SI	High	80.6 (0.4)	6.0 (0.5)	120.3 (2.7)	10.2 (0.6)	6.2 (0.7)
SI	Low	80.5 (0.5)	5.9 (0.4)	120.9 (1.1)	10.0 (0.6)	6.1 (0.7)
DS	High	70.5 (1.9)	10.0 (0.6)	120.9 (1.0)	10.2 (0.6)	6.1 (0.6)
DS	Low	70.5 (1.7)	9.8 (0.6)	120.1 (2.6)	10.1 (0.6)	6.1 (0.6)
EC	–	879.8 (90.8)	109.2 (22.3)	120.7 (1.1)	10.3 (0.5)	6.0 (0.7)
Control	None	–	–	120.3 (1.0)	10.3 (0.5)	6.0 (0.6)
$F_{7,152}$	–	–	–	0.62	1.22	0.40
p	–	–	–	0.74	0.30	0.90

For the same intruder, the difference in SVL and BM between different frequencies of intrusion is insignificant. For ML, the differences of SVL, BM, and CS among eight experimental treatments were also non-significant with the results of the F-test and p-values given at the bottom of the table

significant, indicating that there is no difference between the experimental treatments for SVL, BM and CS. For invaders, including *Oligodon formosanus* (OF), *Sphenomorphus incognitus* (SI) and *Diploderma swinhonis* (DS), the mean SVL and BM of female ML did not differ significantly between high and low frequency invasion treatments using t-tests.

7.4.1 Female Long-Tailed Skinks React Quickly to Intruders

Females in nests with eggs typically reacted strongly to both low-frequency and high-frequent invasions by defendable invaders like *O. formosanus* (Fig. 7.7). While *D. swinhonis* and *S. incognitus* invaders were present at both low and high frequencies of intrusion, the majority of long-tailed skinks chose to ignore them. However, the observing *E. longicaudata* immediately fled and never came back to the nest after coming into contact with the predatory snake *E. carinata*. Within 2 h of the invasion test, every escaped female—except from the *E. carinata* treatment—returned to the nest. Female ML responses to invaders are species-specific. The

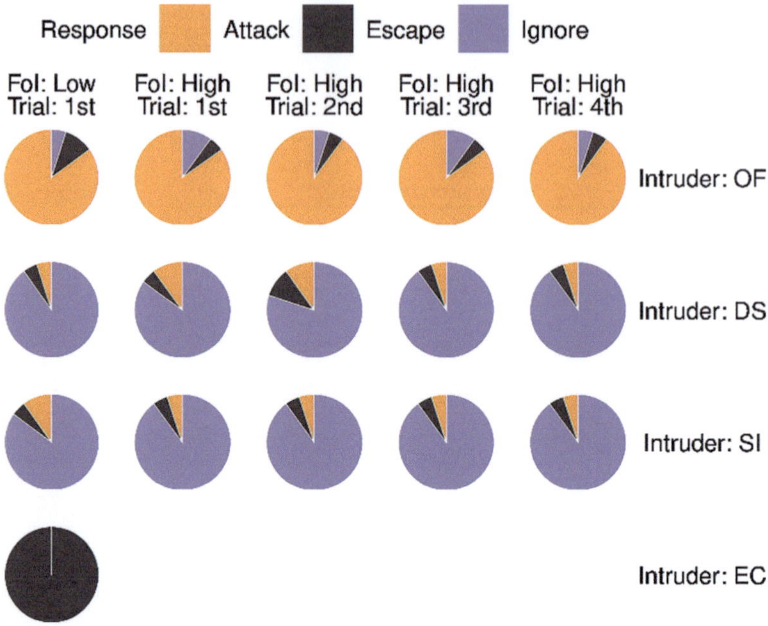

Fig. 7.7 Response of *Eutropis longicaudata* to intruders with low- and high-frequency intrusion. Abbreviation: *FoI* frequency of intruding, *OF Oligodon formosanus*, *DS Diploderma swinhonis*, *SI Sphenomorphus incognitus*, *EC Elaphe carinata*

Fisher-exact test results also demonstrated that EL's response to particular invaders was constant regardless of the frequency and quantity of encounters.

7.4.2 Relationship Between Time Spent on Parental Care and Invasion Frequency

The amount of time spent providing parental care for each experimental treatment is depicted in Fig. 7.8. Since the female long-tailed skink always escaped and never returned to the nest when the invader was *E. carinata*, this treatment was not included in the analysis because parental data were not available. The results of the one-way ANOVA showed significant differences in time spent on parental care ($F_{6,133} = 26.73$, $p < 0.001$). The results of the Tukey HSD showed that the *p*-values for pairwise comparisons for all treatments of *O. formosanus* involving high-frequency invasion were smaller than 0.001, and the *p*-values for the rest of the comparisons were larger than 0.5. Therefore, the significant result of ANOVA was due to the prolonged egg protection time of EL in the high frequency invasive *O. formosanus* treatment. Next, we conducted a two-way ANOVA to test the effects of species and invasion frequency and their interactions. Both species and invasion frequency were significant, $\chi^2 = 22.8$ (d.f. = 3) and $\chi^2 = 31.2$ (d.f. = 1), $p < 0.001$

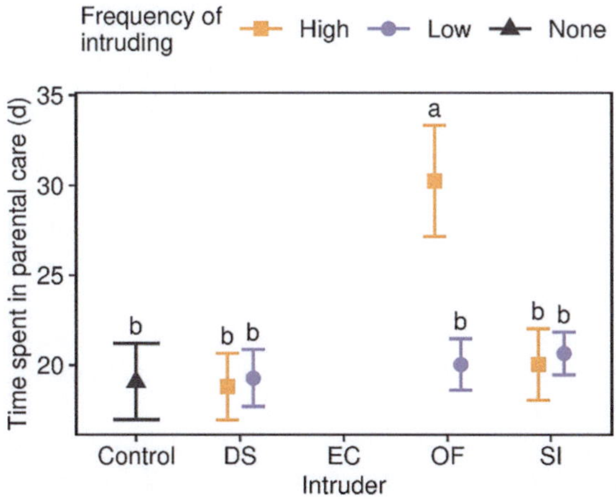

Fig. 7.8 The female *Eutropis longicaudata*, while guarding her eggs, primarily targeted *Oligodon formosanus* (OF) for attacks and paid little attention to other coexisting lizard species. In the control group, the time spent on parental care significantly differed from the high-frequency intrusion by *O. formosanus* treatment. However, no such differences were observed with the low-frequency intrusion or with other sympatric lizard treatments, namely *Diploderma swinhonis* (DS) and *Sphenomorphus incognitus* (SI). Whenever I introduced the snake species *Elaphe carinata* (EC) into the nest hole, the female *E. longicaudata* promptly fled and never returned. Different letters above the bars represent statistically significant distinctions. (Redrawn from Huang and Wang (2009), Ethology)

and $p < 0.001$, respectively. In addition, the interaction term between species and invasion frequency was also significant ($\chi^2 = 30.4$ (d.f. $= 2$), $p < 0.001$).

7.4.3 Discussion

At both the high and low frequency invasions, *Eutropis longicaudata* reacted very aggressively to the egg-eating snakes (*O. formosanus*), attacking and driving them out of the nest, whereas long-tailed skinks typically do not react aggressively to the presence of harmless sympatric lizards (*D. swinhonis* or *S. incognitus*) in the nest. The female *E. longicaudata*, however, always managed to escape the nest after we introduced the stinky snake *E. carinata*, and the long-tailed skink never went back to the nest after that (Fig. 7.8). According to our findings, the advantages of defense or abandonment may frequently depend on the females' capacity to protect their nests or the likelihood that abandoned females will at least make it out alive. Our findings also suggest that *E. longicaudata* responds differently to non-predatory lizards, predatory snakes (*E. carinata*), and snakes that prey on eggs(*O. formosanus*). The maternal responses to particular predators in an earlier study by are fully consistent with the experimental results presented here for the high frequency treatment (Huang, 2006c).

Surprisingly, compared to the other treatments, parental care time was longer in the *O. formosanus* in the high-frequency invasion treatment. This suggests that *E. longicaudata* may, in the presence of egg-eating snakes, modify its expenditures in accordance with the benefits of its offspring, which is an incidental relationship. Since the *E. longicaudata* population on Orchid Island (which exhibits maternal care) is from the mainland Taiwan [where long-tailed skinks do not exhibit care (Huang, 2006c)]. We may be able to trace the evolutionary pathways that led from shorter (part-time) to longer (full-time) parental care from the current study, which may be significant in revealing the evolutionary transition of parental care within a population. Our findings imply that time spent on parental care (whether long- or short-term) may be affected by predators, and that the length of this behavior is influenced by maternal defensiveness.

But why does parental care take longer in *E longicaudata* when *O. formosanus* invasion treatments are more common? One hypothesis is that the longer parents invest in defending offspring, (1) the greater the risk to future reproductive potential; (2) the greater the value of present offspring compared to the parents' expected future contribution; and (3) the expected increase in offspring welfare resulting from this parental expenditure (Pressley, 1981). Thus, natural selection should favor an optimal level of defense based on the risks borne by the parent and the expected benefits received by the offspring (Shine, 1988).

A dramatic drop in hatching success (70%) after the experimental removal of females in the previous chapter revealed the adaptive significance of egg guarding expansion in *E. longicaudata* (Huang, 2006c). The fact that few females abandon their eggs in the wild provides more evidence to the ideas that guarding eggs fundamentally enhances parental fitness and that parental care extensions are primarily intended to reduce the risk of predation by predators.

Parents of multi-nesting species may experience conflicts between reproduction, incubation, and territory establishment due to time constraints, energy constraints, or both (Sanz & Tinbergen, 1999). In birds, the size of the nest is typically thought to be constrained by either (1) the parents' capacity to raise their young (Lack, 1947), depending on their conditions and the quality of their territory or (2) the costs associated with raising an excessive number of young (subsequent fecundity and/or survival) (Daan et al., 1996). However, unlike birds, *E. longicaudata* does not raise or guard its young. Longer care may be a better strategy because there is no obvious correlation between the cost of parental care and the length of care (Huang, 2007a).

We draw the conclusion that long-tailed skinks' parental care and nest abandonment strategies appear to be adaptive and precisely calibrated to the threats posed by other reptiles. The extended duration of parental care in the wild by female skinks further supports the notion that egg guarding fundamentally improves parental fitness.

Since only the long-tailed skinks on Orchid Island exhibit parental care, pushing snakes out of their burrows after biting them, does this indicate that the biting force of this population is greater than that of other populations? Because the care period lasts about a month, the mother lizards must go out to forage, so does the distance to their home affect the success rate of their return? If the eggs are exposed, will the mother lizards return home faster? Since geckos also lay eggs in concrete holes, are there differences in nesting sites between these two species of lizards? The next chapter will provide a detailed discussion.

References

Balshine-Earn, S., & Earn, D. J. D. (1998). On the evolutionary pathway of parental care in mouth-brooding cichlid fish. *Proceedings of the Royal Society B: Biological Sciences, 265*, 2217–2222.

Blumstein, D. T. (2006). The multipredator hypothesis and the evolutionary persistence of antipredator behavior. *Ethology, 112*, 209–271.

Brown, J. R., Ye, H. R., Bronson, T., Dikkes, P., & Greenberg, M. E. (1996). A defect in nurturing in mice lacking the immediate early gene fosB. *Cell, 86*, 297–309.

Burley, N. T., & Johnson, K. (2002). The evolution of avian parental care. *Philosophical Transactions of the Royal Society of London Series B: Biological Sciences, 357*, 241–250.

Calsbeek, R., & Smith, T. B. (2003). Ocean currents mediate evolution in Island lizards. *Nature, 426*, 552.

Clutton-Brock, T. H. (1991). *The evolution of parental care*. Princeton University Press.

Coleman, K., Rothfuss, L. A., Ota, H., & Kardong, K. V. (1993). Kinematics of egg-eating in the specialized Taiwan snake, *Oligodon formosanus* (Colubridae). *Journal of Herpetology, 27*, 320–327.

Cooper, J. W. E., & Burghardt, G. M. (1990). A comparative analysis of scoring methods for chemical discrimination of prey by squamate reptiles. *Journal of Chemical Ecology, 16*, 45–65.

Daan, S., Deerenberg, C., & Dijkstra, C. (1996). Increased daily work precipitates natural death in the kestrel. *Journal of Animal Ecology, 65*, 539–544.

Downes, S. J., & Adams, M. (2001). Geographic variation in antisnake tactics: The evolution of scent-mediated behavior in a lizard. *Evolution, 55*, 605–615.

Downes, S. J., & Bauwens, D. (2002). Does reproductive state affect a lizard's behavior toward predator chemical cues? *Behavioral Ecology and Sociobiology, 52*, 444–450.

Dunn, P. O., & Robertson, R. J. (1992). Geographic variation in the inportance of male parental care and mating systems in tree swallows. *Behavioral Ecology, 3*, 291–299.

Dynesius, M., & Jansson, R. (2000). Evolutionary consequences of changes in species' geographical distributions driven by Milankovitch climate oscillations. *Proceedings of the National Academy of Sciences, 97*, 9115–9120.

Ferrari, M. C. O., Trowell, J. J., Brown, G. E., & Chivers, D. P. (2005). The role of learning in the development of threat-sensitive predator avoidance by fathead minnows. *Animal Behaviour, 70*, 777–784.

Ferrari, M. C. O., Messier, F., & Chivers, D. P. (2008). Can prey exhibit threat-sensitive generalization of predator recognition? Extending the predator recognition continuum hypothesis. *Proceedings of the Royal Society B: Biological Sciences, 275*, 1811–1816.

Forester, D. C. (1979). Homing to the nest by female mountain dusky salamanders (*Desmognathus ochrophaeus*) with comments on the sensory modalities essential to clutch regnition. *Herpetologica, 35*, 330–335.

Forester, D. C., Harrison, K., & McCall, L. (1983). The effects of isolation, the duration of brooding, and non-egg olfactory cues on clutch recognition by salamanders, *Desmognathus ochrophaeus*. *Journal of Herpetology, 17*, 308–314.

Greene, H. W., May, P. G., Hardy, D. L., Sr., Sciturro, J. M., & Farrell, T. M. (2002). Parental behavior by vipers. In G. W. Schuett, M. Hoggren, M. E. Douglas, & H. W. Greene (Eds.), *Biology of the vipers* (pp. 179–205). Eagle Mountain Publications.

Gross, M. R. (2005). The evolution of parental care. *The Quarterly Review of Biology, 80*, 37–46.

Gübitz, T., Thorpe, R. S., & Malhotra, A. (2000). Phylogeography and natural selection in the Tenerife gecko *Tarentola delalandii*: Testing historical and adaptive hypotheses. *Molecular Ecology, 9*(9), 1213–1221.

Hairston, J. N. G., Ellner, S. P., Geber, M. A., Yoshida, T., & Fox, J. A. (2005). Rapid evolution and the convergence of ecological and evolutionary time. *Ecology Letters, 8*, 1114–1127.

Harrison, S. P., Yu, G., Takahara, H., & Prentice, I. C. (2001). Diversity of temperate plants in East Asia. *Nature, 413*, 129–130.

Head, M. L., Doughty, P., Blomberg, S. P., & Keogh, J. S. (2008). Chemical mediation of reciprocal mother-offspring recognition in the southern water skink *(Eulamprus heatwoli)*. *Austral Ecology, 33*, 20–28.

Heulin, B., Surget-Groba, Y., Guiller, A., Guillaume, C. P., & Deunff, J. (1999). Comparisons of mitochondrial DNA (mtDNA) sequences (16S rRNA gene) between oviparous and viviparous strains of *Lacerta vivipara*: A preliminary study. *Molecular Ecology, 8*, 1627–1631.

Hinton, H. E. (1981). *Biology of insect eggs*. Pergamon Press.

Huang, W. S. (1994). Report on egg clutch size of the long-tailed skink, *Mabuya longicaudata* from Taiwan. *Journal of taiwan museum, 47*, 45–47.

Huang, W. S. (2004). *Reptile ecology and the evolution of parental care on a tropical Asian Island*. PH. D. Cornell University.

Huang, W. S. (2006a). Ecological characteristics of the skink, *Mabuya longicaudata*, on a tropical east Asian Island. *Copeia, 2006*, 293–300.

Huang, W. S. (2006b). Ecology and reproductive patterns of the grass lizard, *Takydromus sauteri*, in a tropical rain forest of an East Asian Island. *Journal of Herpetology, 40*, 267–273.

Huang, W. S. (2006c). Parental care in the long-tailed skink, *Mabuya longicaudata,* on a tropical Asian Island. *Animal Behaviour, 72*, 791–795.

Huang, W. S. (2007a). Costs of egg caring in the skink, *Mabuya longicaudata*. *Ecological Research, 22*, 659–664.

Huang, W. S. (2007b). Ecology and reproductive patterns of the agamid lizard, *Japalura swinhonis*, on an East Asian Island, with comments on the small clutch sizes of Island lizrds. *Zoological Science, 24*, 181–188.

Huang, W. S. (2008a). Predation risk of whole-clutch filial cannibalism in a tropical skink with maternal care. *Behavioral Ecology, 19*, 1069–1074.

Huang, W. S. (2008b). Resources exploitation by ants facilitates lizard egg survival. *Ecological Entomolology, 33*, 555–559.

Huang, W. S., & Pike, D. A. (2011a). Climate change impacts on fitness depend on nesting habitat. *Functional Ecology, 25*, 1125–1136.

Huang, W. S., & Pike, D. A. (2011b). Determinants of homing in nest-guarding females: Balancing risks while travelling through unfamiliar landscapes. *Animal Behaviour, 82*, 263–270.

Huang, W. S., & Wang, H. Y. (2009). Predation risks and anti-predation parental care behavior: An experimental study in a tropical skink. *Ethology, 115*, 273–279.

Huang, W. S., Greene, H. W., Chang, T. J., & Shine, R. (2011). Territorial behavior in Taiwanese kukrisnakes *(Oligodon formosanus)*. *Proceedings of the National Academy of Science, 108*, 7455–7459.

Huang, W. S., Lin, S. M., Dubey, S., & Pike, D. A. (2013). Predation drives interpopulation differences in parental care expression. *Journal of Animal Ecology, 82*(2), 429–437.

Klug, H., & Bonsall, M. B. (2009). Life history and the evolution of parental care. *Evolution, 64*, 823–835.

Lack, D. (1947). The significance of clutch size. *Ibis, 89*, 302–352.

Lack, D. (1968). *Ecological adaptations for breeding in birds*. Methuen and.

Lavery, R. J., & Keenleyside, M. H. A. (1990). Filial cannibalism in the biparental fish *Cichlasoma nigrofasciatum* (Pisces: Cichlidae) in response to early brood reductions. *Ethology, 86*, 326–338.

Lena, J. P., & de Fraipont, M. (1998). Kin recognition in the common lizard. *Behavioral Ecology and Sociobiology, 42*, 341–347.

Lin, H.-C., Li, S. H., Fong, J., & Lin, S. M. (2008). Ventral coloration differentiation and mitochondrial sequences of the Chinese cobra (*Naja atra*) in Taiwan. *Conservation Genetics, 9*, 1089–1097.

Lindström, K. (1998). Effects of costs and benefits of brood care on filial cannibalism in the sand goby. *Behavioral Ecology and Sociobiology, 42*, 101–106.

Lindström, K., & Sargent, R. C. (1997). Food access, brood size and filial cannibalism in the fantail darter, *Etheostoma flabellare*. *Behavioral Ecology and Sociobiology, 40*, 107–110.

Lourdais, O., Brischoux, F., DeNardo, D., & Shine, R. (2004). Protein catabolism in pregnant snakes (*Epicrates cenciria maurus*, Boidae) compromises musculature and performance after reproduction. *Journal of Comparative Physiology, 174*, 383–391.

Lourdais, O., Brischoux, F., Shine, R., & Bonnet, X. (2005). Adaptive maternal cannibalism in snakes (Epicrates cenchria maurus, Boidae). *Biological Journal of the Linnean Society, 84*, 767–774.

Macedo, R. H. F., Cariello, M., & Muniz, L. (2001). Context and frequency of infanticide in communally breeding Guira cuckoos. *The Condor, 103*, 170–175.

Magnhagen, C. (1992). Parental care and predation risk in fish. *Annales Zoologici Fennici, 29*, 227–232.

Main, A. R., & Bull, C. M. (1996). Mother-offspring recognition in two Australian lizards, *Tiliqua rugosa* and *Egernia stokesii*. *Animal Behaviour, 52*, 193–200.

Manica, A. (2002a). Alternative strategies for a father with a small brood: Mate, cannibalise or care. *Behavioral Ecology and Sociobiology, 51*, 319–323.

Manica, A. (2002b). Filial cannibalism in teleost fish. *Biological Reviews, 77*, 261–277.

Manica, A. (2004). Parental fish change their cannibalistic behaviour in response to the cost-to-benefit ratio of parental care. *Animal Behaviour, 67*, 1015–1021.

Manica, A., & Johnstone, R. A. (2004). The evolution of paternal care with overlapping broods. *The American Natualist, 164*, 517–530.

Martins, M., Pombal, J. P., & Haddad, C. F. B. (1999). Escalated aggressive behaviour and falculative parental care in the nest building gladiator frog, *Hyla faber*. *Amphibia-Reptilia, 19*, 65–73.

Mehlis, M., Bakker, T. C., Enggvist, L., & Frommen, J. G. (2010). To eat or not to eat: Egg-based assessment of paternity triggers fine-tuned decisions about filial cannibalism. *Proceedings of the Royal Society B: Biological Sciences, 277*, 2627–2635.

O'Connor, D., & Shine, R. (2004). Parental care protects against infanticide in the lizard *Egernia saxatilis* (Scincidae). *Animal Behaviour, 68*, 1361–1369.

Okuda, N., & Yanagisawa, Y. (1996a). Filial cannibalism by mouthbrooding males of the cardinal fish, *Apogon doederleini*, in relation to their physical condition. *Environmental Biology of Fishes, 45*, 397–404.

Okuda, N., & Yanagisawa, Y. (1996b). Filial cannibalism in a paternal mouthbrooding fish in relation to mate availability. *Animal Behaviour, 52*, 307–314.

O'Riain, M. J., Bennett, N. C., Brotherton, P. N. M., McIlrath, G., & Clutton-Brock, T. H. (2000). Reproductive suppression and inbreeding avoidance in wild populations of co-operatively breeding meerkats (Suricata *suricatta*). *Behavioral Ecology and Sociobiology, 48*, 471–477.

Paulo, O. S., Dias, C., Bruford, M. W., Jordan, W. C., & Nichols, R. A. (2001). The persistence of Pliocene populations through the Pleistocene climatic cycles: Evidence from the phylogeography of an Iberian lizard. *Proceedings of the Royal Society of London. Series B: Biological Sciences, 268*(1476), 1625–1630.

Perrone, M., & Zaret, T. M. (1979). Parental care patterns of fishes. *The American Naturalist, 113*, 351–361.

Persson, O., & Öhrström, P. (1989). A new avian mating system: Ambisexual polygamy in the penduline tit *Remiz pendulinus*. *Ornis Scandinavica, 20*, 105–111.

Petersen, C. W. (1990). The occurrence and dynamics of clutch loss and filial cannibalism in two Caribbean damselfishes. *Journal of Experimental Marine Biology and Ecology, 135*, 117–133.

Peterson, M. G. (2000). Nest, but not egg, fidelity in a territorial salamander. *Ethology, 106*, 781–794.
Pressley, P. H. (1981). Parental effort and the evolution of nest-guarding tactics in the threespine stickleback, *Gasterosteus aculeatus* L. *Evolution, 35*, 282–295.
Reynolds, J. D., Goodwin, N. B., & Freckleton, R. P. (2002). Evolutionary transitions in parental care and live bearing in vertebrates. *Philosophical Transactions of the Royal Society of London, Series B, 357*, 269–281.
Roff, D. (2002). *Life history evolution*. Sinauer Associates, Inc.
Rohwer, S. (1978). Parent cannibalism of offspring and egg raiding as a courtship strategy. *The American Naturalist, 112*, 429–440.
Rosenblatt, J. S. (1994). Psychobiology of maternal behavior: Contribution to the clinical understanding of maternal behavior among human. *Acta Paediatrica, Supplement, 83*, 3–8.
Rothstein, S. I. (1975). Mechanisms of avian egg-recognition: Do birds know their own eggs? *Animal Behaviour, 23*, 268–278.
Rothstein, S. I. (1982a). Mechanisms of avian egg recognition: Which egg parameters elict responses by rejecter species. *Behavioral Ecology and Sociobiology, 11*, 229–239.
Rothstein, S. I. (1982b). Success and failures in avian egg and nestling recognition with comments on the utility of optimality reasoning. *American Zoologist, 22*, 547–560.
Salthe, R., & Mechan, J. S. (1974). Reproductive and courtship patterns. In B. Lofts (Ed.), *Phisiology of the Amphibian* (pp. 309–522). Academic Press.
Sanz, J. J., & Tinbergen, J. M. (1999). Energy expenditure, nestling age, and brood size: An experimental study of parental behavior in the great tit *Parus major*. *Behavioral Ecology, 10*, 598–606.
Sargent, R. C. (1992). Ecology of filial cannibalism in fish: Theoretical perspective. In M. A. Elgar & B. J. Crespi (Eds.), *Cannibalism: Ecology and evolution among diverse taxa* (pp. 38–62). Oxford University Press.
Schwank, E. J. (1986). Filial cannibalism in *Tilapia mariae*. *Journal of Applied Ichthyology, 2*, 65–74.
Shang, G. (2001). *Natural portraits of lizards of Taiwan*. Big Tree Press.
Sherman, P. T. (2003). Intergroup infanticide in cooperatively polyandrous white-winged trumpeters (*Psophia leucoptera*). *The Wilson Bulletin, 115*, 339–342.
Shine, R. (1988). Parental care in reptiles. In C. Gans & R. B. Huey (Eds.), *Biology of the Reptilia Vol. 16 Ecology B. Defense and life history* (pp. 275–330). Alan R. Liss.
Sinn, D. L., While, G. M., & Wapstrap, E. (2008). Maternal care in a social lizard: Links between female aggression and offspring fitness. *Animal Behaviour, 76*, 1249–1257.
Somma, L. A. (2003). *Parental behavior in lepidosaurian and testudinian reptiles: A literature survey*. Krieger Publishing Company.
Stahlschmidt, Z. K. (2011). Taxonomic chauvinism revisted: Insight from parental care research. *PLoS One, 6*, e24192.
Székely, T., & Cuthill, I. C. (1999). Brood desertion in Kentish plover: The value of parental care. *Bchavioral Ecology, 10*, 191–197.
Tallamy, D. W. (1984). Insect parental care. *Bioscience, 34*, 20–24.
Tallamy, D. W., & Wood, T. K. (1986). Convergence patterns in subsocial insects. *Annual Review of Entomology, 31*, 20–24.
Thomas, L. K., & Manica, A. (2003). Filial cannibalism in an assassin bug. *Animal Behaviour, 66*, 205–210.
Vidal, N., Azvolinsky, A., Cruaud, C., & Hedges, S. B. (2008). Origin of tropical American burrowing reptiles by transatlantic rafting. *Biology Letters, 4*, 115–118.
Vitt, L. J., & Cooper, W. E., Jr. (1989). Maternal care in skinks (*Eumeces*). *Journal of Herpetology, 23*, 29–34.
Webb, J. N., Houston, A. I., McNamara, J. M., & Szekely, T. (1999). Multiple patterns of parental care. *Animal Behaviour, 58*, 983–993.

Webb, J. K., Du, W. G., Pike, D. A., & Shine, R. (2009). Chemical cues from both dangerous and nondangerous snakes elicit antipredator behaviours from a nocturnal lizard. *Animal Behaviour, 77*, 1471–1478.

Webb, J. K., Du, W.-G., Pike, D. A., & Shine, R. (2010). Generalization of predator recognition: Velvet geckos display anti-predator behaviours in response to chemicals from non-dangerous elapid snakes. *Current Zoology, 56*, 337–342.

Wesolowski, T. (1994). On the origin of parental care and the early evolution of male and female parental-roles in birds. *The American Naturalist, 143*, 39–58.

Wheelwright, N. T., Schultz, C. B., & Hodum, P. J. (1992). Polygyny and male parental care in Savannah sparrowa: Effects on female fitness. *Behavioral Ecology and Sociobiology, 31*, 279–289.

While, G. M., Uller, T., & Wapstra, E. (2009). Family conflict and the evolution of sociality in reptiles. *Behavioral Ecology, 20*, 245–250.

Wiewandt, T. A. (1982). Evolution of nesting patterns in iguanine lizards. In G. M. Burghardt & A. S. Rand (Eds.), *Iguanas of the world: Their behavior, ecology and conservation* (pp. 119–141). Noyes Publications.

Williams, G. C. (1966). Natural selection, the costs of reproduction, and a refinement of Lack's principle. *The American Naturalist, 100*, 687–690.

8 Parental Care and Nesting Selection

Abstract

Environmental variation can drive the evolution of morphology through selection on performance traits such as bite force, which is critical for prey use, competition, defense, and mating in lizards. On Orchid Island, long-tailed skinks (*Eutropis longicaudata*) face intense egg predation from (*Oligodon formosanus*) and uniquely exhibit parental care. We tested whether these skinks have stronger bite force than populations without egg guarding. Three hypotheses were examined: (1) Orchid Island skinks have higher bite force due to parental care; (2) males show greater bite force than females due to sexual selection; and (3) head morphology predicts bite force. Results show Orchid Island skinks possess stronger bites, likely linked to predation pressure and parental behavior. Males exhibit higher bite force, and body mass and head size positively correlate with performance.

Keywords

Bite force · Predation · Long-tailed skink · Kukri snake · Parental care · Homing behavior · Nest site selection

8.1 Great Bite Force in a Skink Population Is Driven by Parental Behavior

8.1.1 Introduction

Differences in selective power and traits across populations have an impact on and modify phenotypic performance, which is frequently related to survival and fitness. Such behavior might be an indication of local adaptation to different selection pressures resulting from the abiotic environment (Dudley, 1996) or biotic factors

W.-S. Huang, *Reptiles in the Changing Climate*, Fascinating Life Sciences,
https://doi.org/10.1007/978-981-92-1887-5_8

like the species community composition (Reznick et al., 1990; Reznick et al., 2001). As a result, morphological characteristics like cryptic or conspicuous colors for predator recognition (Ruxton et al., 2004; Chouteau & Angers, 2011), feeding devices that are used to compete for food (Herrel et al. 2001; Grant & Grant, 2006), and songs or colors that are used to indicate mate preference (Jiggins et al., 2001; Seddon, 2005), evolved as an adaptive strategy. One of the best-known cases of morphological trait adaptation is the surprising coloration of poisonous frogs, which functions as an indication of aversion and mating signals, providing local adaptation advantages, warning predators (Saporito et al., 2007; Noonan & Comeault, 2009), and attracting mates (Summers et al., 1999; Reynolds & Fitzpatrick, 2007). Insularity has been demonstrated to have an impact on a variety of organismal traits, including morphological traits (Clegg & Owens, 2002; Lomolino, 2005), behavior (Huang, 2006c; Donihue et al., 2016), reproductive strategies and life histories (Covas, 2012; Novosolov et al., 2013). Some island populations of lizards exhibit stronger bite force than mainland populations, possibly as a result of adaptation to a particular prey species or population density (Sagonas et al., 2014). Changes in bite force are therefore viewed as indicators of adaptation because they are used to address ecologically relevant challenges. Additionally, bite force is a functional trait linked to both intra- and interspecific competitions. The former includes prey composition (Herrel et al. 2001; Measey et al., 2011) and predator defense (Herrel et al., 2009; de Barros et al., 2010), whereas the latter is primarily associated with sexual selection (Lailvaux et al., 2004; Huyghe et al., 2005; Husak et al., 2006; Cameron et al., 2013).

Multiple studies have shown that a stronger bite improves food handling efficiency (Verwaijen et al., 2002), territoriality (Lappin & Husak, 2005), and predator defense (de Barros et al., 2010). Additionally, lizards frequently engage in sexual selection, and the strength of the bite is essential for both mate choice and territorial defense. Greater bite size increases the likelihood of successful mating and reproduction in an individual (Lappin & Husak, 2005; Lailvaux & Irschick, 2007; Husak et al., 2009). In general, males exhibit larger bites than females due to different trade-offs between selection pressures (Herrel et al., 2010; Vanhooydonck et al., 2010; Cameron et al., 2013). No studies have looked at how parental care affects bite performance, despite the fact that many have focused on variation in bite performance between populations or species.

According to some morphological characteristics, particularly head size and shape, bite force is closely correlated with these characteristics. (Kaliontzopoulou et al., 2012; Sagonas et al., 2014; Žagar et al., 2017). As a result of adaptation to predation pressure and use of microhabitats (such as small crevices), the head shape of some rock-dwelling lizards may change (Herrel et al. 2001; Žagar et al., 2017). In comparison to flatter heads, larger and taller heads had larger bite forces. Because of the greater available space and physiological cross-sectional area of the jaw adductors, head height correlates with bite force (Herrel et al. 2001). On the other hand, in other lizards, the morphology of the head may restrict bite performance (Kaliontzopoulou et al., 2012; Broeckhoven & Mouton, 2014).

Fig. 8.1 (**a**) Study locations of Kenting (south Taiwan), Green Island, and Orchid Island. (**b**) Long-tailed skink (*Eutropis longicaudata*) staying at the entrance of a concrete hole. (Adapted from Tseng et al. (2018), Journal of Zoology, with permission. Map redrawn; photo modified)

Eutropis longicaudata, the long-tailed skink, is found in southern China, Vietnam, Laos, Cambodia, Thailand, and Taiwan (Das, 2010). Long-tailed skinks consume a variety of insect species, including Orthoptera, Coleoptera, and Hemiptera (Huang, 2006a). Because the temperature inside the holes was warmer than the outside, approximately half of the skinks were discovered inside holes in concrete walls (Huang, 2006c). Sometimes in the field, two skinks would engage in violent chases. Orchid Island, off the southeast coast of Taiwan, is home to a single species group that uses unusual parental care techniques to shield its eggs from the kukri snake's (*Oligodon formosanus*) hungry predation (Huang, 2006c; Huang & Wang, 2009). When a long-tailed skink's nest is invaded by a kukri snake, the female skink defends the eggs by biting the intruder fiercely until it flees. This behavior helps eggs survive more successfully (Huang, 2006c), and it may be caused by the high predation pressure found in island habitats(Huang et al., 2013).

This chapter compared three populations of long-tailed skinks that live on mainland Taiwan (Kenting), Green Island, and Orchid Island in terms of their morphological traits and ability to bite (Fig. 8.1a, b). Our hypothesis was that the Orchid Island population's unique maternal care led to a stronger bite force than that of the Green Island and Kenting populations. Additionally, we proposed that males in all three populations had more powerful bite forces than females as a result of competition between sexes. Finally, we explored which morphological features influence bite force most significantly and hypothesized that bite force has a positive correlation with head features.

8.1.2 Density of Long-Tailed Skinks in Three Populations

Findings between 1997 and 2001, there were three populations of long-tailed skinks: Kenting (mean SD, 0.0138 ± 0.00160 #/m2; $n = 5$), Orchid Island (0.00149 ± 0.00146 #/m2; $n = 5$), and Green Island (mean SD, 0.0160 ± 0.002 #/m2; $n = 5$). There were no significant variations in mean population density (type III Wald test, $\chi^2 = 1.590$, DF = 2, $p = 0.452$). Population density was not significantly impacted by survey year as a random factor (type III Wald test, $\chi^2 = 1.770$, DF = 1, $p = 0.183$).

8.1.3 Bite Force Variation Across Populations and Between Sexes

On Orchid Island, Green Island, and Kenting, a total of 44 long-tailed skinks (27 females and 17 males), 66 (36 females and 30 males), and 48 (32 females and 16 males) were caught. Significant variations in bite force between sexes and populations were discovered. The marginal difference in bite force between males and females was roughly 5.787 N (males > females, $p < 0.001$; Table 8.1A; Fig. 8.2a), indicating that males exhibited greater bite force than females in all populations after excluding the insignificant interaction term [sex location, $F_{2, 152} = 0.082$, $p = 0.921$]. Several comparisons among the sites revealed that Orchid Island lizards had significantly more force in their bites than Kenting and Green Island lizards did. There was no significant difference in the bite force in Green Island and Kenting populations(Table 8.1A; Fig. 8.2a).

All five morphological covariates were included in the model to account for how morphological traits affected bite force. Marginal differences in sex and location were similar to the model without covariates after removing insignificant interactions and quadric terms (Table 8.1B; Fig. 8.2b). The model also indicated that BW and HL had a positive correlation with bite force, explaining 29% of the variance, while sex and location only contributed 18%. (Table 8.1B; Fig. 8.2c, d). A low correlation (13%; Table 8.1B; Fig. 8.2e, f) was found between lizard length, width, and height and bite force. Additionally, we discovered that the HW quartiles were significant, showing that extremely high HW may be a factor in increased bite force (Table 8.1B; Fig. 8.2g).

8.1.4 Morphological Variations Among the Three Populations and the Sexes

PERMANOVA analysis of morphological comparisonsrevealed significant differences between locations and between sexes after excluding the non-significant interaction term [sex location, $F_{2, 152} = 0.082$, $p = 0.921$]. (Table 8.2A). There were no significant differences in length and body weight between sexes, according to the univariate model (Table 8.2B, C; Fig. 8.3a, b). But it was discovered that there was sexual dimorphism, with males having larger

Table 8.1 Results of the general linear model fitting bite force (N) (A) against sex and location and (B) against additional morphological covariates. Relative importance (RI) describes the total R^2 partitioned by variable

Parameters/comparisons	Estimate	95% CIs	p	RI
(A)				
Intercept	16.811	(15.34, 18.28)	<0.001	–
Sex (male–female)	5.787	(3.954, 7.619)	<0.001	0.201
Location	Type III test: F (2, 154) = 20.193		<0.001	0.174
KT–GI	−1.104	(−3.130, 0.922)[a]	0.402[a]	–
OI–GI	5.308	(2.626, 7.990)[a]	<0.001[a]	–
OI–KT	6.412	(4.028, 8.797)[a]	<0.001[a]	–
(B)				
Intercept	17.00	(15.98, 18.02)	< 0.001	–
Sex (male–female)	2.342	(0.513, 4.172)	0.012	0.068
Location	Type III test: F (2, 148) = 18.662		<0.001	0.111
KT–GI	1.107	(−0.614, 2.828)[a]	0.282[a]	–
OI–GI	4.437	(2.717, 6.158)[a]	<0.001[a]	–
OI–KT	3.330	(1.316, 5.345)[a]	<0.001[a]	–
Body weight (g)	0.274	(0.154, 0.394)	<0.001	0.155
SVL (mm)	−0.045	(−0.196, 0.105)	0.551	0.056
Head length (mm)	0.921	(0.311, 1.532)	0.003	0.130
Head width (mm)	0.715	(−0.024, 1.453)	0.058	0.115
Head width (mm^2)	0.372	(0.183, 0.560)	< 0.001	0.033
Head height (mm)	−0.016	(−0.759, 0.727)	0.965	0.070

[a]Family-wise error rates for multiple comparisons were controlled to 0.05 using the single-step method

Abbreviations: *KT* Kenting, *GI* Green Island, *OI* Orchid Island

heads than females in terms of head size (height, width, and length) (Table 8.2D–F; Fig 8.3c–e).

Lizards from Green Island did not significantly differ from the other two populations, while those from Orchid Island had higher body weights than those from Kenting (Table 8.2B; Fig. 8.3a). All sites showed a general trend in SVL and head length, width, and height, with Kenting lizards having smaller heads than the other two populations (Table 8.2C–F; Fig. 8.3b–e).

8.1.5 Discussion

8.1.5.1 Parental Care Behaviors Shape the Bite Force of Long-Tailed Skinks

Individuals with stronger bites may have heightened aggression toward potential threats because bite force plays a significant role in determining how aggressive interactions when faced with predators (Herrel et al., 2009). Our research suggests that while Green Island and Kenting skinks both have similar bite forces, Orchid

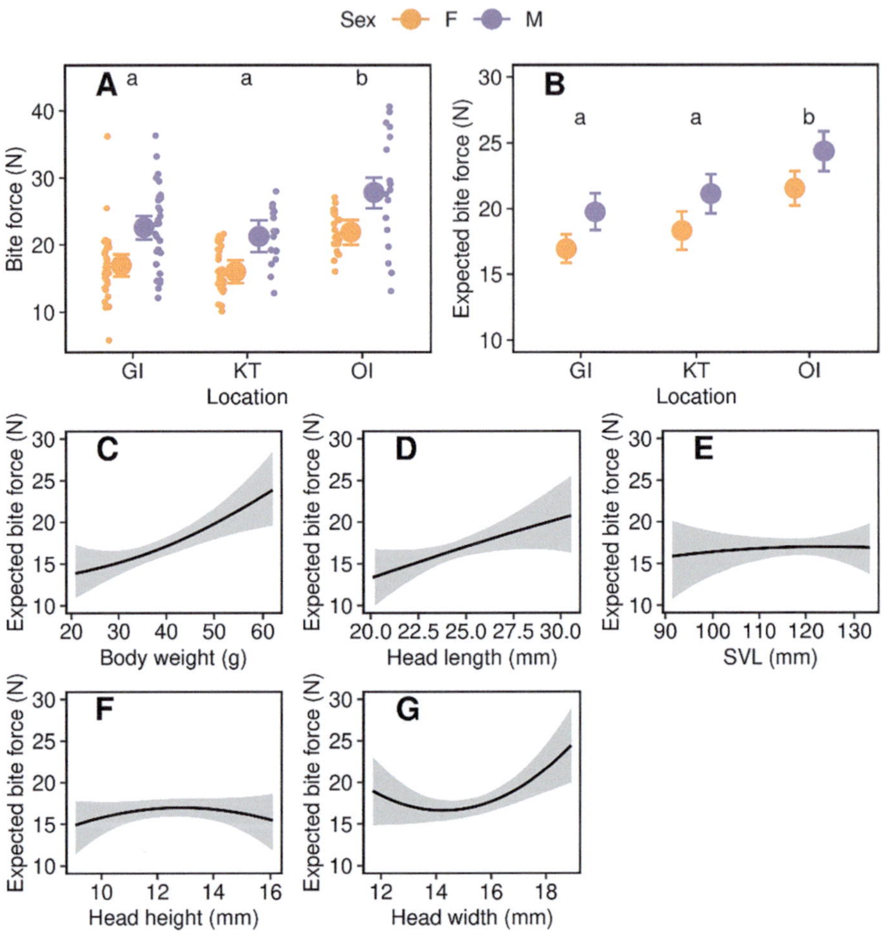

Fig. 8.2 Differences in bite force between sexes and among the three locations, in long-tailed skinks (*Eutropis longicaudata*). (**a**) Original bite force between sexes and among locations. Panel B-G demonstrates the conditional effects of (**b**) population and sex, (**c**) body weight, (**d**) head length, (**e**) SVL, (**f**) head height, and (**g**) head width. Lowercase letters in panels A and B show the results of multiple comparisons when morphological covariates are excluded and included to fit bite force, respectively. Abbreviation: *GI* Green Island, *KT* Kenting, *OI* Orchid Island. (Redrawn from Tseng et al. (2018), Journal of Zoology)

Island skinks have a stronger bite force. Parental care behavior is very widespread in endothermic vertebrates, but extremely rare in reptiles (Somma, 1990), where egg guarding behavior has been documented in only a few species(Greene et al., 2002; Huang, 2006c). Predator deterrence is critical for egg survival in long-tailed skinks. Orchid Island lizards are more vulnerable to predation than the other two populations because kukri snakes are 9–20 times more prevalent there than on Green Island and Kenting (Huang et al., 2013). This suggests that parents may take extra precautions to ensure their offspring's survival. In contrast to other populations, this unusual

Table 8.2 Results of permutational multivariate analysis of variance comparing: (A) five morphological dependent variables between sexes and among locations, and general linear models with corresponding post hoc tests fitted for three locations, (B) body weight, (C) SVL, (D) head length, (E) head width, and (F) head height. Relative importance (RI) describes the total R^2 partitioned by variables

	Comparisons	Estimate	95% CIs	*p*	RI
(A)	Sex	Type III test: F (1, 154) = 24.537		<0.001	0.148
	Location	Type III test: F (1, 154) = 14.809		<0.001	0.151
(B)	Body weight				
	Sex (male–female)	1.921	(−0.770, 4.613)	0.321	0.015
	Location	Type III test: F (1, 154) = 9.39		<0.001	0.111
	KT–GI	−3.157	(−6.586, 0.274)	0.078[a]	–
	OI–GI	3.905	(0.104, 7.706)	0.213[a]	–
	OI–KT	7.063	(3.205, 10.921)	< 0.001[a]	–
(C)	SVL				
	Sex (male–female)	−0.878	(−3.286, 1.531)	0.473	0.002
	Location	Type III test: F (1, 154) = 7.61		< 0.001	0.089
	KT–GI	−5.010	(−8.072, −1.948)	< 0.001[a]	–
	OI–GI	−1.371	(−4.669, 1.928)	1.000[a]	–
	OI–KT	3.639	(0.073, 7.205)	0.044[a]	–
(D)	Head length				
	Sex (male–female)	2.035	(1.487, 2.582)	<0.001	0.280
	Location	Type III test: F (1, 154) = 10.15		<0.001	0.098
	KT–GI	−1.293	(−1.987, −0.599)	<0.001[a]	–
	OI–GI	−0.279	(−1.030, 0.472)	1.000[a]	–
	OI–KT	1.014	(0.218, 1.810)	0.017[a]	–
(E)	Head width				
	Sex (male–female)	1.115	(0.706, 1.523)	<0.001	0.158
	Location	Type III test: F (1, 154) = 10.02		<0.001	0.113
	KT–GI	−1.010	(−1.573, −0.448)	<0.001[a]	–
	OI–GI	−0.073	(−0.645, 0.498)	1.000[a]	–
	OI–KT	0.937	(0.306, 1.568)	0.005[a]	–
(F)	Head height				
	Sex (male–female)	1.246	(0.837, 1.655)	<0.001	0.184
	Location	Type III test: F (1, 154) = 36.93		<0.001	0.250
	KT–GI	−1.680	(−2.160, −1.201)	<0.001[a]	–
	OI–GI	−0.350	(−0.953, 0.254)	1.000[a]	–
	OI–KT	1.330	(0.738, 1.923)	<0.001[a]	–

[a]Type I error rates for multiple comparisons among locations were controlled to 0.05 using the single-step method

Abbreviations: *KT* Kenting, *GI* Green Island, *OI* Orchid Island

behavior might lead to selection for larger bites against dense populations of egg-eating predators in Orchid Island. Only one lizard predator, *Elaphe carinata*, was discovered on Orchid Island in comparison to the high number of predators discovered in Taiwan, including *Protobothrops mucrosquamatus, Viridovipera*

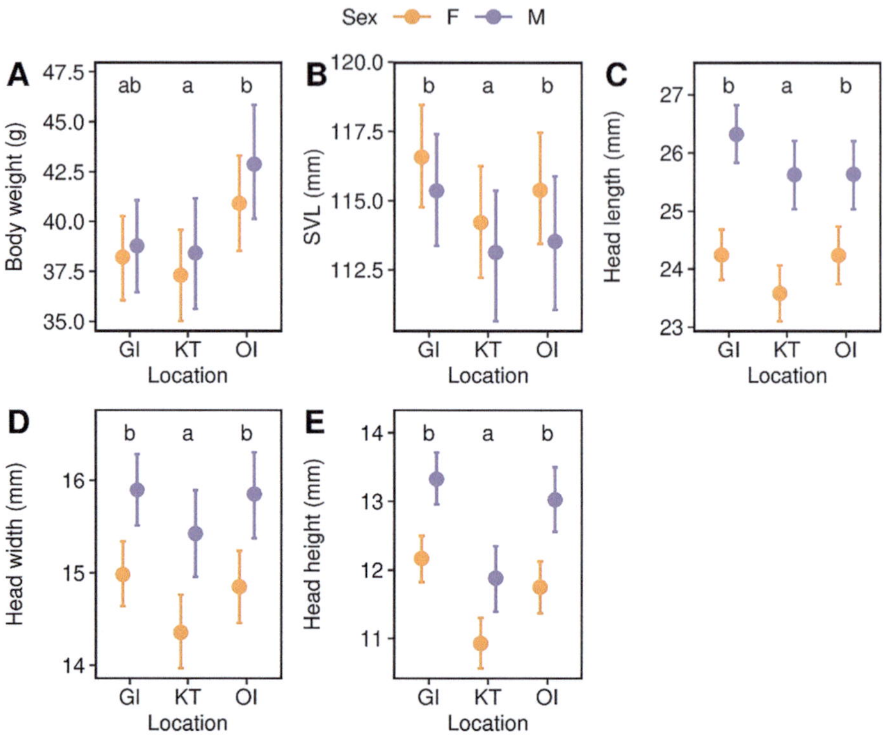

Fig. 8.3 Comparison of morphological variables between sexes and among three locations, in long-tailed skinks (*Eutropis longicaudata*). The variations of (**a**) body weight, (**b**) SVL, (**c**) head length, (**d**) head width, and (**e**) head height between sexes and among populations are shown in box plots. Lowercase letters show the results of post hoc tests comparing the marginal differences among the populations. Abbreviation: *GI* Green Island, *KT* Kenting, *OI* Orchid Island. (Redrawn from Tseng et al. (2018), Journal of Zoology)

stejnegeri, Bungarus multicinctus, and *Naja atra* (Huang et al., 2013). This means that predation pressure is unlikely to be the cause of the Orchid Island long-tailed skink's bite. Although egg guarding was performed by only females on Orchid Island, maternal inheritance may be to blame for males' stronger bites when compared to populations on Green Island and Kenting.

Higher population density may increase intraspecific resource or mate competition (Knell, 2009), which would increase bite force capacity and could explain differences in bite force between populations (Donihue et al., 2016). Our research refuted the hypothesis that population density influences long-tailed skink bite force because it revealed that the Orchid Island population had a stronger bite despite having similar densities to the other two populations. The evolution of bite force was influenced by the unusual parent-child relationship displayed by Orchid Island skinks, though.

Sexual Selection Affects Bite Force in All Populations

Our findings imply that males bite with greater force than females in all three populations, which may reflect different selection pressures on males and females. Sexual selection (intrasexual male competition or intersexual mating bites) and food position allocation are two hypotheses for the sexual dimorphism. Many lizard species exhibit a male advantage in territorial defense and mating success due to sexual selection (Lappin & Husak, 2005; Herrel et al., 2010), in which case the male typically prevails through a stronger bite (Lailvaux et al., 2004; Huyghe et al., 2005; Husak et al., 2006). Wild chases between long-tailed skinks have been seen to be quite violent. This implies that long-tailed skinks engage in antagonistic or competitive behavior.

Several studies have focused on the connection between bite force and foraging apparatus in relation to the delineation of food ecological niches, demonstrating that prey composition affects the bite force of lizards (Verwaijen et al., 2002; Measey et al., 2011; Sagonas et al., 2014). Different sexes may have different diets in the same habitat because males can handle harder prey due to their stronger bite force than females (Verwaijen et al., 2002; Sagonas et al., 2014). Although dietary differences may have an impact on the different bite forces between the sexes, long-tailed skinks consume a variety of prey, including both soft (such as crickets, cicadas, and worms) and hard (such as beetles) prey (Huang, 2006a). Prey were found to be significantly softer than male and female long-tailed skinks when their width and hardness were compared with HW and bite force (Herrel et al. 2001; Wang, 2017) (Fig. 8.4 and Table 8.3). Therefore, prey hardness doesn't appear to be the primary factor influencing the bite force of male and female long-tailed skinks in the current study.

Contribution of Morphological Characteristics to Bite Force

Bite force and head characteristics are positively correlated in some lizard species (Herrel et al., 1996, 1999). Long-tailed skinks' bite force was primarily influenced by BW and HL in the current study, and BW may be a good indicator of the general condition of the muscles that directly influence bite force. However, our findings indicated that bite force correlated positively with HL in long-tailed skinks, but not with HW and HH. This result conflicts with other studies that claim HH and HW are the strongest predictors of bite force in lizards(Herrel et al. 2001, 2004; Lappin et al., 2006), but it is consistent with findings in female tuataras (*Sphenodon punctatus*), where HL partially explains the bite force variation (Herrel et al., 2010). The bite distance between the upper and lower jaws is increased with a longer HH, which may contribute to predator deterrence. The head's shape is not the only factor in bite force. The strength of the bite force in various species is also influenced by a variety of other muscular factors, such as muscle-skeletal attachments, muscle quality, and the distribution of different muscle fiber types (Herrel et al., 1999; Anderson et al., 2008). However, head shape may be limited by cranial systems, and behaviors may limit (Herrel et al. 2001); as a result, additional research into other long-tailed skink systems is required.

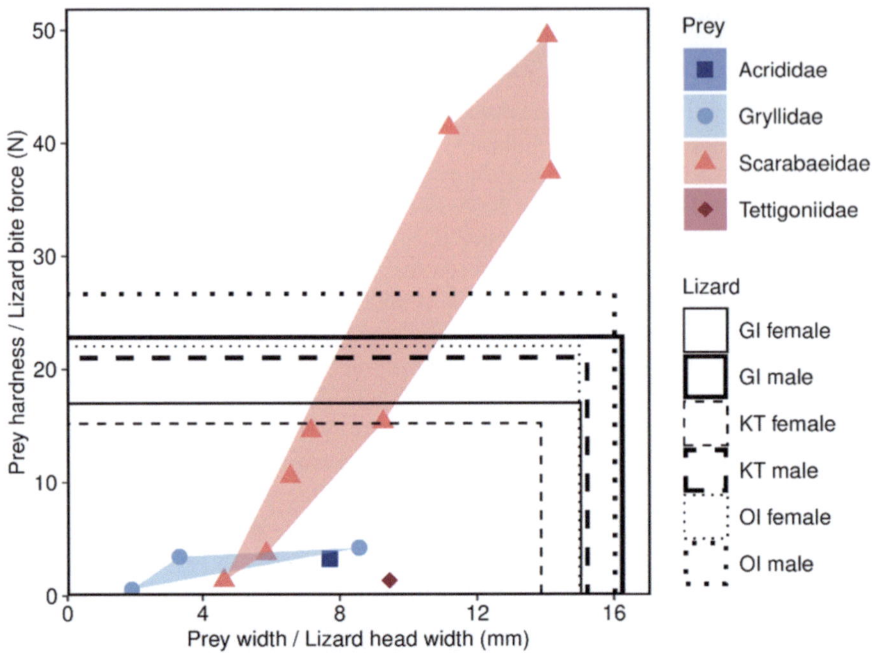

Fig. 8.4 The prey hardness, head width, and bite force of the long-tailed skinks. The potential prey was recorded from Huang (2006a, b, c). The prey hardness is based on Wang (2017). The vertical and horizontal lines represent head width and bite force of the skinks, respectively. Different types of lines show the male or female from different populations. Abbreviation: *GI* Green Island

Morphological Differences Between Sexes and Populations

In many species, sexual dimorphism in morphology is widespread (Herrel et al., 1996, 1999, 2010; Sagonas et al., 2014; Thomas et al., 2015). The head morphology of long-tailed skinks, including HL, HW, and HH, varies between the sexes. However, SVL and BW were similar between sexes. This finding suggests that the head is the source of sexual dimorphism in long-tailed skinks, and that a larger head may increase a male's fitness by increasing bite force (e.g., male competition, mating success, and territorial defense).

Comparisons among the populations of Kenting, Orchid Island, and Green Island show that, with the exception of BW, the Kenting population was smaller in the majority of the studied features. The Island Rule and Island Syndrome hypotheses suggest that island and mainland relatives differ in certain phenotypic or life history traits (Lomolino, 2005; Novosolov et al., 2013). On islands, body gigantism may result from ecological changes like reduced intraspecific competition, resource competition, or predation pressure (Runemark et al., 2015). Compared to the small body size of the Kenting population, the large body size of Orchid Island and Green Island may be consistent with island syndrome. Additionally, Orchid Island's size may be a result of its population of *Oligodon formosanus* having the highest density (Huang et al., 2013).

Table 8.3 Independent sample Wilcoxon rank sum tests were used to test whether (A) head width and (B) bite force of the long-tailed skink were greater than the body width and hardness of potential prey, respectively. *p*-values were adjusted based on the Holm–Bonferroni method to control the family-wise error rate

	Group	Head width (mm) Mean (95% CIs)	Prey width (mm) Mean (95% CIs)	*p*-value
(A)	Orchid Island male	16.06 (15.26–16.85)	8.03 (5.78–10.29)	< 0.001
	Orchid Island female	14.96 (14.55–15.37)		<0.001
	Green Island male	16.12 (15.65–16.59)		<0.001
	Green Island female	15.05 (14.68–15.42)		<0.001
	Kenting male	15.20 (14.56–15.83)		<0.001
	Kenting female	13.99 (13.50–14.48)		<0.001
	Group	Bite force (N) Mean (95% CIs)	Prey hardness (N) Mean (95% CIs)	*p*-value
(B)	Orchid Island male	28.22 (23.75–32.69)	14.41 (4.15–24.67)	<0.001
	Orchid Island female	21.92 (20.80–23.05)		<0.01
	Green Island male	22.59 (20.30–24.88)		<0.01
	Green Island female	16.82 (15.17–18.46)		<0.01
	Kenting male	21.18 (18.9–23.45)		0.01
	Kenting female	15.87 (14.69–17.04)		0.02

In conclusion, after accounting for head and body size, our study offers the first convincing evidence that the evolution of lizard bite force may have been influenced by parental care under strong predation pressure. Bite force differences between the sexes may be significantly influenced by male competition and mating bites. Therefore, males display stronger bites than females across all species. The larger head size that is seen in the morphological traits of males may increase biting power during competition. Additionally, our findings imply that HL and BW are accurate measures of bite force in long-tailed skinks. The F1 survival and growth rates in the laboratory were quite low despite our attempts to hybridize among three populations (40 Orchid Island females mated with 20 Kenting and 20 Green Island males) (only two females survived at each site). The four F1 females exhibited bite force similar to that of the population of Orchid Island at 21 N. This raises the possibility that the stronger bite force has a genetic origin. To test this genetically based hypothesis, additional cross-mating studies are necessary because the small numbers of samples limit this conclusion.

8.2 Determinants of Nest-Guarding Females Returning to the Nest: Trade-Offs Between Risk and Travel of Unfamiliar Environments

Because long-tailed skinks leave the nest to forage during the parental care period, it is important to return to their nest to continue the process of protecting their eggs intact. The ability to successfully return home requires the ability to, first, return home as quickly as possible to shorten the time spent on unfamiliar trails and avoid being eaten by predators; second, avoid wasting too much energy during the return process; and third, gain access to reproductive and food domains if the animal is able to return home, especially for females. Many reptiles have been reported to have successfully returned home using mechanisms such as compass guidance, geomagnetic guidance, sensory mechanisms (e.g., smell, vision, etc.), and e-vector auroral recognition. However, while previous papers have focused on male animals or other factors, this chapter takes the long-tailed skink, which has been cared for by its parents, as the target for its homing time and mechanism. Therefore, the time required to return to the nest and the mechanism of returning to the nest when the lizards are placed at different distances from the nest will be investigated, and whether the time required for the lizards to return to the nest is related to the egg predator, the kukri snake *Oligodon formosanus*, compared to other species of lizards. In other words, this chapter focuses on what factors shorten the time required for nesting and the mechanism of nesting, and whether this is related to the invasion behavior of predators. What is the mechanism of returning to the nest for the long-tailed skink, and does it rely on smell, vision, or other mechanisms to return to its nest? We also tested various homing hypotheses for the animals. The main objective of this study is to test the homing ability of the long-tailed skinks that are caring for eggs.

The majority of animals rely on spatially diverse resources at various stages of their life cycles, such as wintering grounds (Southwood & Avens, 2010), refugia (Webb & Shine, 1997), nesting microhabitats (Brown & Shine, 2007), or mates (Bull et al., 1993). Given that many animals are capable of orienting complex spatial environments, once these resources are discovered, they might later be relatively simple to return to the locations (Pike, 2005; Russel et al., 2005; Gould, 2006). However, finding resources becomes challenging when animals move through unfamiliar environments, such as when they leave their birthplace (Lohmann et al., 2004), nesting (Guyer, 1991), or under involuntary circumstances (e.g., during floods or after attempted predation). Traveling in an unfamiliar area is dangerous because individuals could get lost and wind up in unsuitable habitats (Massot et al., 2002), encounter predators (Gruber & Henle, 2004), or be exposed to extreme conditions (e.g., unsuitable temperatures) where they can't find shelter. According to theory, animals won't travel through strange environments unless the risks are exceeded by the advantages of the available resources.

Suitable nest sites are essential for oviparous species to reproduce and improve fitness by boosting egg survival (Huang & Pike, 2011a). However, there is typically a lot of predation at the nest site, so some parents choose to stay with the eggs while they are incubating (Shine, 1988; Huang, 2008). Some species allow one parent to leave the nest (for example, to go foraging) while the other protects the eggs (parental care: nest guarding). However, only one parent typically watches over the young during incubation in many ectotherms (Shine, 1988; Clutton-Brock, 1991). As a result, when the guardian parent departs from the nest, the eggs are at risk of predators until the parent comes back. In order to minimize the risk of nest predation, parents should spend as little time away from the nest as possible.

However, in some circumstances, parents might be forcibly taken away from the nest (for example by a predator, or during a flood), and they might not be able to return in time. In these situations, parents should only go back to the nest if the fitness benefits of egg protection outweigh the risks of doing so (such as confusion, potential predator encounters, exposure to extreme environmental conditions, etc.). In the worst case, females might become lost or preyed upon and be unable to find their way back to the nest. Females may also go back and discover the nest has been attacked, wasting their time and energy in the process.

The mobility of the organisms under study may affect their ability to build nests; for instance, birds and medium- to large-sized mammals typically have larger home ranges and are more mobile than the majority of ectotherms. Because of this, they have the benefit of being able to travel across long distances in a short amount of time and are frequently familiar with a broad geographic area. Reptiles, on the other hand, are generally less mobile, but a large number of them always make their homes in significant resources, like hibernacula or nesting sites (Brown & Shine, 2007). However, evidence from lizards suggests that after experimental translocation, returning home through unfamiliar environments is not typical. Only 69% of lizards that successfully went back to their homes in nesting experiments with 16 lizard species from five families required several days and traveled no more than a few hundred meters (Table 8.4). Five of the 16 species (31.3%) didn't make any attempt to return home (Table 8.4). Despite these variations, there hasn't been much research done on how individuals balance the advantages of a successful return with the risks of going home. These initiatives can provide insight into how and why certain species decide to go home.

Pregnant females of the long-tailed skink (*Eutropis* [formerly *Mabuya] longicaudata*) deposit eggs beneath rocks or in the ground throughout a large portion of its geographic range. After that, the females leave the nest and stop acting as parental care (Huang, 2006c). On Orchid Island, however, some females build nests in holes through concrete retaining walls. Invading egg-eating snakes (*Oligodon formosanus*), a significant reptile predator, are actively repelled by these females as they guard their nests (Huang, 2006c, 2007a). The eggs are still exposed and therefore vulnerable to predators when long-tailed skinks go foraging away from the nest (Huang, 2007a, b). The longer the female is absent from the nest, the more likely it is that egg predators will discover and eat the unguarded eggs because the lizards' nest guarding efforts are very effective in discouraging egg predators (Huang, 2006c, 2007a). Females should stay as close to the nest as possible to reduce the likelihood of egg predation. Additionally, precise nesting behavior is critical to ensure that the mother returns to her own nest rather than another nest of the same species because there are many nearly identical holes in a small area with nests. In this chapter, we used field experiments to examine the long-tailed skink's ability to nest and make the following discoveries: (1) whether sex differences exist in nesting ability; (2) how displacement distance, body size, and clutch size affect nesting success and return latency; (3) how egg age affects nesting; (4) how threats of egg predation influence nesting speed; and (5) whether visual or olfactory cues are mechanisms of nesting. We predicted that (1) females guarding nests would gain

Table 8.4 Review of lizard displacement distances and the time needed to return home (if successful) . Data are presented as ranges with means in parentheses when available. *Eutropis longicaudata* was the only species tested that guards eggs during incubation

Species	Distance displaced (m)	Days to return	Source
Gekkonidae			
Cyrtodactylus philippinicus	100–150	2–51	Marek et al. (2010)
Iguanidae			
Anolis lineatopus	~180	–	Rand (1967)
A. cristatellus	11–62	(3)	Jenssen (2002)
Dipsosaurus dorsalis	50–274	10–40 (20)	Krekorian (1977)
Sauromalus obesus tumidus	50–500	1–2	Prieto and Ryan (1978)
Phrynosomatidae			
Phrynosoma douglassi	~148	–	Guyer (1978, 1991)
Sceloporus graciosus	~280	–	Guyer (1978, 1991)
S. jarrovii	50–200	1–44 (20)	Ellis-Quinn and Simon (1989)
S. occidentalis	<100	–	Fitch (1940)
S. orcutti	~150	~20	Mahew (1963)
S. undulatus	~270	7–18	Noble (1934)
Uta stansburiana	49–122	–	Spoecker (1967)
Lacertidae			
Takydromus tachydromoides	<180	3	Ishihara (1969)
Scincidae			
Eutropis longicaudata	50–300	0.25–6 (1.4)	This study
Oligosoma grande	25–75	2–34	Stanley (1998)
Tiliqua rugosa	~800	>3	Freake (1998)

more benefit from returning to protect their eggs and should thereforedisplay higher tendencies to do so than males or females without eggs. (2) Lizards would be less likely to return as distance increased, either because it became more difficult for them to find their way home or because the cost of returning to a familiar resource increased. (3) Females should return relatively older eggs more frequently because much time and effort have been put into protecting these eggs, which are also closer to hatching. (4) Females should return relatively larger clutch sizes more frequently than those with relatively smaller clutch sizes because protecting more eggs can have a relative fitness effect. (5) The immediate predation threat at the nest will reduce the time to return due to the increased risk of egg loss.

8.2.1 Homing Success After Displacement

Except for two, all successful homing individuals in all experiments made their way home at night, which was later than the species' typical day/night activity cycle. Males and females who didn't guard the nest had a low homing success rate (only one female successfully returned home from a distance of 50 m and no males returned home). While males and non-nest-guarding females were significantly less likely to return home than nest-guarding females, 90% of nest-guarding females successfully returned home from a distance of 50 m ($\chi^2 = 29.17$, df = 2, $p < 0.00001$).

Female nest-guarding lizards' chances of successfully homing varied according to their migration distance ($\chi^2 = 32.0$, df = 3, $p < 0.00001$); those who traveled 50 and 100 m experienced the highest success rates (90% and 70%, respectively), while those who traveled 300 m experienced much lower success rates (10%; see Fig. 8.5a). Long-tailed skinks took longer to return as migration distance increased ($F_{3,36} = 245.94$, $p < 0.00001$; all distances significantly different, Tukey's HSD post hoc, all $p < 0.0001$; Fig. 8.5b); this effect was also seen for homing time (Fig. 8.5b). Homing success was independent of body size (homing success. $F_{1,72} = 1.43$, $p = 0.24$; distance. $F_{3,72} = 2.11$, $p = 0.11$; interaction. $F_{3,72} = 1.60$, $p = 0.20$; Fig. 8.6a) and clutch size (homing success. $F_{1,72} = 3.80$, $p = 0.06$; distance. $F_{3,72} = 0.74$, $p = 0.53$; interaction: $F_{3,72} = 0.63$, $p = 0.60$; Fig. 8.6b) We excluded data from individuals that migrated to 50 m and re-analyzed them because there was a tendency for females with larger clutch sizes to successfully return home after migrating to distances of 100, 200, and 300 m (Fig. 8.6b). When they had larger clutches, females that migrated to 100, 200, or 300 m were more likely to successfully return home (homing success). (homing success. $F_{1,54} = 9.87$, $p = 0.017$; distance. $F_{2,54} = 1.34$, $p = 0.44$; interaction. $F_{2,54} = 0.20$, $p = 0.82$; Fig. 8.6b).

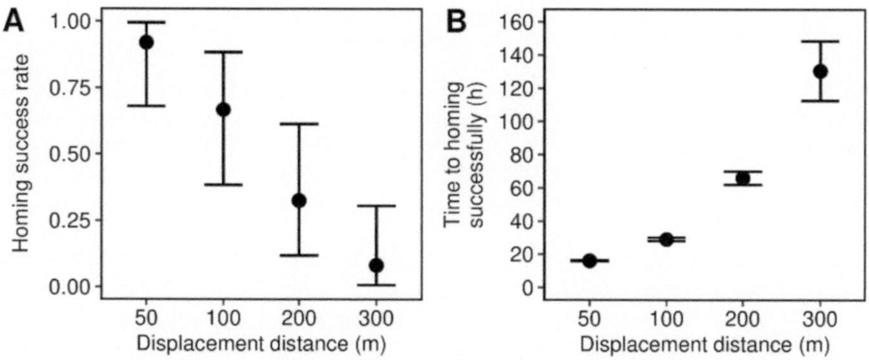

Fig. 8.5 Displacement success of female nest-guarding long-tailed skinks (*Eutropis longicaudata*) by distance, showing the (**a**) ratio of individuals ($N = 12$ females/treatment) successfully returning to the nest and (**b**) time it took successfully-homing individuals to return (hours). (Redrawn from Huang and Pike (2011a, b), Animal Behaviour)

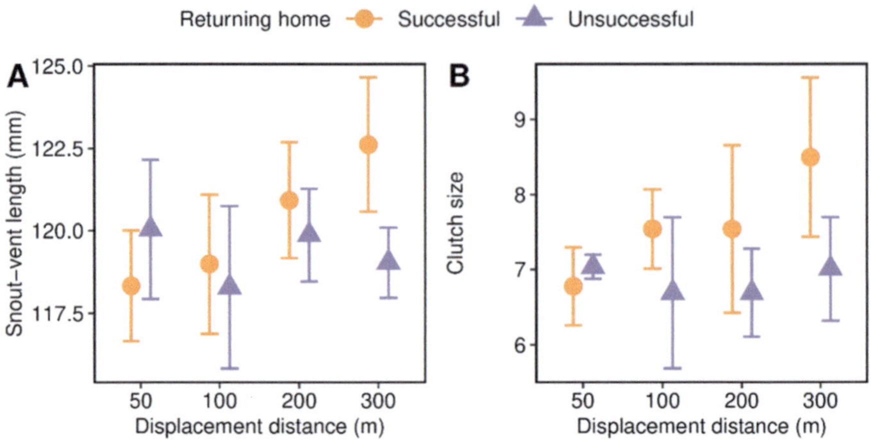

Fig. 8.6 Displacement of female nest-guarding long-tailed skinks (*Eutropis longicaudata*) by distance, showing comparisons of mean (**a**) body sizes (snout-vent length) and (**b**) clutch sizes of individuals that successfully homed vs. never returned home ($N = 12$ lizards/treatment). Error bars represent ±SE. (Redrawn from Huang and Pike (2011a, b), Animal Behaviour)

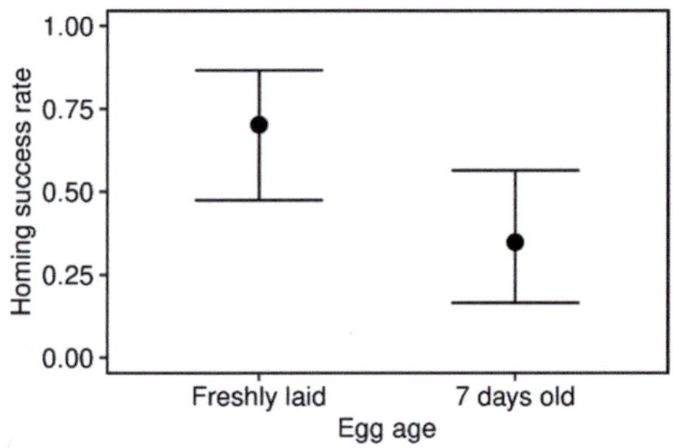

Fig. 8.7 Percentage of displaced female nest-guarding long-tailed skinks (*Eutropis longicaudata*) which successfully returned home vs. never returned home in relation to the timing of displacement (immediately upon laying eggs vs. 1 week later). In each treatment, $N = 20$ females. (Redrawn from Huang and Pike (2011a, b), Animal Behaviour)

Mean clutch size (± SE) was 7.8 ± 0.4 for successfully homing females and 6.8 ± 0.2 for not returned females.

Egg age (<1 day vs. 7 days) strongly influenced homing success ($\chi^2 = 4.9$, df $= 1$, $p = 0.026$; Fig. 8.7). Compared to nest-guarding females of 7 days, the number of freshly laid eggs returning home was twice as high as the latter (Fig. 8.7). Homing success was independent of body size (nesting success. $F_{1,36} = 0.38$, $p = 0.54$; egg age. $F_{1,36} = 0.01$, $p = 0.90$; interaction. $F_{1,36} = 0.43$, $p = 0.52$).

Long-tailed skinks that guard nests had a mean time to home that was significantly shorter than most reported values for other lizard species (mean time over all distances = 1.38 days) (Table 8.4). Most species that successfully returned home took much longer than the long-tailed skink to do so (e.g., up to 44 days; Table 8.4).

8.2.2 Does the Threat of Predation Impact Homing Behavior?

Exposure to egg-eating predators prior to migration did not affect homing success ($\chi^2 = 0.44$, df = 1, $p = 0.51$); in both cases, most females were successful returned home (Fig. 8.8a). The time it took for females to successfully return home, however, was significantly shortened by exposure to egg-eating predators ($F_{1,24} = 0.31$, $p < 0.00001$); exposed females typically returned 12 h faster than unexposed females (Fig. 8.8b). Homing success was independent of body size (homing success. $F_{1,36} = 0.38$, $p = 0.54$; exposure to predators. $F_{1,36} = 0.02$, $p = 0.89$; interaction. $F_{1,36} = 0.06$, $p = 0.81$). Despite differences in clutch size between treatments (mean for control and predator-exposed females 6.0 ± 2 and 7.1 ± 0.2, respectively), clutch size did not affect homing success (homing success. $F_{1,36} = 2.09$, $p = 0.16$; predator exposure. $F_{1,36} = 14.11$, $p < 0.001$; interaction: $F_{1,36} = 2.09$, $p = 0.16$).

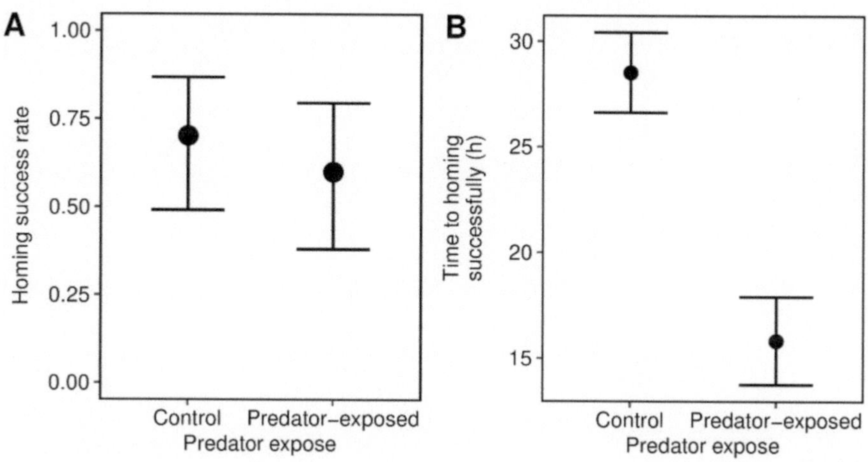

Fig. 8.8 Responses of nest-guarding long-tailed skinks (*Eutropis longicaudata*) either with or without exposure to an egg predator prior to displacement, showing (**a**) the proportion of females which successfully returned home vs. never returned home and (**b**) the time it took successfully homing females to return to the nest site. In each treatment, $N = 20$ females. (Redrawn from Huang and Pike (2011a, b), Animal Behaviour)

8.2.3 Mechanisms of Homing

Both the time to successful homing ($F_{1,38} = 0.31$, $p = 0.58$) and homing success ($\chi^2 = 0.23$, df $= 1$, $p = 0.63$; Fig. 8.9a) were unaffected by altering the nest site's visual appearance. In contrast, homing success were strongly influenced by the presence of chemical traces in the nest site (washed vs. unwashed) ($\chi^2 = 25.9$, df $= 1$, $p < 0.0001$); significantly fewer female lizards successfully homing after both nests and eggs were washed with chemical traces (Fig. 8.9b). However, the presence of chemical traces did not affect the time to successful homing ($F_{1,16} = 0.07$, $p = 0.93$). Finally, the presence of chemical traces on eggs and in nest sites strongly influenced homing success ($\chi^2 = 19.80$, df $= 1$, $p < 0.0001$); significantly more female lizards successfully returned to the nests when only eggs were cleaned compared to when nest sites were cleaned (Fig. 8.9c). However, the presence of chemical cues on the eggs and in the nest did not affect the time to successful homing ($F_{1,16} = 0.16$, $p = 0.70$). In these experiments, all females had a clutch of 6 eggs, and body size did not affect homing success ($p > 0.14$ for all treatments, homing success, and interactions).

8.2.4 Discussion

Homing is a crucial behavior that enables individuals to return to particular resources scattered across the landscape. Compare available data of other lizard species, long-tailed skinks typically homing faster (Table 8.4). Additionally, we discovered that

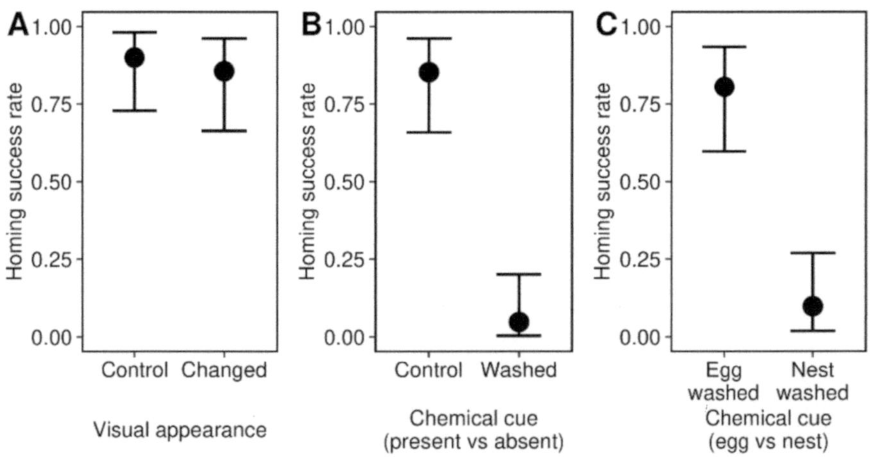

Fig. 8.9 Percentage of displaced female nest-guarding long-tailed skinks (*Eutropis longicaudata*) which successfully returned home vs. never returned home in relation to whether (**a**) the physical appearance of the nest site was altered vs. unaltered, (**b**) the nest and eggs were washed of chemical cues vs. unwashed, and (**c**) either the nest or eggs were washed of chemical cues. In each treatment, $N = 20$ females. (Redrawn from Huang and Pike (2011a, b), Animal Behaviour)

male and female homing behavior varied significantly. While non-nest-guarding males and females (i.e., 5% of those tested) rarely succeeded in returning, nest-guarding females did so in a variety of circumstances. These variations may be the result of nest site defense adaptations, in which females defend incubating eggs from nest predators, thereby improving egg survival (Huang, 2006c). In contrast, for non-guarding skinks, there are many other nearly identical retreat sites within the concrete wall, and the fitness benefit of returning to a specific retreat site when a nest is absent is not apparent. In our study, we discovered that nest-guarding long-tailed skinks: (1) homed occurred at night, when the species was typically inactive; (2) were more successful at homing over shorter distances; (3) returned faster from distances >50 m when nest size was relatively large; (4) were more likely to home when eggs were recently laid; (5) reduced nearly half homing time after exposure to egg predators; and (6) homed through olfactory cues associated with the nest sites, rather than egg-based or visual cues. In conclusion, these findings suggest a trade-off between the risks of returning through unfamiliar environment and the significant fitness benefits of protecting eggs from predators.

Because nest-guarding long-tailed skinks discrete in separate nests, we can predict when particular lizards will come home. Contrarily, many other lizard species are secretive or live in difficult-to-access microhabitats, making it difficult for researchers to observe them (e.g., high in trees; Ellis-Quinn & Simon, 1989; Stanley, 1998). This could partially account for the fact that other species typically have longer homing times than the species we studied (Table 8.4). Long-tailed skinks are the only species that have been observed to exhibit parental care during incubation period. In addition, females of this species aggressively pursue invasive egg predators, which increases the likelihood that eggs will hatch (e.g., 81% in the presence and 18% in the absence; Huang, 2006a, b, c). Females adapt their nest-guarding behavior, extending the time spent with eggs under the high predator abundance (Huang & Wang, 2009) and shortening the time spent back home under the high predation threats (this chapter). As a result, nest-guarding species have a higher level of investment in a single discrete microhabitat (the nest), which could account for why these species are more likely to return to their homes in a variety of circumstances than non-nest-guarding species.

When traveling through an unfamiliar landscape, there is a high predation risk because it may be difficult for individuals to find essentials like shelter when they come across potential predators or when they are exposed to harsh weather (Huang, 2004), and this is also true of their only known natural enemies on Orchid Island (Stinky snake, *Elaphe carinata*). Long-tailed skinks, on the other hand, altered their behavior and now only visit their nests at night. This change in behavior can greatly decrease the risk of predation on the way back to the nest (Lima & Dill, 1990; Mathis & Smith, 1993) and the risk of exposure to fatal high temperatures [which may be high in the tropical environment of our study site; (Kearney et al., 2009)]. The only risk for displaced females is predation upon return to the nest because nocturnal nesting eliminates the possibility of overheating while traveling in tropical sunlight and the immediate risk of predation. The success of homing declines with increasing migration distance for two distinct reasons. One explanation is that females may have simply been moved too far away from home to navigate. This is unlikely, though, as some individuals manage to navigate home no matter the distance we test (Fig. 8.5a). The second possibility entails a trade-off between the advantages of a return trip and the risks related to a lengthy journey. Animals should only use energy when doing so will not incur high costs. Reptiles spend more time moving, which significantly increases their daily energy expenditure (Huey & Pianka, 1981). The energy required to travel home and the likelihood that eggs will be eaten while traveling back both significantly rise as the distance of migration increases. For instance, it takes female skinks 130 h (or roughly 5.5 days) to return from a distance of 300 m before egg-eating snakes can locate and consume the entire clutch of eggs. Even if the nest has already been attacked by a predator, for these females, the relative fitness benefits of having more eggs hatch successfully may exceed the energy costs of going back to the nest. Larger clutch sizes also increase the likelihood that nest-guarding females will attack intruders, whereas smaller clutch sizes increase the likelihood that predators will flee the nest after attempting to eat the eggs(Huang, 2006c). This suggests a trade-off between the energy of reproductive investment and maternal care efforts, with females adjusting their maternal care strategies (including returning home from a greater distance) based on the advantages larger clutch size provides for fitness.

Age-related directional differences may have an impact on how successfully an individual returns home (Weintraub, 1970; Ellis-Quinn & Simon, 1989). Older females may have spent more time exploring the area than younger females, which could shorten the time it takes to get home (Guyer, 1991; Chelazzi, 1992; Hein & Whitaker, 1997). Because clutch size increases with body size in long-tailed skinks, it is challenging to determine whether young adults have smaller clutch sizes than older females (Huang, 2006a). However, in any of our experiments, body size had no impact on the success of homing, which could be attributed to the relatively narrow range of female body sizes we used (Fig. 8.6a).

Cost-benefit analyses suggest that since the eggs approach hatching, females who devote more time (and thus energy) to raising young should be more likely to return to their nests. Contrary to what was expected, nest return success was strongly influenced by the the duration of guarding eggs before migration (24 h vs. 7 days): females exhibited greater return probability when the pre-migration maternal investment was short (i.e., 24 h). Approximately 1 week after oviposition, however, long-tailed skinks exhibit a significant decline in nest-guarding rates (Huang, 2006c), which may be related to a decreased risk of egg predation. Fresh eggs may provide stronger odor cues that attract predators than older eggs (e.g., (Congdon et al. 1983; 1987), which may reduce the encounter probability of successful predator with older eggs. Once the peak period of egg predation has passed, the low risk of nest predation may compensate for the cost of returning home after this period.

After exposure to egg-specific predators that did not endanger females, the time it took females to return to the nest was reduced by almost half. This was perhaps the most exciting finding of our study (Fig. 8.8b). The parental care efforts of long-tailed skinks can be adjusted to the prevalence of predators; when exposed to egg predators, females increase the duration of staying on the nest (Huang & Wang,

2009). Our finding extends this work by showing that females also reduce their time away from the nest in response to the high risk of egg predation. This may be achieved by moving rapidly through the landscape.

It is possible for olfactory and visual cues to help discriminate between their own nests and those of conspecifics (e.g., Fleishman, 1992, Peterson, 2000, Zuri & Bull, 2000). Olfactory cues are used by long-tailed skinks instead of visual cues to locate their nests (Fig. 8.9). The fewer proportion of individuals that returned after cleaning the site compared to after cleaning the eggs suggests that chemical cues from the nest site itself (rather than the eggs) appear to be the main signals as retuning cues during nesting. Despite the absence of a unified hypothesis for the operation of olfactory-based homing (Chelazzi, 1992), this supports our previous research that females lack the ability to discriminate between their eggs and those of conspecifics (Previous Chapter). Similarly, female terrestrial salamanders can use non-nest scents to navigate their nests (Forester et al., 1983, Peterson, 2000).

Our study suggests that by adjusting the homing behavior, nest-guarding females could minimize the risk of crossing unfamiliar environments, ultimately, successfully homing females gain fitness benefits by egg protection from predation. This suggests that parental care behaviors exhibit significant plasticity, but that in some cases the risks of return exceed the benefits of nest protection .

8.3 The Influence of Intraguild Predators on Prey Nest Site Selection

Due to the trade-off between nest site quality (microenvironment) and nest site predation risk (for female lizards and/or lizard nests), suitable nest sites are frequently scarce. The choice of a nest site can be significantly influenced by interactions with "classical" predators (i.e., those that do not fight over shared resources), but it is unknown whether predation within the animal itself also affects this behavior. We investigated whether the diurnal Scincid lizard *Eutropis* [*Mabuya*] *longicaudata*, an intraguild predator, influences the prey's choice of nest site (the nocturnal gecko *Gekko hokouensis*). The larger *E. longicaudata* also preys on *G. hokouensis*, but it is believed that there are resource competitions, such as invertebrate prey and nesting microhabitats (but not its eggs), between these two species. Both species create nests in concrete walls. *E. longicaudata* prefers to nest in holes blocked by debris, preventing water intrusion from damage of the eggs, whereas *G. hokouensis* prefers holes that are open on both ends, allowing rapid escape from predators. In our experiments, when we kept *E. longicaudata* out of its preferred nesting area, *G. hokouensis* not only built more nests, but also changed its nest site choice, e.g., collective nesting was more common and there were similar occupancies of both open and closed holes. Egg survival was slightly lower (7%) in areas where predators were excluded, but hole type selection had no effect on it (probably due to higher local incubation temperatures). Our field experiments suggest that even when diel activity patterns differ between predators and prey and eggs are not at risk of predation, predators can directly influence prey nest density by changing maternal nest site selection behavior.

As a strong evolutionary driver, predation modifies the behavioral ecology of prey species in two ways, depending on the ecological relationships that exist between the predator and the prey: (1) "classical" predation, in which the predator and prey occupy distinct ecological guilds and lack competitions, and (2) intraguild predation, in which the predator and prey share the same guild and resource requirements, resulting in potential competitive relationships (Polis et al., 1989; Arim & Marquet, 2004).

The majority of studies have concentrated on traditional predation; notable examples include the effects of mammals on the ecology, behavior, and evolution of birds and reptiles. Contrarily, despite being more common taxonomically (Arim & Marquet, 2004) and having a stronger impact on population dynamics than either competition or predation itself, intraguild predation has attracted limited research attention. This is due to the fact that intraguild predators reduce interspecific competition by cutting down on the amount of prey they must compete with (Polis et al., 1989). In order to comprehend the causes and effects of prey behavior more generally, it is crucial to understand the relationship between predators and prey within ecological guilds.

Antipredator responses often vary with life stage of prey, reflecting differential susceptibility to predators (e.g., Stamps, 1983; Webb et al. 2009; Pike et al., 2010), as well as population-level impacts of predation on different life stages (e.g., Congdon et al., 1993). The survival of oviparous, sexually reproducing populations depends on females and eggs. Due to the additional strain of carrying eggs or offspring, pregnant females are frequently particularly vulnerable to predation, which causes them to have decreased motility (Shine, 1980). Similarly, eggs are extremely vulnerable to predators during incubation, particularly without parental care (Huang, 2008). Therefore, natural selection favors strategies that minimize the risk of predation during these vulnerable phases, including modifications in the activity patterns of egg-laying females (Cooper et al., 1990), as well as strategies used by females to discourage egg predation, including non-random nest selection (Stoleson & Beissinger, 2001; Relyea, 2003; Spencer & Thompson, 2003) and/or hidden nests (Forstmeier & Weiss, 2004; Huang, 2007a).

In order to ensure reproductive success , oviparous species face the additional challenge of finding nest sites with a microenvironment appropriate to embryonic development. Breeding females of reptiles select nest locations with thermal conditions that optimize the offspring's fitness (Shine & Harlow, 1996; Shine et al., 1997; Shine, 2004; Warner et al., 2009; Huang & Pike, 2011a). Because of their direct effects on development, sex, phenotype, and survival, research to date has concentrated on how microhabitat features affect egg fitness (Kolbe & Janzen, 2002; Spencer, 2002; Spencer & Thompson, 2003; Huang & Pike, 2011a). Little attention has been paid to the maternal risks associated with nest site selection and the role of predators (especially intraspecific predators) in modifying nest site selection behavior.

On Orchid Island, we looked into whether intraguild predation affects the choice of nest sites for the nocturnal gecko *Gekko hokouensis*. The nesting mother and the nest site are generally simultaneously exposed to predation risks in traditional predator-prey relationships; however, in this unique situation, the lizards themselves are susceptible to predation but not their hard-shelled eggs. This allowed us to test directly whether maternal responses are driven by her own predation risk or by the predation risk of her hatched offspring, rather than by the risk of the nest itself. The predator is a large diurnal long-tailed skink (*Eutropis longicaudata*) that coexists with *G. hokouensis* on an invertebrate food web (e.g., (Huang, 2006a). The eggs of lizards and other soft-shelled reptiles are preyed upon by *E. longicaudata*, but not the hard-shelled eggs produced by *G. hokouensis* (Huang, 2006a). Although the two species exhibit different temporal activity patterns, their microhabitat utilization overlaps significantly, and they could compete for nesting sites. For instance, both species build nests in holes through concrete retaining walls; however, *G. hokouensis* rarely nests in those occupied by *E. longicaudata*. We predict that the delineation of such nest sites is being influenced by the intraguild predation risk, and if this is the case, the predator exclusion will lead to higher levels of prey nesting. We also anticipated that this response wouldn't have a significant impact on the egg survival during incubation (i.e., sites with and without predators would provide equally suitable incubation conditions). By comparing the two species' nest site preferences, we were able to test these hypotheses and see whether prey changed their nest site preferences when predators were artificially kept out of the study area.

8.3.1 Nesting Patterns of *E. longicaudata* and *G. hokouensis*

In the southeastern location (SE), we discovered 420 nests of *E. longicaudata* and clutch sizes ranged from 2–13 eggs, and eggs were typically laid on the floor far end of the hole by *G. hokouensis*. Open holes were not preferred by *E. longicaudata* as nesting sites ($n = 402$ closed sites and 268 open sites vs. 390 closed nests and 30 open nests; $\chi^2 = 41.41$, $\text{df} = 1$, $p < 0.0001$; Fig. 8.10). Only 25% ($n = 2$) of the remaining open nests successfully hatched, and 73% ($n = 22$) of the open nests were were lost to rainfall. Thus, choosing the right nest site had a significant impact on lizard nest survival ($\chi^2 = 83.8$, $\text{df} = 1$, $p < 0.0001$); only 6.7% ($n = 2$) of the 30 nests that were spawned in open sites, as opposed to 81% ($n = 316$) of those in closed sites, hatched.

The northwestern location (NW) was where we discovered 320 *G. hokouensis* nests. Each nest contained two eggs that were attached to the top or sides of the hole. *G. hokouensis* favored using open holes as nesting sites ($n = 10$ closed nests and 310 open nests versus 250 closed nests and 280 open nests; $\chi^2 = 47.70$, $\text{df} = 1$, $p < 0.0001$; Fig. 8.10). Despite this habitat selection was non-random, there was no significant difference in survival rates between nest types ($\chi^2 = 1.37$, $\text{df} = 1$, $p = 0.24$); 8 out of 10 open nests and 282 out of 310 closed nests hatched.

At both sites, nest temperatures showed considerable diurnal variation, reaching a maximum temperature at noon and a minimum temperature at midnight. Within each

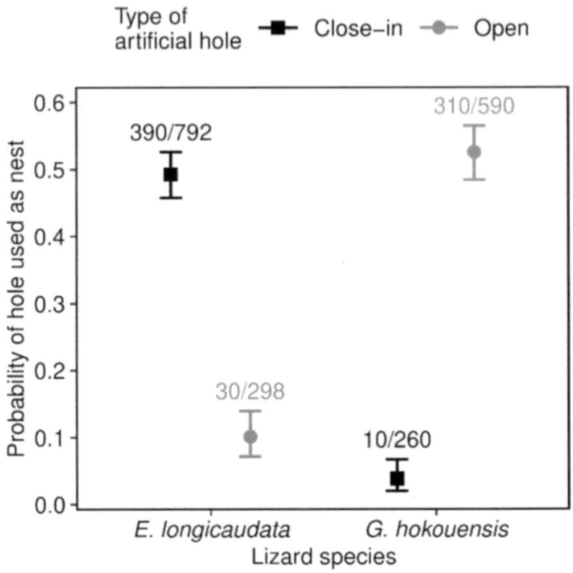

Fig. 8.10 The percentage of artificial holes (either closed-in or open-ended) available vs. used as nest sites by *Eutropis longicaudata* ($n = 420$ nests) and *Gekko hokouensis* ($n = 320$ nests) in the absence of the other species. Note that *E. longicaudata* selected closed-in sites more often than they were available, while *G. hokouensis* selected open sites more often than they were available. The number above the bar indicates the number of holes used as nests and the total observed holes. (Redrawn from Huang and Pike (2013), Animal Behaviour)

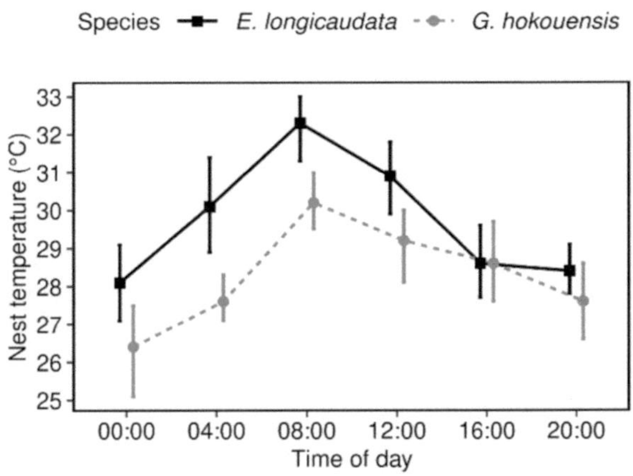

Fig. 8.11 Mean daily thermal regimes (± SE) inside *Eutropis longicaudata* and *Gekko hokouensis* nests averaged over the nesting season from 1999 to 2009. (Redrawn from Huang and Pike (2013), Animal Behaviour)

site, there was no significant difference in temperature between nests laid out in open and closed cavities ($p > 0.74$ for both cases). Therefore, these data were pooled to compare nest temperatures at the northwestern site with those at the southeastern site. Mean temperatures at the northwestern site (where *G. hokouensis* nests) were significantly lower than those at the southeastern site, where *E. longicaudata* prefers to nest (mean daily temperatures of 28.4 ± 2.0 °C vs. 30.5 ± 1.5 °C, respectively; $t = 8.7$; $p < 0.001$; Fig. 8.11).

8.3.2 Does the Exclusion of Predators Result in Altering Prey Nest Site Selection?

The use of snake chemical trails was highly effective in excluding *E. longicaudata* from the study area; only six individuals were observed in the exclusion area during the seven-year period (2001–2007) when chemical trails were used (total of 10 captures). In contrast, we observed 122 and 136 *E. longicaudata* individuals (256 and 202 catches) before and 2 years after the application of chemical nudges, respectively.

There was no difference in the number of eggs laid (*E. longicaudata*: $t_9 = -0.77$, $p = 0.46$; *G. hokouensis*: $t_9 = -1.50$, $p = 0.17$; Fig. 8.12a) or the number of artificial holes used as nest sites (*E. longicaudata*: $t_9 = -0.74$, $p = 0.48$; *G. hokouensis*: $t_9 = -0.49$, $p = 0.63$; Fig. 8.12b) in the unmanipulated control area between years with and without predators. Therefore, any variation in nesting behavior in predator-excluded areas should be the consequence of experimental treatment instead of other factors.

The predator experimental exclusion affected prey nesting behavior; in the years without predators, *G. hokouensis* produced significantly more eggs ($t_9 = -13.78$, $p < 0.0001$; Fig. 8.12c) and used significantly more artificial holes ($t_9 = -11.37$, $p < 0.0001$; Fig. 8.12c). Prior to the exclusion of predators (1999–2000), *G. hokouensis* used few holes and produced few eggs; these numbers increased dramatically during the exclusion of predators (2001–2007), but then declined intensely after predators were allowed to re-colonize the area (2008–2009; Fig. 8.12c).

The prevalence of collective nesting also increased significantly when predators were excluded ($t_9 = -0.84$, $p = 0.01$); geckos produced the average number of nests per hole rose from 1.25 in the presence of predators to 2.1 in their absence. In the northwestern region (where *G. hokouensis* primarily nests), there was collective nesting did not significantly change over time ($t_9 = -1.38$, $p = 0.20$); the average nesting ratio per hole was 2.2.

The presence of predators also influenced nest selection in *G. hokouensis*; when predators were present, a higher proportion of eggs were produced in open nests (87.5%), whereas in the absence of predators, nest selection was the same for both types (49%) ($t_9 = 3.02$, $p = 0.003$). Thus, the exclusion of *E. longicaudata* altered not only the frequency of nesting in the experimental area, but also the type of preferred microhabitat where *G. hokouensis* nested. In contrast, nest site selection behavior was similar in the control area (*E. longicaudata*: $t_9 = -0.44$, $p = 0.67$; *G. hokouensis*: $t_9 = -0.29$, $p = 0.78$); 98.3% of nests of *E. longicaudata* were produced in closed cavities, whereas 96.7% of nests of Geckos were produced in open cavities.

As expected,there was no difference in nest survival of gecko between years with and without predator exclusion ($t_9 = 0.15$, $p = 0.88$); overall nest success was high, averaging 81%. As in the northwestern region, nest survival was not affected by nest type ($t_9 = 0.67$, $p = 0.51$). Nest survival for gecko (*G. hokouensis*) in predator exclusion areas had a slight but significant decrease ($\chi^2 = 5.14$, df $= 1$, $p = 0.023$); 91% of nests hatched at the northwestern (cooler) site compared to 84% at the

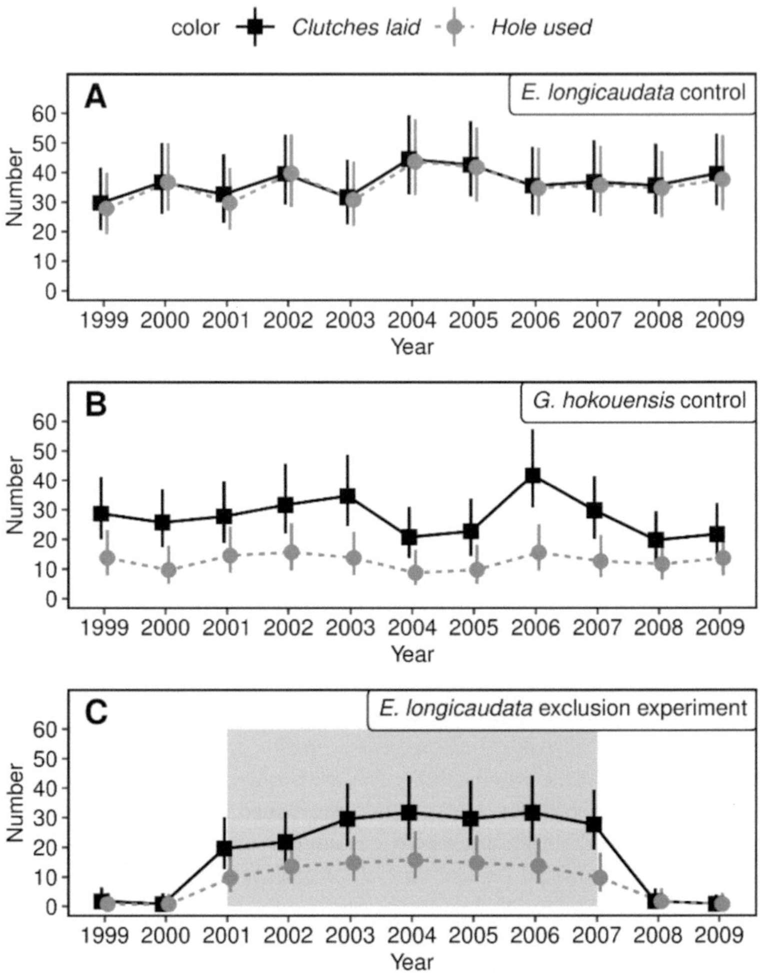

Fig. 8.12 The number of artificial holes used by (**a**) *Gekko hokouensis* and (**b**) *Eutropis longicaudata* as nest sites and the number of clutches laid in these holes over 11 years. (**a**) and (**b**) represent control sites, which were not manipulated. (**c**) Use of artificial holes as nest sites by *G. hokouensis* in the presence (1999–2000 and 2008–2009) or absence (2001–2007) of predatory *E. longicaudata* (years in which the predator was experimentally excluded from the nesting area are shaded). *G. hokouensis* often nests communally, resulting in more than one clutch per hole, while for *E. longicaudata,* instances of more clutches than holes were due to females occasionally using a hole in which a clutch had hatched earlier in the season. (Redrawn from Huang and Pike (2013), Animal Behaviour)

southeastern (warmer) site. This pattern was due to differences in nest survival at open nests ($\chi^2 = 4.75$, df = 1, $p = 0.03$); hatching rate at open nests was 91% at the northwestern site, while 83% hatched with juveniles at the southeastern site. Egg survival for closed nests differs significantly between the two locations (Fisher's exact test, $p = 0.65$).

8.3.3 Discussion

The decision of where to build a nest involves a complex interplay of biotic (predation and/or competition) and abiotic (temperature and nest site quality) trade-offs (Martin, 2001). We looked at two lizard species' choice of nesting locations for eggs laid in concrete wall holes. Both open and closed-in holes offered comparable heat, but predators and prey made non-random decisions about where to build their nests. Because *E. longicaudata* nested in a man-made hole that was enclosed at one end, the destructive effects of heavy rainfall were reduced (which would have washed away eggs laid in the open hole). In contrast, *G. hokouensis* prefers open nests, and both types of nests have high rates of egg survival. The gecko (*G. hokouensis*) prefers to nest in open holes, whereit can produce alive offspring, giving the prey priority access to this resource and potentially lowering predation within the animal (Polis et al., 1989; Schellhorn & Andow, 1999). For example, *E. longicaudata* likely avoids open holes while searching the nests or guarding eggs. Utilizing openings also provides the gecko with an escape path as the additional advantage in the event of predator entry. By contrast, nests of *E. longicaudata* exhibit higher survival rates in only one type of nest, which suggests that intraspecific competition for this type of nest should be particularly intense.

Our field studies offer more evidence that gecko nest sites are directly influenced by the presence of intraguild predators, even when the former's eggs are not in danger of being eaten. Following the exclusion of predators, prey altered their selection of nest sites by (1) increasing the number of nests they built; (2) increasing the number of artificial holes they used as nests; (3) increasing the prevalence of collective nest building [further evidence of nest site limitation; reviewed in (Doody et al., 2009)]; and (4) changing the type of hole they used. Importantly, these patterns changed when predators were permitted to return to the study area (see, for example, Fig. 8.12), offering compelling evidence of a connection between the presence of intraguild predators and prey nesting behavior.

We do not support the hypothesis that *G. hokouensis* simply avoids the warmer microhabitats favored by *E. longicaudata* because egg incubation is limited by temperature; despite the fact that nest survival in warmer and predator-free areas is only significantly lower by 7%. Furthermore, if thermal constraints are driving this pattern, then rapid colonization of predator-excluded areas didn't happen due to predator removal (e.g., Fig. 8.12). Although this pattern is consistent well with the theoretical models predicting that intraguild predators often displace their prey (e.g., Holt & Polis, 1997, Sergio et al., 2007), it is unclear whether the reduced abundance of prey in the presence of predators reflects predator avoidance behavior or population suppression due to predation. This is a persistent challenge in studies of intraguild predation (Sergio et al., 2003; Sergio & Hiraldo, 2008; Webb et al. 2009).

However, *E. longicaudata* consumes primarily invertebrate, with lizards making up only a small fraction of its dietary intake [<2%; (Huang, 2006a)], suggesting that intense predation pressure is unlikely an explanation; instead, predator avoidance and/or competitive exclusion are more plausible drivers. Although there is some uncertainty, the significant movement of *G. hokouensis* into areas where predators were excluded suggests that the risk of predation is a key to determining the behavior of gravid females, as given the shortage of appropriate nest sites. The fact that *G. hokouensis* nests thrived both in the presence and absence of predators is significant because it proves that *E. longicaudata* does not feed on hard-shelled eggs. This finding supports the idea that females react to immediate predatory risk to themselves or their young rather than to incubating eggs.

The temporal and/or spatial separation between the predator and the prey has been attributed to intraguild predation [reviewed in (Polis et al., 1989, Sergio & Hiraldo, 2008)]. Although the temporal activity patterns in our two study species are different (predators are diurnal and prey are nocturnal), this is unlikely to be the result of intraguild predation. Due to evolutionary constraints, most long-tailed skink species are nocturnal, while most gecko species are diurnal long-tailed skinks species are predominantly nocturnal, whereas gecko species are primarily diurnal. These activity times are broadly conserved across phylogeny (Underwood, 1954). Although it may initially appear strange that daytime predators can affect the location of nocturnal prey's nest, but this phenomenon may offer a coexistence mechanism (Amarasekare, 2007). Moreover, since less energy is spent on capturing inactive prey than active prey, differences in activity may actually improve *E. longicaudata*'s capacity to find and ingest geckos during their inactive phase,. Last but not least, despite the fact that warmer nests appear to have little impact on gecko egg survival, The significance of thermal variation for more subtly functional traits of the nest site warrants consideration, such as locomotor function. This may help to explain why *G. hokouensis* typically occurs in regions with poorer thermal habitat quality and why predators and prey are frequently conspecific on a local scale (e.g., habitat-mediated predator avoidance; Sergio et al., 2007).

Our study utilized a simulated environment where the abundance of predators and prey, as well as the availability of good nesting sites, may be unusually high, likely exceeding typical natural conditions. Predator density can modify the nature, strength, and likelihood of interspecific interactions (Burley et al., 2006), so it remains important to determine whether the patterns we documented persist in natural habitats (where abundance is low). In our study area, artificial walls may increase the frequency of interactions between predator and prey because in natural habitats, the two species select distinct nesting microhabitats; *E. longicaudata* deposits its eggs in the soil under rocks, while *G. hokouensis* favors to use loose bark or rock crevices.

This suggests that the presence of artificial habitats may create interaction opportunities that are rare in nature. Because predator avoidance often involves a set of trade-offs with other activities (e.g., reduced activity levels) or selection of lower-quality habitat (Lima & Dill, 1990; Webb et al., 2010), we predict that *G. hokouensis* exhibit reduced selectivity in avoiding *E. longicaudata* at low levels of predator abundance. Despite this restriction, habitat modification is becoming more widespread, continuing to alter potential species interactions and habitat availability.

Finally, the sensory mechanism by which *G. hokouensis* detects predation risks is unknown. For example, whether geckos rely on visual signals from long-tailed skinks or eggs, or whether chemosensory signals from feces, nests, or the lizards themselves provide sufficient information to trigger predator avoidance? Chemosensory-based predator detection is well documented in various invertebrates (e.g., Walzer et al., 2006; Turner et al., 2007; Putra et al., 2009; Choh et al., 2010), and geckos are capable of utilizing both visual and chemosensory cues to avoid predators (e.g., (Webb et al. 2009). In areas of high lizard abundance (e.g., artificial walls), these cues become more prevalent (and therefore easier to detect), which may increase more opportunities for prey to facilitate predator avoidance. Although the mechanisms driving these patterns and their potential divergence between natural and artificial habitats are not understood, it is clear that even if eggs are not at predation threat, predator and prey activity levels do not overlap and the threat of intraguild predation may influence prey nest site selection. Investigating whether comparable patterns occur in other intraguild predator-prey dichotomies across a wider range of taxonomic groups is an important avenue for future research.

Long-tailed skinks lay their eggs in concrete holes, and there happen to be two types of ants that utilize these eggs. One type feeds on the eggs, while the other only sucks the water droplets that condense on the surface of the eggs. Eventually, these two types of ants fight each other. Which one will win? We will see in the next chapter.

References

Amarasekare, P. (2007). Trade-offs, temporal variation, and species coexistence in communities with intraguild predation. *Ecology, 88*, 2720–2728.

Anderson, R. A., McBrayer, L. D., & Herrel, A. (2008). Bite force in vertebrates: Opportunities and caveats for use of a nonpareil whole-animal performance measure. *Biological Journal of the Linnean Society, 93*, 709–720.

Arim, M., & Marquet, P. A. (2004). Intraguild predation: A widespread interaction related to species biology. *Ecology Letters, 7*, 557–564.

Broeckhoven, C., & Mouton, P. l. F. N. (2014). Under pressure: Morphological and ecological correlates of bite force in the rock-dwelling lizards *Ouroborus cataphractus* and *Karusasaurus polyzonus* (Squamata: Cordylidae). *Biological Journal of the Linnean Society, 111*, 823–833.

Brown, G. P., & Shine, R. (2007). Like mother, like daughter: Inheritance of nest-site location in snakes. *Biology Letters, 3*, 131–133.

Bull, C. M., Bedford, G. S., & Schulz, B. A. (1993). How to sleepy lizards find each other? *Herpetologica, 49*, 294–300.

Burley, L. A., Moyer, A. T., & Petranka, J. W. (2006). Density of an intraguild predator mediates feeding group size, intraguild egg predation, and intra- and interspecific competition. *Oecologia, 148*, 641–649.

Cameron, S. F., Wynn, M. L., & Wilson, R. S. (2013). Sex-specific trade-offs and compensatory mechanisms: Bite force and sprint speed pose conflicting demands on the design of geckos (*Hemidactylus frenatus*). *Journal of Experimental Biology, 216*, 3781–3789.

Chelazzi, G. (1992). Reptiles. In F. Papi (Ed.), *Animal homing* (pp. 235–261). Springer.

Choh, Y., van der Hammen, T., Sabelis, M. W., & Janssen, A. (2010). Cues of intraguild predators affect the distribution of intraguild prey. *Oecologia, 163*, 335–340.

Chouteau, M., & Angers, B. (2011). The role of predators in maintaining the geographic organization of aposematic signals. *The American Naturalist, 178*, 810–817.

Clegg, S. M., & Owens, P. (2002). The 'Island rule' in birds: Medium body size and its ecological explanation. *Proceedings of the Royal Society B: Biological Sciences, 269*, 1359–1365.
Clutton-Brock, T. H. (1991). *The evolution of parental care*. Princeton University Press.
Congdon, J. D., Tinkle, D. W., Breitenbach, G. L., & Van Loben Sels, R. C. (1983). Nesting ecology and hatching success in the turtle *Emydoidea blandingi*. *Herpetologica, 39*, 417–429.
Congdon, J. D., Breitenbach, G. L., Van Loben Sels, R. C., & Tinkle, D. W. (1987). Reproduction and nesting ecology of snapping turtle (*Chelydra serpentina*) in southeastern Michigan. *Herpetologica, 43*, 39–54.
Congdon, J. D., Dunham, A. E., & Van Loben Sels, R. C. (1993). Delayed sexual maturity and demographics of Blanding's turtles (*Emydoidea blandingii*): Implications for conservation and management of long-lived organisms. *Conservation Biology, 7*, 826–833.
Cooper, W. E. J., Vitt, L. J., Hedges, R., & Huey, R. B. (1990). Locomotor impairment and defense in gravid lizards (*Eumeces laticeps*): Behavioral shift in activity may offset costs of reproduction in an active forager. *Behavioral Ecology and Sociobiology, 27*, 153–157.
Covas, R. (2012). Evolution of reproductive life histories in Island birds worldwide. *Proceedings of the Royal Society B: Biological Sciences, 279*, 1531–1537.
Das, I. (2010). *A field guide to the reptiles of Southeast Asia*. New Holland.
de Barros, F. C., de Carvalho, J. E., Abe, A. S., & Kohlsdorf, T. (2010). Fight versus flight: The interaction of temperature and body size determines antipredator behaviour in tegu lizards. *Animal Behaviour, 79*, 83–88.
Donihue, C. M., Brock, K. M., Foufopoulos, J., & Herrel, A. (2016). Feed or fight: Testing the impact of food availability and intraspecific aggression on the functional ecology of an Island lizard. *Functional Ecology, 30*, 566–575.
Doody, J. S., Freedberg, S., & Keogh, J. S. (2009). Communal egg-laying in reptiles and amphibians: Evolutionary patterns and hypotheses. *The Quarterly Review of Biology, 84*, 229–252.
Dudley, S. A. (1996). The response to differing selection on plant physiological traits: Evidence for local adaptation. *Evolution, 50*, 103–110.
Ellis-Quinn, B. A., & Simon, C. A. (1989). Homing behavior of the lizard *Sceloporus jarrovi*. *Journal of Herpetology, 23*, 146–152.
Fitch, H. S. (1940). A field study of the growth and behavior of the fence lizard. *University of California Publications in Zoology, 44*, 151–172.
Fleishman, L. J. (1992). The influence of the sensory system and the environment on motion patterns in the visual displays of anoline lizards and other vertebrates. *The American Naturalist, 139*, 36–61.
Forester, D. C., Harrison, K., & McCall, L. (1983). The effects of isolation, the duration of brooding, and non-egg olfactory cues on clutch recognition by salamanders, *Desmognathus ochrophaeus*. *Journal of Herpetology, 17*, 308–314.
Forstmeier, W., & Weiss, I. (2004). Adaptive plasticity in nest-site selection in response to changing predation risk. *Oikos, 104*, 487–499.
Freake, M. J. (1998). Variation in homeward orientation performance in the sleepy lizard (*Tiliqua rugosa*): Effects of sex and reproductive period. *Behavioral Ecology and Sociobiology, 43*, 339–344.
Gould, J. L. (2006). Homing behavior: Decisions, dominance and democracy. *Current Biology, 16*, R920–R921.
Grant, P. R., & Grant, B. R. (2006). Evolution of character displacement in Darwin's finches. *Science, 313*, 224–226.
Greene, H. W., May, P. G., Hardy, D. L., Sr., Sciturro, J. M., & Farrell, T. M. (2002). Parental behavior by vipers. In G. W. Schuett, M. Hoggren, M. E. Douglas, & H. W. Greene (Eds.), *Biology of the vipers* (pp. 179–205). Eagle Mountain Publications.
Gruber, B., & Henle, K. (2004). Linking habitat structure and orientation in an arboreal species *Gehyra variegata* (Gekkonidae). *Oikos, 107*, 406–414.

Guyer, C. (1978). *Comparative ecology of the short-horned lizard (Phrynosoma douglassi) and the sagebrush lizard (Sceloporus graciosus) in southeastern Idaho*. Idaho State University.

Guyer, C. (1991). Orientation and homing behaviour as a measure of affinity for the home range in two species of iguanid lizards. *Amphibia-Reptilia, 12*, 373–384.

Hein, E. W., & Whitaker, S. J. (1997). Homing in eastern fence lizards (*Sceloporus undulatus*) following short-distance translocation. *Great Basin Nat., 57*, 348–351.

Herrel, A., Van Damme, R., & De Vree, F. (1996). Sexual dimorphism of head size in *Podarcis hispanica atrata*: Testing the dietary divergence hypothesis by bite force analysis. *Netherlands Journal of Zoology, 46*, 253–262.

Herrel, A., Spithoven, L., Van Damme, R., & De Vree, F. (1999). Sexual dimorphism of head size in *Gallotia galloti*: Testing the niche divergence hypothesis by functional analyses. *Functional Ecology, 13*, 289–297.

Herrel, A., Damme, R. V., Vanhooydonck, B., & Vree, F. D. (2001). The implications of bite performance for diet in two species of lacertid lizards. *Canadian Journal of Zoology, 79*, 662–670.

Herrel, A., De Grauw, E., & Lemos-Espinal, J. A. (2001). Head shape and bite performance in xenosaurid lizards. *Journal of Experimental Zoology, 290*, 101–107.

Herrel, A., Vanhooydonck, B., & Van Damme, R. (2004). Omnivory in lacertid lizards: Adaptive evolution or constraint? *Journal of Evolutionary Biology, 17*, 974–984.

Herrel, A., Andrade, D. V., De Carvalho, J. E., Brito, A., Abe, A., & Navas, C. (2009). Aggressive behavior and performance in the tegu lizard *Tupinambis merianae*. *Physiological and Biochemical Zoology, 82*, 680–685.

Herrel, A., Moore, J., Bredeweg, E., & Nelson, N. (2010). Sexual dimorphism, body size, bite force and male mating success in tuatara. *Biological Journal of the Linnean Society, 100*, 287–292.

Holt, R. D., & Polis, G. A. (1997). A theoretical framework for intraguild predation. *The American Naturalist, 149*, 745–764.

Huang, W. S. (2004). *Reptile ecology and the evolution of parental care on a tropical Asian Island*. PH. D. Cornell University.

Huang, W. S. (2006a). Ecological characteristics of the skink, *Mabuya longicaudata*, on a tropical east Asian Island. *Copeia, 2006*, 293–300.

Huang, W. S. (2006b). Ecology and reproductive patterns of the grass lizard, *Takydromus sauteri*, in a tropical rain forest of an East Asian Island. *Journal of Herpetology, 40*, 267–273.

Huang, W. S. (2006c). Parental care in the long-tailed skink, *Mabuya longicaudata*, on a tropical Asian Island. *Animal Behaviour, 72*, 791–795.

Huang, W. S. (2007a). Costs of egg caring in the skink, *Mabuya longicaudata*. *Ecological Research, 22*, 659–664.

Huang, W. S. (2007b). Ecology and reproductive patterns of the agamid lizard, *Japalura swinhonis*, on an East Asian Island, with comments on the small clutch sizes of Island lizrds. *Zoological Science, 24*, 181–188.

Huang, W. S. (2008). Predation risk of whole-clutch filial cannibalism in a tropical skink with maternal care. *Behavioral Ecology, 19*, 1069–1074.

Huang, W. S., & Pike, D. A. (2011a). Climate change impacts on fitness depend on nesting habitat. *Functional Ecology, 25*, 1125–1136.

Huang, W. S., & Pike, D. A. (2011b). Determinants of homing in nest-guarding females: Balancing risks while travelling through unfamiliar landscapes. *Animal Behaviour, 82*, 263–270.

Huang, W. S., & Pike, D. A. (2013). Testing cost-benefit models of parental care evolution using lizard populations differing in the expression of care. *PLoS One, 8*, e54065.

Huang, W. S., & Wang, H. Y. (2009). Predation risks and anti-predation parental care behavior: An experimental study in a tropical skink. *Ethology, 115*, 273–279.

Huang, W. S., Lin, S. M., Dubey, S., & Pike, D. A. (2013). Predation drives interpopulation differences in parental care expression. *Journal of Animal Ecology, 82*, 429–437.

Huey, R. B., & Pianka, E. R. (1981). Ecological correlates of foraging mode. *Ecology, 62*, 991–999.

Husak, J. F., Kristopher Lappin, A., Fox, S. F., & Lemos-Espinal, J. A. (2006). Bite-force performance predicts dominance in male venerable collared lizards (*Crotaphytus antiquus*). *Copeia, 2006*, 301–306.
Husak, J. F., Lappin, A. K., & Van Den Bussche, R. A. (2009). The fitness advantage of a high-performance weapon. *Biological Journal of the Linnean Society, 96*, 840–845.
Huyghe, K., Vanhooydonck, B., Scheers, H., Molina-Borja, M., & Van Damme, R. (2005). Morphology, performance and fighting capacity in male lizards, *Gallotia galloti*. *Functional Ecology, 19*, 800–807.
Ishihara, S. (1969). Homing behavior of the lizard, *Tachydromus tachydromoides* (Schlegel). *Bulletin of the Kyoto University of Education Series, 36*, 11–23.
Jenssen, T. A. (2002). Spatial awareness by the lizard *Anolis cristatellus*: Why should a non-range species demonstrate homing behavior? *Herpetologica, 58*, 364–371.
Jiggins, C. D., Naisbit, R. E., Coe, R. L., & Mallet, J. (2001). Reproductive isolation caused by colour pattern mimicry. *Nature, 411*, 302–305.
Kaliontzopoulou, A., Adams, D. C., van der Meijden, A., Perera, A., & Carretero, M. A. (2012). Relationships between head morphology, bite performance and ecology in two species of *Podarcis* wall lizards. *Evolutionary Ecology, 26*, 825–845.
Kearney, M. R., Porter, W., & Shine, R. (2009). The potential for behavioral thermoregulation to buffer "cold-blooded" animals against climate warming. *Proceedings of the National Academy of Science, 106*, 3835–3840.
Knell, R. J. (2009). Population density and the evolution of male aggression. *Journal of Zoology, 278*, 83–90.
Kolbe, J. J., & Janzen, F. J. (2002). Impact of nest-site selection on nest success and nest temperature in natural and disturbed habitats. *Ecology, 83*, 269–281.
Krekorian, C. O. (1977). Homing in the desert iguana, *Dipsosaurus dorsalis*. *Herpetologica, 33*, 123–127.
Lailvaux, S. P., & Irschick, D. J. (2007). The evolution of performance-based male fighting ability in Caribbean *Anolis* lizards. *The American Naturalist, 170*, 573–586.
Lailvaux, S. P., Herrel, A., VanHooydonck, B., Meyers, J. J., & Irschick, D. J. (2004). Performance capacity, fighting tactics and the evolution of life–stage male morphs in the green anole lizard (*Anolis carolinensis*). *Proceedings of the Royal Society B: Biological Sciences, 271*, 2501–2508.
Lappin, A. K., & Husak, J. F. (2005). Weapon performance, not size, determines mating success and potential reproductive output in the collared lizard (*Crotaphytus collaris*). *The American Naturalist, 166*, 426–436.
Lappin, A. K., Hamilton, P. S., & Sullivan, B. K. (2006). Bite-force performance and head shape in a sexually dimorphic crevice-dwelling lizard, the common chuckwalla [*Sauromalus ater* (= *obesus*)]. *Biological Journal of the Linnean Society, 88*, 215–222.
Lima, S., & Dill, L. M. (1990). Behavioral decisions made under the risk of predation: A review and prospectus. *Canadian Journal of Zoology, 68*, 619–640.
Lohmann, K. J., Lohmann, C. M. F., Ehrhart, L. M., Bagley, D. A., & Swing, T. (2004). Geomagnetic map used in sea-turtle navigation. *Nature, 428*, 909–910.
Lomolino, M. V. (2005). Body size evolution in insular vertebrates: Generality of the Island rule. *Journal of Biogeography, 32*, 1683–1699.
Mahew, W. W. (1963). Reproduction in the granite spiny lizard, *Sceloporus orcutti*. *Copeia, 1963*, 144–152.
Marek, C., Bissantz, N., Curio, E., Siegert, A., Tacud, B., & Ziggel, D. (2010). Spatial orientation of the Philippine bent-toed gecko (*Cyrtodactylus philippincus*) in relation to its home range. *Salamandra, 46*, 93–97.
Martin, T. E. (2001). Abiotic vs. biotic influences on habitat selection of coexisting species: Climate change impacts? *Ecology, 82*, 175–188.

Massot, M., Clobert, J., Lorenzon, P., & Rossi, J. M. (2002). Condition dependent dispersal and ontogeny of the dispersal behavior: An experimental approach. *Journal of Animal Ecology, 71*, 235–261.

Mathis, A., & Smith, J. F. (1993). Fathead minnows, *Pimephales promelas*, learn to recognize northern pike, *Esox lucius*, as predators on the basis of chemical stimuli from monnows in the pike's diet. *Animal Behaviour, 46*, 645–656.

Measey, G., Rebelo, A., Herrel, A., Vanhooydonck, B., & Tolley, K. (2011). Diet, morphology and performance in two chameleon morphs: Do harder bites equate with harder prey? *Journal of Zoology, 285*, 247–255.

Noble, G. K. (1934). Experimenting with the courtership of lizards. *Natural History, 34*, 5–15.

Noonan, B. P., & Comeault, A. A. (2009). The role of predator selection on polymorphic aposematic poison frogs. *Biology Letters, 5*, 51–54.

Novosolov, M., Raia, P., & Meiri, S. (2013). The Island syndrome in lizards. *Global Ecology and Biogeography, 22*, 184–191.

Peterson, M. G. (2000). Nest, but not egg, fidelity in a territorial salamander. *Ethology, 106*, 781–794.

Pike, D. A. (2005). Spatial learning in a hatchling Chelonia, *Gopherus polyphemus*. *Florida Scientist, 68*, 267–271.

Pike, D. A., Croak, B. M., Webb, J. K., & Shine, R. (2010). Context-dependent avoidance of predatory centipedes by nocturnal geckos (*Oedura lesueurii*). *Behaviour, 147*, 397–412.

Polis, G. A., Myers, C. A., & Holt, R. D. (1989). The ecology and evolution of intraguild predation: Potential competitors that eat each other. *Annual Review of Ecology and Systematics, 20*, 297–330.

Prieto, A. A., & Ryan, M. J. (1978). Some observations of the social behavior of the Arizona chuckwalla, *Sauromalus obesus tumidus* (Reptilia, Lacertilia, Iguanidae). *Journal of Herpetology, 12*, 327–336.

Putra, N. S., Yasuda, H., & Sat, S. (2009). Oviposition preference of two hoverfly species in response to risk of intraguild predation. *Applied Entomology and Zoology, 44*, 29–36.

Rand, A. S. (1967). Ecology and social organization in the iguanid lizard *Anolis lineatopus*. *Proceedings of the United States National Museum, 122*, 1–79.

Relyea, R. A. (2003). Predators come and predators go: The reversibility of predator-induces traits. *Ecology, 84*, 1840–1848.

Reynolds, R. G., & Fitzpatrick, B. M. (2007). Assortative mating in poison-dart frogs based on an ecologically important trait. *Evolution, 61*, 2253–2259.

Reznick, D. A., Bryga, H., & Endler, J. A. (1990). Experimentally induced life-history evolution in a natural population. *Nature, 346*, 357–359.

Reznick, D., Butler, M. J., IV, & Rodd, H. (2001). Life-history evolution in guppies. VII. The comparative ecology of high-and low-predation environments. *The American Naturalist, 157*, 126–140.

Runemark, A., Sagonas, K., & Svensson, E. I. (2015). Ecological explanations to Island gigantism: Dietary niche divergence, predation, and size in an endemic lizard. *Ecology, 96*, 2077–2092.

Russel, A. P., Bauer, A. M., & Johnson, M. K. (2005). Migration in amphibians and reptiles: An overview of patterns and orientation mechanisms in relation to life history strategies. In A. M. T. Elewa (Ed.), *Migration of organisms: Climate, geography, ecology* (pp. 152–203). Springer.

Ruxton, G. D., Sherratt, T. N., & Speed, M. P. (2004). *Avoiding attack: The evolutionary ecology of crypsis, warning signals and mimicry* (1st ed.). Oxford University Press.

Sagonas, K., Pafilis, P., Lymberakis, P., Donihue, C. M., Herrel, A., & Valakos, E. D. (2014). Insularity affects head morphology, bite force and diet in a Mediterranean lizard. *Biological Journal of the Linnean Society, 112*, 469–484.

Saporito, R. A., Zuercher, R., Roberts, M., Gerow, K. G., & Donnelly, M. A. (2007). Experimental evidence for aposematism in the dendrobatid poison frog *Oophaga pumilio*. *Copeia, 2007*, 1006–1011.

Schellhorn, N. A., & Andow, D. A. (1999). Cannibalism and interspecific predation: Role of oviposition behaviour. *Ecological Applications, 9*, 418–428.

Seddon, N. (2005). Ecological adaptation and species recognition drives vocal evolution in neotropical suboscine birds. *Evolution, 59*, 200–215.

Sergio, F., & Hiraldo, F. (2008). Intraguild predation in raptor assemblages: A review. *Ibis, 150*, 132–145.

Sergio, F., Marchesi, L., & Pedrini, P. (2003). Spatial refugia and the coexistence of a diurnal raptor with its intraguild owl predator. *Journal of Animal Ecology, 72*, 232–245.

Sergio, F., Marchesi, L., Pedrini, P., & Penteriani, V. (2007). Coexistence of a generalist owl with its intraguild predator: Distance-sensitive or habitat-mediated avoidance? *Animal Behaviour, 74*, 1607–1616.

Shine, R. (1980). "Costs" of reproduction in reptiles. *Oecologia, 46*, 92–100.

Shine, R. (1988). Parental care in reptiles. In C. Gans & R. B. Huey (Eds.), *Biology of the Reptilia Vol. 16 Ecology B. Defense and life history* (pp. 275–330). Alan R. Liss.

Shine, R. (2004). Incubation regimes of cold-climate reptiles: The thermal consequences of nest-site choice, viviparity and maternal basking. *Biological Journal of the Linnean Society, 83*, 145–155.

Shine, R., & Harlow, P. S. (1996). Maternal manipulation of offspring phenotypes via nest-site selection in an oviparous lizard. *Ecology, 77*, 1808–1817.

Shine, R., Elphick, M. J., & Harlow, P. S. (1997). The influence of natural incubation environments on the phenotypic traits of hatchling lizards. *Ecology, 78*, 2559–2568.

Somma, L. A. (1990). A categorization and bibliographic survey of parental behavior in lepidosaurian reptiles. *Smithsonian Herpetological Information Service, 81*, 1–53.

Southwood, A., & Avens, L. (2010). Physiological, behavioral, and ecological aspects of migration in reptiles. *Journal of Comparative Physiology, 180*, 1–23.

Spencer, R. J. (2002). Experimentally testing nest site selection: Fitness trade-offs and predation risk in turtles. *Ecology, 83*, 2136–2144.

Spencer, R.-J., & Thompson, M. B. (2003). The significance of predation in nest site selection of turtles: An experimental consideration of macro- and microhabitat preferences. *Oikos, 102*, 592–600.

Spoecker, P. D. (1967). Movements and seasonal activity cycles of the lizard *Uta stansburiana stejnegeri*. *The American Midland Naturalist, 77*, 484–494.

Stamps, J. A. (1983). Sexual selection, sexual dimorphism, and territoriality. In R. B. Huey, E. R. Pianka, & T. W. Schoener (Eds.), *Lizard ecology: Studies of a model organism* (pp. 169–204). Harvard University Press.

Stanley, M. C. (1998). Homing in the skink, *Oligosoma grande*, within a fragmented habitat. *Journal of Herpetology, 32*, 461–464.

Stoleson, S. H., & Beissinger, S. R. (2001). Does risk of nest failure or adult predation influence hatching patterns of the green-rumped parrotlet. *The Condor, 103*, 85–97.

Summers, K., Symula, R., Clough, M., & Cronin, T. (1999). Visual mate choice in poison frogs. *Proceedings of the Royal Society B: Biological Sciences, 266*, 2141–2145.

Thomas, P., Pouydebat, E., Hardy, I., Aujard, F., Ross, C. F., & Herrel, A. (2015). Sexual dimorphism in bite force in the grey mouse lemur. *Journal of Zoology, 296*, 133–138.

Tseng, H.-Y., Huang, W.-S., Jeng, M.-L., Villanueva, R. J. T., Nuñeza, O. M., & Lin, C.-P. (2018). Complex inter-island colonization and peripatric founder speciation promote diversification of flightless *Pachyrhynchus* weevils in the Taiwan–Luzon volcanic belt. *Journal of Biogeography, 45*, 89–100.

Turner, A. M., Turner, R. R., & Ray, S. R. (2007). Competition and intraguild egg predation among freshwater snails: Re-examining the mechanism of interspecific interactions. *Oikos, 116*, 1895–1903.

Underwood, G. (1954). On the classification and evolution of geckos. *Proceedings of the Zoological Society of London, 124*, 469–492.

Vanhooydonck, B., Cruz, F. B., Abdala, C. S., Azócar, D. L. M., Bonino, M. F., & Herrel, A. (2010). Sex-specific evolution of bite performance in *Liolaemus* lizards (Iguania: Liolaemidae): The battle of the sexes. *Biological Journal of the Linnean Society, 101*, 461–475.

Verwaijen, D., Van Damme, R., & Herrel, A. (2002). Relationships between head size, bite force, prey handling efficiency and diet in two sympatric lacertid lizards. *Functional Ecology, 16*, 842–850.

Walzer, A., Paulus, H. F., & Schausberger, P. (2006). Oviposition behavior of interacting predatory mites: Response to the presence of con- and heterospecific eggs. *Journal of Insect Behavior, 19*, 305–320.

Wang, L.-Y. (2017). *Too hard to swallow: A secret secondary defence of an aposematic insect*. National Taiwan Normal University.

Warner, D. A., Uller, T., & Shine, R. (2009). Fitness effects of the timing of hatching may drive the evolution of temperature-dependent sex determination in short-lived lizards. *Evolutionary Ecology, 23*, 281–294.

Webb, J. K., & Shine, R. (1997). A field study of spatial ecology and movements of a threatened snake species, Hoplocephalus bungaroides. *Biological Conservation, 82*, 203–217.

Webb, J. K., Pringle, R. M., & Shine, R. (2009). Intraguild predation, thermoregulation, and microhabitat selection by snakes. *Behavioral Ecology, 20*, 271–277.

Webb, J. K., Du, W.-G., Pike, D. A., & Shine, R. (2010). Generalization of predator recognition: Velvet geckos display anti-predator behaviours in response to chemicals from non-dangerous elapid snakes. *Current Zoology, 56*, 337–342.

Weintraub, J. D. (1970). Homing in the lizard *Sceloporus orcutti*. *Animal Behaviour, 18*, 132–137.

Žagar, A., Carretero, M. A., Vrezec, A., Drašler, K., & Kaliontzopoulou, A. (2017). Towards a functional understanding of species coexistence: Ecomorphological variation in relation to whole-organism performance in two sympatric lizards. *Functional Ecology, 31*, 1780–1791.

Zuri, I., & Bull, C. M. (2000). The use of visual cues for spatial orientation in the sleepy lizard (*Tiliqua rugosa*). *Canadian Journal of Zoology, 78*, 515–520.

Lizard Eggs and Ants

9

Abstract

A trade-off between dominance and resource discovery may promote species coexistence, yet evidence from natural systems remains limited. We tested this using a field experiment on two ants: the dominant *Paratrechina longicornis* and the exploiting *Pheidole taivanensis*, competing for lizard eggs. When *P. longicornis* encountered eggs occupied by *P. taivanensis* for 1 day, it displaced them. However, after 2 days, it ignored the eggs because *P. taivanensis* had damaged the shells, reducing resource value. At Sandimen, lower resource use led to shorter exploitation time, but higher ant density increased displacement rates. These results suggest lizard eggs are ephemeral resources used both spatially and temporally, and support a context-dependent trade-off between resource discovery and dominance.

Keywords

Coexistence of species · Ant · Resource overlap · Food competition

Mutualism refers to a symbiotic relationship between two species in the biological world where both parties depend on each other and benefit mutually. The most well-known example is the relationship between ants and aphids: ants protect aphids from predators and help them migrate to better plants to obtain more food, while the honeydew secreted by aphids serves as an important food source for ants. Another example is the relationship between clownfish and sea anemones. Clownfish have mucus on their bodies that prevents the stinging cells on the sea anemone's tentacles from attacking them, allowing them to use the sea anemone for protection. In return, the sea anemone benefits from the clownfish cleaning its tentacles.

The interesting story I am about to tell you below is that, although one of these two species of ants is a predator of lizard eggs, the other is a mutualistic symbiont. In

W.-S. Huang, *Reptiles in the Changing Climate*, Fascinating Life Sciences,
https://doi.org/10.1007/978-981-92-1887-5_9

a high-humidity environment, the condensation on the eggshell attracts the mutualistic symbiont, which helps defeat the predatory ant and protects the egg so that it can hatch.

9.1 Response of Two Sympatric Ants Foraging on Lizard Eggs

Mutualistic and antagonistic interactions frequently intersect within ecological communities, particularly when organisms compete for resources that are temporary and subject to rapid transformation. Rather than viewing competition as a fixed hierarchy determined solely by aggression or numerical dominance, this study emphasizes how the temporal dynamics of natural resources can alter interaction outcomes. Using lizard eggs in lowland forests of Taiwan as a focal system, this work examines how functionally distinct ant species respond to changing resource conditions and how those responses influence broader patterns of coexistence and community structure.

Lizard eggs represent a transient and vulnerable substrate embedded within a humid forest floor environment. Shortly after deposition, condensation on the eggshell surface and the diffusion of organic cues attract opportunistic ants searching for protein-rich resources. Some species behave strictly as predators, penetrating the shell and consuming embryonic contents whenever access is possible. Others, however, may occupy the eggs for extended periods, feeding gradually and in the process altering the shell's physical integrity and surface characteristics. These modifications do not merely reflect consumption; they transform the resource itself.

In interactions between *Pheidole taivanensis* and *Paratrechina longicornis*, such behavioral contrasts generate shifting competitive outcomes across time. The former rapidly discovers and exploits newly available eggs, modifying their structural properties through sustained use. The latter is an efficient encounter species that commonly displaces competitors from intact and profitable resources. Yet once the eggshell has been sufficiently altered, its nutritional value, accessibility, or structural suitability may decline from the perspective of the dominant competitor, thereby reducing the likelihood of successful displacement. Competitive advantage, therefore, becomes contingent upon the stage of resource transformation.

This sequence illustrates how resource-mediated change can reshape competitive asymmetries. Traditional interpretations of the dominance–discovery trade-off largely derive from artificial bait experiments that assume resource constancy. In natural environments, however, resources deteriorate, are restructured by consumers, and fluctuate in attractiveness across time. Such temporal variability creates sequential windows of opportunity, allowing different species to exploit the same substrate at different stages without continuous direct confrontation.

By foregrounding resource plasticity, this study suggests that coexistence may emerge from shifting temporal niches rather than stable competitive equilibria. The interaction between behavioral strategy, population density, and resource state determines whether antagonism, avoidance, tolerance, or indirect facilitation occurs.

Ephemeral substrates such as reptile eggs thus function as dynamic ecological arenas in which relationships are continually renegotiated, offering insight into how environmental variability and organismal flexibility together structure ecological communities.

9.2 Introduction

Ecological communities are structured through continual interactions among species competing for finite resources. These interactions influence not only patterns of resource acquisition but also behavioral decisions that scale up to affect population dynamics and community organization (Pleasants, 1980; Connell, 1983; Armbruster & Herzig, 1984; Hölldobler & Wilson, 1990; Davidson, 1998; Stone et al., 1998; Sanders & Gordon, 2000; Raine et al., 2007). Empirical research across a wide range of taxa demonstrates that interspecific competition can alter diet breadth, spatial distribution, activity schedules, and foraging strategies (Connell, 1961; Fellers, 1987; Hölldobler & Wilson, 1990; Heske et al., 1994; Davidson, 1998; Brouat et al., 2000; McGlynn & Kirksey, 2000; Bogran et al., 2002; Eccard & Ylonen, 2002; Bonesi et al., 2004; Alder & Silverman, 2005; Huang, 2008b). For instance, experimental removal of *Myrmecocystus* colonies caused neighboring *Aphaenogaster cockerelli* to shift their dietary allocation toward alternative prey, illustrating how competitive release can restructure patterns of resource use (Sarty et al., 2006; Santini et al., 2007). Despite these advances, most studies emphasize competitive presence or absence rather than how species differentially exploit distinct stages or components of the same natural resource.

Ant assemblages provide a particularly useful system for investigating competitive organization because species frequently overlap in diet and microhabitat. Wilson (1971) categorized ants into opportunists, exterminators, and insinuators, highlighting alternative strategies for securing food. Subsequent theoretical and empirical work distinguished between exploitative competition, which emphasizes rapid discovery, and interference competition, which relies on aggression and recruitment. Encounter species are often behaviorally dominant and capable of displacing competitors, whereas exploitative species excel at locating resources quickly but may relinquish them once dominant rivals arrive(Schoener, 1983; Hölldobler & Wilson, 1990; Davidson, 1998). Although these frameworks have clarified mechanisms underlying competitive asymmetry, they frequently assume that resources remain constant throughout interactions.

In natural systems, however, resources are dynamic entities. Organic substrates deteriorate, desiccate, fragment, or are structurally modified by consumers, thereby altering accessibility and energetic value over time. Such transformation may fundamentally influence whether displacement occurs, how long occupation persists, and whether sequential exploitation replaces direct confrontation. Yet relatively few field studies explicitly examine how resource conditions mediate interspecific interference in ants (Sarty et al., 2006; Santini et al., 2007). Moreover, limited research has integrated detailed behavioral observation with experimental manipulation to determine how overlapping resource use shifts under high exploitation intensity (Connell, 1983; Munday et al., 2001). Although strong competitive effects have been demonstrated in diverse organisms (Connell, 1983; Schoener, 1983; Davidson et al., 1984; Munday et al., 2001; Cross & Benke, 2002; Eccard & Ylonen, 2002; Richter-Boix et al., 2004), little is known about how variation in population density and resource state jointly influence foraging duration and displacement dynamics in ants, particularly when encounter species can exclude exploitative species entirely (Huang, 2008b).

To address these issues, this chapter investigates competitive interactions between *P. longicornis* and *P. taivanensis*, two sympatric and ecologically similar species in Taitung County, southeastern Taiwan (Huang, 2008b). Through field experiments using naturally occurring lizard eggs and controlled manipulation of eggshell condition, I evaluate how temporal changes in resource structure alter displacement probability and occupation duration. Comparisons among multiple sites, including SanDiMen in Pingtung County and Orchid Island off Taiwan's southeastern coast, allow assessment of how local population density and resource-use intensity shape interaction outcomes. By integrating behavioral strategy, resource transformation, and spatial context, this study aims to clarify how dynamic natural substrates mediate competitive asymmetries and contribute to coexistence within ant communities.

9.2.1 Study Species and Location

The present study focuses on two ant species that commonly occur in southern Taiwan: *P. longicornis* and *P. taivanensis*. Although these species differ markedly in nesting habits and foraging ecology, they overlap in their use of reptile eggs as a resource (Huang, 2008b).

P. longicornis is a highly mobile, polydomous species that maintains multiple interconnected nest sites and forages across diverse microhabitats. Colonies exploit soil cavities, decomposing wood, spaces beneath stones, vegetation, and artificial structures. In the study region, workers were frequently observed moving through drainage holes embedded in concrete retaining walls, where humidity and surface moisture are relatively stable. In contrast, *P. taivanensis* is primarily subterranean and typically forms visible soil mounds in exposed ground. Its diet is broad and includes arthropods, organic material, and reptile eggs.

Field observations in Pingtung County have documented that both species occur sympatrically in habitats containing nests of the skink *Eutropis longicaudata* (Huang, 2008b). Despite sharing the same resource, their interactions with lizard eggs differ fundamentally. *P. taivanensis* functions as a direct predator: workers use their mandibles to tear the flexible eggshell and access embryonic contents, a behavior that can dramatically reduce hatching success when predation persists (Fig. 9.1a). In contrast, *P. longicornis* does not normally penetrate intact eggs. Instead, it appears to associate with moisture that condenses on the eggshell surface, which likely provides a water source. When surface moisture diminishes, whether

Fig. 9.1 (**a**) *P. taivanensis* use their mandibles to open the elastic lizard eggshell; (**b**) *P. taivanensis* are displaced and killed by *P. longicornis* (small circle), as evidenced by their corpses (large circle) being found around the lizard egg nests. (Adapted from Huang (2010), Zoology, with permission. Photos rearranged and modified)

through drying or structural damage to the shell, the association weakens and colonies often abandon the nest site (Huang, 2008b).

Previous observations further indicate that nest occupancy can shift between the two species over time. Although *P. taivanensis* frequently locates nests rapidly, displacement patterns are not unidirectional. In some cases, colonies of *P. longicornis* were able to establish initial occupancy; however, replacement by *P. taivanensis* occurred in a substantial proportion of nests, and by the end of the observation period, only a subset of sites remained occupied by *P. longicornis* (Fig. 9.1b). These dynamics suggest that competitive outcomes are contingent upon both timing and changes in resource conditions.

The focal reptile, *E. longicaudata*, is the largest skink species in Taiwan and is widely distributed across Taiwan, southeastern China, Indochina, and the Malay Peninsula (Huang, 2006a). In southern Taiwan, it is especially abundant at low elevations. Breeding typically occurs from March through August, with incubation lasting approximately 26–33 days. Clutch size ranges from 2 to 13 eggs, with an average of about 6–7 per nest (Huang, 2006a). During the present study, skink eggs were commonly observed within artificial drainage cavities of concrete retaining walls, creating a semi-enclosed microenvironment in which both ant species and lizard eggs came into close spatial proximity.

9.2.2 Ants' Arrival Time to the Nest and the Hatching Success Rate of Lizard Eggs

A total of 240 lizard nests were monitored across five field sites from oviposition to hatching. Among these, 59 nests were associated with *P. longicornis* and 26 with *P. taivanensis*, all located within artificial concrete cavities.

The two ant species differed markedly in the timing of nest detection. Colonies of *P. taivanensis* were recorded at nests earlier in the incubation period (mean ± SD: 6.2 ± 2.4 days; range 1–14 days; $n = 26$), whereas *P. longicornis* appeared later on

Table 9.1 Arrival time at egg after being laid, frequency of first arrival by an ant species (occupied nests), and daily arrival period to lizard nests by *Pheidole taivanensis* (PT) and *Paratrechina longicornis* (PL) at the five study sites. first, 00:00–06:00; second, 06:00–12:00; third, 1200: 1800; fourth, 18:00–0:00

Locality	Ant	Arrival time (days)	Frequency of first arrival species	Daily arrival period			
				1st	2nd	3rd	4th
SanDiMen	PT	5.4 ± 2.0	22	7	6	3	6
	PL	9.3 ± 3.4	12	3	2	4	3
Orchid Is.	PT	6.8 ± 2.3	10	4	3	2	1
	PL	10.3 ± 4.2	7	2	2	2	1
Napu	PT	7.2 ± 3.2	8	2	3	1	2
	PL	10.2 ± 3.7	6	1	2	1	2
Tapu	PT	7.4 ± 2.0	8	2	3	1	1
	PL	11.3 ± 4.0	5	1	2	1	1
Maja	PT	7.8 ± 3.3	7	1	2	2	2
	PL	10.5 ± 3.5	3	1	1	1	0

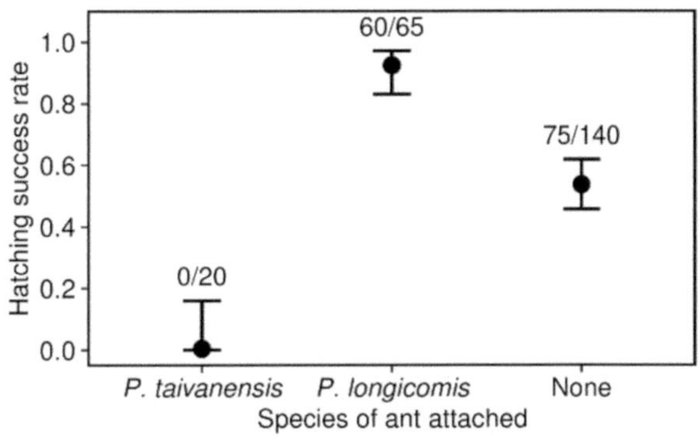

Fig. 9.2 Hatching success of lizard clutches when attended by ants (*Pheidole taivanensis* or *Paratrechina longicornis*) or unattended at the five study sites. All lizard eggs were completely consumed in *P. taivanensis*-attended nests. Numbers indicate the number of attached nests and the total observed nests. (Redrawn from Huang (2010), Zoology)

average (10.2 ± 3.7 days; range 1–27 days; $n = 59$; Table 9.1). Statistical comparison confirmed that the difference in arrival timing between species was significant (unpaired *t*-test: $t = 4.30$, $p < 0.001$; Table 9.1). When comparisons were made within matched site conditions, the earlier appearance of *P. taivanensis* remained evident (paired *t*-test: $p = 0.037$, df = 5; Table 9.1).

Egg survival outcomes varied according to the attending species. Nests occupied by *P. longicornis* exhibited substantially higher hatching success relative to those associated with *P. taivanensis* ($\chi^2 = 53.5$, df = 2, $p < 0.0001$; Fig. 9.2). This contrast is consistent with the distinct ecological roles of the two ants, as one species is primarily associated with moisture use while the other directly consumes egg contents.

Arrival timing within the daily cycle did not differ among the five study sites for either species. Analyses using Fisher's exact probability test indicated no site-specific pattern in the time of day when ants first reached nests ($p = 0.87$ for *P. taivanensis*; $p = 0.95$ for *P. longicornis*; Table 9.1). These results suggest that nest detection is not restricted to a particular diel period but instead reflects broader foraging dynamics.

9.2.3 Duration of Eggshell Damage, Intensity of Food Utilization, and Frequency of Foraging Routes in *P. taivanensis* and *P. longicornis*

Among eggs attended by *P. taivanensis*, surface condensation was recorded only when occupation lasted 1 day ($n = 30$). Eggs occupied for 2–4 consecutive days showed clear shell deterioration and no visible condensation. Nevertheless, damaged eggs were readily reoccupied when *P. taivanensis* was experimentally reintroduced.

At SanDiMen, the mean duration of shell damage was 24 ± 1.4 h ($n = 30$), significantly shorter than at Orchid Island and the other sites ($F_{1,101} = 12.3$, $p < 0.001$; Fig. 9.3). This location also exhibited a greater proportion of nests initially used by *P. taivanensis* that were later replaced by *P. longicornis* ($\chi^2 = 12.7$, df = 4, $p = 0.013$; Fig. 9.3).

Foraging route frequency, reflecting relative population density, was highest at SanDiMen for both species (ANOVA: *P. taivanensis*, $F_{4,20} = 121.1$; *P. longicornis*, $F_{4,20} = 38.6$; both $p < 0.0001$; Fig. 9.3). Collectively, SanDiMen showed the most intensive ant activity and nest displacement, yet the shortest eggshell damage duration.

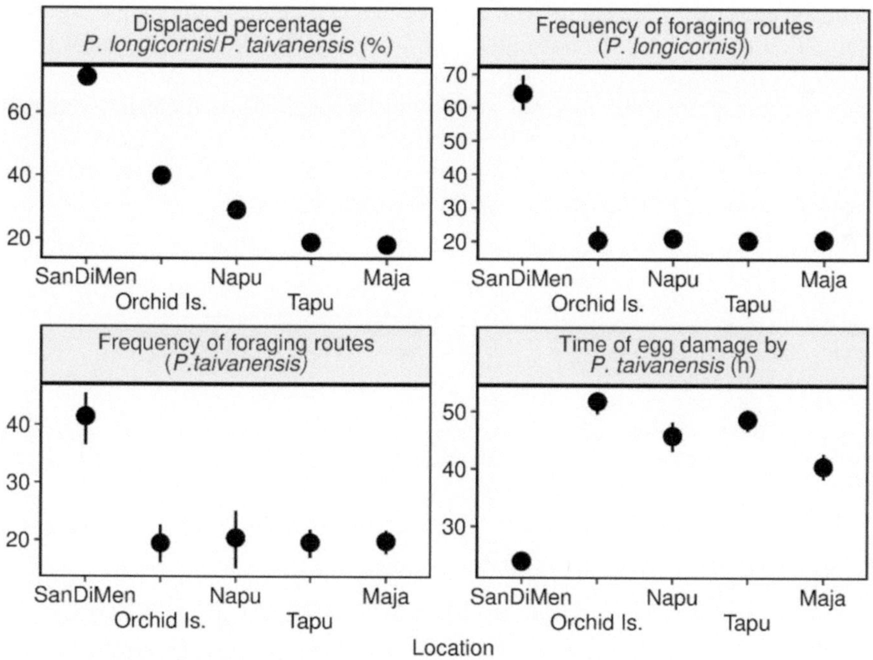

Fig. 9.3 Timeline of eggshell damage by *Pheidole taivanensis*, intensity of food competition (represented by the frequency of egg occupation by *P. taivanensis* and then displacement by *Paratrechina longicornis*), and frequency of foraging routes by *P. taivanensis* and *P. longicornis* at the five study sites. (Redrawn from Huang (2010), Zoology)

9.2.4 Introduction of Ants into Untreated Eggs

When *P. longicornis* was experimentally introduced into nests already occupied by *P. taivanensis*, competitive outcomes depended strongly on the condition of the eggs. In nests containing eggs with intact shells (group 1; $n = 29$; SanDiMen), *P. longicornis* frequently displaced *P. taivanensis*. In contrast, such replacement rarely occurred in nests where eggs had already been structurally damaged. Only a few displacement events were recorded in these groups (three on day 2, two on day 3, four on day 3 in another set, and one on day 4), producing a significant difference among egg conditions ($\chi^2 = 88.2$, df $= 3$, $p < 0.0001$; Fig. 9.4). In several cases where displacement occurred, dead workers of *P. taivanensis* were observed near the pipe openings of nests that were subsequently occupied by *P. longicornis*.

9.2.5 Introduction of Ants into Artificially Damaged Eggs

A contrasting pattern emerged when eggs were experimentally damaged before ant introduction. Colonies of *P. longicornis* did not establish on any of the artificially damaged eggs ($n = 15$), and no workers appeared at these nests during the observation period (Fisher's exact probability test, $p < 0.0001$). Unlike the control treatment, *P. taivanensis* later returned to these nests and gradually removed the egg contents, transporting the material away until the eggs were completely emptied ($n = 14$; mean duration $= 3.8 \pm 0.5$ days; $\chi^2 = 26.3$, df $= 1$, $p < 0.0001$). This pattern suggests that the shell condition strongly influences resource suitability for the two species and shapes their competitive interactions.

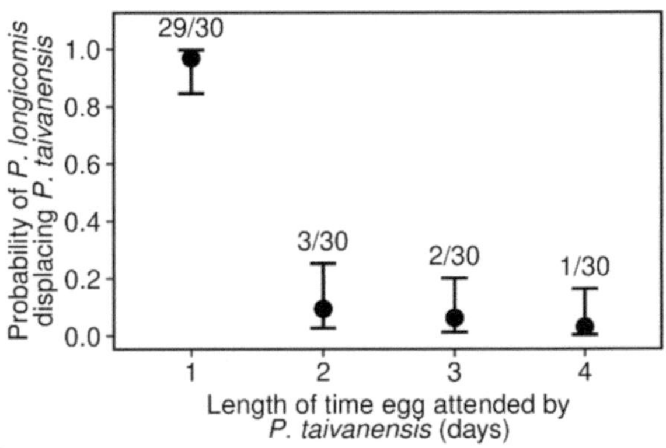

Fig. 9.4 Responses of *Paratrechina longicornis* when introduced to eggs occupied by *Pheidole taivanensis* for four different lengths of time. Only the group occupied by *P. taivanensis* for 1 day was significantly displaced by *P. longicornis*. (Redrawn from Huang (2010), Zoology)

9.2.6 Discussion

Interactions between ant species competing for the same food source often depend on differences in behavioral strategy. In this system, the aggressive and behaviorally dominant *P. longicornis* tends to exclude the subordinate *P. taivanensis* from lizard eggs through interference or disturbance. Nevertheless, *P. taivanensis* appears to compensate for its lower competitive dominance by locating food more rapidly. By discovering nests earlier, this species can begin exploiting egg resources before *P. longicornis* arrives, allowing it temporary access to food that might otherwise be monopolized by the dominant competitor.

Differences among study sites further influenced these outcomes. At SanDiMen, eggshell destruction by *P. taivanensis* occurred over a shorter period than at the other four locations, while displacement events involving *P. longicornis* were comparatively frequent (Fig. 9.3). One possible explanation is the relatively high ant population density at this site. A fruit orchard located near the retaining wall may increase food availability during the mango-falling season (May–July), which in turn supports larger ant populations. Greater population density could intensify competition between the two species and accelerate resource exploitation.

Temporal differences in resource use also appear to influence how these ants interact. Two contrasting foraging patterns were observed: the encounter-type strategy of *P. longicornis* and the exploitative strategy of *P. taivanensis*. The former tends to interact with eggs primarily when the eggs provide a direct reward, while the latter locates nests earlier and rapidly damages eggshells. Because *P. longicornis* does not typically establish permanent nests directly at egg sites, its presence may fluctuate according to the availability of rewards such as moisture condensation on the eggshell surface (Huang, 2008b). Ant attendance is therefore likely to concentrate on eggs that currently provide these benefits, as has been reported in other ant-resource interactions (Oliveira, 1997; Giusto et al., 2001). When rewards decline, for example, after eggs hatch or when shells become damaged, *P. longicornis* colonies tend to abandon the nest area.

In contrast, *P. taivanensis* adopts a strategy of early arrival and rapid exploitation. By reaching nests soon after eggs are deposited, it can begin feeding before the dominant species establishes control. However, if eggs remain intact, later arrival by *P. longicornis* often results in displacement of *P. taivanensis*. This competitive pressure likely explains the rapid shell-breaking behavior observed in southern Taiwan. Once the eggshell structure is sufficiently damaged, condensation disappears and the resource becomes less attractive to *P. longicornis*, reducing the likelihood of takeover.

Several mechanisms may therefore explain why *P. taivanensis* destroys eggshells quickly. First, it lacks the ability to defend egg resources against the more aggressive *P. longicornis*. Second, even when it discovers nests first, intact eggs may still attract the dominant competitor, eventually leading to displacement. Finally, when unattended eggs are encountered, *P. taivanensis* appears to initiate shell damage immediately, regardless of clutch size, thereby accelerating access to the contents while simultaneously decreasing the probability of later interference. Together, these behavioral differences highlight how temporal resource dynamics and contrasting foraging strategies can shape competitive interactions between coexisting ant species.

9.3 Ants' Resource Exploitation Promotes Lizard Egg Survival

Interactions between ants and vertebrate eggs represent a complex ecological system in which predation, commensalism, and competition may occur simultaneously. In southern Taiwan, eggs of the long-tailed skink *E. longicaudata* provide a temporary but valuable resource that attracts multiple ant species. Among these, the tramp ant *P. longicornis* and the egg-predatory ant *P. taivanensis* display contrasting ecological roles. The former often associates with lizard nests without consuming the eggs directly, whereas the latter feeds on egg contents. Examining how these species interact with the same resource helps clarify how behavioral strategies influence ecological outcomes.

Field observations showed that both ant species actively searched for newly laid lizard eggs. However, their arrival patterns differed markedly. Colonies of *P. taivanensis* generally discovered nests earlier than *P. longicornis*, allowing them to exploit eggs soon after oviposition. When undisturbed by other ants, this predatory species could substantially reduce egg survival through direct feeding. In contrast, the later arrival of *P. longicornis* often altered these outcomes. Despite reaching nests later, this species frequently displaced *P. taivanensis*, resulting in significantly higher egg survival in nests attended by *P. longicornis* (0.95 ± 0.04 eggs) compared with nests lacking ants (0.65 ± 0.09 eggs) or nests occupied by *P. taivanensis* (0.07 ± 0.02 eggs).

Competition among ants may occur through different mechanisms. One involves direct interference, in which aggressive encounters lead to injury or displacement. Another involves exploitative competition, where species gain an advantage by rapidly locating and consuming resources before competitors arrive (Hölldobler, 1986). Many ants exhibit aggressive interference behavior, yet some species rely primarily on rapid discovery of food. Workers of *P. longicornis*, for example, are highly mobile and possess strong chemosensory abilities that allow them to detect newly available resources quickly. After locating food, workers recruit nestmates through pheromone trails. Despite this efficiency, the species often retreats when confronted by more aggressive ants and may abandon contested resources (Wilson, 1971).

Most studies of ant competition have focused on interactions among ants themselves, particularly in relation to survival or dominance hierarchies. Comparatively fewer studies have examined how ants interact with the resources that structure these relationships. In many mutualistic systems, ants protect plants or insects in exchange for nutritional rewards such as nectar or honeydew (Peterson, 1995; Moya-Raygoza & Nault, 2000). These mutualisms are generally stabilized by a balance of costs and benefits between partners (Bronstein, 1994; Giusto et al., 2001). Experiments have demonstrated that increased nectar production can attract more ants and enhance defensive services against herbivores (Giusto et al., 2001). However, relatively little attention has been given to what occurs when the resource that initially attracts ants' changes or disappears.

Water availability is another factor influencing ant distribution and colony persistence (Hölldobler & Wilson, 1990). Because ants are sensitive to desiccation, moisture strongly affects colony establishment and foraging behavior (Hölldobler & Wilson, 1990). Reptile eggs can represent an unusual source of moisture. Under conditions of very high humidity (90–100%), condensation forms on the eggshell surface, producing small droplets of water (Huang, 2004, 2008b, 2010). Field observations in southern Taiwan showed that *P. longicornis* frequently collected these droplets, suggesting that the ants may exploit reptile eggs primarily as a water resource rather than as food. The ecological consequences of this association, however, remained largely unexplored.

The biology of the two ant species suggests contrasting strategies. *P. taivanensis* is mainly subterranean and commonly constructs soil mounds in exposed ground. It consumes various animal materials, including reptile eggs. In contrast, *P. longicornis* is an extremely adaptable species that nests in decaying wood, beneath rocks, within soil, or inside artificial structures such as concrete cavities. Workers forage both above and below ground and can exploit resources across a wide area. Because these species frequently coexist in southern Taiwan, they provide an opportunity to examine how behavioral differences influence access to shared resources.

The lizard *E. longicaudata* also plays a critical role in this ecological system. Females guard their eggs, a form of parental care that significantly increases hatching success (Huang, 2006b). When guarding females are experimentally removed, egg survival declines sharply. One major predator of these eggs is the snake *O. formosanus*, which frequently attacks nests. On Orchid Island, where lizards often deposit eggs inside concrete retaining walls, nests experience relatively warm incubation temperatures but are also more exposed to predators (Huang & Wang, 2009; Huang & Pike, 2011a, b). Under natural conditions, parental care maintains hatching success near 85%, whereas removal of the guarding female can reduce survival to about 65% (Huang, 2006b). However, *If lizards* are absent and various species of ants coexist *near the eggs, would egg survival vary depending on which ant species is present?*

This chapter describes unusual interactions between the ant-accompanying lizard egg (*E. longicaudata*) and its temporal variation in the foraging between *P. longicornis* and *P. taivanensis*. Specifically, we experimentally removed water and *P. longicornis* from the egg surface to evaluate the extent to which egg survival was directly influenced by predators, to estimate the fitness effect on eggs, and to test ant-egg interactions.

When lizards are absent and eggs remain unattended, ants become important participants in nest dynamics. To investigate these interactions, 139 nests were monitored from the time of egg laying until hatching during the 2003 breeding season. Among these nests, 18 were attended by *P. taivanensis* and 31 by *P. longicornis*. The average time required for *P. taivanensis* to locate nests after

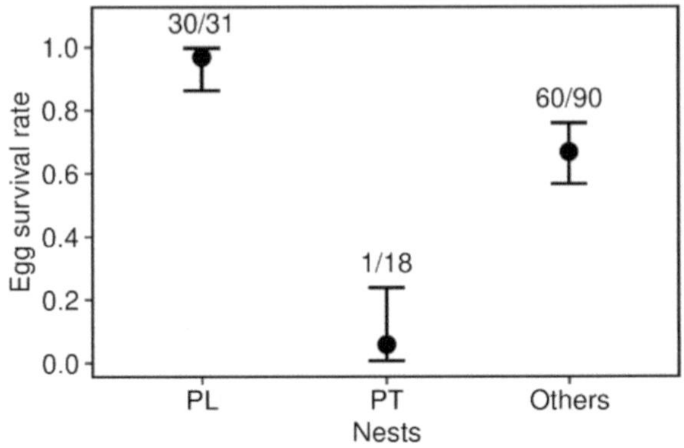

Fig. 9.5 Proportions of eggs of *Eutropis longicaudata* hatching in the presence and absence of *Paratrechina longicornis* (PL). Survival rates were higher when eggs were accompanied by *P. longicornis* than by *Pheidole taivanensis* (PT) and others. (Redrawn from Huang (2008a, b), Ecological Entomology)

oviposition was 5 ± 4.5 days (range 1–14 days; $n = 18$), significantly earlier than the arrival of *P. longicornis* (12.3 ± 6.3 days; range 1–27 days; $t = -4.75$, $n = 31$; $p < 0.0001$). However, once present at nests, *P. longicornis* remained considerably longer than *P. taivanensis* (mean, 20.3 ± 4.6 days; range, 13 ~ 27 days; $n = 31$ in *P. longicornis*; mean, 5.3 ± 2.0 days; range, 3 to 8 days; $n = 18$ in *P. taivanensis*; $t = 15.8$; $p < 0.0001$).

These differences had important consequences for egg survival. Hatching success was highest in nests attended by *P. longicornis*, intermediate in nests without ants, and extremely low in nests occupied by *P. taivanensis*. Statistical analysis confirmed significant differences among these groups (95.5% ± 3.5% mean ± 1 SD, $n = 31$ in nests with *P. longicornis*; 6.5% ± 2%, $n = 18$ in nests with *P. taivanensis*: other nests: 66.5% ± 9%, $n = 90$; $\chi^2 = 269.4$, df = 2, $p < 0.0001$; Fig. 9.5). Evidence of aggressive displacement was also observed: dead workers of *P. taivanensis* were found outside 10 of the 31 nests later occupied by *P. longicornis*, suggesting that the predatory species had initially discovered the nests but was subsequently driven away (mean, 1.2 days; range 1–3 days; $n = 10$).

Additional experiments demonstrated that the association between *P. longicornis* and lizard eggs depended strongly on the presence of moisture. When water was experimentally removed from eggs that had been attended by *P. longicornis* for 2 days, the ants gradually abandoned the nests within 3–6 days (mean 4.7 ± 1.5 days; $n = 12$). In untreated nests, by contrast, *P. longicornis* colonies remained for much longer periods (mean 20.3 ± 4.6 days; range, $n = 31$, $t = 17.2$, $p < 0.0001$). After the ants disappeared, *P. taivanensis* frequently returned and consumed the eggs, removing their contents within approximately 3.2 ± 0.3 days ($n = 8$).

Consistent with this pattern, egg mortality was significantly higher in nests from which *P. longicornis* had been experimentally excluded. Egg survival averaged 65% ± 5% ($n = 66$) in nests where the species was present but declined to 31% ± 4% ($n = 22$) when it was removed. Exclusion of *P. longicornis* also resulted in a substantial increase in predation by *P. taivanensis* ($\chi^2 = 46.3$, df = 4, $p < 0.0001$; Fig. 9.6).

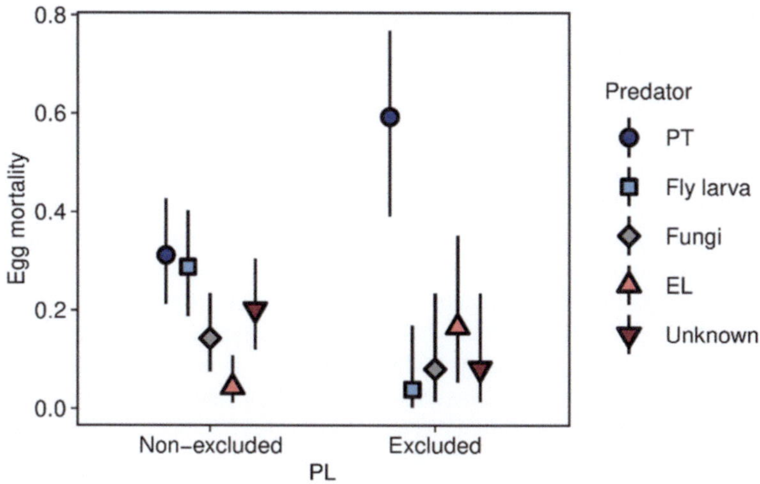

Fig. 9.6 Proportions of eggs consumed between the non-excluded and excluded nests from which *Paratrechina longicornis* had been removed. Higher predation rates by *Pheidole taivanensis* (PT) were noted when *P. longicornis* had been excluded. EL: *Eutropis longicaudata*. (Redrawn from Huang (2008a, b), Ecological Entomology)

Taken together, these findings suggest that the relationship between *P. longicornis* and lizard eggs represents an indirect protective interaction. Although the ants do not consume the eggs themselves, their presence deters predatory ants and thereby increases egg survival. At the same time, this association appears to depend on a transient environmental resource, condensed water on the eggshell surface. When this moisture disappears, *P. longicornis* abandons the nest, leaving the eggs vulnerable to predation by *P. taivanensis*.

9.3.1 Discussion

The present chapter provides evidence that the presence of *P. longicornis* increases the survival rate of long-tailed skink (*Eutropis longicaudata*) eggs (Fig. 9.6). Previous studies have shown that when either *P. longicornis* or *P. taivanensis* invades long-tailed skink nests, the females abandon their eggs (Huang, 2004). In nests invaded by *P. taivanensis*, nearly all eggs are consumed. In contrast, nests occupied by *P. longicornis* exhibited a hatching success rate of approximately 95% (Fig. 9.6), which is even higher than the 85% success rate observed in nests with maternal care (Huang, 2006b). Furthermore, *P. longicornis* eliminates the invasive *P. taivanensis*, thereby protecting the lizard eggs.

Although *P. taivanensis* clearly feeds on lizard eggs, the benefits gained by *P. longicornis* from guarding the eggs are less obvious. One possible explanation is related to water acquisition. High relative humidity in nests may facilitate the condensation of water on the eggshell surface, providing a potential moisture source for ants in otherwise dry shelters (mean relative humidity 98.5 ± 2.2%; range 95–100%). The experimental removal of water from lizard eggs in this study indirectly supports this hypothesis. A comparable mechanism has been documented in the ant *Diacamma rugosum*, in which workers decorate nest entrances with hygroscopic materials such as dead ants and bird feathers to collect dew that later forms on these objects (Moffett, 1985).

In addition to predation by *P. taivanensis*, egg loss in the Pingtung region has also been attributed to parasitic fly larvae and fungal infections on eggshells. Numerous studies have demonstrated that *E. longicaudata* provides parental care and actively defends nests against snake predators on Orchid Island (Huang & Wang, 2009; Huang et al., 2013; Huang & Pike, 2013). However, females are unable to protect nests from fungal infection and typically abandon nests once they are invaded by either *P. longicornis* or *P. taivanensis*, suggesting that maternal presence becomes less effective once ants or pathogens occupy the nest (Huang, 2006b). Because *P. longicornis* eliminates *P. taivanensis* and deters other potential predators, its presence appears to benefit egg survival. Moreover, by extracting water from eggshell surfaces, *P. longicornis* may also reduce the likelihood of lethal fungal infections and thereby promote higher hatching success.

Environmental conditions also influence offspring development. Eggs of *E. longicaudata* incubated in concrete holes exhibit significantly higher survival rates and larger hatchling body sizes compared with those incubated under natural conditions (Huang & Pike, 2011a). Moist nesting environments during warmer seasons may promote faster embryonic development, resulting in shorter incubation periods and higher survival rates. Larger hatchlings may gain advantages in capturing larger prey during periods of food scarcity, potentially enhancing their future survival compared with smaller individuals that hatch later in the season. Thus, nest-site selection by females can influence offspring phenotypic traits and ultimately affect their fitness (Huang & Pike, 2011a).

The structural organization of most squamate eggshells is relatively simple and uniform, consisting of calcium carbonate calcification arranged in various patterns on the shell surface (Packard & Packard, 1980; Schleich & Kastle, 1988). Eggshell characteristics are shaped by the ecological conditions of the nesting environment, and their functions are generally associated with adaptations to local environmental constraints, such as gas exchange and water regulation(Packard & Packard, 1980, Schleich & Kastle, 1988). In the moist nests of *E. longicaudata*, the highly hygroscopic surface of reptile eggshells may facilitate moisture capture. *P. longicornis* may initially be attracted to condensed moisture on the eggshells, and once established within the nest, the ants may reduce parasitism and predation on the eggs, thereby increasing hatching success.

Ant populations often increase markedly when eggs are deposited in concrete holes, sometimes completely surrounding the nesting site. A comparable ecological association has been reported for the Amazonian lizard *Tupinambis teguixin*, which lays its eggs in termite nests of *Nasutitermes* (Beebe, 1945). In this system, termites maintain the nest structure and create a warm, humid microenvironment that protects eggs from predators and provides favorable incubation conditions (Pianka & Vitt, 2003).

In opportunistic systems, ant activity may decline when rewards become scarce. In this study, colonies of *P. longicornis* disappeared within approximately 4.6 days after the lizards hatched. A remaining question is why *P. longicornis* does not directly consume the lizard eggs when water is unavailable. One possible explanation is that the soldiers of *P. longicornis* lack the strong mandibles required to break open eggshells, unlike the soldiers of *P. taivanensis*. Nevertheless, the precise mechanism underlying this interaction remains unclear and warrants further field observations and controlled laboratory experiments.

While *P. taivanensis* adopts an opportunistic strategy by arriving early at nests and directly consuming eggs, *P. longicornis* establishes colonies around lizard eggs. The presence of *P. longicornis* reduces the likelihood that *P. taivanensis* will occupy the nest and is associated with increased egg survival.

More broadly, highly territorial animals are generally expected to exhibit more frequent conflicts when population density increases. This raises further ecological questions: do individuals return successfully to their territories primarily because of food availability or mating opportunities? How do lizards respond evolutionarily when their eggs face intense predation pressure from snakes? Under such conditions, could the evolution of viviparity be favored? These questions will be examined in the next chapter using tree lizards as a model system.

References

Alder, P., & Silverman, J. (2005). Effects of interspecific competition between two urban ant species, *Linepithema humile* and *Monomorium minimum*, on toxic bait performance. *Journal of Economic Entomology, 98*, 493–501.

Armbruster, W. S., & Herzig, A. L. (1984). Partitioning and sharing of pollinators by four sympatric species of *Dalechampia* (Euphorbiaceae) in Panama. *Annals of the Missouri Botanical Garden, 71*, 1–16.

Beebe, W. (1945). Field notes on the lizards of Kartabo, British Guiana, and Caripito, Venezuela. Part 3. Teiidae, Amphisbaenidae, and Scincidae. *Zoologica: Scientific Contributions of the New York Zoological Society, 30*, 7–32.

Bogran, C. E., Heinz, K. M., & Ciomperlik, M. A. (2002). Interspecific competition among insect parasitoids: Field experiments with whiteflies as hosts in cotton. *Ecology, 83*, 653–668.

Bonesi, L., Chanin, P., & Macdonald, D. W. (2004). Competition between eurasian otter *Lutra lutra* and American mink *Mustela vison* probed by niche shift. *Oikos, 106*, 19–26.

Bronstein, J. L. (1994). Our current understanding of mutualism. *The Quarterly Review of Biology, 69*, 31–51.

Brouat, C., Mckey, D., Bessiere, J.-M., Pascal, L., & Hossaert-Mckey, M. (2000). Leaf volatile compounds and the distribution of ant patrolling in an ant-plant protection mutualism: Preliminary results on *Leonardoxa* (Fabaceae: Caesalpinioideae) and *Petalomyrmex* (Formicidae: Formicinae). *Acta Oecologica, 21*, 349–357.

Connell, J. H. (1961). The influence of interspecific competition and other factors on the distribution of the barnacle *Chthamalus stellatus*. *Ecology, 42*, 710–723.

Connell, J. H. (1983). On the prevalence and relative importance of interspecific competition: Evidence from field experiments. *The American Naturalist, 122*, 661–696.

Cross, W. F., & Benke, A. C. (2002). Intra- and interspecific competition among coexisting lotic snails. *Oikos, 96*, 251–264.

Davidson, D. W. (1998). Resource discovery versus resource domination in ants: A functional mechanism for breaking the trade-off. *Ecological Entomology, 23*, 484–490.
Davidson, D. W., Inouye, R. S., & Brown, J. H. (1984). Granivory in a desert ecosystem: Experimental evidence for indirect facilitation of ants by rodents. *Ecology, 65*, 1780–1786.
Eccard, J. A., & Ylonen, H. (2002). Direct interference or indirect exploitation? An experimental study of fitness costs of interspecific competition in voles. *Oikos, 99*, 580–590.
Fellers, J. H. (1987). Interference and exploita tion in a guild of woodland ants. *Ecology, 68*, 1466–1478.
Giusto, B. D., Anstett, M.-C., Dounias, E., & Mckey, D. (2001). Variation in the effectiveness of biotic defence: The case of an opportunistic ant-plant protection mutualism. *Oecologia, 129*, 367–375.
Heske, E. J., Brown, J. H., & Mistry, S. (1994). Long-term experimental study of a chihuahuan desert rodent community: 13 years of competition. *Ecology, 75*, 438–445.
Hölldobler, R. (1986). Food robbing in ants, a form of interference competition. *Oecologia, 69*, 12–15.
Hölldobler, R., & Wilson, E. O. (1990). *The ants*. Harvard University Press.
Huang, W. S. (2004). *Reptile ecology and the evolution of parental care on a tropical Asian Island*. PH. D. Cornell University.
Huang, W. S. (2006a). Ecological characteristics of the skink, *Mabuya longicaudata*, on a tropical east Asian Island. *Copeia, 2006*, 293–300.
Huang, W. S. (2006b). Parental care in the long-tailed skink, *Mabuya longicaudata,* on a tropical Asian Island. *Animal Behaviour, 72*, 791–795.
Huang, W. S. (2008a). Predation risk of whole-clutch filial cannibalism in a tropical skink with maternal care. *Behavioral Ecology, 19*, 1069–1074.
Huang, W. S. (2008b). Resources exploitation by ants facilitates lizard egg survival. *Ecological Entomolology, 33*, 555–559.
Huang, W. S. (2010). Foraging behaviors of two sympatric ant species in response to lizard eggs. *Zoology, 113*, 85–90.
Huang, W. S., & Pike, D. A. (2011a). Climate change impacts on fitness depend on nesting habitat. *Functional Ecology, 25*, 1125–1136.
Huang, W. S., & Pike, D. A. (2011b). Determinants of homing in nest-guarding females: Balancing risks while travelling through unfamiliar landscapes. *Animal Behaviour, 82*, 263–270.
Huang, W. S., & Pike, D. A. (2013). Testing cost-benefit models of parental care evolution using lizard populations differing in the expression of care. *PLoS One, 8*, e54065.
Huang, W. S., & Wang, H. Y. (2009). Predation risks and anti-predation parental care behavior: An experimental study in a tropical skink. *Ethology, 115*, 273–279.
Huang, W. S., Lin, S. M., Dubey, S., & Pike, D. A. (2013). Predation drives interpopulation differences in parental care expression. *Journal of Animal Ecology, 82*, 429–437.
McGlynn, T. P., & Kirksey, S. E. (2000). The effects of food presentation and microhabitat upon resource monopoly in a ground-foraging ant (hymenoptera: Formicidae) community. *Revista de Biología Tropical, 48*, 629–642.
Moffett, M. W. (1985). An Indian ant's novel method for obtaining water. *National Geographic Research, 1*, 146–149.
Moya-Raygoza, G., & Nault, L. R. (2000). Obligatory mutualism between *Dalbulus quinquenotatus* (Homoptera: Cicadellidae) and attendant ants. *Annals of the Entomological Society of America, 93*, 929–940.
Munday, P. L., Jones, G. P., & Caley, M. J. (2001). Interspecific competition and coexistence in a guild of coral-dwelling fishes. *Ecology, 82*, 2177–2189.
Oliveira, P. S. (1997). The ecological function of extrafloral nectaries: Herbivore deterrence by visiting ants and reproductive output in *Caryocar brasiliense* (Caryocaraceae). *Functional Ecology, 11*, 323–330.
Packard, G. C., & Packard, M. J. (1980). Evolution of the cleidoic egg among reptilian antecedents of birds. *American Zoologist, 20*, 351–362.

Peterson, M. A. (1995). Unpredictability in the facultative association between larvae of *Euphilotes enoptes* (Lepidoptera: Lycaenidae) and ants. *Biological Journal of the Linnean Society, 55*, 209–223.
Pianka, E. R., & Vitt, L. J. (2003). *Lizards: Windows to the evolution of diversity*. University of California Press.
Pleasants, J. M. (1980). Competition for bumblebee pollinators in rocky mountain plant communities. *Ecology, 61*, 1446–1459.
Raine, N. E., Pierson, A. S., & Stone, G. N. (2007). Plant-pollinator interactions in a Mexican Acacia community. *Arthropod-Plant Interactions, 1*, 101–117.
Richter-Boix, A., Liorente, G. A., & Montori, A. (2004). Responses to competition effects of two anuran tadpoles according to life-history traits. *Oikos, 106*, 39–50.
Sanders, N. J., & Gordon, D. M. (2000). The effects of interspecific interactions on resource use and behavior in a desert ant. *Oecologia, 125*, 436–443.
Santini, G., Tucci, L., Ottonetti, L., & Frizzi, F. (2007). Competition trade-offs in the organisation of a Mediterranean ant assemblage. *Ecological Entomology, 32*, 319–326.
Sarty, M., Abbott, K. L., & Lester, P. J. (2006). Habitat complexity facilitates coexistence in a tropical ant community. *Oecologia, 149*, 465–473.
Schleich, H. H., & Kastle, W. (1988). *Reptile egg-shells SEM atlas*. Gustav Fischer Verlag.
Schoener, T. W. (1983). Field experiments on interspecific competition. *The American Naturalist, 122*, 240–285.
Stone, G. N., Willmer, P. G., & Rowe, J. A. (1998). Partitioning of pollinators during flowering in an African *Acacia* community. *Ecology, 79*, 2808–2827.
Wilson, E. O. (1971). *The insect society*. Belknap Press of Harvard University Press.

Tree Lizard Territory and Adaptation 10

Abstract

Territoriality, shaped by intraspecific competition, often depends on population density, yet empirical evidence for this relationship is inconsistent. Using eight populations of Swinhoe's tree lizards, we examined how male density influences multiple combat-related traits. Males in high-density populations displayed less aggressive tactics, lower immobilization and wounding rates, weaker bite forces, and smaller body sizes, whereas the opposite pattern occurred at low density. Additionally, the association between body size and bite force weakened as density increased. These coordinated shifts in behavior, performance, and morphology along a density gradient demonstrate how density-dependent selection can drive adaptive polymorphism within species.

Keywords

Density dependence · Territority · Tree lizard · Intraspecific competition · Homing behavior · Evolution of viviparity

10.1 Low-Density Moderated Competition Performance Facilitates Conflict Resolution

Wang Han's Liangzhou poem from the Tang Dynasty: Grape wine in a nightlight cup, ready to drink, but the pipa on horseback urges us on. Drunk and lying on the battlefield, don't laugh at me, for how many have returned from ancient battles?

Supplementary Information The online version contains supplementary material available at https://doi.org/10.1007/978-981-92-1887-5_10.

W.-S. Huang, *Reptiles in the Changing Climate*, Fascinating Life Sciences,
https://doi.org/10.1007/978-981-92-1887-5_10

The meaning of this poem is roughly as follows: A wine cup made of jade, filled with fine wine made from grapes, is about to be drunk, but the sound of the pipa on horseback urges everyone to the battlefield. Despite the urging, everyone still drank to their heart's content. Even if they were drunk and lying on the battlefield, do not mock me, for throughout history, how many soldiers have returned alive from battle? The emotions of the soldiers described in this poem are a complex mix of both grandeur and sorrow.

Human territorial desires can be understood as a pursuit of resources, security, and social status, stemming from the combined influence of biological instincts, social structures, and historical factors. The following are some possible reasons: (1) Biological instincts and survival competition: From a biological perspective, territory is an important guarantee for obtaining resources (such as food, water, and mates) and reproducing offspring. As animals, humans are also driven by survival instincts and have a natural desire to possess territory. In early human societies, territorial disputes were often closely tied to survival. Possessing more territory meant access to more resources and a stronger survival advantage. (2) Social structure and power struggles: As society develops, territory is no longer merely a carrier of survival resources but also a symbol of power, status, and identity. Territorial disputes between nations and ethnic groups are often the result of intertwined political, economic, and cultural factors. Possessing more territory signifies greater influence and international standing. (3) Historical and Cultural Factors: Wars and conflicts throughout history have left deep imprints on different nations and ethnic groups, shaping unique emotions and values toward territory. In certain cultures, a strong sense of belonging and honor toward territory is deeply ingrained, often becoming an integral part of ethnic identity. (4) Psychological Factors: The desire to possess territory is also linked to the psychological needs of individuals and groups. Possessing territory can bring a sense of security, belonging, and superiority. The fear of losing territory may trigger anxiety and unease. This fear may also serve as a driving force behind territorial desires.

Even so, are lizards with highly territorial behavior really the same as humans? Perhaps animals engage in fights similar to humans, but most are not fatal. After all, the purpose of animals is survival and reproduction, and if they die, neither survival nor reproduction is possible. This chapter uses research on tree lizards to test the territorial behavior hypothesis, the true reasons for returning to their territory, and a novel hypothesis about the evolution from oviparity to viviparity.

Territorial behavior is an adaptive strategy that emerges when individuals compete for limited resources within a population (Kaufmann, 1983; Raffestin, 2012; Toobaie & Grant, 2013; Deitloff et al., 2014). The intensity of such competition is often closely linked to population density (Knell, 2009; Lamb et al., 2017). When environmental resources remain relatively stable, higher densities typically lead to increased competition among conspecifics (Tomkins & Brown, 2004; Kokko & Rankin, 2006; Knell, 2009). Despite this intuitive expectation, empirical studies attempting to evaluate density-dependent influences on territorial behavior frequently encounter difficulties (Moczek, 2003). One reason is that territorial contests

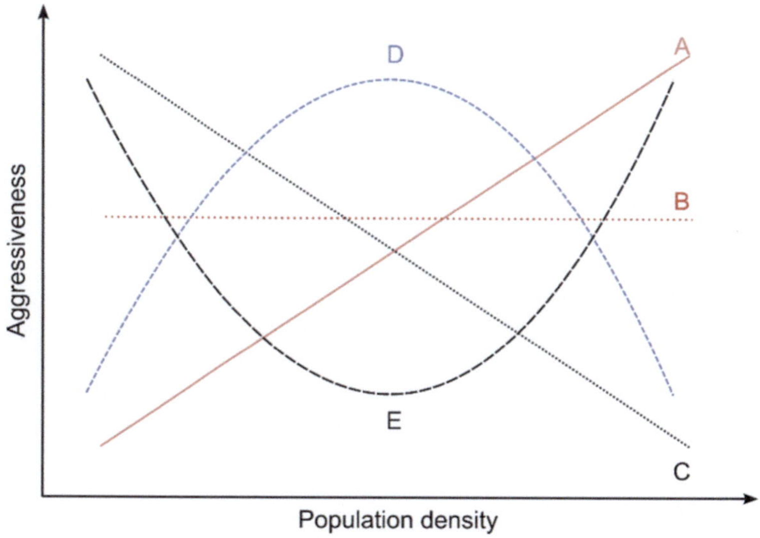

Fig. 10.1 Five hypothetical relationships between density and aggressiveness. The level of aggressiveness would increase (**a**), remain constant (**b**), or decrease (**c**) with population density. The level of aggressiveness may also reach the highest level in a medium-density population with population density (**d**), or reach the lowest level in a medium-density population (**e**). (Redrawn from Hsu et al. (2023), Functional Ecology)

involve a complex interaction among behavioral strategies, morphological traits and performance abilities.

Animals often exhibit flexible behavioral tactics during territorial conflicts (Mitani & Rodman, 1979), employing a variety of combat-related traits such as status signals, weapon-like morphological structures, and ritualized behaviors to resolve disputes (Stuart-Fox, 2006; Arnott & Elwood, 2009; Camerlink et al., 2015). These traits are shaped by different selective pressures and evolutionary trade-offs (Kilgour et al., 2018), which may obscure straightforward predictions regarding density-dependent effects.

Previous research has proposed several potential relationships between population density and aggressive interactions (Fig. 10.1a–e). These include positive (Donihue et al., 2016; Wegener et al., 2019), negative (Pomfret & Knell, 2007; Knell, 2009), unimodal (Jones, 1983), bimodal (Mayer et al., 2020), and neutral associations (Niemelä et al., 2012). Such patterns can be interpreted through the evolutionary balance between the costs and benefits of aggression as population density changes. For example, aggression may increase with density when the benefits of dominance rise more rapidly than the associated energetic or injury costs. Under such conditions, dominant individuals in dense populations may gain disproportionate access to valuable resources, including mates and high-quality habitat (Holbrook & Schmitt, 2002; Abesamis & Russ, 2005). Although empirical support exists for each of these patterns, the diversity and occasional contradiction among them highlight the complexity of density-dependent behavioral evolution and the lack of a unified theory explaining how density shapes territorial strategies (Stamps & Buechner, 1985; O'Neil et al., 2020). A clearer understanding of these mechanisms would improve our knowledge of how behavioral interactions contribute to population regulation.

Besides population density, the spatial distribution and quality of resources, such as food availability and habitat structure, can also influence the effectiveness and value of territorial defense (Grant & Guha, 1993; Jensen et al., 2005). According to the resource dispersion hypothesis, when resources are abundant but spatially dispersed, individuals may share overlapping areas while still maintaining defendable resource patches (Macdonald & Johnson, 2015). In such situations, animals are expected to invest more effort in defending sites that contain valuable resources (Monaghan & Metcalfe, 1985; Grant & Guha, 1993; Jensen et al., 2005) or when resources are sufficiently concentrated to be economically defended (Grant & Guha, 1993, Jensen et al., 2005). Consequently, variation in the distribution and abundance of resources may strongly influence where territories are established and how individuals defend them (Diego-Rasilla & Perez-Mellado, 2003; Kokko & Rankin, 2006; Macdonald & Johnson, 2015).

In this study, we examine how population density influences territorial behavior in the Swinhoe's tree lizard (*Diploderma swinhonis*). Specifically, we address three questions: (i) how population density affects the outcome and escalation of territorial contests, (ii) whether injury frequency varies with density, and (iii) how morphology and performance traits associated with fighting change across density gradients. To address these questions, we quantified population density, habitat characteristics, food availability, and dietary composition across eight lizard populations. Behavioral experiments were conducted in each population, and several traits associated with fighting performance, including snout–vent length (SVL), bite force and injury frequency, were measured.

Territorial encounters in *D. swinhonis* typically involve a sequence of aggressive displays that range from ritualized non-contact behaviors to direct physical combat. These behavioral characteristics, together with the species' clear territorial system and measurable performance traits, make it a suitable model organism for investigating density-dependent variation in territorial behavior.

Our fieldwork was carried out from March to October in 2013–2014 and 2018–2020. The tree lizard, *Diploderma swinhonis*, is a lizard endemic to Taiwan that typically inhabits forested areas and shrub-dominated habitats on the main island as well as nearby offshore islands. Eight populations were examined in this study. Six of these populations occurred on the main island of Taiwan, distributed from northern to southern regions: Taipei (25°01′04.1″N 121°33′08.6″E), Taoyuan (25°00′09.3″N 121°19′41.9″E), Yilan (24°36′51.8″N 121°51′18″5″E), Taichung (24°08′44.3″ N 120°33′25.5″ E), Hualien (23°36′54.2″N 121°25′05.4″E), and Kenting (22°08′29.70″N, 120°42′11.77″E). Two additional populations were located on offshore islands east of Taiwan: Green Island (22°39′33″N, 121°29′ 15E) and Orchid Island (22°3′18″N, 121°32′41″E) (Fig. 10.2a).

Diploderma swinhonis is the largest member of the genus *Diploderma* occurring in Taiwan, with adults typically reaching a snout–vent length (SVL) of

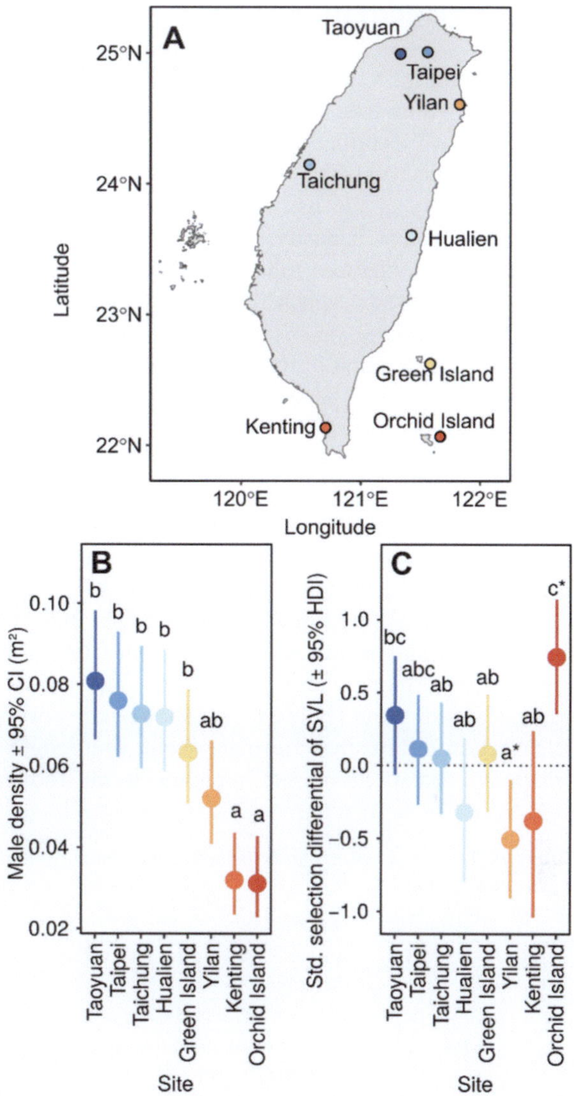

Fig. 10.2 Map showing the sampling locations of the eight populations studied at low altitude tropical forests in Taiwan (**a**), the corresponding male densities (**b**), and the selection differential of male snout–vent length (SVL) (**c**). Lowercase letters in panel B-C summarize the results of multiple comparisons; the same letter indicates an insignificant difference in density or selection differentials and vice versa. Asterisk signs in panel C indicate selection differentials that are significantly different from zero. (Adapted from Hsu et al. (2023), Functional Ecology, with permission. (**a**) Reproduced. (**b**) and (**c**) Redrawn)

approximately 65–80 mm (Lin & Lu, 1982; Huang, 2007). Males are known to compete for territories, each generally centered on a primary tree that serves as the focal point of the defended area (Lin & Lu, 1982, Huang, 2007). Resident males commonly perform conspicuous displays while occupying their territory tree. These behaviors are especially frequent during the reproductive season, which extends from March to October. The species also exhibits pronounced sexual dimorphism: males generally have a larger body size and display a conspicuous yellow stripe along each side of the body(Baird et al., 1996; Huang, 2007).

10.1.1 Categorization of Fight

Video recordings were used to examine the fighting performance of both resident males and intruding individuals (Supplementary Video 10.1). From these recordings, we identified eight types of aggressive behaviors that occurred during contests (Table 10.1). These behaviors were grouped into three categories according to the intensity of the interaction. Encounters typically started with non-contact displays (i.e., low intensity, such as push-ups, standoffs, and approach). These displays were sometimes followed by chasing behavior (i.e., medium intensity). In some cases, contests escalated further into direct physical combat (i.e., high intensity, such as pecking, crashing, and biting). We documented the frequency of these behaviors, the duration of the contest, and the eventual outcome of the fight.

10.1.2 Density of Males in Eight Lizard Populations

Across the eight study populations, we captured a total of 600 male *Diploderma swinhonis*. Sample sizes varied among locations, including 39 individuals from Orchid Island, 40 from Kenting, 65 from Yilan, 79 from Green Island, 90 from Hualien, 91 from Taichung, 95 from Taipei and 101 from Taoyuan. Population densities differed significantly among sites ($\chi^2 = 61.359$, df $= 7$, $p < 0.001$; Fig. 10.2b). Among these populations, Taoyuan exhibited the highest density, whereas Kenting and Orchid Island showed the lowest densities.

Selection differentials for male snout–vent length (SVL) generally declined as population density decreased. In particular, the Taoyuan population, which had the

Table 10.1 A series of agonistic behaviors exhibited by *Diploderma swinhonis* during staged dyadic encounters

Level	Behavior	Description
Lowest	Retreat	An act of moving back or withdrawing toward the opposite direction of the opponent
Low—no involving physical contact between opponents		
	Push-up	Vertical raising and lowering of the anterior part of the body using rhythmic motion
	Approach	Movement directly toward the agonistic male can be with a fast or slow speed
	Standoff	Stand facing each other with or without an intimidating motion
Medium—escalating the aggressive level but still not involving physical contact		
	Chase	Rapid running toward the retreated opponent
High—physical contact and fighting; injuries may occur		
	Crash	A blow or a jolting collision
	Pecking	Pecks at another individual with its snout the mouth closed
Highest—violent fighting; injuries usually occur		
	Bite	An act to bite the opponent, mainly in the regions of neck, body, and tail base in order to expel it from the territory

highest density, exhibited a significantly larger selection differential than the Yilan population, where density was comparatively lower (Fig. 10.2c). In Yilan, the selection differential was negative (posterior median $\delta = -0.51$ SD; 95% highest density interval [HDI]: −0.91 to −0.10 SD), indicating that territorial resident males were smaller than the average male body size within the population.

Interestingly, Orchid Island displayed an exceptional pattern. Despite having the lowest population density, the selection differential there was significantly larger than in most other populations (Fig. 10.2c). The estimated value was positive and significantly greater than zero (posterior median $\delta = 0.74$ SD; 95% HDI: 0.35–1.14), suggesting that resident males on Orchid Island were larger than the population mean.

10.1.3 Relationship Between Male Density and Aggressiveness

A total of 176 staged contests were conducted across the eight populations (Fig. 10.3). Model comparisons based on Bayesian Information Criterion (BIC)

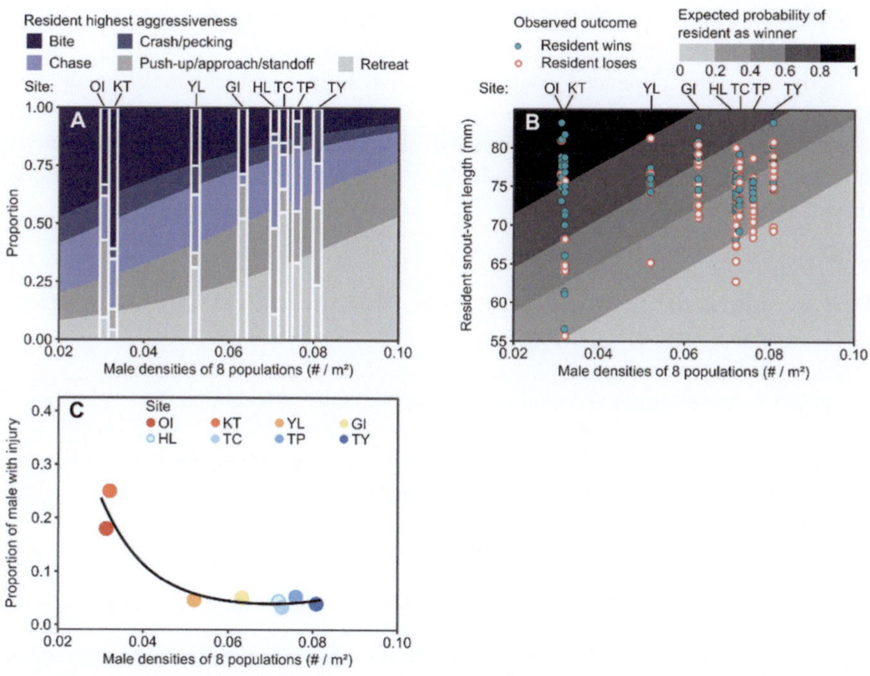

Fig. 10.3 The contest results include the composition of the resident lizards' most aggressive behaviors (**a**), the outcome of residents as winners (**b**), and the proportions of residents with injuries (**c**) in eight populations exhibiting varying male densities. The color patches in panels A and B, as well as the regression line in panel C, depict the expected probability/proportion based on the most BIC-favored models. (Redrawn from Hsu et al. (2023), Functional Ecology)

among four candidate models indicated that the "intrasexual competition model" best explained variation in fight intensity, contest outcomes and the proportion of scarred males.

Within this model, male population density emerged as a strong predictor negatively associated with all three indicators of combat performance. Specifically, higher density populations exhibited lower fight intensity ($\beta \pm SE = -31.85 \pm 12.02$ per m^2, $Z = -2.649$, $p = 0.008$), reduced probability of resident victory ($\beta \pm SE = -44.10 \pm 13.00$ per m^2, $Z = -3.393$, $p < 0.001$), and a lower proportion of scarred males ($\beta \pm SE = -131.3 \pm 33.13$ per mm^2, $Z = -3.962$, $p < 0.001$; Fig. 10.3).

In contrast, our analyses provided little evidence that habitat composition, food resource availability, or dietary composition contributed substantially to explaining variation in these behavioral patterns.

Behavioral observations further revealed that males in low-density populations tended to escalate territorial disputes into high-intensity physical combat. For instance, in Kenting, 65% of encounters involved high-intensity combat, whereas only 8.7% involved low-intensity displays. Conversely, populations with higher densities tended to resolve conflicts primarily through lower-intensity interactions. Correspondingly, the proportion of males bearing scars declined with increasing population density, decreasing from approximately 0.2 in Kenting to ~0.04 in Taoyuan.

Body size also influenced contest outcomes. Larger resident males were more likely to defeat intruders (odds ratio ≈ 1.15 per mm increase in body size). However, this resident size advantage was reduced in populations with high male density, indicating that the influence of body size on territorial success weakens under crowded conditions.

10.1.4 Relationship Among Density, Morphological Characteristics, and Bite Performance

To quantify morphological variation, we calculated a standardized morphological index (SMI) for each individual (male: $N = 398$; female: $N = 401$). Five log-transformed morphological variables were reduced using principal component analysis (PCA). The resulting composite index was:

$$\text{SMI} = 0.061 \text{ body weight} + 0.150 \text{ head height} + 0.066 \text{ head length} + 0.028 \text{ SVL} + 0.110 \text{ body weight} - 7.067$$

This index accounted for 90.7% of the total variance among morphological measurements. Statistical analyses indicated that sex, male density, sex ratio, and the interaction between male density and sex significantly influenced morphology associated with fighting ability.

Overall, males possessed larger combat-related morphological traits than females. Additionally, individuals from populations characterized by lower male density and

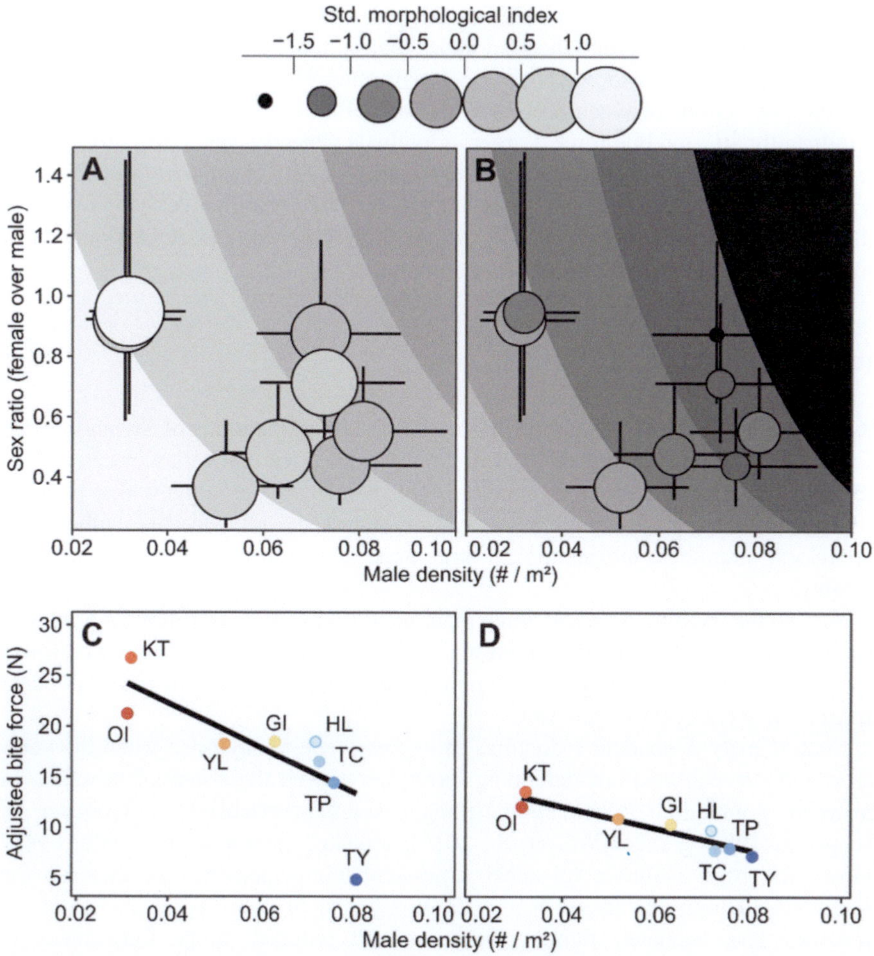

Fig. 10.4 Relationship between male density and std. morphological index in males (**a**) and females (**b**) across eight populations. Std. The morphological indices were the standardized scores of first principle component after downscaling snout–vent length (SVL), head width, head length, head height, and body weight and explained 90.7% variance. Color patches present the expected value according to the most BIC-favored model. Relationships between males' (**c**) and females' (**d**) adjusted bite force (e.g., bite force when snout–vent length (SVL) constrained at 70.3 mm) and male density among eight populations. Black lines present the M-robust regression lines. All error bars present the 95% CI. (Redrawn from Hsu et al. (2023), Functional Ecology)

lower sex ratios tended to have larger body dimensions (Fig. 10.4a, b). The interaction between sex and male density further revealed that increases in male density were associated with a stronger reduction in combat-related morphology in females than in males (Fig. 10.4a, b).

Adjusted bite force was negatively related to male population density. This pattern was observed in both sexes (male: Kendall's $\tau = -0.714$, $p = 0.014$; female:

Kendall's $\tau = -0.857$, $p = 0.002$; Fig. 10.4c, d). However, the magnitude of this relationship was more pronounced in males than in females.

Additional analyses showed that log-transformed SVL, sex, population identity, and the interactions between SVL and sex, as well as between sex and population, significantly affected bite force. Larger individuals generally generated stronger bite forces, and males displayed higher values than females. Lizards from populations with lower male densities also exhibited greater bite performance. Moreover, the allometric relationship between body size and bite force was especially pronounced among males inhabiting high-density populations.

10.1.5 Discussion

By comparing multiple phenotypic traits across eight populations of Swinhoe's tree lizard, this study demonstrates that variation in male territorial behavior can largely be explained by intrasexual competition. Our results further suggest that density-dependent competitive interactions may simultaneously influence the evolution of behavioral strategies, morphological structures, and performance traits related to fighting.

Across the examined populations, male territorial responses followed a density gradient: individuals from areas with higher male densities generally showed lower levels of aggression. This pattern supports the idea that territorial aggressiveness declines as local density increases.

Such density-dependent reductions in aggressiveness may arise when the energetic and survival costs of defending territories exceed the potential benefits.. In crowded populations where home ranges overlap extensively (Schoener & Schoener, 1982; Vander Wal et al., 2014), repeated intense confrontations may reduce the time available for other fitness-enhancing activities or increase the likelihood of injury (Vervust et al., 2009). Consistent with this explanation, we observed less escalated fighting behavior and reduced injury frequencies in populations with higher densities(Fig. 10.3a, b). These results indicate that the escalating costs of territorial disputes may weaken territorial defense in this species.

We also detected a reduced resident advantage in populations with high densities (Fig. 10.3c). This suggests that occupying a territory does not necessarily confer a substantial fighting advantage under crowded conditions, which may further contribute to the observed decline in aggressiveness.

Male density appeared to influence not only behavior but also morphology and performance traits associated with combat(Fig. 10.4). Lizards from dense populations displayed smaller combat-related morphological traits and weaker bite forces. These patterns correspond with behavioral observations showing reduced escalation during contests, fewer injuries, and a diminished resident effect.

Although bite force increased positively with body size (SVL), the strength of this allometric relationship differed among populations and between sexes. In particular, males inhabiting high-density populations exhibited comparatively weaker bite forces. One possible explanation is that individuals in these environments may allocate energetic resources toward alternative morphological features rather than bite performance due to the elevated costs of intrasexual competition.

Sexual dimorphism in bite force further suggests that male-specific selective pressures play a key role in shaping this trait. Interestingly, sex ratio did not strongly explain variation in bite force among populations, implying that its evolution may not be directly linked to mate competition.

The development of combat-related morphology appears to be influenced by multiple evolutionary forces. Our results show that individuals from both high-density populations and populations with less male-biased sex ratios tend to exhibit reduced combat morphology. Because these conditions do not necessarily occur simultaneously as density increases, they may represent two opposing selective pressures affecting trait evolution.

Although male density emerged as the strongest predictor of territorial behavior in our analyses, other ecological variables may also play important roles. For instance, shelter availability has been recognized as a key ecological resource influencing territoriality (Chapman, 1966; Tricarico & Gherardi, 2010). We did not explicitly include refuge abundance in our analyses because field observations indicated that Swinhoe's tree lizards can use a wide variety of temporary shelters, including tree canopies, grasses, and leaf litter. Nevertheless, for species that depend strongly on specific refuge sites, this factor could substantially influence territorial behavior and should be examined in future studies.

Our findings are consistent with several earlier studies demonstrating negative density-dependent territorial behavior. For example, populations of island lizards, birds, and mammals with high densities often show reduced aggression toward conspecifics, resulting in smaller territories, increased boundary overlap, greater tolerance of subordinate individuals, and sometimes even the absence of territorial defense (Smith et al., 1980; Stamps & Buechner, 1985).

Unlike the pronounced density-dependent patterns observed across all three phenotypic aspects, variation in sex ratio significantly influenced only morphological traits, particularly body size measures such as SVL (Fig. 10.4a, b), while having minimal impact on territorial behavior or bite force performance. Male body size tended to increase when the operational sex ratio was biased toward males. This pattern suggests that reduced access to mates may intensify intrasexual competition, thereby favoring individuals with larger body size under sexual selection. Although strong competition for environmental resources at high male densities may favor smaller males, intensified rivalry for mating opportunities could impose the opposite selective pressure. The contrasting directions of morphological responses under different resource limitations highlight how ecological variables, including population density and sex ratio, can exert opposing influences on phenotypic evolution. Such antagonistic selective pressures, together with the evolutionary linkage among multiple traits, emphasize the importance of developing integrated frameworks to understand the complex interplay between ecological dynamics and evolutionary processes.

Furthermore, a meta-analysis of 483 lizard populations revealed an inverse relationship between body size and population density (Buckley et al., 2008).

Although different taxa display a wide variety of interactions between density and aggression (Jones, 1983; Knell, 2009; Niemelä et al., 2012; Wegener et al., 2019; Mayer et al., 2020), our study highlights the importance of examining multiple populations and integrating several phenotypic traits to better understand the ecological mechanisms underlying territoriality.

Future research could further clarify these mechanisms by incorporating individual-level environmental measurements and applying path analysis approaches. Such methods would help disentangle the complex relationships among ecological variables, behavior, morphology, and performance traits.

In summary, our results provide strong support for the hypothesis that negative density dependence is an important driver of eco-evolutionary processes. The density gradient across populations produced male-specific variation in fighting behavior, morphology, and performance, suggesting that territorial strategies may reach different evolutionary equilibria depending on local costs and benefits.

This phenotypic variation underscores the influence of both ecological conditions and social interactions on ontogenetic growth patterns and life-history strategies, as reflected in the differences in body size among populations (Calsbeek & Smith, 2007; Buckley et al., 2008). These findings also provide a foundation for future studies examining pace-of-life syndromes (Wright et al., 2019).

By investigating how population density influences territorial behavior and the integration of multiple phenotypic traits, this study contributes to a deeper understanding of how phenotypic adaptation may shape population structure and evolutionary dynamics.

10.2 Homing Behavior of a Tree Lizard: Influences of Mating Resource, and Habitat Structure

Homing behavior refers to an animal's ability to repeatedly relocate and return to a particular spatial area. This behavior is influenced by an individual's physiological state as well as environmental conditions operating at both individual and population scales. Two commonly used indicators of homing behavior are site fidelity. and homing success. Site fidelity describes the tendency of an individual to remain at or repeatedly occupy a previously used location, whereas homing success represents the probability that an individual can relocate its original territory after being experimentally displaced. Understanding how ecological conditions shape these behaviors across different populations provides insights into how animals adjust navigation strategies under variable environmental contexts.

In this study, we performed a translocation experiment across eight populations of the Swinhoe's tree lizard (*Diploderma swinhonis*) to investigate the ecological determinants of homing behavior. Specifically, we evaluated how resource availability (mates and food), morphological traits, habitat characteristics, and population density influence both site fidelity and homing success.

Our results showed pronounced variation in these behaviors at both within-population and among-population scales. Males displayed stronger site fidelity when they were surrounded by more female neighbours but experienced lower heterogeneity among nearby males. Patterns of homing success further revealed that males returned to their territories more quickly when those territories offered greater mating opportunities and reduced levels of male–male competition. Habitat structure also played an important role: homing success was higher in populations dominated by smaller trees, or when males occupied territories located on smaller host trees.

Taken together, these findings highlight substantial ecological and social influences on homing behavior. Our study underscores the importance of examining both intra- and inter-population variation in order to better understand the evolutionary and ecological mechanisms shaping spatial navigation in animals.

Movement is a fundamental aspect of animal behavior and is strongly influenced by the spatial and temporal distribution of environmental resources (Willems et al., 2009; Willems & Hill, 2009). Patterns of movement can profoundly affect both ecological interactions and evolutionary fitness, depending on the behavioral context and spatial scale involved, such as migration, dispersal, foraging, or mate searching (Hooten et al., 2017; Lewis et al., 2021). In environments where resources fluctuate in space and time, animals must efficiently detect and access key resources, including food, mates, and shelter, to survive and reproduce successfully (Williams & Safi, 2021). Achieving this requires the integration of environmental information with locomotor decisions and orientation behaviors (Åkesson et al., 2014; Breed & Moore, 2016). Consequently, studying movement behavior not only provides insights into the functioning of animal sensory systems and cognitive abilities (Fagan et al., 2013; Lewis et al., 2021), but also contributes to the development of effective conservation strategies for threatened species (Allen & Singh, 2016; Hromada et al., 2023).

Many animals repeatedly utilize particular locations or objects in their environment, including shelters, nesting sites, and territories. Returning to these sites after displacement or temporary absence requires homing behavior (Gerking, 1959; Papi, 1992; Breed & Moore, 2016). Homing is typically defined as the ability of an animal to navigate back to a previously occupied location through unfamiliar surroundings, distinguishing it from large-scale movements such as dispersal or migration (Humphries & Sisson, 2012; Breed & Moore, 2016; Galib et al., 2022). Animals employ a variety of orientation mechanisms when navigating home, integrating multiple sensory cues and cognitive processes such as landmark navigation, path integration, and cognitive mapping (Papi, 1992; Pašukonis et al. 2014a; Breed & Moore, 2016). For example, pigeons are known to rely on solar positioning and geomagnetic signals during homing (Grüter & Wiltschko, 1990; Armstrong et al., 2013), whereas lizards appear to orient primarily using visual landmarks and solar compass cues (Zuri & Bull, 2000; Freake, 2001; Foà et al., 2009; LaDage et al., 2012; Gagliardo, 2013). Although previous research has extensively examined the sensory and cognitive mechanisms underlying homing behavior, the ecological drivers explaining why homing behavior varies across contexts remain less well understood. In particular, evaluating the balance between the potential costs and benefits of returning to a familiar location may provide key insights into variation in homing outcomes.

Two metrics are commonly used to describe homing behavior: site fidelity and homing success (Jreidini & Green, 2022). Site fidelity refers to the tendency of an individual to remain in, or repeatedly occupy, a previously used location and is therefore closely related to homing tendencies (Switzer, 1993; Jiménez-Franco et al., 2013; Richardson et al., 2017). In contrast, homing success is experimentally measured as the probability that an animal returns to its original location after being displaced to an unfamiliar site (Grüter & Wiltschko, 1990; Papi, 1992). Both measures are influenced by a combination of intrinsic characteristics and environmental conditions. From an evolutionary perspective, animals should be more likely to return to their original site when the fitness benefits of doing so exceed the associated costs (Ellis-Quinn & Simon, 1989). For instance, territories containing abundant mating opportunities may promote higher site fidelity and stronger homing tendencies because they directly enhance reproductive success (Hoover, 2003; Navarro-Salcedo et al., 2022). Individual characteristics can also influence homing performance. Larger individuals, which are often older or more experienced, may possess better knowledge of their local surroundings, enabling them to relocate familiar areas more effectively (Weintraub, 1970; Ellis-Quinn & Simon, 1989). Their greater body size may also provide advantages during territorial defense, allowing them to deter intruders and thereby increase both site fidelity and homing success (McEvoy et al., 2013).

Comparative studies among populations are particularly valuable for understanding how environmental heterogeneity influences homing behavior and navigation strategies. Differences in ecological conditions, such as resource distribution, habitat structure, and population density, may lead to substantial variation in homing outcomes across populations. For example, research on the wall lizard *Podarcis muralis* revealed intraspecific variation in homing behavior among populations, although the ecological mechanisms driving this variation remained unclear (Scali et al., 2012). Resource distribution may strongly influence movement patterns. In White's skinks (*Liopholis whitii*), individuals exhibited smaller home ranges when high-quality resources were evenly distributed, suggesting that uneven resource availability promotes greater movement (Halliwell et al., 2017). Such patterns imply that spatial variation in resources may also influence homing tendencies and experimental homing outcomes.

Habitat structure may also play a critical role. Environments characterized by dense vegetation or numerous vertical obstacles can reduce locomotor efficiency (Losos & Irschick, 1996; Neel et al., 2021) and limit the use of visual landmarks required for navigation (Papi, 1992; Janson & Bitetti, 1997; Freake, 2001; Auburn et al., 2009). Therefore, structural complexity of the habitat may significantly influence homing success (Bélisle et al., 2001). In addition, social conditions such as population density and resource distribution may alter the intensity of social interactions (Fletcher Jr, 2007) and levels of resource competition (Harada et al., 1995), thereby modifying the benefits associated with returning to a previously occupied site. However, theoretical expectations regarding homing behavior can

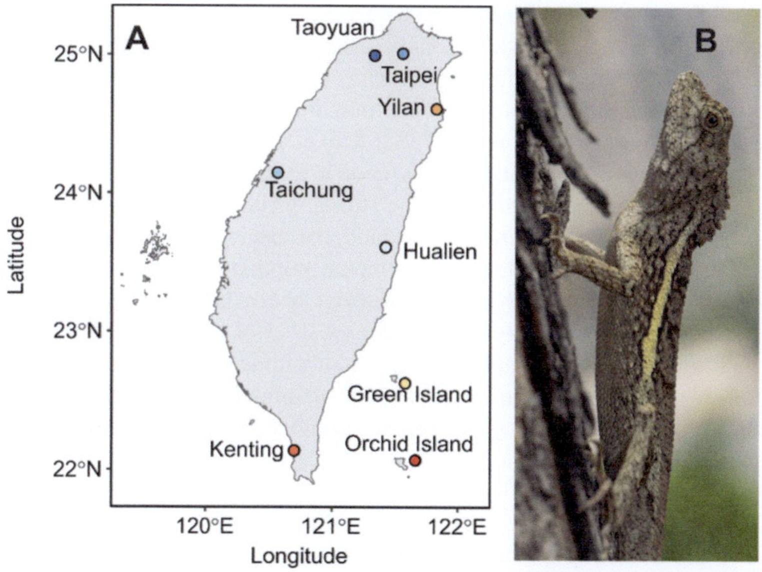

Fig. 10.5 The study system for our homing experiment. (**a**) A map of the eight sample populations ranging from urban parks to natural forests across Taiwan. (**b**) An adult male Swinhoe's tree lizard, the largest agamid in Taiwan, in typical territorial display posture. (Adapted from Zhang et al. (2024), Behavioral Ecology. (**a**) Redrawn. (**b**) Photo by Yu-De Zhang)

sometimes be contradictory. In resource-rich populations, individuals might exhibit stronger site fidelity because defending high-quality territories provides clear fitness benefits (Carpenter, 1984; Hoover, 2003; Wasko & Sasa, 2012; Cram et al., 2013). Conversely, individuals may be less motivated to return after displacement if comparable resources are readily available at the new location (Rueger et al., 2016). These alternative predictions highlight the importance of conducting comparative studies across multiple populations.

In this study, we examined variation in site fidelity and homing success within and among populations of a territorial vertebrate, the Swinhoe's tree lizard (*Diploderma swinhonis*) (Fig. 10.5a, b). This species is the largest endemic arboreal agamid in Taiwan and inhabits a wide range of environments, from natural forests to urban parks. Swinhoe's tree lizards exhibit pronounced sexual dimorphism, with males significantly larger than females and possessing a conspicuous yellow lateral stripe (Kuo et al., 2009; Norval et al., 2011). Adult males are strongly territorial and typically occupy a single tree, referred to as a territory tree, which they use for resting, basking, and territorial displays. When confronted with intruders, males perform push-up displays and extend their dewlaps to deter competitors (Hsu et al., 2023). Male territories rarely overlap, and individuals seldom share the same territory tree (Lin & Lu, 1982). In contrast, multiple females are often observed within the same tree, and their home ranges frequently overlap with those of males, indicating a polygynous mating system(Lin & Lu, 1982). In arboreal lizards, the degree of polygyny often increases with tree size (Manzur & Fuentes, 1979; Cuadrado, 2001). Previous studies have also shown that dominant males perch higher and occupy thicker trunks or branches than subordinate males and females, suggesting that perch height and tree size may influence both territorial dominance and mating opportunities(Lin & Lu, 1982; Kuo et al., 2007; Kuo et al., 2009).

Several characteristics make the Swinhoe's tree lizard an excellent model for investigating homing behavior. First, the strong territoriality of males allows clear identification of territorial boundaries. Second, the species occupies a wide variety of habitats, providing an opportunity to compare populations experiencing different environmental conditions. Third, our previous work has demonstrated negative density-dependent territoriality in this species (Hsu et al., 2023), suggesting that behavioral traits can evolve rapidly among populations. Building on these findings, we conducted a translocation experiment to examine how morphological traits, resource availability, habitat structure, and population density influence both site fidelity and homing success in this species (Table 10.2).

10.2.1 PCA of Morphological Traits

Principal component analysis (PCA) of morphological variables accounted for 95% of the total variation in the dataset. The first three principal components (PC1–PC3) explained 53%, 21%, and 21% of the variance, respectively. PC1 primarily represented overall body size, with the largest loadings contributed by SVL, HL, HH, and body mass. PC2 mainly reflected body condition, whereas PC3 was largely associated with head width (HW) (Table 10.3).

10.2.2 Site Fidelity

Site fidelity of male Swinhoe's tree lizards differed among the eight study populations. Across all surveys, males were observed occupying their territory trees in 54.17% of observations.

During model selection, inclusion of four individual-level predictors reduced the AIC by 3.77, whereas the addition of two population-level variables further reduced the AIC by 6.27, indicating that population-level conditions were stronger predictors of variation in site fidelity.

At the individual level, males were more frequently detected within their territories when they had a greater number of female neighbors (OR = 1.438 per SD, $Z = 1.980$, $p = 0.048$; Fig. 10.6a). Perch height also showed a positive association with site fidelity (OR = 1.362 per SD, $Z = 1.921$, $p = 0.055$). However, neither of these variables explained variation among populations, as their random slopes were excluded from the final model.

No significant effects were detected for average number of male neighbors ($p = 0.766$) or for the interaction between perch height and male neighbor number ($p = 0.109$). Likewise, PCA-derived morphological variables and territory tree size

Table 10.2 The potential (a) individual-level and (b) population-level factors that shape homing behavior

Explanatory variable		Description
(a)	Resource availability	
	Average female neighbors	Mean of non-resident (i.e., neighboring) females observed in a territory across monitoring sessions
	Average male neighbors	Mean of non-resident (i.e., neighboring) males observed in a territory across monitoring sessions
	Morphological traits	
	PCA.M1	Longer SVL, HL, and HH, and heavier weight
	PCA.M2	Greater body condition
	PCA.M3	Longer HW
	Habitat structure	
	Territory tree size	DBH of the territory tree
	Perch height	Height where each male is located
	Treatment	
	Displacement distance	Distance from the territory tree to the translocation site
(b)	Resource availability	
	Heterogeneity of female neighbors	CV of average female neighbors
	Heterogeneity of male neighbors	CV of average male neighbors
	Food abundance	Mean weight of prey samples caught in pitfall traps
	Population density	
	Male density	Number of males within 1250 m^2
	Female density	Number of females within 1250 m^2
	Habitat structures	
	Habitat tree size	Mean DBH of trees nearest to survey points
	Heterogeneity of the territory tree size	CV of territory tree size
	PCA.H1	More coverage of stone, rock, and bare land and a longer distance to the nearest tree
	PCA.H2	More number of stems and trunks
	PCA.H3	More coverage of manmade construction and less coverage of mulch
	PCA.H4	More coverage of grass and less coverage of the herb
	Site fidelity	
	Average site fidelity	Mean site fidelity of each population

Abbreviations: *SVL* snout–vent length, *HL* head length, *HH* head height, *HW* head width, *DBH* diameter of breast height, *CV* coefficient of variation

were absent from the best-supported model, suggesting that these factors did not substantially influence site fidelity.

At the population level, males exhibited stronger site fidelity in populations where variation in the number of male neighbors was lower, indicating a more homogeneous distribution of competitors (OR = 0.617 per SD, $Z = -2.443$,

Table 10.3 Summary of PCA loadings and proportion of variation explained by downscaling morphological variables among eight locations

	Component		
Variable	PCA.M1	PCA.M2	PCA.M3
SVL	0.93	0.27	−0.10
HL	0.82	0.46	0.05
HW	0.44	0.89	0.12
HH	0.86	0.30	0.20
Weight	0.84	0.26	0.43
Body condition	0.08	0.08	0.99
Proportion variance	0.53	0.21	0.21
Cumulative variance	0.53	0.74	0.95

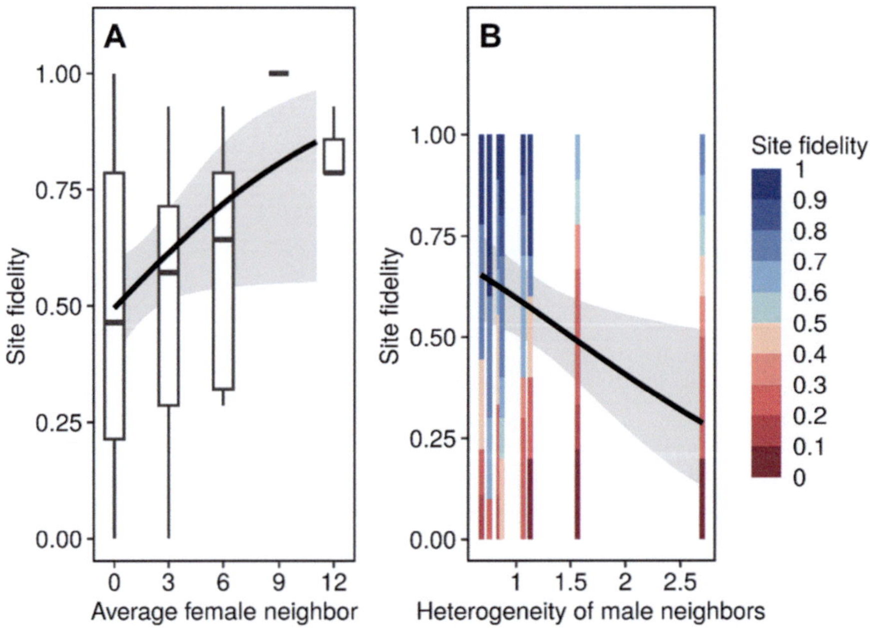

Fig. 10.6 The conditional effects of (**a**) number of female neighbors in the territory and (**b**) the heterogeneity of male neighbors on site fidelity. The solid black line and the gray area indicate the average effect of each factor and its 95% confidence band, respectively. (Redrawn from Zhang et al. (2024), Behavioral Ecology)

$p = 0.015$; Fig. 10.6b) . Additionally, rock cover within a habitat tended to be negatively related to site fidelity (OR = 0.697 per SD, $Z = -1.806$, $p = 0.071$).

Other environmental variables, including population density, heterogeneity in female neighbors, food availability, average tree size, variation in territory tree size, nearest-tree distance, number of trunks or stems, and other habitat components, were excluded from the top model and therefore showed no detectable association with site fidelity (Table 10.4).

Table 10.4 Summary of PCA loadings and proportion of variation explained by downscaling habitat structures among eight locations

Variable	Component			
	PCA.H1	PCA.H2	PCA.H3	PCA.H4
Grass	0.12	−0.19	−0.04	0.83
Herb	−0.24	−0.04	−0.22	−0.89
Stone/rock	0.90	−0.29	−0.03	0.14
Mulch	−0.07	0.35	−0.90	0.06
Bare land	0.66	0.24	−0.19	0.46
Artificial	0.09	0.06	0.92	0.23
Others	−0.45	−0.38	0.54	−0.40
Number of stems	−0.13	0.98	−0.01	−0.06
Number of trunks	−0.10	0.92	−0.30	−0.07
Distance to the nearest tree	0.85	−0.17	0.34	0.12
Proportion variance	0.23	0.23	0.22	0.20
Cumulative variance	0.23	0.46	0.68	0.88

10.2.3 Homing Success

Homing success differed markedly among populations. Overall, 167 of 240 individuals (69.5%) successfully returned to their territories following translocation. Return probability declined with increasing displacement distance. Among males displaced 20 m, 68 of 80 (85.0%) successfully homed, returning after 1.441 ± 1.300 days on average. For those displaced 40 m, 57 of 80 (74.0%) returned within 1.737 ± 1.433 days, whereas only 39 of 80 (48.8%) individuals translocated 60 m successfully homed, requiring 1.949 ± 1.445 days on average. Model selection showed that three individual-level predictors reduced AIC by 4.31, while the addition of five population-level predictors or interactions produced an additional decrease of 8.86, again suggesting that population-level factors played a stronger role in shaping homing success.

At the individual level, the likelihood of returning to the territory increased when translocation distance was shorter (HR = 0.614 per 20 m, Z = −5.92, $p < 0.001$). Homing success was also higher for males with more female neighbors within their territories (HR = 1.257 per SD, Z = 2.71, $p = 0.007$; Fig. 10.7a). In contrast, individuals occupying smaller territory trees were more likely to return successfully (HR = 0.808 per SD, Z = −2.37, $p = 0.018$; Fig. 10.7b). Despite these effects at the individual level, neither female neighbor number nor territory tree size explained population-level variation because their random slopes were removed during model selection. Morphological PCs and perch height were also absent from the final model.

Among population-level predictors, the average number of male neighbors, female density, and their interaction significantly influenced homing success (female density: HR = 1.155 per SD, Z = 1.52, $p = 0.130$; average male neighbors: HR = 0.676 per SD, Z = −2.88, $p = 0.004$; female density × average male

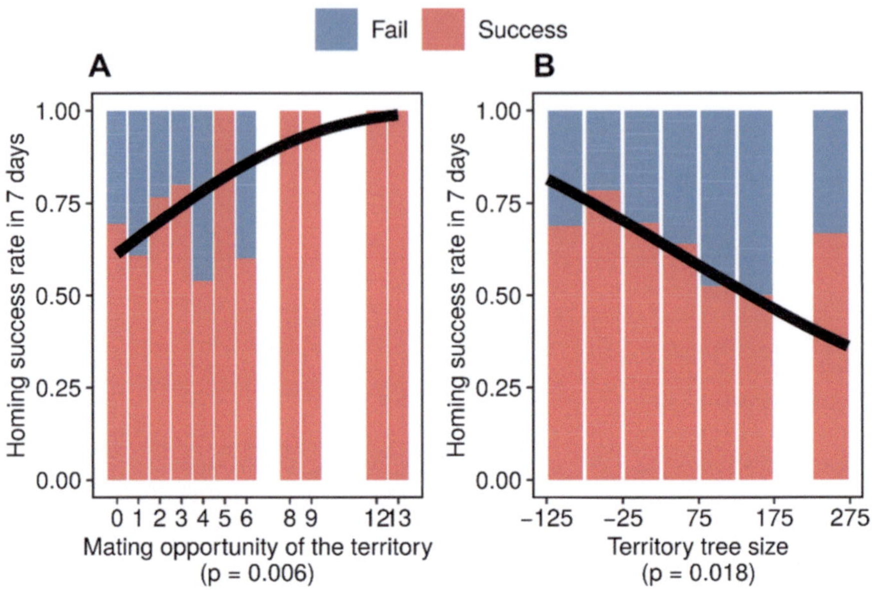

Fig. 10.7 The conditional effect of (**a**) number of female neighbors in the territory (i.e., mating opportunity) and (**b**) territory tree size on homing success. The solid black lines and the gray areas indicate the average effect and their confidence bands respectively. The blue circles present successful homing events, and the red circles represent failed homing events. (Redrawn from Zhang et al. (2024), Behavioral Ecology)

neighbors: HR = 0.955 per SD, Z = 2.52, $p = 0.012$). Males from populations with higher female density tended to return to their territories regardless of the number of male competitors. Conversely, in populations with lower female density, males were less likely to return when many competing males were present (Fig. 10.8). Homing success was also greater in populations characterized by smaller trees (HR = 0.706 per SD, Z = −3.42, $p < 0.001$; Fig. 10.9). The interaction between male neighbor number and habitat tree size was not significant ($p = 0.140$). Other environmental variables, including heterogeneity in female or male neighbors, male density, food abundance, variation in territory tree size, nearest-tree distance, number of stems, habitat cover components, and population-level mean site fidelity, were removed during model selection.

10.2.4 Discussion

Our results demonstrate that both site fidelity and homing success can vary substantially both within populations and among populations. By examining multiple ecological and environmental variables across eight populations of male Swinhoe's tree lizards, we identified several key drivers of homing behavior r, most of which were linked to resource availability and habitat characteristics. Previous research on

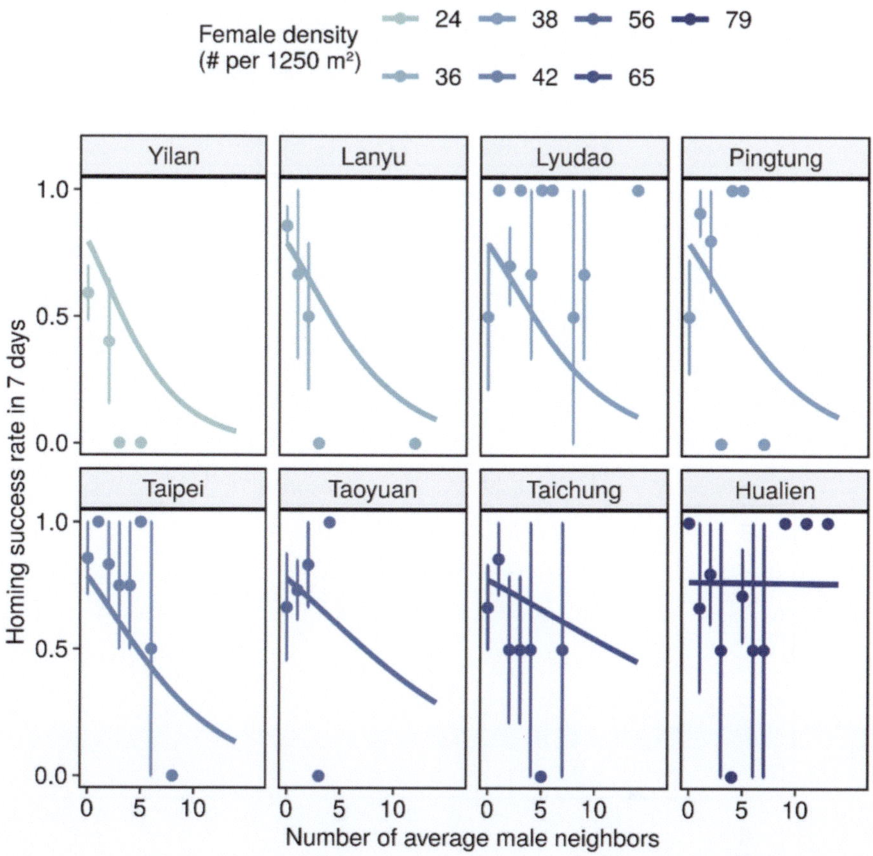

Fig. 10.8 The conditional effect of the interaction between female population density and number of male neighbors (i.e., male-male competition pressure). (Redrawn from Zhang et al. (2024), Behavioral Ecology)

homing behavior has largely focused on mechanistic questions, addressing how animals navigate unfamiliar environments (Freake, 2001; Hansson & Åkesson, 2014; Pašukonis et al. 2014b; Breed & Moore, 2016; Mandal, 2018). Studies examining the adaptive significance of homing, why animals return to particular locations, have typically concentrated on individual-level variation (Freake, 1998; Scali et al., 2012; Streit & Bellwood, 2017; Nothacker et al., 2018). In contrast, our study integrates both individual- and population-level factors, allowing us to identify ecological processes that may shape the evolution of homing behavior.

Regarding site fidelity, males were more frequently observed within their territories when they had a greater number of neighboring females. Because female abundance likely reflects increased mating opportunities, this pattern suggests that the reproductive value of a territory promotes stronger site fidelity (Navarro-Salcedo et al., 2022; Rebstock et al., 2022). At the population scale, site fidelity was also

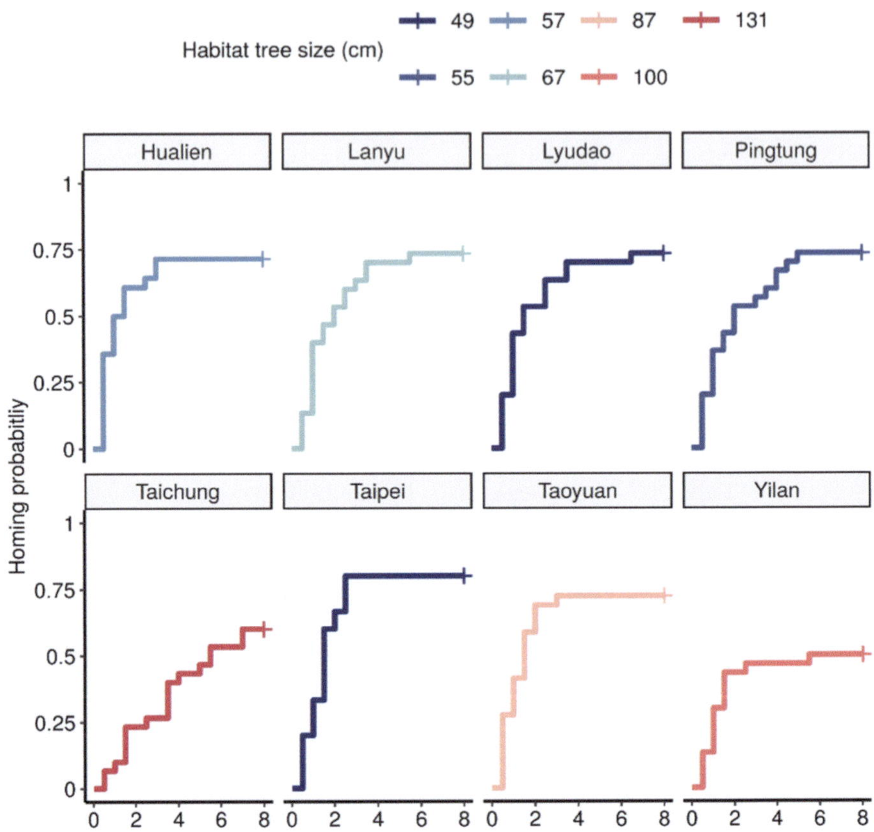

Fig. 10.9 The conditional effect of habitat tree size on homing success. (Redrawn from Zhang et al. (2024), Behavioral Ecology)

higher in populations where the number of neighboring males was more evenly distributed. Given the strong territorial behavior of this species, a more uniform competitor distribution may arise when resources are evenly distributed or when individuals reach a balance between territory expansion and the costs of territorial conflict. Under such conditions, abandoning a territory would be risky because other males could rapidly occupy it, forcing the displaced individual to compete for a new territory of similar quality.

Patterns in homing success similarly suggest that males evaluate the costs and benefits of returning home (Ellis-Quinn & Simon, 1989). Our analyses indicate that homing decisions are influenced by mating opportunities, intrasexual competition, and habitat structure. Males were generally more likely to return when their territories contained more females and fewer male competitors. Furthermore, the interaction between female density and male neighbors indicates that males tend to return when their territories provide abundant mating opportunities, even when competition from other males is high. These results highlight the importance of evaluating multiple ecological factors simultaneously across populations.

Habitat structure also influenced homing outcomes. Males were more likely to return successfully when they came from populations characterized by smaller trees or when their territories were located on relatively small trees. One possible explanation is that larger trees create greater visual obstruction, potentially interfering with landmark-based navigation. Another possibility is that individuals occupying large trees are often dominant males with strong resource-holding ability. Such individuals might not need to return to their original territories because they could establish new territories elsewhere following translocation. However, our morphological analyses do not support this hypothesis: if dominance were responsible, larger males should show lower homing success, but body size and other morphological traits were not associated with homing outcomes. These results suggest that habitat structure may influence homing success primarily through navigation ability rather than motivation to return.

Site fidelity and homing success are commonly used to quantify homing behavior, with homing success representing the combined outcome of homing ability and homing tendency (Jreidini & Green, 2022). Both site fidelity and homing tendency are likely influenced by how individuals assess the costs and benefits of territory defense (Ellis-Quinn & Simon, 1989). Interestingly, although previous work showed that males from high-density populations exhibit weaker territorial behavior and reduced bite force (Hsu et al., 2023), we found no evidence that male density or morphology influenced site fidelity or homing success.

Our findings suggest that site fidelity and homing tendency may be shaped by partially independent ecological drivers. For example, both territory tree size and average tree size within a population strongly influenced homing success but had no effect on site fidelity. In addition, population-level site fidelity did not predict individual homing success, implying that navigation ability may contribute more strongly to homing success than motivation to return. However, our study did not directly test whether individual site fidelity correlates with homing success. Addressing this question would require longitudinal data in which both traits are measured repeatedly for the same individuals.

In summary, our results demonstrate that homing behavior in Swinhoe's tree lizards is shaped by a complex combination of mating opportunities, competitive interactions, habitat structure, and environmental conditions, leading to substantial variation both within and among populations. These findings highlight the importance of understanding not only the mechanisms that allow animals to navigate home, but also the ecological motivations that drive them to do so. Such perspectives are essential for understanding variation in animal movement strategies and may provide broader insights into the evolution of dispersal and migration across species.

10.3 The Path of Viviparous Evolution-Predation-Risk Hypothesis

Predators are among the strongest ecological forces shaping life-history evolution, yet their role in the emergence of viviparity in tropical reptiles remains poorly understood. We integrated field surveys, manipulative experiments, and evolutionary modeling to examine how contrasting predation pressures on adults and eggs influence reproductive variation in the Swinhoe's tree lizard, *Diploderma swinhonis*, across eight populations in Taiwan.

On Orchid Island, where egg-specialist snakes (*Oligodon formosanus*) are common, females attain larger body sizes, reproduce later, and frequently deposit two clutches on the same day despite differences in embryonic stage. This pattern effectively shortens the period during which eggs remain exposed in the environment. By contrast, in Kenting, where avian predators impose high mortality on adult lizards, females mature earlier and produce several clutches per season with lower energetic investment per egg, consistent with a rapid reproduction strategy. Selection-gradient models indicate that the optimal duration of external egg development and maternal investment per offspring depends strongly on the relative intensity of adult versus egg predation. Our results demonstrate that spatial variation in predation risk can generate distinct reproductive strategies within a species and may promote evolutionary trajectories toward viviparity in tropical squamates. These findings highlight asymmetric mortality across life stages as a key driver of life-history diversification.

Predation has long been recognized as a major selective force influencing the evolution of prey behavior and reproduction (Lima & Dill, 1990; Kats & Dill, 1998). Throughout their lives, most prey species encounter multiple predator types that impose different levels of mortality risk. Such variation can trigger plastic or evolutionary responses in morphology, physiology, behavior, and reproductive allocation (McCauley et al., 2011; Skelhorn et al., 2011; Zanette et al., 2011; Clinchy et al., 2013; Wei & Zhang, 2022).

Adaptive responses to predators require accurate risk assessment. Excessive defensive responses may waste time or energy, reducing fitness, whereas underestimating danger can lead to mortality or reproductive failure (Walzer & Schausberger, 2011). Consequently, prey species are expected to adjust behavior and life-history strategies in proportion to perceived risk levels. These adjustments may be particularly pronounced during reproduction, when adults face increased vulnerability and offspring are often defenseless (Lima & Dill, 1990; Magnhagen, 1991). Despite extensive research on predator-induced behavioral plasticity, little work has examined how distinct predator types, those targeting adults versus those targeting eggs, shape reproductive strategies in lizards.

During the reproductive cycle, animals may experience different categories of predation risk. Species facing high adult mortality are expected to reduce exposure to predators by accelerating reproduction or maturing earlier (Norrdahl & KorpimÄki, 2000). In contrast, species subjected to high offspring or egg predation may reduce reproductive investment or modify nesting behavior to minimize losses(Martin et al.,

2000; Fontaine & Martin, 2006). Adaptive maternal effects can also influence embryonic development, for example by accelerating growth to shorten vulnerable developmental stages (Coslovsky & Richner, 2011).

Viviparity represents another potential response to offspring predation. By retaining embryos internally, females can shield developing offspring from external hazards and maintain favorable developmental conditions (Webb et al., 2006; Blackburn & Stewart, 2011; Whittington et al., 2022). Similarly, species may evolve parental care behaviors or defensive nesting strategies when predation pressure on offspring increases (Huang & Pike, 2013; Pike et al., 2016). These responses reflect evolutionary trade-offs between reproductive investment and survival, with optimal strategies depending on the balance between risks to adults and their offspring.

Life-history theory predicts that mortality acting on different life stages should shape reproduction in contrasting ways (Hua et al., 2014). Predators may target either juveniles or adults, generating distinct selective pressures on age at maturity, reproductive output, and offspring investment (Abrams & Rowe, 1996; Belk, 1998). Adult mortality primarily affects survival costs, whereas offspring predation affects fecundity costs (Magnhagen, 1991). Despite these predictions, most empirical studies have focused on a single type of predation risk or have combined multiple risks within a single experimental treatment (Fontaine & Martin, 2006; Lasley-Rasher & Yen, 2012; Creel et al., 2007). A simultaneous evaluation of these risk types across natural populations remains rare.

In squamate reptiles, the transition from egg laying (oviparity) to live birth (viviparity) has traditionally been attributed to cold climates. However, this explanation fails to account for viviparous species in tropical regions. In this study, we investigate whether predation pressure could represent an alternative driver of reproductive evolution in tropical reptiles.

Using eight Taiwanese populations of the arboreal lizard *Diploderma swinhonis*, we examine how variation in egg predation and adult predation influences reproductive timing, clutch frequency, egg retention, and maternal energy allocation. By integrating empirical data with theoretical models, we evaluate whether asymmetric predation risk can promote reproductive traits associated with early stages of viviparity.

10.3.1 Predation on Individuals

To assess predation pressure in each study population, we deployed 20 artificial lizard models per site and monitored them for a period of 10 days. Each model was attached to a tree trunk at a height of approximately 1.5–2 m to simulate the typical basking posture of live lizards (Fig. 10.10). The models were arranged along a transect roughly 100 m in length and inspected once daily throughout the observation period. Evidence of predation was identified from attack marks on the models, usually appearing as deep scratches or puncture-like cuts on the head, body, or tail, damage patterns commonly associated with attacks by birds. In cases where a model

Fig. 10.10 A model adult lizard *Diploderma swinhonis* is on a tree trunk

could no longer be located, it was recorded as having been removed and presumed consumed by a predator.

10.3.2 Variation of Life-History Traits Across the Eight Lizard Populations

10.3.2.1 Breeding Age at Maturity

To estimate the age at sexual maturity, we employed skeletochronology. Age determination was based on the examination of annual growth rings present in cross-sections of phalanges. Thin sections of the phalanx were prepared and observed under a microscope at magnifications ranging from 20× to 40×. Age was inferred by counting the Lines of Arrested Growth (LAGs), which are reliable indicators for determining individual age (Fig. 10.11).

10.3.2.2 Predation on Eggs and Adults

When egg clutches were discovered by predators, they were typically consumed entirely. Egg predation was extremely high on Orchid Island, where 53.6% of buried clutches (Supplementary Videos 10.2 and 10.3) were attacked by snakes within 5 days (Fig. 10.12). In contrast, predation rates were much lower in Taichung, Taipei, and Taoyuan populations (2.4–7.4%; Fig. 10.13).

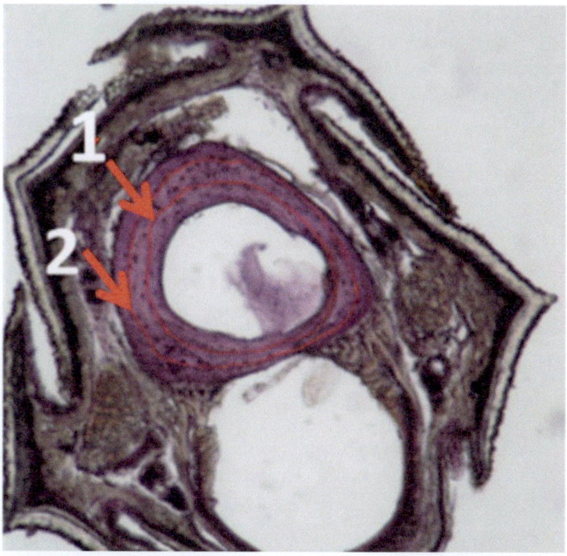

Fig. 10.11 Cross section of the third phalanx of a specimen. Two lines of arrested growth (LAGs) of phalanges can be counted

Fig. 10.12 Kukri snake attempts to eat tree lizard's freshly buried egg nest

Predation risk to adult lizards showed a different geographic pattern. Kenting exhibited the highest attack rate on lizard models, with 54.8% being attacked within 7 days. Other populations experienced considerably lower rates (7.1–11.9%), except Yilan, which showed intermediate predation (21.4%) (Fig. 10.13).

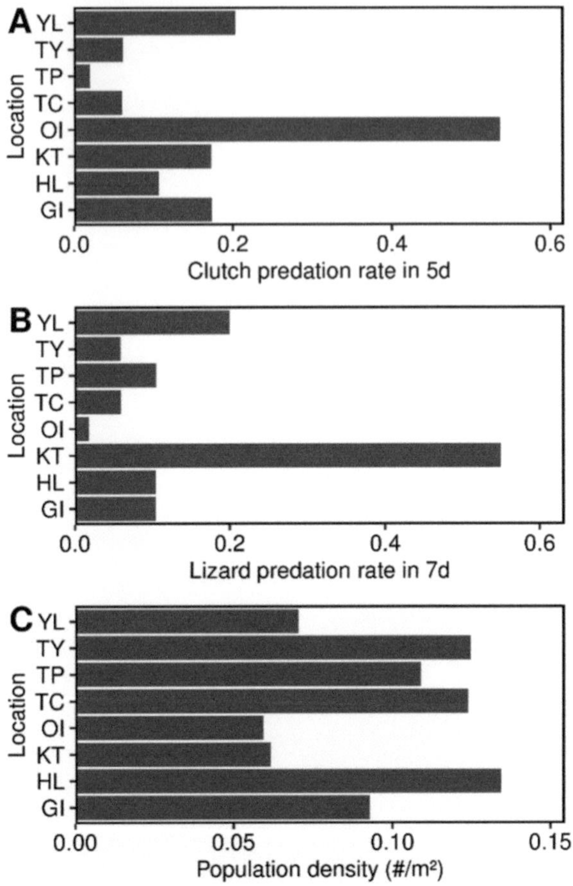

Fig. 10.13 Population-level covariates, including egg predation risk (**a**), individual predation risk (**b**), and lizard density (**c**), were tested in this study

10.3.2.3 Reproductive Strategies

Statistical models examining age at sexual maturity revealed that female reproductive timing was strongly associated with adult predation risk but only weakly related to egg predation. In Kenting, the site with the highest adult predation, more than 50% of reproductive females were 1 year old. In other populations, however, fewer than 20% of females reproduced at that age (Fig. 10.14).

Across all populations, females displayed three reproductive modes. Some produced a single clutch per breeding season, whereas others laid multiple clutches separated by 17–71 days (mean = 32.6 days; Fig. 10.15a). Remarkably, about 20% of females from Orchid Island deposited two clutches on the same day, a phenomenon we describe as the simultaneous reproductive mode (Fig. 10.15a). In these cases, one clutch completed external incubation in approximately 9 days instead of the usual ~50 days (Fig. 10.15b).

Dissection of a gravid female from Orchid Island revealed pronounced asymmetry between oviducts. Eggs in one oviduct were in advanced developmental stages

Fig. 10.14 Age structures of adult females correlated with individual predation risks and were modulated by island type

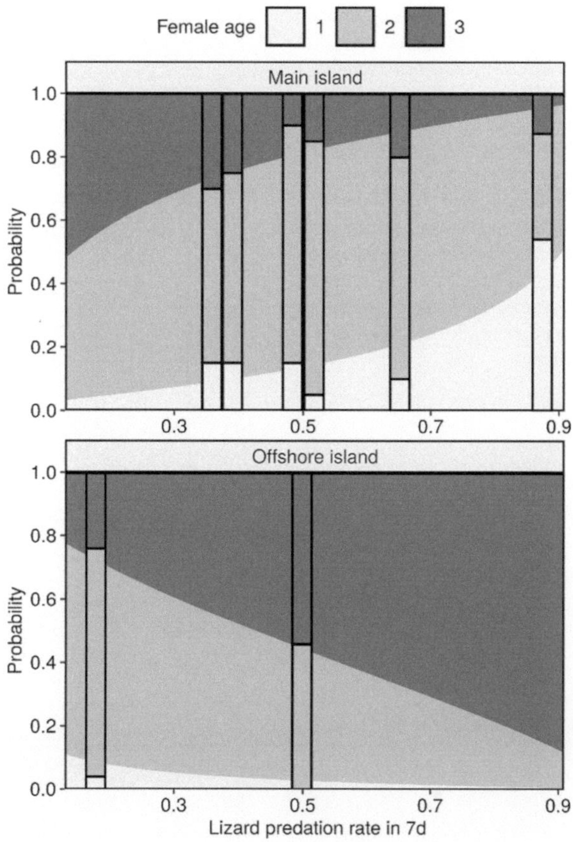

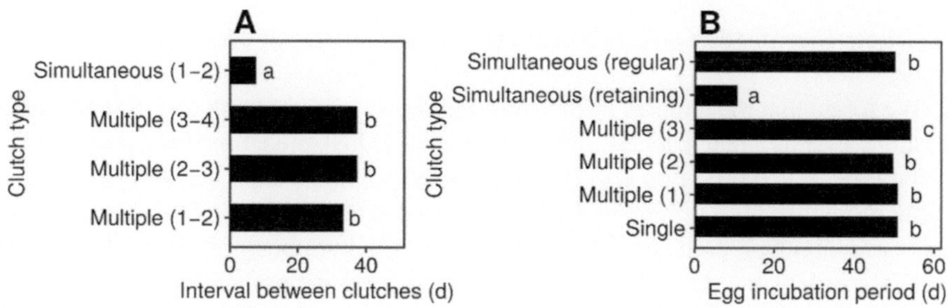

Fig. 10.15 Comparisons of interval between clutches per female (**a**) and incubation period (**b**) among reproduction strategies

with calcified shells and larger size, whereas eggs in the other oviduct were smaller, softer, and at earlier developmental stages.

Model predictions indicated that this simultaneous clutch-laying behavior was associated with environments characterized by high egg predation but low adult

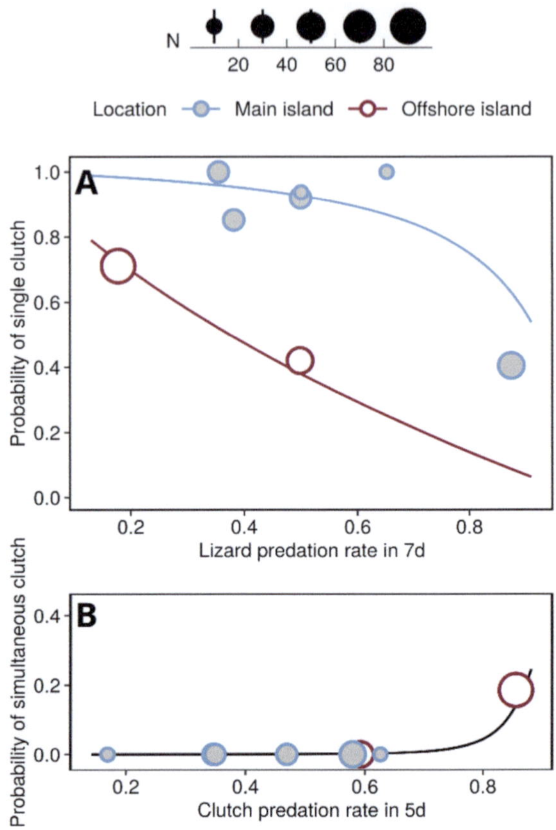

Fig. 10.16 The probabilities of females laying a single clutch negatively correlated with lizard predation rates (**a**) and those laying a simultaneous clutch positively correlated with clutch predation rates (**b**). Females on the main island exhibited a greater tendency to produce a single clutch of eggs during a reproductive season compared to those on the offshore islands

predation (Fig. 10.16). Only the Orchid Island population exhibited this reproductive mode.

We found that the females from lower individual predation risk and females from offshore islands tended to lay only clutch (individual predation risk, $p < 0.001$; island type, $p < 0.001$; Fig. 10.16a). Among the populations on the main island, we found more than 60% of females laid more than one clutch in the population with the highest individual predation risk (i.e., Kenting), but most females from other populations laid only one clutch (Fig. 10.16a). Besides the probability of only laying a single clutch, the total clutch number is also positively associated with individual predation risk ($p < 0.001$) and island type ($p < 0.001$). Females from the populations with higher individual predation risk and/or from the offshore islands tended to lay more clutches (Fig. 10.17).

The model suggested that females laid two clutches on the same day was associated with the higher egg predation rate ($p = 0.013$; Fig. 10.16b). Only females from the population with the highest egg predation risk exhibited (i.e., Orchid Island) such a simultaneous reproductive mode (Fig. 10.16b).

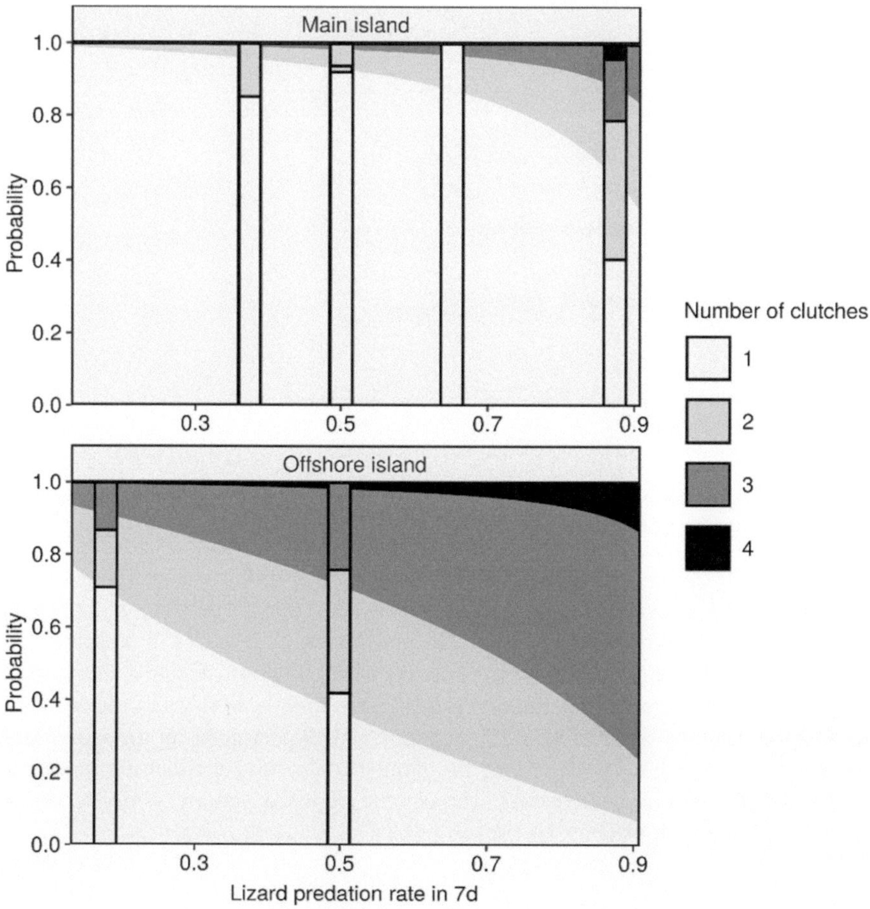

Fig. 10.17 The number of clutches per female per breading season among eight populations positively correlated with lizard predation rate. Females on the main island laid fewer clutches of eggs during a reproductive season compared to those on the offshore islands

10.3.2.4 Reproductive Efforts

Analysis showed that the clutch size of a female's first clutch within a breeding season was positively associated with snout–vent length (SVL) in populations from both the main island and the offshore islands ($p < 0.001$; Fig. 10.18). In addition, females from the main island generally produced more eggs than females from offshore populations. This difference became especially pronounced among larger individuals, as indicated by a significant interaction between island type and SVL ($p < 0.001$; Fig. 10.18).

Considering the overall reproductive investment of females across an entire breeding season, we found that the cumulative number of eggs produced increased with individual predation risk ($p = 0.005$; Fig. 10.19a). In contrast, total egg mass

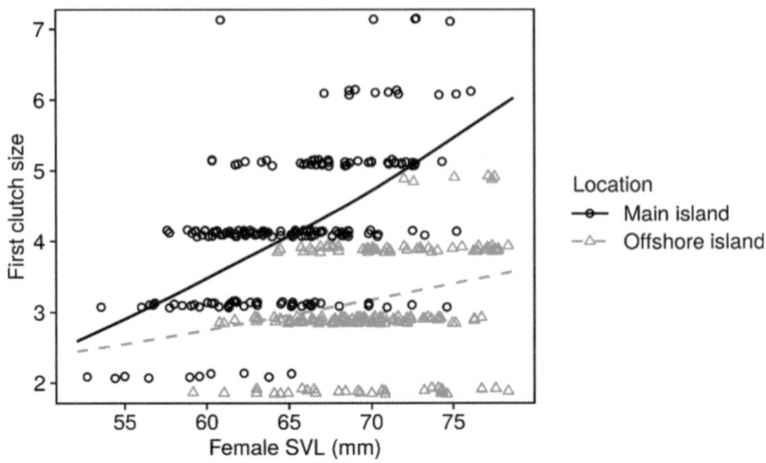

Fig. 10.18 The first clutch size of a female positively correlated with female SVL, especially in those from the main island

was positively associated with female snout–vent length (SVL) ($p = 0.009$; Fig. 10.19b), indicating that larger females invested more biomass in reproduction. When both clutch output and hatching success were taken into account, the number of offspring surviving from a female during a breeding season was influenced by both island category and the level of predation risk experienced by the individual. Specifically, in offshore populations, the number of surviving offspring tended to decline as predation risk increased, whereas the opposite pattern was observed in main-island populations (Fig. 10.19c).

10.3.3 Discussion

Our results demonstrate that contrasting predation pressures can generate markedly different reproductive strategies within a single species. On Orchid Island, the reptile egg-eating snake *Oligodon formosanus* imposes intense predation on lizard eggs (Lin et al., 2023), whereas adult predation remains relatively low. Under these conditions, females retain embryos longer and frequently deposit two clutches simultaneously, effectively reducing the duration that eggs remain exposed to predators.

By contrast, populations in Kenting face strong predation from migratory raptors such as *Accipiter soloensis*. In this environment, females mature earlier and reduce per-egg investment while producing multiple clutches each season. Such traits are consistent with life-history strategies that emphasize rapid reproduction under high adult mortality.

The traditional explanation for the evolution of viviparity, the cold climate hypothesis, suggests that prolonged egg retention evolved to stabilize embryonic

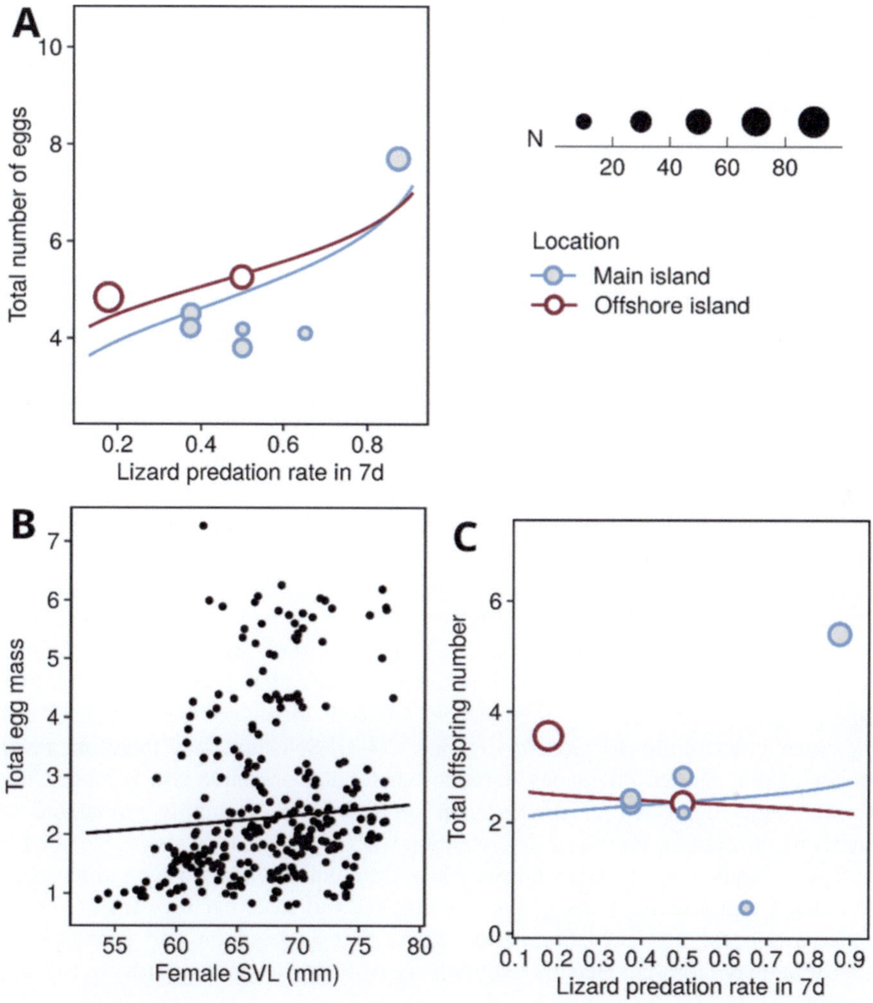

Fig. 10.19 The reproductive investment of each female per breeding season on two types of isands. Total number of eggs was positively associated with the lizard predation rate in 7 day (**a**). The total egg mass was positively associated with SVL of female (**b**). Total offspring number was positively associated with the lizard predation rate in 7 day on main island, but those was negatively associated on offshore island (**c**)

temperature in cool environments (Qualls & Shine, 1995; Smith et al., 2001). Several associated traits have been predicted, including extended egg retention, shorter external incubation periods, and larger female body size. Interestingly, these characteristics also occur in the Orchid Island population of Diploderma swinhonis despite the island's tropical climate.

Embryonic development in lizards consists of 40 stages (Dufaure & Hubert, 1961). Most oviparous species lay eggs at stages 27–28 and hatch at stage 40. On Orchid Island, however, the second clutch laid simultaneously often reaches stages

36–39 before deposition, leaving only 9–13 days of external incubation. These eggs therefore represent a transitional developmental state approaching viviparity.

Our selection-gradient analyses further reveal that egg exposure duration behaves as a threshold trait rather than a continuously varying one. Evolution tends to favor either minimal exposure (nearly internal development) or maximal exposure, depending on the relative strengths of egg versus adult predation. Such discrete outcomes highlight the importance of asymmetric mortality in shaping reproductive strategies..

Adult predation risk also strongly influences reproductive timing. Migratory raptors passing through Kenting during autumn can number in the hundreds of thousands within a 15–20 day period, feeding heavily on lizards and amphibians during migration (Chen et al., 2022). This intense predation pressure likely explains the early maturation and increased reproductive frequency observed in the Kenting population.

Energy allocation patterns provide further evidence for life-history trade-offs. When adult mortality risk is low but egg predation risk is high, females invest heavily in each egg, producing fewer but more robust offspring. Conversely, under high adult mortality, females reduce investment per egg and reproduce repeatedly, spreading reproductive risk across multiple clutches, a classic "live fast, reproduce often" strategy (Reznick & Bryga, 1987).

The ecological context of this species also shapes these responses. As a diurnal arboreal lizard that often occupies exposed substrates such as tree trunks (Huang, 2007), gravid females may experience reduced locomotor performance and increased vulnerability to predators (Shine, 2003). Selection may therefore favor strategies that shorten pregnancy duration when adult predation risk is high. Conversely, eggs deposited in shallow soil or leaf litter are highly vulnerable to predators, promoting increased egg retention when egg predation risk is elevated.

Taken together, our findings illustrate how differential predation across life stages can shape reproductive timing, clutch structure, and maternal investment. These patterns suggest that predation-driven shifts in egg retention and reproductive allocation may represent an early evolutionary pathway toward viviparity in tropical reptiles.

References

Abesamis, R. A., & Russ, G. R. (2005). Density-dependent spillover from a marine reserve: Long-term evidence. *Ecological Applications, 15*, 1798–1812.

Abrams, P. A., & Rowe, L. (1996). The effects of predation on the age and size of maturity of prey. *Evolution, 50*(3), 1052–1061.

Åkesson, S., Boström, J., Liedvogel, M., & Muheim, R. (2014). Animal navigation. In L.-A. Hansson & S. Åkesson (Eds.), *Animal movement across scales* (pp. 151–178). Oxford University Press.

Allen, A. M., & Singh, N. J. (2016). Linking movement ecology with wildlife management and conservation. *Frontiers in Ecology and Evolution, 3*, 155.

Armstrong, C., Wilkinson, H., Meade, J., Biro, D., Freeman, R., & Guilford, T. (2013). Homing pigeons respond to time-compensated solar cues even in sight of the loft. *PLoS One, 8*, e63130.

Arnott, G., & Elwood, R. W. (2009). Assessment of fighting ability in animal contests. *Animal Behaviour, 77*, 991–1004.

Auburn, Z. M., Bull, C. M., & Kerr, G. D. (2009). The visual perceptual range of a lizard, *Tiliqua rugosa*. *Journal of Ethology, 27*, 75–81.

Baird, T. A., Acree, M. A., & Sloan, C. L. (1996). Age and gender-related differences in the social behavior and mating success of free-living collared lizards, *Crotaphytus collaris*. *Copeia, 1996*, 336–347.

Belk, M. C. (1998). Predator-induced delayed maturity in bluegill sunfish (*Lepomis macrochirus*): Variation among populations. *Oecologia, 113*(2), 203–209.

Bélisle, M., Desrochers, A., & Fortin, M.-J. (2001). Influence of forest cover on the movements of forest birds: A homing experiment. *Ecology, 82*, 1893–1904.

Blackburn, D. G., & Stewart, J. R. (2011). Viviparity and placentation in snakes. In R. Aldridge & D. Sever (Eds.), *Reproductive biology and phylogeny of snakes* (pp. 119–181). CRC Press.

Breed, M. D., & Moore, J. (2016). Movement: Search, navigation, migration, and dispersal. In M. D. Breed & J. Moore (Eds.), *Animal behavior* (pp. 253–292). Academic Press.

Buckley, L. B., Rodda, G. H., & Jetz, W. (2008). Thermal and energetic constraints on ectotherm abundance: A global test using lizards. *Ecology, 89*, 48–55.

Calsbeek, R., & Smith, T. B. (2007). Probing the adaptive landscape using experimental islands: Density-dependent natural selection on lizard body size. *Evolution, 61*, 1052–1061.

Camerlink, I., Turner, S. P., Farish, M., & Arnott, G. (2015). Aggressiveness as a component of fighting ability in pigs using a game-theoretical framework. *Animal Behaviour, 108*, 183–191.

Carpenter, R. C. (1984). Predator and population density control of homing behavior in the Caribbean echinoid *Diadema antillarum*. *Marine Biology, 82*, 101–108.

Chapman, D. W. (1966). Food and space as regulators of salmonid populations in streams. *The American Naturalist, 100*, 345–357.

Chen, C.-C., Lin, J.-G., Cheng, Y.-J., Chen, T.-C., Tsai, J.-S., & Sun, Y.-H. (2022). Autumn migration of diurnal raptors across Kenting National Park, Taiwan, along the east-Asian oceanic flyway. *Journal of Raptor Research, 56*, 95–100.

Clinchy, M., Sheriff, M. J., & Zanette, L. Y. (2013). Predator-induced stress and the ecology of fear. *Functional Ecology, 27*, 56–65.

Coslovsky, M., & Richner, H. (2011). Predation risk affects offspring growth via maternal effects. *Functional Ecology, 25*, 878–888.

Cram, J. M., Torgersen, C. E., Klett, R. S., Pess, G. R., May, D., Pearsons, T. N., & Dittman, A. H. (2013). Tradeoffs between homing and habitat quality for spawning site selection by hatchery-origin Chinook salmon. *Environmental Biology of Fishes, 96*, 109–122.

Creel, S., Christianson, D., Liley, S., & Winnie, J. A., Jr. (2007). Predation risk affects reproductive physiology and demography of elk. *Science, 315*, 960–960.

Cuadrado, M. (2001). Mate guarding and social mating system in male common chameleons (*Chamaeleo chamaeleon*). *Journal of Zoology, 255*, 425–435.

Deitloff, J., Alcorn, M. A., & Graham, S. P. (2014). Variation in mating systems of salamanders: Mate guarding or territoriality? *Behavioural Processes, 106*, 111–117.

Diego-Rasilla, F. J., & Perez-Mellado, V. (2003). Home range and habitat selection by *Podarcis hispanica* (Squamata, Lacertidae) in Western Spain. *Folia Zoologica-Praha, 52*, 87–98.

Donihue, C. M., Brock, K. M., Foufopoulos, J., & Herrel, A. (2016). Feed or fight: Testing the impact of food availability and intraspecific aggression on the functional ecology of an Island lizard. *Functional Ecology, 30*, 566–575.

Dufaure, J. P., & Hubert, J. (1961). Table de développement du lézard vivipare: *Lacerta* (*Zootoca*) *vivipara* Jacquin. *Archives d'Anatomie Microscopique et de Morphologie Expérimentale, 50*, 309–327.

Ellis-Quinn, B. A., & Simon, C. A. (1989). Homing behavior of the lizard *Sceloporus jarrovi*. *Journal of Herpetology, 23*, 146–152.

Fagan, W. F., Lewis, M. A., Auger-Méthé, M., Avgar, T., Benhamou, S., Breed, G., Ladage, L., Schlägel, U. E., Tang, W.-W., Papastamatiou, Y. P., Forester, J., & Mueller, T. (2013). Spatial memory and animal movement. *Ecology Letters, 16*, 1316–1329.

Fletcher, R. J., Jr. (2007). Species interactions and population density mediate the use of social cues for habitat selection. *Journal of Animal Ecology, 76*, 598–606.

Foà, A., Basaglia, F., Beltrami, G., Carnacina, M., Moretto, E., & Bertolucci, C. (2009). Orientation of lizards in a Morris water-maze: Roles of the sun compass and the parietal eye. *Journal of Experimental Biology, 212*, 2918–2924.

Fontaine, J. J., & Martin, T. E. (2006). Parent birds assess nest predation risk and adjust their reproductive strategies. *Ecology Letters, 9*, 428–434.

Freake, M. J. (1998). Variation in homeward orientation performance in the sleepy lizard (*Tiliqua rugosa*): Effects of sex and reproductive period. *Behavioral Ecology and Sociobiology, 43*, 339–344.

Freake, M. J. (2001). Homing behaviour in the sleepy lizard (*Tiliqua rugosa*): The role of visual cues and the parietal eye. *Behavioral Ecology and Sociobiology, 50*, 563–569.

Gagliardo, A. (2013). Forty years of olfactory navigation in birds. *Journal of Experimental Biology, 216*, 2165–2171.

Galib, S. M., Sun, J., Twiss, S. D., & Lucas, M. C. (2022). Personality, density and habitat drive the dispersal of invasive crayfish. *Scientific Reports, 12*, 1114.

Gerking, S. D. (1959). The restricted movement of fish populations. *Biological Reviews, 34*, 221–242.

Grant, J. W., & Guha, R. T. (1993). Spatial clumping of food increases its monopolization and defense by convict cichlids, *Cichlasoma nigrofasciatum*. *Behavioral Ecology, 4*, 293–296.

Grüter, M., & Wiltschko, R. (1990). Pigeon homing: The effect of local experience on initial orientation and homing success. *Ethology, 84*, 239–255.

Halliwell, B., Uller, T., Wapstra, E., & While, G. M. (2017). Resource distribution mediates social and mating behavior in a family living lizard. *Behavioral Ecology, 28*, 145–153.

Hansson, L.-A., & Åkesson, S. (2014). *Animal movement across scales*. Oxford University Press.

Harada, Y., Ezoe, H., Iwasa, Y., Matsuda, H., & Sato, K. (1995). Population persistence and spatially limited social interaction. *Theoretical Population Biology, 48*, 65–91.

Holbrook, S. J., & Schmitt, R. J. (2002). Competition for shelter space causes density-dependent predation mortality in damselfishes. *Ecology, 83*, 2855–2868.

Hooten, M. B., Johnson, D. S., McClintock, B. T., & Morales, J. M. (2017). *Animal movement: Statistical models for telemetry data* (1st ed.). CRC Press.

Hoover, J. P. (2003). Decision rules for site fidelity in a migratory bird, the prothonotary warbler. *Ecology, 84*, 416–430.

Hromada, S. J., Esque, T. C., Vandergast, A. G., Drake, K. K., Chen, F., Gottsacker, B., Swart, J., & Nussear, K. E. (2023). Linear and landscape disturbances alter Mojave desert tortoise movement behavior. *Frontiers in Ecology and Evolution, 11*, 971337.

Hsu, J.-Y., Chou, C.-C., Liao, C.-P., Lin, J.-W., Cheng, R.-C., & Huang, W.-S. (2023). Negative density-regulated contest performance promotes conflict resolution in a tree lizard. *Functional Ecology, 37*, 3040–3051.

Hua, F., Sieving, K. E., Fletcher Jr, R. J., & Wright, C. A. (2014). Increased perception of predation risk to adults and offspring alters avian reproductive strategy and performance. *Behavioral Ecology, 25*(3), 509–519.

Huang, W. S. (2007). Ecology and reproductive patterns of the agamid lizard, *Japalura swinhonis*, on an East Asian Island, with comments on the small clutch sizes of Island lizrds. *Zoological Science, 24*, 181–188.

Huang, W. S., & Pike, D. A. (2013). Testing cost-benefit models of parental care evolution using lizard populations differing in the expression of care. *PLoS One, 8*, e54065.

Humphries, W. J., & Sisson, M. A. (2012). Long distance migrations, landscape use, and vulnerability to prescribed fire of the gopher frog (*Lithobates capito*). *Journal of Herpetology, 46*, 665–670.

Janson, C. H., & Bitetti, M. S. D. (1997). Experimental analysis of food detection in capuchin monkeys: Effects of distance, travel speed, and resource size. *Behavioral Ecology and Sociobiology, 41*, 17–24.

Jensen, S. P., Gray, S. J., & Hurst, J. L. (2005). Excluding neighbours from territories: Effects of habitat structure and resource distribution. *Animal Behaviour, 69*, 785–795.

Jiménez-Franco, M. V., Martínez, J. E., Pagán, I., & Calvo, J. F. (2013). Factors determining territory fidelity in a migratory forest raptor, the booted eagle *Hieraaetus pennatus*. *Journal of Ornithology, 154*, 311–318.

Jones, G. P. (1983). Relationship between density and behaviour in juvenile *Pseudolabrus celidotus* (Pisces: Labridae). *Animal Behaviour, 31*, 729–735.

Jreidini, N., & Green, D. M. (2022). Artificial displacement alters movement behavior of a terrestrial amphibian. *Herpetologica, 78*, 154–160.

Kats, L. B., & Dill, L. M. (1998). The scent of death: Chemosensory assessment of predation risk by prey animals. *Ecoscience, 5*, 361–394.

Kaufmann, J. H. (1983). On the definitions and functions of dominance and territoriality. *Biological Reviews, 58*, 1–20.

Kilgour, R. J., McAdam, A. G., Betini, G. S., & Norris, D. R. (2018). Experimental evidence that density mediates negative frequency-dependent selection on aggression. *Journal of Animal Ecology, 87*, 1091–1101.

Knell, R. J. (2009). Population density and the evolution of male aggression. *Journal of Zoology, 278*, 83–90.

Kokko, H., & Rankin, D. J. (2006). Lonely hearts or sex in the city? Density-dependent effects in mating systems. *Philosophical Transactions of the Royal Society B: Biological Sciences, 361*, 319–334.

Kuo, C. Y., Lin, Y. S., & Lin, Y. K. (2007). Resource use and morphology of two sympatric *Japalura* lizards (Iguania: Agamidae). *Journal of Herpetology, 41*, 713–723.

Kuo, C. Y., Lin, Y. S., & Lin, Y. K. (2009). Sexual size and shape dimorphism in an agamid lizard, *Japalura swinhonis* (Squamata: Lacertilia: Agamidae). *Zoological Studies, 48*, 351–361.

LaDage, L. D., Roth, T. C., Cerjanic, A. M., Sinervo, B., & Pravosudov, V. V. (2012). Spatial memory: Are lizards really deficient? *Biology Letters, 8*, 939–941.

Lamb, J. S., Satge, Y. G., & Jodice, P. G. R. (2017). Influence of density-dependent competition on foraging and migratory behavior of a subtropical colonial seabird. *Ecology and Evolution, 7*, 6469–6481.

Lasley-Rasher, R. S., & Yen, J. (2012). Predation risk suppresses mating success and offspring production in the coastal marine copepod, *Eurytemora herdmani*. *Limnology and Oceanography, 57*, 433–440.

Lewis, M. A., Fagan, W. F., Auger-Méthé, M., Frair, J., Fryxell, J. M., Gros, C., Gurarie, E., Healy, S. D., & Merkle, J. A. (2021). Learning and animal movement. *Frontiers in Ecology and Evolution, 9*, 681704.

Lima, S., & Dill, L. M. (1990). Behavioral decisions made under the risk of predation: A review and prospectus. *Canadian Journal of Zoology, 68*, 619–640.

Lin, Y. S., & Lu, K. H. (1982). Population ecology of the lizard *Japalura swinhonis formosensis* (Sauria: Agamidae) in Taiwan. *Copeia, 1982*, 425–434.

Lin, J.-W., Liao, C.-P., Chou, C.-C., Clark, R. W., Tseng, H.-Y., Hsu, J.-Y., & Huang, W.-S. (2023). Loss of sea turtle eggs drives the collapse of an insular reptile community. Science. *Advances, 9*, eadj7052.

Losos, J. B., & Irschick, D. J. (1996). The effect of perch diameter on escape behaviour of *Anolis* lizards: Laboratory predictions and field tests. *Animal Behaviour, 51*, 593–602.

Macdonald, D. W., & Johnson, D. D. P. (2015). Patchwork planet: The resource dispersion hypothesis, society, and the ecology of life. *Journal of Zoology, 295*, 75–107.

Magnhagen, C. (1991). Predation risk as a cost of reproduction. *Trends in Ecology & Evolution, 6*, 183–186.

Mandal, S. (2018). How do animals find their way back home? A brief overview of homing behavior with special reference to social hymenoptera. *Insectes Sociaux, 65*, 521–536.

Manzur, M. I., & Fuentes, E. R. (1979). Polygyny and agonistic behavior in the tree-dwelling lizard *Liolaemus tenuis* (Iguanidae). *Behavioral Ecology and Sociobiology, 6*, 23–28.

Martin, T. E., Scott, J., & Menge, C. (2000). Nest predation increases with parental activity: Separating nest site and parental activity effects. Proceedings of the Royal Society of London. *Series B: Biological Sciences, 267*, 2287–2293.

Mayer, M., Frank, S. C., Zedrosser, A., & Rosell, F. (2020). Causes and consequences of inverse density-dependent territorial behaviour and aggression in a monogamous mammal. *Journal of Animal Ecology, 89*, 577–588.

McCauley, S. J., Rowe, L., & Fortin, M. J. (2011). The deadly effects of "nonlethal" predators. *Ecology, 92*, 2043–2048.

McEvoy, J., While, G. M., Sinn, D. L., & Wapstra, E. (2013). The role of size and aggression in intrasexual male competition in a social lizard species, *Egernia whitii*. *Behavioral Ecology and Sociobiology, 67*, 79–90.

Mitani, J. C., & Rodman, P. S. (1979). Territoriality: The relation of ranging pattern and home range size to defendability, with an analysis of territoriality among primate species. *Behavioral Ecology and Sociobiology, 5*, 241–251.

Moczek, A. P. (2003). The behavioral ecology of threshold evolution in a polyphenic beetle. *Behavioral Ecology, 14*, 841–854.

Monaghan, P., & Metcalfe, N. B. (1985). Group foraging in wild brown hares: Effects of resource distribution and social status. *Animal Behaviour, 33*, 993–999.

Navarro-Salcedo, P., Arcila-Pérez, L. F., Pérez-González, J. L., Rueda-Solano, L. A., Rada, M., & Vargas-Salinas, F. (2022). Sex difference in homing: Males but not females return home despite offspring mortality in *Ikakogi tayrona*, a glassfrog with prolonged maternal care. *Behavioral Ecology and Sociobiology, 76*, 18.

Neel, L. K., Logan, M. L., Nicholson, D. J., Miller, C., Chung, A. K., Maayan, I., Degon, Z., DuBois, M., Curlis, J. D., Taylor, Q., Keegan, K. M., McMillan, W. O., Losos, J. B., & Cox, C. L. (2021). Habitat structure mediates vulnerability to climate change through its effects on thermoregulatory behavior. *Biotropica, 53*, 1121–1133.

Niemelä, P. T., Vainikka, A., Lahdenperä, S., & Kortet, R. (2012). Nymphal density, behavioral development, and life history in a field cricket. *Behavioral Ecology and Sociobiology, 66*, 645–652.

Norrdahl, K., & KorpimÄki, E. (2000). The impact of predation risk from small mustelids on prey populations. *Mammal Review, 30*, 147–156.

Norval, G., Mao, J. J., & Slater, K. (2011). Notes on the reproduction of the Swinhoe's tree lizard, *Japalura swinhonis* Günther, 1864, (Squamata: Agamidae) from southwestern Taiwan. *Herpetol Notes, 4*, 319–324.

Nothacker, J. A., Neu, C. P., Mayer, M., Wagner, N., & Lötters, S. (2018). Homing behavior in the Neotropical poison frog *Ameerega trivittata*. *Salamandra, 54*, 30–36.

O'Neil, S. T., Vucetich, J. A., Beyer, D. E., Jr., Hoy, S. R., & Bump, J. K. (2020). Territoriality drives preemptive habitat selection in recovering wolves: Implications for carnivore conservation. *Journal of Animal Ecology, 89*, 1433–1447.

Papi, F. (1992). General aspects. In F. Papi (Ed.), *Animal homing* (pp. 1–18). Springer.

Pašukonis, A., et al. (2014a). Homing trajectories and initial orientation in a Neotropical territorial frog, *Allobates femoralis* (Dendrobatidae). *Frontiers in Zoology, 11*, 29.

Pašukonis, A., et al. (2014b). Poison frogs rely on experience to find the way home in the rainforest. *Biology Letters, 10*, 20140642.

Pike, D. A., Clark, R. W., Manica, A., Tseng, H. Y., Hsu, J. Y., & Huang, W. S. (2016). Surf and turf: Predation by egg-eating snakes has led to the evolution of parental care in a terrestrial lizard. *Scientific Reports, 6*, 22207.

Pomfret, J. C., & Knell, R. J. (2007). Crowding, sex ratio and horn evolution in a South African beetle community. *Proceedings of the Royal Society B: Biological Sciences, 275*(1632), 315.

Qualls, C. P., & Shine, R. (1995). Maternal body-volume as a constraint on reproductive output in lizards: Evidence from the evolution of viviparity. *Oecologia, 103*, 73–78.

Raffestin, C. (2012). Space, territory, and territoriality. *Environment and Planning D: Society and Space, 30*, 121–141.

Rebstock, G. A., Abrahms, B., & Boersma, P. D. (2022). Site fidelity increases reproductive success by increasing foraging efficiency in a marine predator. *Behavioral Ecology, 33*, 868–875.

Reznick, D. N., & Bryga, H. (1987). Life-history evolution in guppies (*Poecilia reticulata*): 1. Phenotypic and genetic changes in an introduction experiment. *Evolution, 41*, 1370–1385.

Richardson, T., Giuggioli, L., Franks, N., & Sendova-Franks, A. (2017). Measuring site fidelity and spatial segregation within animal societies. *Methods in Ecology and Evolution, 8*, 965–975.

Rueger, T., Gardiner, N. M., & Jones, G. P. (2016). Homing is not for everyone: Displaced cardinalfish find a new place to live. *Journal of Fish Biology, 89*, 2182–2188.

Scali, S., Sacchi, R., Azzusi, M., Daverio, S., Oppedisano, T., & Mangiacotti, M. (2012). Homeward bound: Factors affecting homing ability in a polymorphic lizard. *Journal of Zoology, 289*, 196–203.

Schoener, T. W., & Schoener, A. (1982). Intraspecific variation in home-range size in some *Anolis* lizards. *Ecology, 63*, 809–823.

Shine, R. (2003). Locomotor speeds of gravid lizards: Placing 'costs of reproduction' within an ecological context. *Functional Ecology, 17*(4), 526–533.

Skelhorn, J., Rowland, H. M., Delf, J., Speed, M. P., & Ruxton, G. D. (2011). Density-dependent predation influences the evolution and behavior of masquerading prey. *Proceedings of the National Academy of Sciences, 108*, 6532–6536.

Smith, J. N., Montgomerie, R. D., Taitt, M. J., & Yom-Tov, Y. (1980). A winter feeding experiment on an Island song sparrow population. *Oecologia, 47*, 164–170.

Smith, S. A., Austin, C. C., & Shine, R. (2001). A phylogenetic analysis of variation in reproductive mode within an Australian lizard (*Saiphos equalis*, Scincidae). *Biological Journal of the Linnean Society, 74*, 131–139.

Stamps, J. A., & Buechner, M. (1985). The territorial defense hypothesis and the ecology of insular vertebrates. *The Quarterly Review of Biology, 60*, 155–181.

Streit, R. P., & Bellwood, D. R. (2017). High prevalence of homing behaviour among juvenile coral-reef fishes and the role of body size. *Coral Reefs, 36*, 1083–1095.

Stuart-Fox, D. (2006). Testing game theory models: Fighting ability and decision rules in chameleon contests. *Proceedings of the Royal Society B: Biological Sciences, 273*, 1555–1561.

Switzer, P. V. (1993). Site fidelity in predictable and unpredictable habitats. *Evolutionary Ecology, 7*, 533–555.

Tomkins, S. L., & Brown, G. S. (2004). Population density drives the local evolution of a threshold dimorphism. *Nature, 431*, 1099–1103.

Toobaie, A., & Grant, J. W. A. (2013). Effect of food abundance on aggressiveness and territory size of juvenile rainbow trout, *Oncorhynchus mykiss*. *Animal Behaviour, 85*, 241–246.

Tricarico, E., & Gherardi, F. (2010). Past ownership makes crayfish more aggressive. *Behavioral Ecology and Sociobiology, 64*, 575–581.

Vander Wal, E., Laforge, M. P., & McLoughlin, P. D. (2014). Density dependence in social behaviour home range overlap and density interacts to affect conspecific encounter rates in a gregarious ungulate. *Behavioral Ecology and Sociobiology, 68*, 383–390.

Vervust, B., Van Dongen, S., Grbac, I., & Van Damme, R. (2009). The mystery of the missing toes: Extreme levels of natural mutilation in Island lizard populations. *Functional Ecology, 23*, 996–1003.

Walzer, A., & Schausberger, P. (2011). Threat-sensitive anti-intraguild predation behaviour: Maternal strategies to reduce offspring predation risk in mites. *Animal Behaviour, 81*, 177–184.

Wasko, D. K., & Sasa, M. (2012). Food resources influence spatial ecology, habitat selection, and foraging behavior in an ambush-hunting snake (Viperidae: *Bothrops asper*): An experimental study. *Zoology, 115*, 179–187.

Webb, J. K., Shine, R., & Christian, K. A. (2006). The adaptive significance of reptilian viviparity in the tropics: Testing the maternal manipulation hypothesis. *Evolution, 60*, 115–122.
Wegener, J. E., Mulder, K. P., Pringle, R. M., Losos, J. B., & Kolbe, J. J. (2019). Head size of male and female lizards increases with population density across Island populations in The Bahamas. *Breviora, 566*, 1–9.
Wei, X., & Zhang, Z. Q. (2022). Level-dependent effects of predation stress on prey development, lifespan and reproduction in mites. *Biogerontology, 23*, 515–527.
Weintraub, J. D. (1970). Homing in the lizard *Sceloporus orcutti*. *Animal Behaviour, 18*, 132–137.
Whittington, C. M., Van Dyke, J. U., Liang, S. Q. T., Edwards, S. V., Shine, R., Thompson, M. B., & Grueber, C. E. (2022). Understanding the evolution of viviparity using intraspecific variation in reproductive mode and transitional forms of pregnancy. *Biological Reviews, 97*, 1179–1192.
Willems, E. P., & Hill, R. A. (2009). Predator-specific landscapes of fear and resource distribution: Effects on spatial range use. *Ecology, 90*, 546–555.
Willems, E. P., Barton, R. A., & Hill, R. A. (2009). Remotely sensed productivity, regional home range selection, and local range use by an omnivorous primate. *Behavioral Ecology, 20*, 985–992.
Williams, H. J., & Safi, K. (2021). Certainty and integration of options in animal movement. *Trends in Ecology & Evolution, 36*, 990–999.
Wright, J., Bolstad, G. H., Araya-Ajoy, Y. G., & Dingemanse, N. J. (2019). Life-history evolution under fluctuating density-dependent selection and the adaptive alignment of pace-of-life syndromes. *Biological Reviews, 94*, 230–247.
Zanette, L. Y., White, A. F., Allen, M. C., & Clinchy, M. (2011). Perceived predation risk reduces the number of offspring songbirds produce per year. *Science, 334*, 1398–1401.
Zhang, Y. D., Chou, C. C., Liao, C. P., Lin, J. W., Clark, R. W., Hsu, J. Y., & Huang, W. S. (2024). Homing behavior of a tree lizard: Influences of mating resource and habitat structure. *Behavioral Ecology, 35*(6), arae081.
Zuri, I., & Bull, C. M. (2000). The use of visual cues for spatial orientation in the sleepy lizard (*Tiliqua rugosa*). *Canadian Journal of Zoology, 78*, 515–520.

Aposematic Signals to Sympatric and Allopatric Lizard Predators

11

Abstract

Aposematic coloration allows toxic prey to signal unprofitability and serves as a key model in evolutionary biology. Using a visually oriented predator, the Swinhoe's tree lizard (*Diploderma swinhonis*), we tested whether coloration enhances learning and memory of aversive prey. Lizards were offered quinine-treated crickets (*Gryllus bimaculatus*) colored red (aposematic) or green (non-aposematic). After a single exposure, lizards rejected both prey types, but showed greater hesitation toward red crickets in later trials. Memory of both prey types persisted for up to 60 days, with strong generalization from colored to uncolored prey. These results reveal nuanced learning and long-term memory in visually hunting lizards, supporting their value as models for studying aposematism.

Keywords

Tree lizard · Aposematism · Learning · Memory · Prey · Sympatry · Allopatry

Generally speaking, "disguise" is not a positive term. However, in Chinese history, "disguise" has been regarded as a form of great wisdom, giving rise to various tactics such as feigning ignorance, concealing one's abilities, and enduring humiliation in the presence of opponents. Each of these tactics is rooted in profound strategic thinking.

The most famous example in history is the 32nd strategy of the Thirty-Six Stratagems: the Empty City Strategy.

Zhuge Liang had only 5000 soldiers with him in Xicheng, busy transporting supplies. Suddenly, he heard that Sima Yi was leading an army of 150,000 soldiers toward them. Zhuge Liang realized the situation was dire, as the main general was absent, and half of the soldiers had already left with the supplies. After some thought,

W.-S. Huang, *Reptiles in the Changing Climate*, Fascinating Life Sciences,
https://doi.org/10.1007/978-981-92-1887-5_11

he decided to take a bold gamble. Zhuge Liang ordered the soldiers to remove the military flags, open all four city gates, and station 20 soldiers at each gate, dressed as civilians, leisurely sweeping the streets. No one else was allowed to enter or exit freely, and no loud noise was permitted; otherwise, they would be beheaded. Then, two attendants followed Zhuge Liang to the city wall with a zither. After burning incense, Zhuge Liang began to play the zither. When the Wei army arrived, they saw the scene before them and dared not act rashly. When Sima Yi went to the city gate to observe, he saw Zhuge Liang smiling and playing the zither intently, while the civilians at the city gate were quietly sweeping the streets. There was no sense of the restless atmosphere that usually precedes battle. The sound of the zither was also calm and harmonious. Sima Yi ordered the army to retreat immediately and told his son, who opposed the retreat, "Zhuge Liang is a cautious man who would not take such a great risk. There must be ambush troops, which is why he is so leisurely playing the zither."

There are numerous examples of animal camouflage or aposematic colors. For instance, the non-venomous milk snake (*Lampropeltis triangulu*) mimics the highly venomous coral snake (*Micrurus*), while the aposematic colors of venomous poison dart frogs serve to deter predators. Of course, these camouflage or aposematic strategies are not the result of human ingenuity but rather the outcome of natural selection. This chapter will describe how weevils use aposematic coloration to avoid attacks from lizards.

11.1 Aposematic Coloration of Prey Improves Memory in the Tree Lizard

11.1.1 Introduction

Predation is a major natural selection factor that influences both behavioral and morphological traits across animal taxa. To avoid the risk of attack, many species employ deterrent signaling strategies that convey reduced profitability to potential predators. These defenses may be mediated through chemical (Eisner & Grant, 1981), visual (Endler & Mappes, 2004), or acoustic (Guilford, 1990) signals. Aposematism, which characterized by conspicuous visual cues advertising unpalatability, represents one of generally documented antipredator adaptations (Guilford, 1990). To date, the majority of studied on aposematism has focused on species that signal unpalatability through vivid, highly conspicuous coloration. Such warning displays have been documented across many groups, including insects (Cott, 1940; Edmunds, 1974), amphibians (Howard & Brodie, 1973), reptiles (Cooper & Greenberg, 1992; Britson & Gutzke, 1993), as well as birds and mammals (Cott, 1940; Dumbacher et al., 1992; Caro, 2009). Longer wavelength conspicuous colors, particularly orange, red, or yellow, are typically associated with aposematism in these various systems (Lindström et al., 2001). These colors are typically create chromatic contrast against natural background (Gamberale-Stille & Guilford, 2003; Polo-Cavia et al., 2016), enhancing detectability, learning and memory retention of predator relative to concealed colors, and keep them in mind for longer periods of time (Siddall & Marples, 2008). By promoting rapid recognition and avoidance by experienced predators, conspicuous coloration may reduce mortality among conspecifics, thereby conferring a selective advantage (Gittleman & Harvey, 1980; Roper & Redston, 1987). The adaptive of persistent conspicuous coloration is not universal. For instance, Lee et al. (2018) reported that lizards relied more heavily on prey body shape than coloration when avoiding distasteful weevils, while Ham et al. (2006) found that great tits exhibited limited color bias and were equally capable of learning to discriminate between gray and colored prey.

Avoidance of conspicuously colored prey by predators is generally attributed to two mechanisms. One is innate avoidance: inexperienced predators may refrain from attacking certain visual targets because of innate avoidance to particular colors or objects (Smith, 1975; Roper, 1990). The other depends on associative learning, whereby predators form links between warning color patterns (conditioned stimuli) and negative experiences such as distastefulness (unconditioned stimuli) (Roper & Wistow, 1986; Bowers, 1992). The aversive experiences of unpalatable prey generate fitness benefits for both interacting parties. Predators that rapidly acquire and retain avoidance associations reduce the risk of toxin ingestion, while prey benefit from diminished attack rates in future. Given that the tendency of predators to avoid repeated negative experiences encounters, aposematism has become a model system for investigating the mechanisms and evolution of learning and memory (Roper & Wistow, 1986; Ruxton et al., 2004; Pegram & Rutowski, 2014).

Following the description of aposematism by (Poulton, 1890), many experiments demonstrated that particular colors and/or patterns improve predator retention of distasteful prey (e.g., Cott, 1940; Evans & Schmidt, 1990; Guilford, 1990). For instance, naive quails (*Callipepla gambelii*) exhibit a stronger innate aversion against orange relative to blue, and this facilitates stronger learning and broader generalization following negative experiences (Pegram & Rutowski, 2014). Despite fewer studies in reptiles, it has been demonstrated that garter snakes (*Thamnophis radix*) exhibit stronger aversive learning toward yellow and black prey than green prey (Terrick et al., 1995).

Lizards function simultaneously as predators and prey within the terrestrial consumer-resource network and occur worldwide. Their trophic ecology, including diet composition, habitat use, activity patterns, and foraging strategies has been extensively studied and serves as the foundation for several "classic" ecological studies [reviewed in (Vitt et al., 2003)]. Typically, behavioral studies of lizard foraging divide species into two strategies, which are primarily related to the major clades "Iguania"and "Scleroglossa"(Cooper, 1994). Iguania is known for its "sit-and-wait" foraging strategy, relying primarily on visual cues to identify and consume a variety of highly mobile prey. In contrast, scleroglossa is known for its aggressive foraging, which involves the use of visual and chemical cues to locate and consume low-mobility prey (Vitt et al., 2003). Even with a substantial ecological literature, a predominantly insectivorous diet, global distribution, and abundance, lizards are underrepresented in aposematism studies relative to birds and mammals (Roper, 1994; Siddall & Marples, 2008; Caro, 2009). Such underrepresentation may stem from historical biases that the reptile cognitive abilities.. However, squamates have complex cognitive, learning, and problem-solving abilities, according to recent research (Manrod et al., 2008; Leal & Powell, 2012). Moreover, lizards are often more easily captured and maintained under laboratory conditions than other terrestrial vertebrates, making them well-suited for learning and memory control behavioral experiments.

Diploderma (Agamidae, Iguania) is an agamid genus found in Pakistan, India, Burma, China, Vietnam, Taiwan, and Japan. Five *Diploderma* species occur in Taiwan, among which *D. swinhonis* is the most widespread and abundant. This species is only found in Taiwan and is common across the lowlands (Huang, 2007). *D. swinhonis*, like other ambush-feeding tree lizards, is primarily a sit-and-wait predator that uses visual cues to detect and identify prey, according to a previous study (Shen, 2009). Prey identification in *D. swinhonis* is also aided by movement (Woo et al., 2009; Woo et al., 2017), and most species are considered capable of detecting the full spectrum of visible (Barbour et al., 2002) and ultraviolet light (LeBas & Marshall, 2000).

Despite coexisting with a diverse range of palatable and toxic prey, the role of warning color in shaping foraging behavior of *D. swinhonis* is unknown (but see Lee et al., 2018). Using *D. swinhonis* as a model, we investigated the impacts of warning color on learning and memory in lizards in this chapter. We colored crickets red (a color associated with mimicry in many systems) or green for prey (a color rarely associated with mimicry). We first looked into whether *D. swinhonis* had a color preference, and then we tested whether aposematic coloration improved *D. swinhonis*' ability to the memory of toxic prey and the generalization of aversive experience to non-colored crickets.

To investigate the effects of a particular color, we modified the color of crickets using "tapioca," a refined starch processed into a relatively flavorless edible white powder. Aposematic (red), non-aposematic (green), and control (black) colors were generated by blending tapioca (1:1 w/w) with the addition of red (Food coloring Red No.6), green (Food coloring Yellow No.4, 57% w/w and Food coloring Blue No.1, 43%), and black (Food coloring Black) food coloring powders, respectively, to produce color toners. At the beginning of the experiment, the crickets were immersed in the specified color toner until their surfaces were evenly coated (Fig. 11.1a). Red crickets simulated aposematic coloration consistent with other experimental designs, green crickets were used as a non-aposematic prey, and black crickets were used as a control for the color method, as our black toner closely resembled their natural body coloration.

11.1.2 Learning and Memory/Generalization Tests

Following the determination of initial color preferences in the first experiment, the second experiment was conducted to quantify the memory associated with noxious prey and generalization of aversion to non-colored prey in *D. swinhonis*. A total of 268 new lizards (different individuals than in the color preference experiment) were

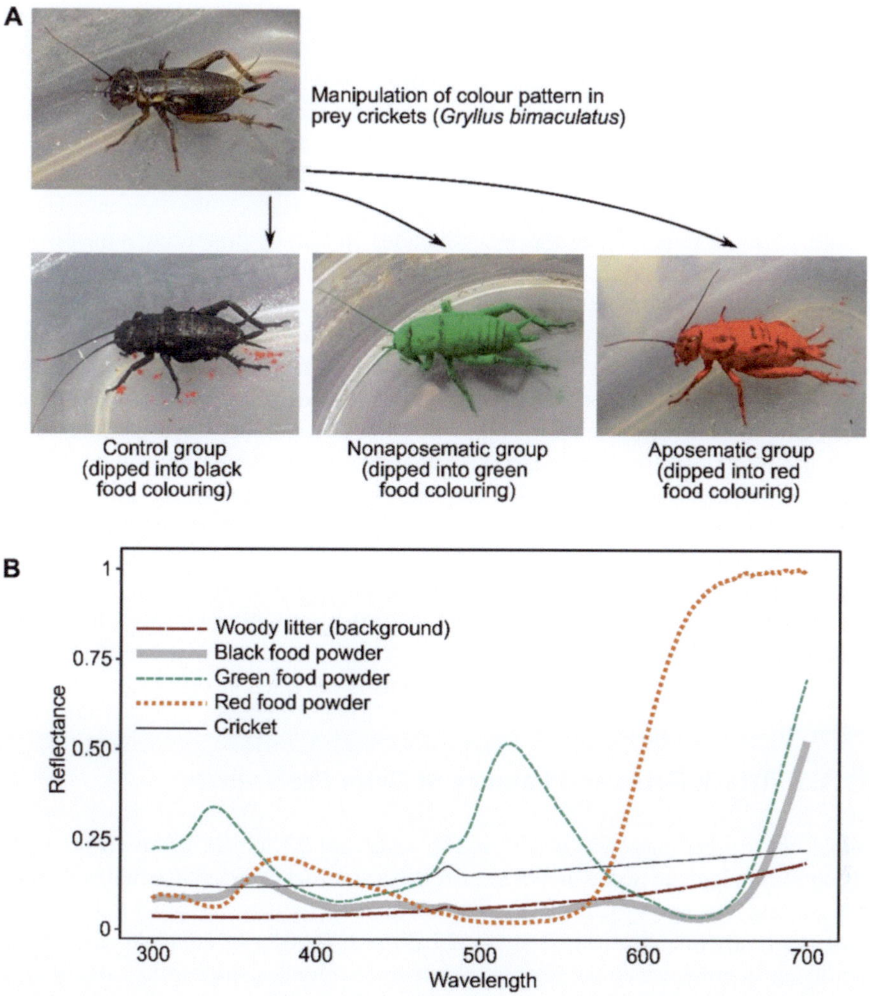

Fig. 11.1 Schematic of experimentally colored prey (**a**) and the average reflectance spectrum of color powders (**b**). (Adapted from Ko et al. (2020), Animal Behaviour, with permission. Photos modified)

randomly grouped to either an aposematic (red) or non-aposematic (green) treatment. Both groups included male and female lizards. Following 2 days of fasting, each lizard was presented with a cricket matching their assigned color treatment. These prey items were rendered unpalatable by coating them with a solution containing 20% (w/w) quinine and 10% (w/w) mustard. For both groups, the latency to attack was documented using video recordings.. Lizards in both experimental groups displayed rejection responses of taste aversion, such as spitting out the cricket or tongue protrusion. If the lizard bit the cricket and exhibited aversive reactions, then after 30 min, a palatable cricket with the same color but without quinine was offered to determine whether it learned an aversion due to not attacking the second cricket. Failure to attack this similarly colored cricket was interpreted as evidence of successful learning. Lizards that consumed both crickets were classified as non-learners and excluded from further experiment. Consequently, only individuals who attacked the first unpalatable cricket exhibited aversive reactions and subsequently refused the second palatable cricket with a similar color were advanced to the memory retention experiment. If a lizard did not attack the first unpalatable cricket within 2 min, hunger was assessed after 30 min by presenting a mealworm. Individuals who also refused the mealworm were retested after 2 days; only one retest was permitted before being discarded from the experiment.

Following successful aversion trials, lizards were fasted for 2 days, and then were allocated to 7, 20, or 60 days to test their memory retention. Lizards were maintained on a diet of two mealworms per day during these periods. Lizards in each group were offered a palatable cricket that matched the color of the initial test after the period.

If the lizard abstained from attacking the colored cricket within 2 min, a black palatable cricket was offered. This step aimed to determine if the aversion to a specific color was generalized to a prey with similar morphology and behavior but lacking the learned color cue. If the black cricket was also ignored for 2 min, the hunger of the lizard was tested by providing a mealworm after 30 min. If the lizard did not still attack the mealworm, we repeated the generalization test 2 days later, with only a single retest permitted per lizard. The full experimental workflow of the learning trials and memory/generalization trials is provided in Fig. 11.2.

11.1.3 Attack Rates and Latency of Color Preference

Initial color preference was tested in 116 male and 93 female lizards (Fig. 11.1). Using Firth's logistic regression demonstrated a significant difference in lizard attack rate among the aposematic (red crickets), non-aposematic (green crickets), and control groups (black crickets) ($\chi^2 = 33.29$, $p < 0.001$). In contrast, sex did not significantly influence attack likelihood (males vs. females, odds ratio [OR] = 1.14, Z = 0.831, $p = 0.702$). Post hoc comparisons revealed that red crickets were attacked at a significantly lower rate than green crickets (OR = 0.164, Z = 4.71, $p = 0.001$) or the control black cricket (OR = 0.134, Z = 4.65, $p < 0.001$; Fig. 11.3a). Attack rates did not significantly differ between green and black crickets (OR = 0.819, Z = 0.346, $p = 0.729$; Fig. 11.3a).

According to the analysis of the forward selection process, there were no significant interactions between sex and handling treatment ($p = 0.632$), and the morphological covariates were not significant to attack rate (head width, $p = 0.882$; head height, $p = 0.778$; head length, $p = 0.839$; SVL, $p = 0.260$; body mass, $p = 0.218$).

Similarly, attack latency showed a significant difference between treatment groups ($\chi^2 = 23.00$, $p < 0.001$). Lizards took longer latencies on the red crickets than green or black crickets (red vs. green, hazard ratio [HR = 0. 359, Z = 5.236, $p < 0.001$; red vs. black, HR = 0.360, Z = 4.831, $p < 0.001$; green versus black, HR = 1.002, Z = 0.994, $p = 0.419$; Fig. 11.3b). The forward selection process indicated except for SVL, either the interaction term or most morphological covariates were not significant (interaction between sex and treatment, $p = 0.273$;

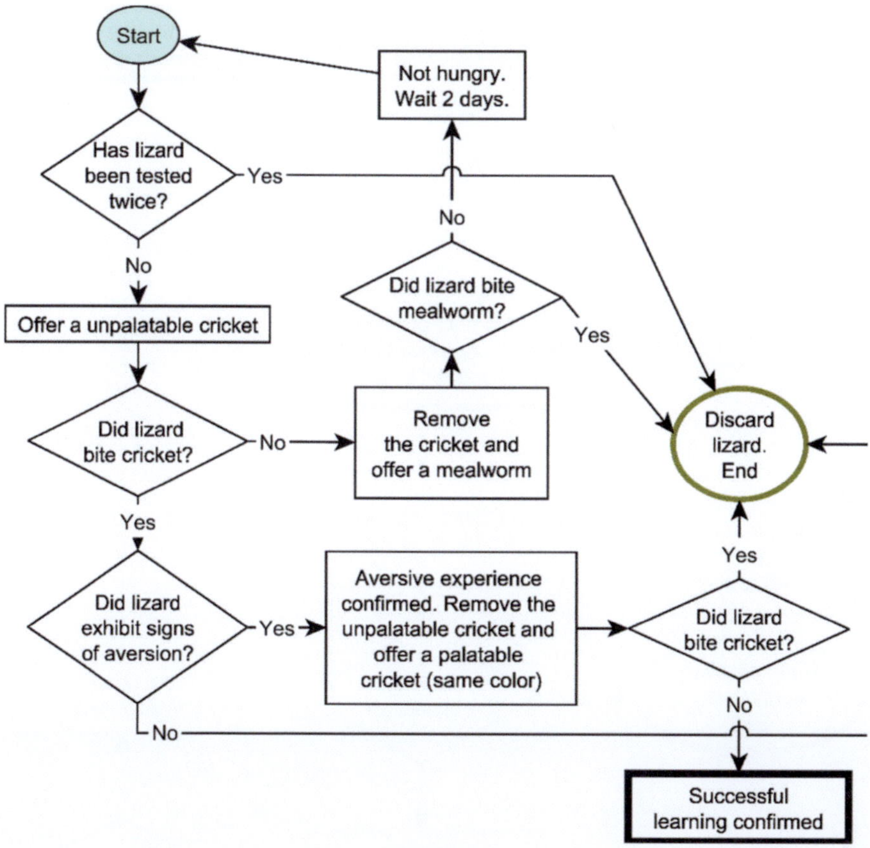

Fig. 11.2 Flowchart of memory retention and generalization test for lizards. Rectangles with thicker borders indicate states determining learning, memory retention, and generalizations. (Redrawn from Ko et al. (2020), Animal Behaviour)

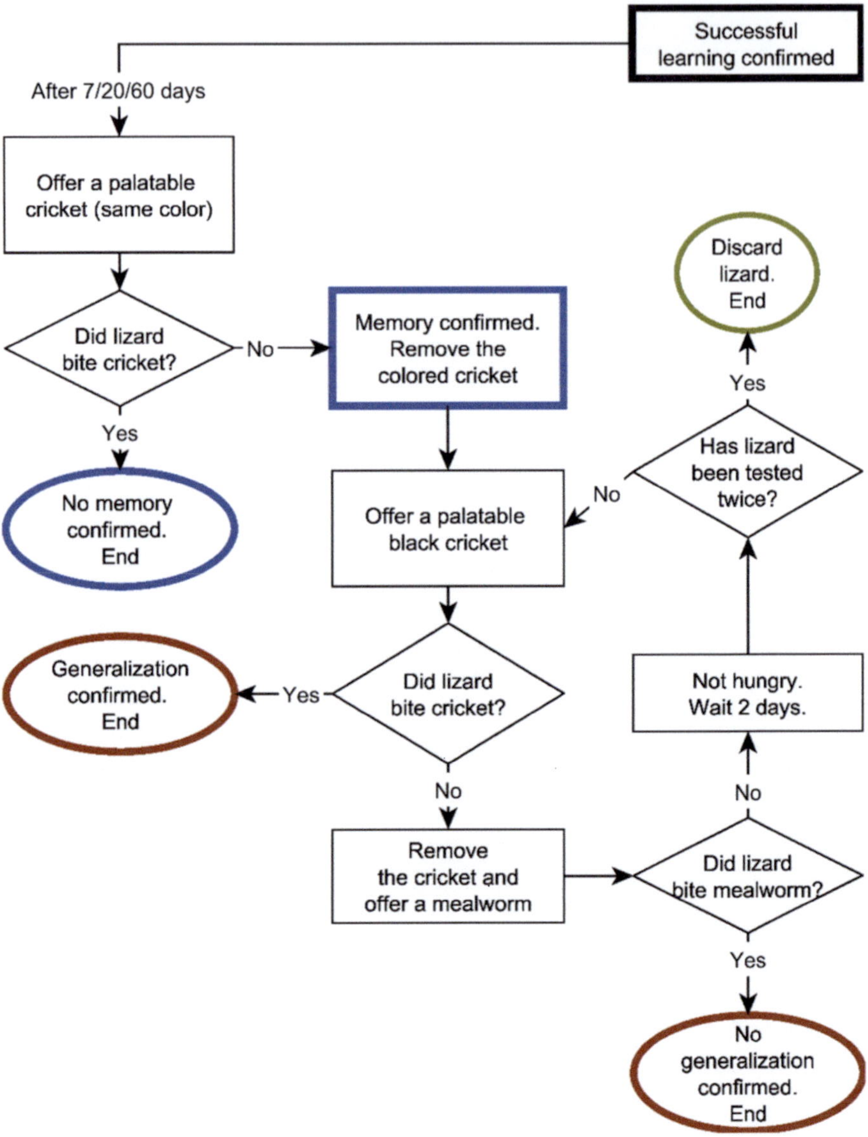

Fig. 11.2 (continued)

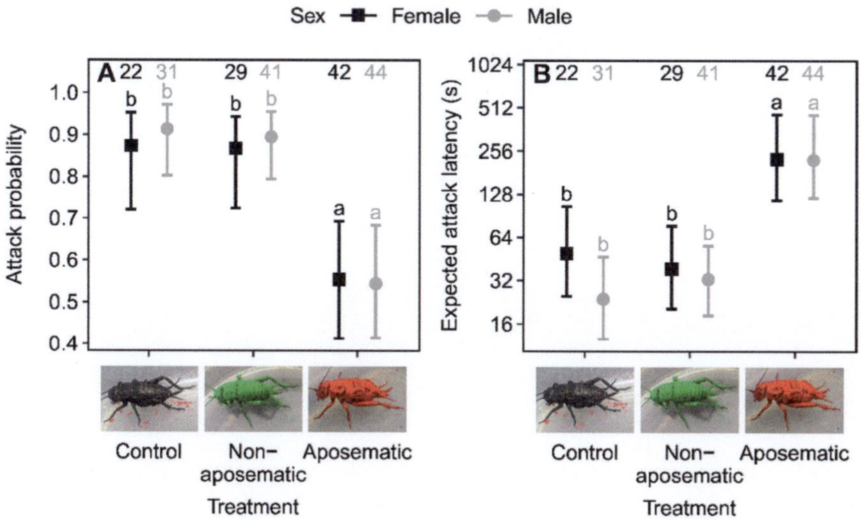

Fig. 11.3 Probability of lizard attacking and latency to attack crickets of different colors. (**a**) In our initial color preference experiment, the results showed that the proportion of lizards that attacked aposematic (red colored) crickets was statistically lower than the proportion attacking either non-aposematic or control. (**b**) The expected attack latency indicates that lizards in the aposematic group showed a longer attack latency than those in the non-aposematic and control groups. The lowercase letters indicate the significant differences based on multiple comparisons. Numbers above bars indicate the sample sizes. (Adapted from Ko et al. (2020), Animal Behaviour, with permission. Photos modified; plots redrawn)

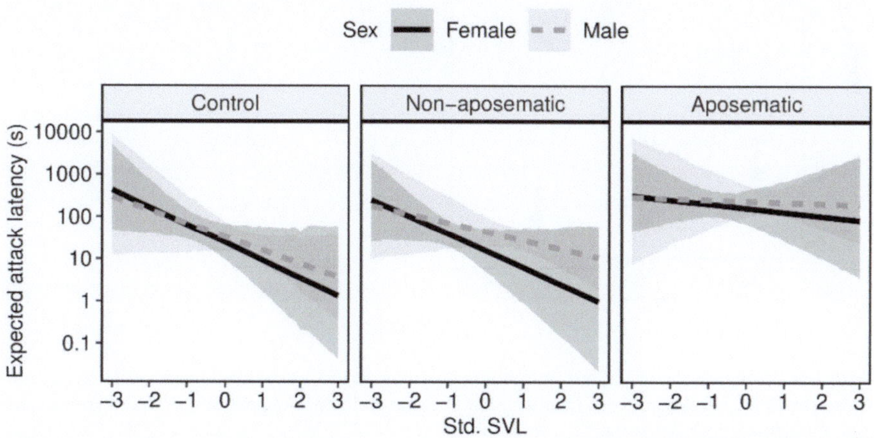

Fig. 11.4 Larger lizards attack more readily. Survival analysis of attack latency also indicated that larger lizards showed shorter attack latencies than smaller lizards

head width, $p = 0.871$; head height, $p = 0.986$; head length, $p = 0.234$; body mass, $p = 0.059$). Lizards with longer SVL attacked more rapidly (HR = 1.182/SVL SD, $Z = 1.992$, $p = 0.046$) (Fig. 11.4). There were no morphological

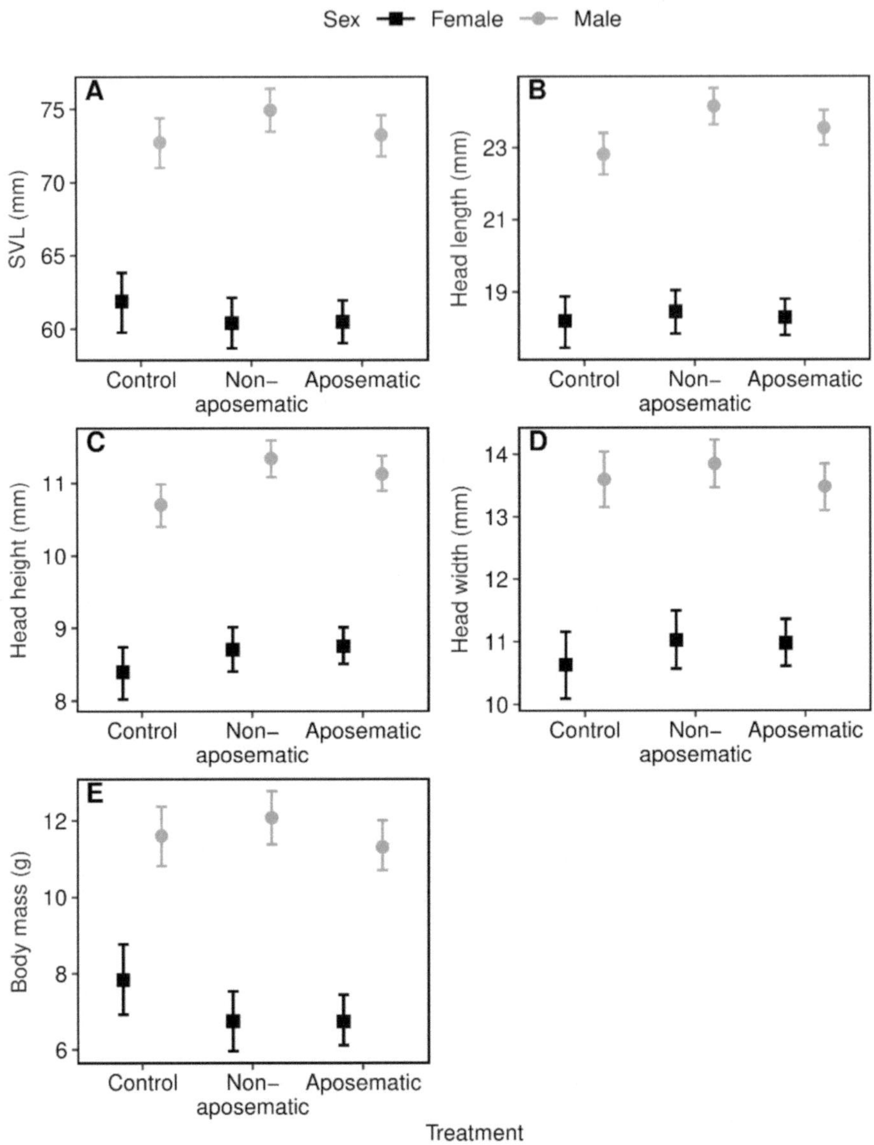

Fig. 11.5 Comparisons of various morphological traits. (**a**) SVL; (**b**) head length; (**c**) head height; (**d**) head width; (**e**) body mass of lizards tested in color preference experiments. (Redrawn from Ko et al. (2020), Animal Behaviour)

differences in handling between the sexes (permutational MANOVA; males, $R^2 = 0.041$, $p = 0.063$; females, $R^2 = 0.035$, $p = 0.152$; Supplementary Fig. 11.5).

11.1.4 Memory for Unpalatable Prey

In our second experiment, 350 lizards were successfully fed on inedible red or green crickets. Consistent with the results of our color preference experiment, roughly 75% of the lizards attacked tasty (red) crickets and roughly 93% attacked tasteless (green) crickets (confirmed by a weak interaction between the two experiments, lizard sex and prey color; $\chi^2 = 1.229$, $p = 0.746$). Lizards in the red mimic group exhibited avoidance more frequently to the second colored cricket than lizards in the green non-mimic group despite the most of lizards (93%) did not attack unpalatable crickets during re-testing (Fig. 11.6a) (OR = 2.139, $p = 0.041$; Fig. 11.6a). The result also showed that higher body mass lizards were lower attack rate for toxic prey during re-testing (OR = 1.576, $p = 0.018$ for SD per body mass; Fig. 11.6b). There was no relationship with memory period ($\chi2[2] = 4.985$, $p = 0.083$), sex (males vs. females, OR = 1.433, $p = 0.329$; Fig. 11.6a), interaction terms (sex × prey color, $p =$ 0. 779; sex × memory period, $p = 0.523$; memory period × prey color, $p = 0.068$), or any morphological covariates except for body weight (head width, $p = 0.098$; head height, $p = 0.554$; head length, $p = 0.452$; SVL, $p = 0.707$).

We also looked at how the two cricket colors affected the probability of an attack on the first cricket during training and memory tests. By pooling the two sexes and three memory periods, the probability of lizards attacking a red mimic cricket decreased significantly from 65% in the training test to 8.8% in the memory test ($p < 0.0001$; Fig. 11.6a), as did the probability of attacking a green non-mimic cricket ($p < 0.0001$; Fig. 11.6a). Thus, while lizards trained with red crickets were less likely to attack them overall, lizards trained with green crickets showed a greater decrease in attack probability after training because the initial attack probability for green crickets was much higher ($p < 0.001$).

Using the envelope MANOVA to assess morphological variations between treatments within the two sexes, the results showed no significant differences in morphological characteristics within females ($R^2 = 0.040$, $p = 0.563$; Fig. 11.7). In contrast, there were significant differences within males ($R^2 = 0.089$, $p = 0.004$), due to the males in the 7- and 60-day non-aposematic treatments were smaller size (Fig. 11.7).

11.1.5 Generalization

Of the 268 lizards exposed to unpalatable colored crickets, 234 individuals did not attack them again upon reexposure, and this avoidance also occurred in generalizations to black crickets. The memory period significantly influenced this generalization to black crickets, according to our logistic regression model ($\chi2[2] = 11.059$, $p = 0.004$). According to post hoc comparisons, the lizards in the 7-day group exhibited lower generalization (OR = 4.498, Z = 3.323, $p = 0.003$), but none of the other comparisons were significant (7-day group vs. 20-day group OR = 1.927, Z = 1.417, $p = 0.157$; 20-day group vs. 60-day group OR = 0.429, Z = 1.473, $p = 0.157$). We also discovered generalization was lower in male lizards

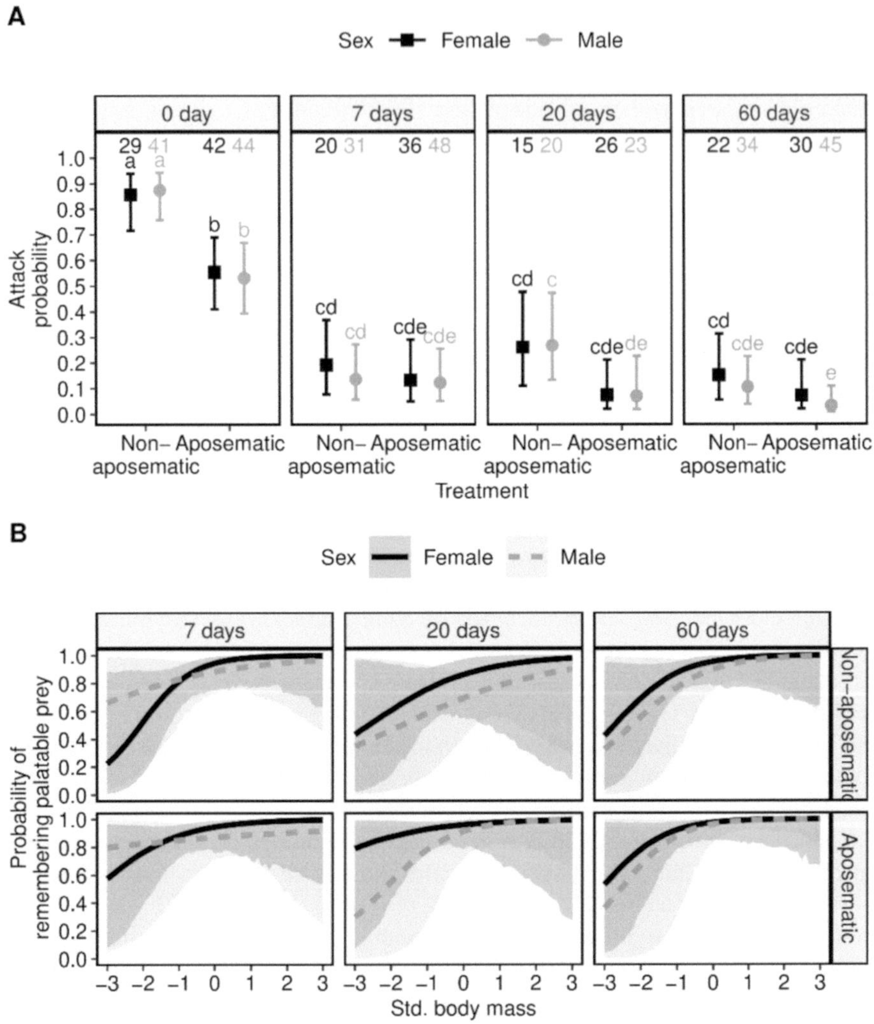

Fig. 11.6 Comparisons of lizard responses to noxious prey at learning phase (0 day) after different memory periods (7/20/60 days) by prey color (**a**) and lizard body mass (**b**). Lizards exposed to noxious red crickets were significantly more likely to avoid the colored cricket in the next trial than those exposed to noxious green crickets. Lizards with higher body mass were also less likely to attack colored crickets upon reexposure. The "0 day" sample indicates the proportion of lizards that initially attacked unpalatable crickets, and the memory retention samples show the proportion of lizards which learned aversion in the initial trial but then attacked colored crickets in the subsequent memory retention test. The lowercase letters indicate the significant differences based on multiple comparisons. Numbers above bars indicate the sample sizes of tested lizards. (Redrawn from Ko et al. (2020), Animal Behaviour)

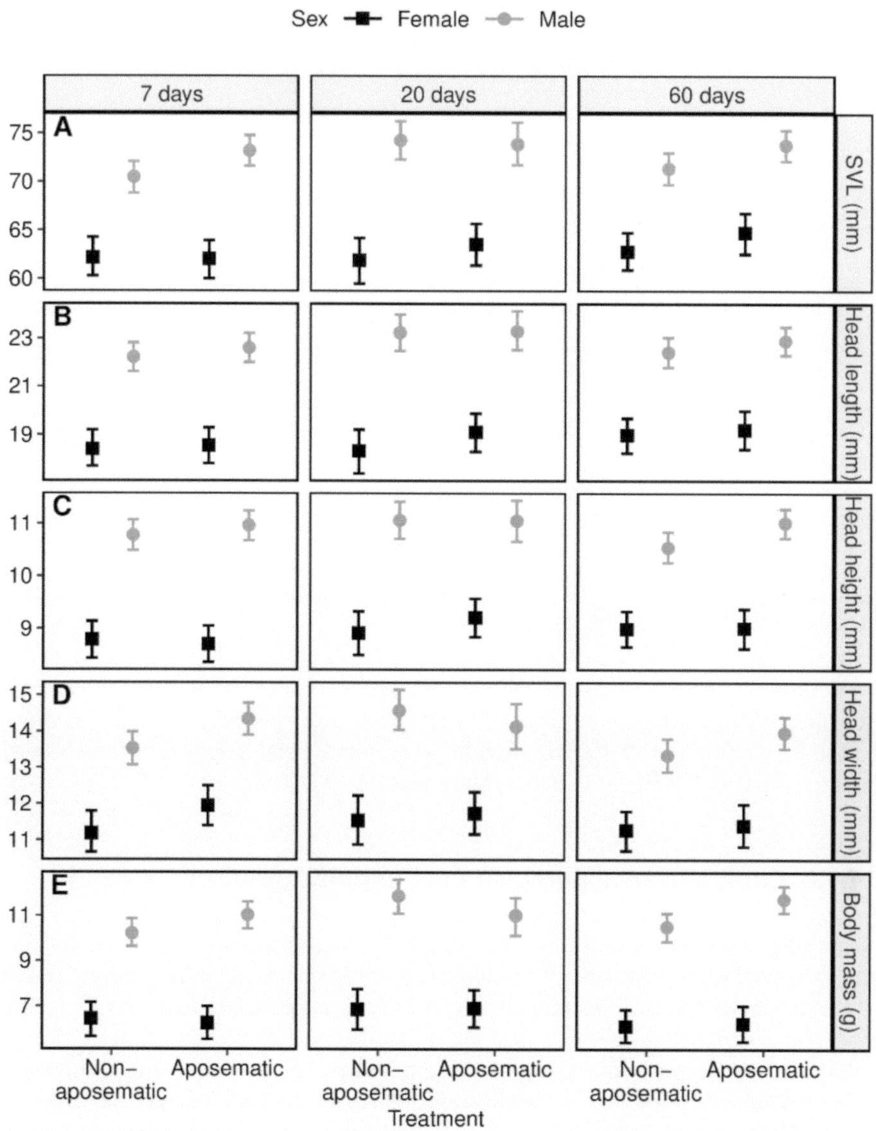

Fig. 11.7 Comparisons of various morphological traits. (**a**) SVL; (**b**) head length; (**c**) head height; (**d**) head width; (**e**) body mass of lizards tested in memory experiments. (Redrawn from Ko et al. (2020), Animal Behaviour)

than females (OR = 3.567, Z = 2.968, $p = 0.003$; Fig. 11.8), but was no significant difference between the green and red groups (OR = 1.301, Z = 1.622, $p = 0.469$; Fig. 11.8).

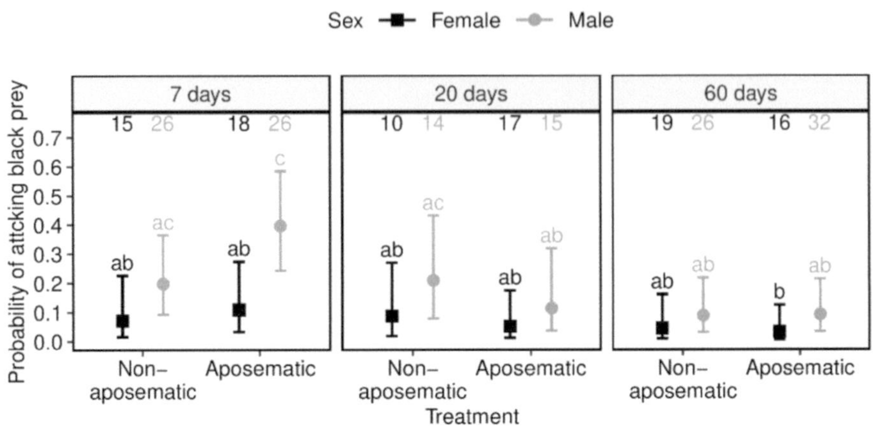

Fig. 11.8 The proportion of lizards that avoided black crickets after exposure to noxious colored crickets compared by latency to reexposure and sex. Lizards in the 7-day group were more likely to attack the black cricket after ignoring the colored one than those in the 60-day group and male lizards were more inclined to attack the black cricket than female lizards, but generalization did not differ between green and red color groups. The lowercase letters indicate the significant differences based on multiple comparisons. Numbers above bars indicate the sample sizes of tested lizards. (Redrawn from Ko et al. (2020), Animal Behaviour)

Neither interaction terms nor morphological covariates were significant in terms of generalization (sex prey color, $p = 0.401$; sex memory period, $p = 0.783$; memory period prey color, $p = 0.060$; head width, $p = 0.860$; head height, $p = 0.351$; head length, $p = 0.947$; SVL, $p = 0.096$; body mass, $p = 0.319$).

11.1.6 Color Distance Between Prey and Background

According to the Kruskal-Wallis rank sum test, the color distances between crickets and powders were different ($\chi^2 = 24.821$, $p < 0.001$; Table 11.1). Figure 11.1b illustrates the reflectance spectra of crickets and food powders used in this study. Post hoc comparisons revealed that the dorsal-ventral of crickets showed similar color distance from woody litter and green powders, respectively, but significantly lower than that of black and red powders (Table 11.1). A single-sample single-tailed exact symbol examination further demonstrated that all types of powders, except crickets, were significantly larger than the 2-JND threshold (Table 11.1). For the color distance of crickets, the red and black powders were significantly larger than 2, but not the green powder (Table 11.1). These findings indicate that the lizards in this study may be capable of visually discriminating between the red/green/black powder and backgrounds, but in terms of lizard chromatic function, the red and black powders may be stronger signals compared to the green powders and local crickets.

Table 11.1 Lizard tetrachromatic color distances among cricket and color powder used in this research. Color distances (JND ± SD) and p-values of one-sample one-tailed exact sign tests to test JND greater than 2 are presented in the lower and upper triangles, respectively

	Cricket	Red powder	Green powder	Black powder	Woody litter
Cricket ($n = 8$)	–	<0.001	0.322	<0.001	0.844
Red powder ($n = 8$)	6.21 ± 1.56	–	<0.001	<0.001	0.004
Green power ($n = 8$)	2.05 ± 0.60	7.23 ± 1.58	–	<0.001	0.004
Black power ($n = 8$)	3.26 ± 0.95	6.53 ± 1.39	4.33 ± 0.83	–	0.004
Woody litter ($n = 1$)	1.82 ± 0.76	6.13 ± 1.60	2.22 ± 0.16	4.97 ± 0.65	–

11.1.7 Discussion

Our findings suggest that different *chromatic* signals affect lizards' *ability to recognize* unpalatable prey. In our initial experiments, lizards in the wild exhibited longer attack delays and lower attack probabilities against red-dyed crickets compared with green- and black-dyed ones. Following an initial encounter with quinine-treated dyed prey, lizards that tasted red crickets (but not green crickets) had lower attack rates on these crickets, with the avoidance persisted for up to 60 days. These patterns support the interpretation that red is a more aposematic color than green for lizards. Consistent with this behavioral pattern, our visual modeling indicates that in a woody substrate context, red is more recognizable than green, potentially enhancing the more robust memory formation and discrimination and the ability to differentiate red prey from similar, non-spontaneous prey.

Our second experiment demonstrated that *D. swinhonis* possesses a remarkable capacity for long-term memory, retaining avoidance of unpalatable prey for at least 60 days following a single aversion experience. This learned aversion could be displayed the generalization to the control (black) cricket. In the memory experiment, Cricket coloration is of low importance. While lizards trained with red crickets displayed marginally higher propensity for avoidance upon retesting many days later, those trained with green crickets also exhibited strong avoidance at retested. This indicated a greater overall reduction in attack probability due to lizards did not initially avoid green crickets. The generalization of aversion to the control color (black) was not influenced by the color of the cricket. These findings are discussed in greater depth below.

11.1.8 Memory

D. swinhonis can maintain long-term memory of coloration avoidance associated with unpalatable prey (including red and green) for at least 60 days after an aversive encounter, and we discovered that this effect is slightly stronger in red crickets than in green crickets. Consistent with in garter snakes (*Thamnophis radix*), where obvious color prey (yellow and black) facilitated stronger aversion learning to toxic prey than unknown color prey (green) (Terrick et al., 1995). Our findings suggest that *D. swinhonis* currently exhibits a unique level of memory retention for aposematic coloration. In garter snakes (*Thamnophis radix*), for example, the aversion was induced through linking to prey color signals and discomfort caused by lithium chloride injections, such as inducing aversion vanished after 22 days (Burghardt et al., 1973; Terrick et al., 1995). In contrast, wild-caught jungle crows (*Corvus macrorhynchos*) are capable of retaining Wild-caught jungle crows' color discrimination tasks for at least 10 months (Bogale et al., 2012). Because the 60 days were the longest memory period during our study, this memory may be retained longer than the period.

In a previous study of *D. swinhonis*, the lizard exhibited strong avoidance of weevil after attacking the hard elytra of the weevil initially (Tseng et al., 2014), but this avoidance persisted for only 23 days, less than half of the 60-day memory period we recorded here. This difference suggests that the memory retention may be based on the type or degree of defensive signal of the prey, as quinine and mustard-treated crickets represent direct toxicity in the current study, whereas hard-shelled weevils may only represent a foraging obstacle rather than a lethal risk. The capacity to retain information about toxic prey is clearly an adaptation of predatory lizard survival and reproduction. Following targeted aversive conditioning against invasive cane toads, the lizard, *Varanus panoptes* survived 482 days longer than untrained lizards in toad-invaded areas (Ward-Fear et al., 2016).

Toxin levels may also play a role in mediating this effect. The tendency of European starlings (*Sturnus vulgaris*) to attack toxic prey is determined by their energy requirements as well as the amount of toxins in the prey. Because of the increased toxin burden, birds stop eating toxic prey earlier (Barnett et al., 2011). The toxicity of quinine-treated crickets in our study may have far exceeded the acceptance threshold for lizards, as most individuals refused to attack and avoid colored crickets after only one encounter.

11.1.9 Color Preference of *D. swinhonis*

Adult *D. swinhonis* in the wild exhibited more tendency to avoid red (aposematic) colored prey compared with green prey in our initial experiments. Lizards can avoid colored prey via three different mechanisms: (1) innate avoidance, (2) learning and generalization, and (3) neophobia. Several previous studies have shown that in a laboratory setting, innate aversion in certain species can be evoked by specific colors.

Naive quail chicks (*Coturnix coturnix*), for example, preferentially attack blue-black rather than orange-blue prey (Pegram & Rutowski, 2014). Likewise, inexperienced blue tits (*Cyanistes caeruleus*) and coal tits (*Periparus ater*) avoid attacking colored fireworms (Exnerová et al., 2007). In contrast, Lee et al. (2018) discovered that when *D. swinhonis* attacks on inedible weevils, prey body shape had a greater effect than color pattern. Although neophobia has been documented in predators, particularly in naive predators, encountering novel prey types (Coppinger, 1970; Marples et al., 1998), it is unlikely that the aversion to red crickets in our lizards is attributable to neophobia.

It is possible that our lizards had prior exposure to prey with various colors, including red insects, such as red soapberry, which is very abundant at our collection sites. Furthermore, due to an unusual bright and uniform coloration that does not naturally occur in wild prey, all of our dyed crickets are equally novel. Despite this, the lizards only exhibit some (but not complete) aversion to red crickets, but readily attack green and black.

Through learning and generalization, color avoidance, at least in avian studies, can be acquired (Kazemi et al., 2014). Colored fire worms are more frequently attacked by naive great tits (*Parus major*) and crowned tits (*Lophophanes cristatus*) than by wild-caught adults of the same species, but they have also learned to color avoidance through experience (Exnerová et al., 2007). In our case, the wild-caught lizards were most likely exposed to difficult red prey prior to our experiments. Soapberry bugs (*Leptocoris augur*) are abundant in the areas where we captured lizards, and they have an aversive coloration (distinct red coloration) as well as toxic defenses (toxic odor) (Ho et al., 2006). We discovered that colored prey induced a longer attack latency than non-colored prey and controls in wild-caught *D. swinhonis*, implying that red elicits lizard vigilance more effectively than green (Gittleman & Harvey, 1980; Gittleman et al., 1980; Roper & Redston, 1987). In previous research on scaled lizards, attack latency is extensively used to measure predatory behavior (reviewed in Cooper & Burghardt, 1990). Aside from cricket coloration, we discovered a moderate effect of body size (SVL), with larger, older lizards thought to be more likely to attack colored crickets. We evaluated several variables, including sex, SVL, head length, head width, head height, and body size, as well as first-order interactions between morphological characteristics and handling, and discovered that SVL showed a sole significant association with attack latency. Other vertebrates, such as adult great tits, the preference for black- and yellow-striped prey was weaker than juvenile individuals (Lindström et al., 2001). Similarly, inexperienced juvenile blackcaps (*Sylvia atricapilla*) favored red artificial fruits over blue, green, yellow, and white, but this preference was absent in wild-caught adults, possibly due to prior experience (Schmidt & Schaefer, 2004). These studies suggest that there is influence of both age and experience on cognition and color preference in birds. Consistent with this pattern, there was an effect of age in our study, with older lizards exhibiting weaker discriminations of colored (red or green) crickets. With time, lizards may develop a more flexible foraging strategy due to learn that the palatability of colored prey varies in nature.

11.1.10 Summarization

When experienced predators are exposed to a new aversive stimulus, avoidance can extend beyond the initially toxic prey through stimulus generalization (Darst & Cummings, 2006). Color and shape can both be effective stimuli for generalization. Predatory birds, for example, easily generalize their experience and avoidancewith toxic prey to prey with similar signals (Ruxton et al., 2004). In our study, lizards exposed to aposematic red prey exhibited greater memory and recognition than those exposed to green prey, but the generalizations of aversion to black prey were similar between both groups. Previous research on gecko generalization discovered a similar pattern. Nocturnal downy geckos (*Amalosia lesueurii*) respond to chemicals emitted by both dangerous and non-dangerous snakes, possibly by generalizing from predatory snakes to other snakes (Webb et al., 2010). According to the authors, the generalized anti-predatory response is adaptive under the condition in which predators or their cues occur at an extreme frequency. Similarly, *D. swinhonis* may have a generalized tendency to avoid crickets with different colors because, compared to palatable prey, toxic prey is uncommon in its natural environment.

Furthermore, the extent of behavioral generalization appears to be less a function of the specific color and more closely related to the intensity of aversion felt or other cues. Generalization may require different cognitive mechanisms than memory recall or discrimination, and it is possible that the shape or overall appearance of aversive prey, rather than color, serves as the primary determinant of salience. Consistent with this interpretation, previous research on weevils found that *D. swinhonis* relied on overall shape and appearance rather than colored markings as the memory cues of aversive encounters (Lee et al., 2018).

We also discovered that the interval between encounters influenced the generalization: lizards in the 60-day group were more likely to exhibit generalization than those in the 7-day group (avoiding black crickets). This finding implies that color signal memory may become blurred over time, making coloration discrimination ability of lizards decline from previous experiences and more likely to promote avoidance of aversive stimulus (i.e., more likely to generalize) to all similar prey items (i.e., more likely to generalize).

11.1.11 Sex

Male and female lizards differed a little in their memory retention and generalization in this study. Males and females had the same proclivity to remember toxic prey regardless of prey color. Females, on the other hand, are more likely to exhibit aversive generalizations to black crickets after encounter to toxic prey in the generalization test. This divergence could be attributed to differences in territory and foraging behavior between males and females. Male *D. swinhonis* defend territories aggressively against other males, have larger home ranges, and prefer to forage from the habitat rather than the ground (Lin & Lu, 1982; Kuo et al., 2009), whereas females move safely across the territories of multiple males. Male lizards may be less risk-averse and exhibit more risky behavior toward new prey to balance the energetic costs of territorial behavior. Female lizards may have more opportunities to exposure a broader diversity prey types because they have a larger foraging range. As a result, female lizards may be more conservative on dietary, preferring familiar foods over unfamiliar or distasteful prey.

11.1.12 Conclusion

As major insectivorous predators in global ecosystems, lizards are an important consumer in terrestrial ecosystems as well as a primary selective force in arthropod behavior and morphology. Despite their widespread presence, little is known about their foraging, learning, and memory abilities. Despite numerous studies demonstrating that lizards can distinguish prey using chemical cues, most sitters rely on visual cues to identify and attack prey, and the role of visual signals in prey identification and discrimination remains comparatively underexplored. In many ways, our experiments indicate that lizards are ideal subjects for such studies: they can be high abundance, are easily tested in experimental settings, and exhibit a previously unknown ability associated with the memory and generalization of aversion. Their experimental operability provides new subjects for research into the costs and benefits of mimicry and imitation, which have previously been used as model systems for microevolutionary and behavioral genetics studies.

11.2 The Role of Different Visual Features in Weevils to Send Aposematic Signals to Sympatric Lizard Predators

Conspicuous or aposematic coloration are frequently used by prey to alert potential predators to their unpalatable status. Although most studies have focused on color, predators can identify unprofitable prey using other features, such as the shape or size of the markings, or general body shape and appearance. These more general characteristics may play a crucial role when predators co-occur with radiating species or populations with different colors or patterns; predators may use these species-consistent features to learn to avoid them. Predators are more likely to learn consistent features to avoid these species rather than learning multiple types of patterns independently.

In this study, we performed feeding trials using weevils (*Pachyrhynchus tobafolius*) bearing conspicuous blue spots as prey, and sympatric insect-eating lizards (*Diploderma swinhonis*) as predators. We experimentally altered spot number, size, shape, and color, as well as modifying the weevils' overall body appearance, to evaluate which traits most effectively convey warning signals. Weevils left unaltered were rarely attacked, and changes to spot presence, color, or size did not significantly affect predator responses. In contrast, prey whose physical appearance was manipulated experienced higher attack frequencies and were targeted more rapidly. These findings indicate that body shape and overall appearance play a greater role than bright spot coloration in deterring lizard predators. This mechanism may help explain the pronounced diversity of colors, stripes, and spot patterns observed among weevil species within the same habitat.

Aposematism is an antipredator mechanism that uses highly visible or distinctive signals to advertise their low profitability to predators (Ruxton et al., 2004). This form of signaling has evolved repeatedly across diverse animal groups, including mollusks (Gosliner, 2001), insects (HotovVá Svádová et al., 2010), amphibians (Toledo & Haddad, 2009), reptiles (Kikuchi & Pfennig, 2010), and mammals (Caro, 2009). Visually conspicuous colors and patterns are advantageous because not only are they readily detected but also they promote predator learning and memory of aversive responses (Terrick et al., 1995; Servedio, 2000). As a result, many predators that rely on vision, such as insects, birds, and lizards, learn to associate distinct signals with aversive or defensive stimuli and subsequently avoid attacking aposematic prey (Exnerová et al., 2007; Prudic et al., 2007; Tseng et al., 2014).

Long-wavelength colors, including red, orange, and yellow, are the most frequently observed aposematic colors, largely because they remain relatively consistent across varying environmental conditions and provide strong chromatic contrast against typical backgrounds (Stevens & Ruxton, 2012; Arenas et al., 2014). Nevertheless, some defended species employ short-wavelength colors, such as green or blue, as part of their warning displays (Summers et al., 2003; Pegram et al., 2013; Fabricant & Smith, 2014). Well-known examples include strawberry poison frogs (*Dendrobates pumilio*), hibiscus hobbits (T*ectocoris diophthalmus*), and pipistrelle swallowtail butterflies (*Battus philenor*). These colors are often produced through structural mechanisms, including iridescence, which can generate pronounced luminance contrast (Doucet & Meadows, 2009; Seago et al., 2009; Umbers, 2013). Experimental studies involving avian and insect predators have shown that increased luminance contrast can improve the effectiveness of aposematic signals and accelerate aversion learning (Prudic et al., 2007; Fabricant et al., 2014).

Beyond color itself, additional properties of warning displays, such as signal size, pattern configuration, and symmetry, can substantially influence aposematic performance. Some evidence indicates that larger aposematic signals tend to be more effective in deterring predators. For instance, domestic chicks (*Gallus gallus domesticus*) acquire avoidance responses more rapidly when exposed to artificial prey bearing relatively large markings (Forsman & Merilaita, 1999). This study, as well as comparative analyses of rubber grasshoppers (Whitman & Vincent, 2008), red firebugs (Prokopová et al., 2010), and artificial cake prey (Remmel & Tammaru, 2011) have shown that both the relative size of warning signals and overall prey body size can strengthen aposematism effectiveness. Moreover, several studies report enhanced predator avoidance of prey displaying more symmetrical patterns [(Forsman & Merilaita, 1999; Forsman & Herrström, 2004); but see (Stevens et al., 2009)]. Beyond color and pattern, some insects have a suite of additional visual cues, including body shape, antennae, hair, legs, and behavior, that can contribute to both aposematism and mimicry (Kauppinen & Mappes, 2003; Dolenská et al., 2009; Veselý et al., 2013).

Pachyrhynchid weevils (Insecta: Coleoptera: Curculionidae) are distributed throughout islands of the Old World tropics, with the highest species diversity and endemism occurring in the Philippines (Schultze, 1923; Starr & Wang, 1992). Extreme body hardness renders pachyrhynchid weevils unprofitable to certain avian predators (Schultze, 1923), and their conspicuous markings have been proposed to function as aposematism advertising this defense (Welch et al., 2007). Experimental evidence supports this interpretation: Tseng et al. (2014) demonstrated that lizard predators from allopatric populations attacked weevils more readily when their bright markings were obscured with black marker, indicating that conspicuous coloration plays a role in aposematic signaling (Dolenská et al., 2009, Veselý et al., 2013).

The goal of this study was to determine the significance of specific features in *P. tobafolius* visual mimicry signals for recognition by a typical lizard predator. Swinhoe's tree lizards (*D. swinhonis*, Agamidae) were used in our study because they are captured from the wild and therefore possessed natural foraging experiences. They are foragers who sit and wait for prey, relying on visual cues (Huang, 2007). Lizards and weevils (*Pachyrhynchus tobafolius*) are sympatric and have been shown to avoid attacking weevils; learning from previous experience is assumed (Tseng et al., 2014). We systematically manipulated the number, size, color, pattern and appearance of *Pachyrhynchus tobafolius* in order to identify the traits most important for lizard cognition. Because the marker patterns of Green Island weevils differ, we predicted that lizard predators would use color and body shape to identify weevils as unpalatable prey.

Weevils were collected from their host plant, *Pipturus arborescens*, through hand or insect net on Green Island (Fig. 11.9). They were housed in plastic containers (13.5 cm × 9.5 cm × 8 cm) and provisioned with fresh *P. arborescens* leaves once every 2 days. The color pattern of this weevil comprises a black base with metallic green spots on the head, pronotum, elytra, and legs. To evaluate which visual

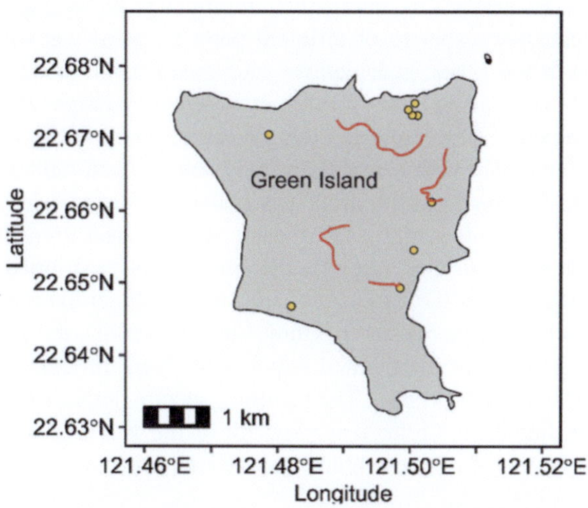

Fig. 11.9 Locations where we collected Swinhoe's tree lizards *Diploderma swinhonis)* and Pachyrhynchid weevils (*Pachyrhynchus tobafolius*) on Green Island, Taiwan. Yellow dots and red lines represent weevils and lizards, respectively. (Redrawn from Lee et al. (2018), Journal of Zoology)

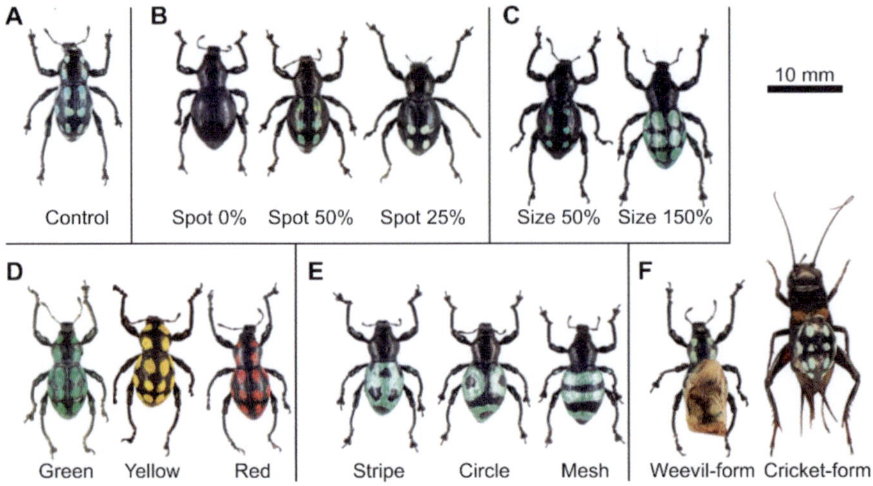

Fig. 11.10 Pachyrhynchid weevils (*Pachyrhynchus tobafolius*) and crickets (*Gryllus bimaculatus*) used in the experiments. (**a**) control treatment, unchanged weevils; (**b**) number of spots treatments retained 8, 4, and 0 spots; (**c**) size of spots treatments increased and decreased 50% area of spots; (**d**) color of spots treatments painted over spots with red, yellow, and green nontoxic acrylic paint; (**e**) pattern of spots treatment changed spots to stripes, circles, or mesh; (**f**) appearance of prey treatments attached a fore wing of crickets to weevils and a weevil elytra to crickets. (Adapted from Lee et al. (2018), Journal of Zoology, with permission)

components mediate lizard recognition of these weevils as unprofitable prey, we experimentally altered specific pattern attributes. Collected weevils were randomly assigned to either a control group or one of five treatment groups (Fig. 11.10) that manipulated: (1) number of spots; (2) size of spots; (3) color of spots; (4) pattern of spots; (5) appearance of weevils. Weevils in the control group retained their natural coloration and pattern.

To verify that our paint mixture was biologically realistic, we quantified the reflectance spectra of different parts of intact weevil patterns and compared them with the black, red, yellow, and green paints used. Spectral measurements were obtained using a Jaz spectrometer (detection range: 250–800 nm, Ocean Optics, Inc., Dunedin, FL) connected to a reflection probe (ZFQ-13101, Ocean Optics, Inc.) and illuminated with a deuterium-tungsten halogen light source (DH-2000-BAL, Ocean Optics, Inc.). Calibration was performed using a reflectance standard (WS-1-SL, Ocean Optics, Inc.). For each weevil measurement (N = 12, 3 patterns or backgrounds per weevil), the probe was positioned vertically 1 mm above the focal part of the cuticle and adjusted to obtain the maximal reflectance spectra. For each paint measurement (three sites per color), the probe was positioned vertically 1 mm above the painted paper. Data were processed using SpectraSuite software (Ocean Optics, Inc.). The resulting spectra demonstrated that the black paint closely resembled the reflectance spectra of the weevil's black body, and the green paint mixture closely resembled the metallic green spots (Fig. 11.11).

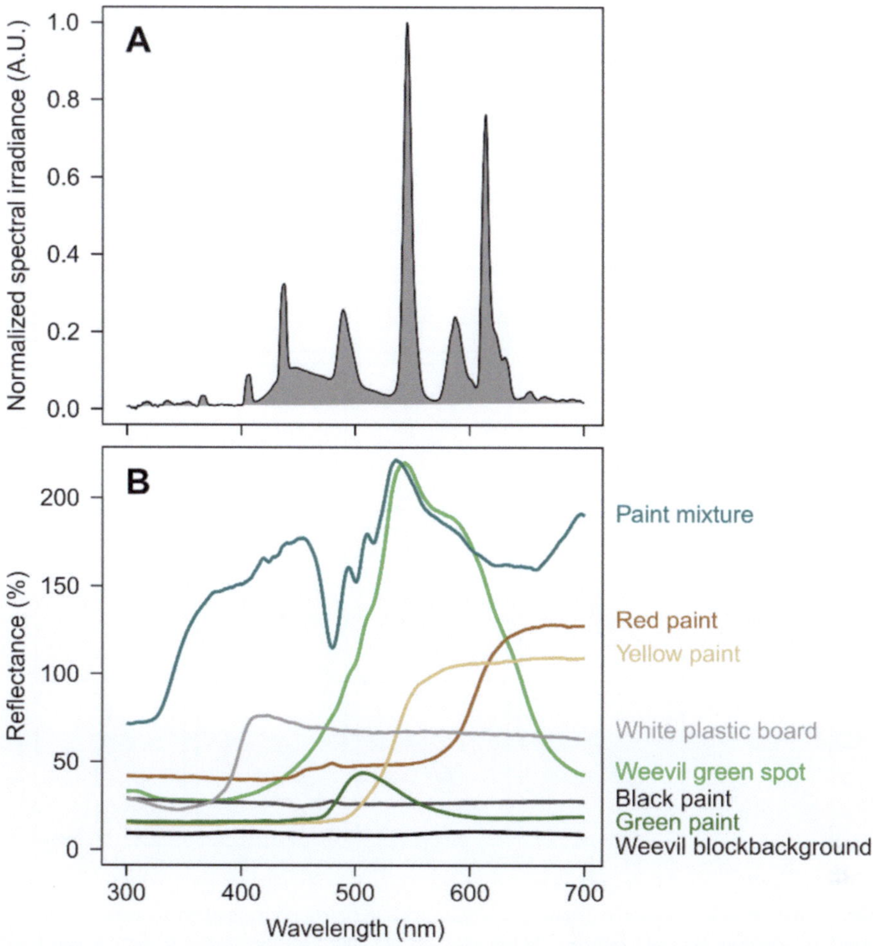

Fig. 11.11 The normalized irradiance spectra of the fluorescent lamb used in the experiments (**a**) and the reflectance spectrum of the bright spots and the black background of *Pachyrhynchus tobafolius*, the paint mixture to enlarge the spot, the black paints to eliminate/shrink the spot, and the red/yellow/green paints to alter the stimuli of spot (**b**). (Redrawn from Lee et al. (2018), Journal of Zoology)

Morphometric variables, such as head length, head width, head depth, snout-vent length (SVL), and body mass, were highly linearly correlated within each sex of lizards (all of Pearson $r > 0.741$; Fig. 11.12). We conducted a principal component analysis (PCA) based on the correlation matrix of these measurements separately and then extracted the first principal component (PC1) for males and females, respectively. PC1 was highly correlated with each body size measurement with high Pearson correlation coefficients (all $r > 0.894$; Fig. 11.12), indicating that it effectively represented body size and justifying its use as a standardized size index.

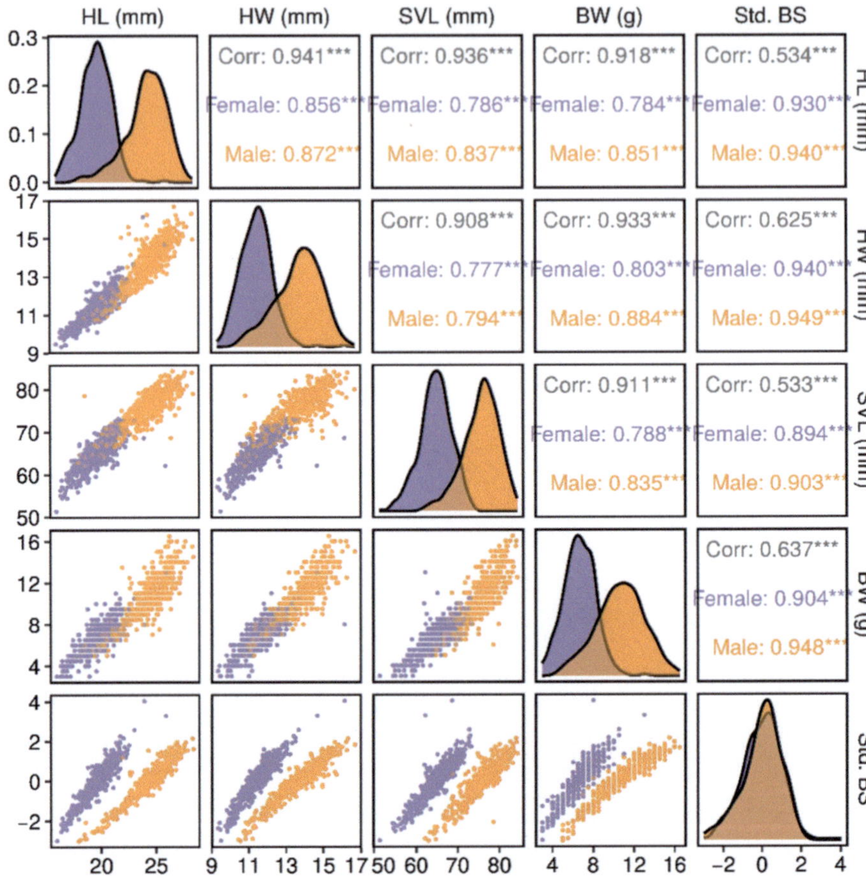

Fig. 11.12 Highly correlated lizard body size measurements were applied to extract the first principal component as standardized body size. Orange and blue dots represent female and male lizard measurements, respectively. Pearson coefficients between measurements are labeled. Abbreviations: *HL* head length, *HW* head width, *SVL* snout-vent length, *BW* body weight, *Std. BS* standardized body size. (Redrawn from Lee et al. (2018), Journal of Zoology)

Of the 1220 lizards initially tested, 373 individuals (30.6%) were excluded from subsequent analyses because they failed to attack the crickets after presentation of the experimental prey.

11.2.1 Aggressive Behavior

Handling, sex, and body weight were found to be significant predictors of attack behavior in lizards (Table 11.2), confirming that male lizards attacked prey significantly more frequently than females (Wald $\chi^2 = 12.126$, df $= 1$, $p < 0.001$), and that larger lizards attacked less frequently than smaller ones (Wald $\chi^2 = 4.775$, df $= 1$, $p = 0.029$).

Table 11.2 Interaction terms were eliminated by comparison to the best-fit logistic model (no interaction) after a forward model selection procedure

Interaction	BIC	ΔBIC	$\Delta\chi^2$	ΔDF	*p*
No interaction	740.73	0	0	0	–
Std. body size × treatments	797.57	56.83	8.00	13	0.843
Std. body size × sex	744.22	3.48	0.01	1	0.927
Sex × treatments	803.70	62.97	20.34	13	0.087

No significant difference in attack frequency was detected between control prey and prey with spots had been removed (spot 50%, $p = 0.147$, odds ratio = 1.584, CI95 = 0.87–3.13; spot 25%, $p = 0.192$, odds ratio = 1.229, CI95 = 0.90–1.68; spot 0, $p = 0.971$, odds ratio = 1.004, CI95 = 0.78–1.25; Table 11.3).

The increase and decrease in spot size did not affect the percentage of lizards attacking prey in the experiment (size 150%, P = 0.707, geometric ratio = 1.032, CI95 = 0.87–1.21; size 50%, P = 0.154, geometric ratio = 0.877, CI95 = 0.71–1.03; Table 11.3).

Modifying the color of the spots also did not affect the proportion of lizards attacking prey (red, $p = 0.096$, geometric ratio = 1.095, CI95 = 0.98–1.21; yellow, $p = 0.736$, geometric ratio = 1.018, CI95 = 0.91–1.12; green, $p = 0.332$, geometric ratio = 0.948, CI95 = 0.84–1.05; Table (Table 11.3).

The proportion of lizards that attacked prey did not differ significantly among control prey, prey with circular patterns, and prey with reticulated patterns. However, the attack rate was significantly higher for striped-patterned weevils than for control weevils ($p = 0.006$, odds ratio = 1.085, CI95 = 1.02–1.15; Table 11.3; Fig. 11.13).

Both weevils and crickets with modified appearance were attacked more frequently than the control weevils (weevil type, $p < 0.001$, odds ratio = 1.125, CI95 = 1.07–1.18; cricket type, $p < 0.001$, odds ratio = 1.227, CI95 = 1.18–1.28; Table 11.3; Fig. 11.13).

11.2.2 Observation Results

Handling, sex, and body weight were all factors in the final Cox hazard regression model that explained the latency of lizard attacks on prey (Table 11.4). Female lizards hesitated longer than males (Wald $\chi2 = 9.545$, df = 1, $p = 0.002$; Table 11.5a; Fig. 11.14), and lizards with larger body masses hesitated longer than lizards with smaller body masses (Wald $\chi2 = 4.100$, df = 1, $p = 0.043$; Table 11.5a).

Lizards attacked *P. tobafolius* individuals with similar latency for all different numbers of spots (spot 50%, $p = 0.140$; spot 25%, $p = 0.233$; spot 0%, $p = 0.920$; Table 11.5b). The size of the spots had no effect on attack latency (size 150%, $p = 0.662$; size 50%, $p = 0.168$; Table 11.5b). For the spot color group, the time to attack prey did not differ between the three colors and control prey (red, $p = 0.087$;

Table 11.3 Results of the best-fit logistic model explaining the probability of prey attacked by lizards (A) and the consequent post hoc comparisons among treatments (B). The intercept term represents the log odds of the reference levels: sex = female and treatment = control

Comparison or covariate	Estimate	SE	Odd ratio (95% C.I.)	*P*
(A)				
Intercept	−2.825	0.509	–	<0.001
Std. body size	−0.207	0.098	0.813 (0.670, 0.986)	0.035
Sex (male vs. female)	0.541	0.198	1.718 (1.167, 2.549)	0.006
Treatments[a]	–	–	–	<0.001
(B)				
Among control and spots[a]	–	–	–	0.331
Spot 50% vs. control	0.888	0.611	2.431 (0.772, 8.800)	0.131
Spot 25% vs. control	0.983	0.603	2.673 (0.868, 9.570)	0.088
Spot 0% vs. control	0.596	0.633	1.815 (0.542, 6.767)	0.336
Among control and sizes[a]	–	–	–	0.299
Size 150% vs. control	0.733	0.611	2.081 (0.661, 7.524)	0.215
Size 50% vs. control	−0.080	0.699	0.923 (0.227, 3.759)	0.908
Among control and colors[a]	–	–	–	0.174
Red vs. control	1.105	0.596	3.019 (0.998, 10.710)	0.051
Yellow vs. control	0.747	0.622	2.111 (0.651, 7.759)	0.216
Green vs. control	0.181	0.670	1.198 (0.321, 4.697)	0.785
Among control and shapes[a]	–	–	–	0.013
Circle vs. control	−0.290	0.743	0.749 (0.160, 3.228)	0.649[b]
Stripe vs. control	1.332	0.576	3.788 (1.320, 13.061)	0.037[b]
Mesh vs. control	0.383	0.649	1.467 (0.418, 5.582)	0.659[b]
Circle vs. stripe	−1.622	0.630	0.198 (0.049, 0.611)	0.023[b]
Circle vs. mesh	−0.673	0.696	0.510 (0.116, 1.907)	0.481[b]
Stripe vs. mesh	0.949	0.514	2.583 (0.985, 7.480)	0.108[b]
Among control and forms[a]	–	–	–	<0.001
Weevil vs. control	2.015	0.564	7.503 (2.706, 25.453)	<0.001[b]
Cricket vs. control	3.403	0.571	30.064 (10.788, 103.589)	<0.001[b]
Weevil vs. cricket	−1.388	0.392	0.250 (0.114, 0.527)	<0.001[b]

[a]Joint hypothesis tested by using a likelihood ratio test
[b]*p*-value adjusted by using the Benjamini-Hochberg procedure to control false discovery rate

yellow, $p = 0.725$; green, $p = 0.366$; Table 11.5b). Lizards were less hesitant to attack striped prey ($p = 0.006$); however, there was no difference in latency to attack control, round-spotted, and reticulated prey (round-spotted, $p = 0.136$; reticulated, $p = 0.638$; Table 11.5b; Fig. 11.14). The latency to attack weevil-type prey and cricket-type prey differed significantly from control prey, respectively (weevil-type, $p < 0.001$; cricket-type, $p < 0.001$; Table 11.5b; Fig. 11.14).

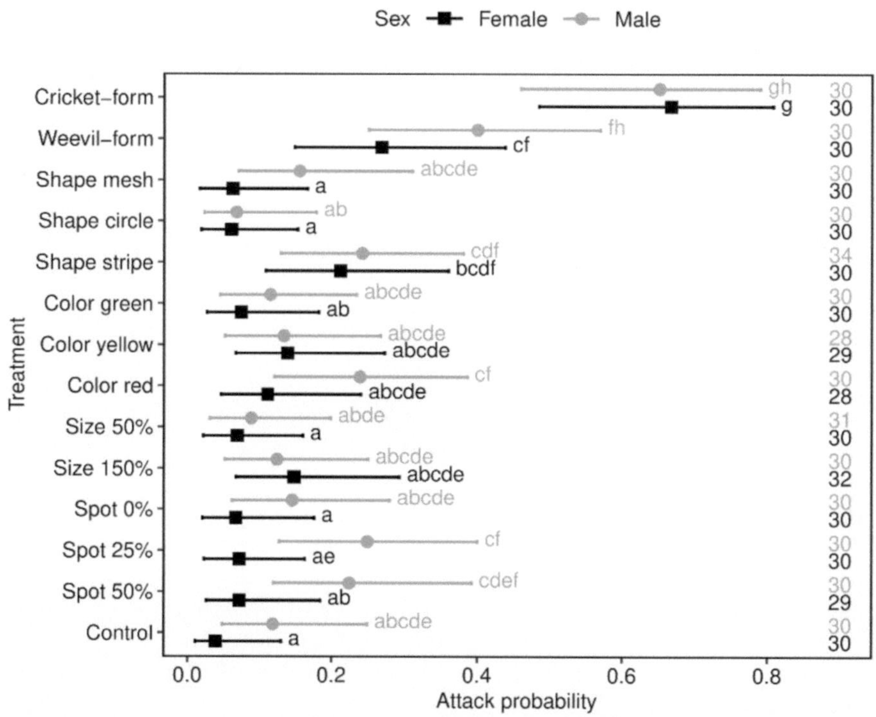

Fig. 11.13 The percentage of male and female lizards that exhibited attack behavior toward controls and prey with altered appearance. Striped patterns, weevil-form, and cricket-form prey were attacked with higher percentages than the control weevils. The lowercase letters indicate the significant differences based on multiple comparisons. Numbers indicate the sample sizes. (Redrawn from Lee et al. (2018), Journal of Zoology)

Table 11.4 Interaction terms were eliminated by comparison to the best-fit exponential-distributed survival regression model (no interaction) after a forward model selection procedure

Interaction	BIC	ΔBIC	$\Delta\chi^2$	ΔDF	p
No interaction	2388.9	0	0	0	–
Std. body size × treatments	2467.6	78.67	8.88	13	0.782
Std. body size × sex	2395.7	6.73	0.01	1	0.970
Sex × treatments	2454.6	65.65	21.90	13	0.057

11.2.3 Discussion

We show that male lizards attack their prey more frequently and with less hesitation than females, and that larger lizards attack their prey more frequently and with more hesitation. This could be because male lizards exhibit strong territorial defensive behaviors over short distances, whereas large lizards have a competitive advantage

Table 11.5 Results of the best-fit exponential-distributed survival regression fitting latency of lizards to attack prey over feeding trial time (A) and the consequent post hoc comparisons among treatments (B). The intercept term represents the log-hazard rate of the reference levels: sex = female and treatment = control

Comparison or covariate	Estimate	SE	Hazard ratio (95% C.I.)	*p*
(A)				
Intercept	8.079	0.509	–	<0.001
Standardized body size	0.156	0.080	0.856 (0.731, 1.001)	0.052
Sex (male vs. female)	−0.406	0.164	1.500 (1.088, 2.069)	0.013
Treatments[a]	–	–	–	< 0.001
(B)				
Among control and spots[a]	–	–	–	0.375
Spot 50% vs. control	−0.924	0.601	2.519 (0.776, 8.18)	0.124
Spot 25% vs. control	−0.969	0.592	2.636 (0.827, 8.404)	0.101
Spot 0% vs. control	−0.642	0.627	1.900 (0.556, 6.493)	0.306
Among control and sizes[a]	–	–	–	0.245
Size 150% vs. control	−0.783	0.707	2.188 (0.673, 7.117)	0.193
Size 50% vs. control	0.063	0.584	0.939 (0.235, 3.755)	0.929
Among control and colors	–	–	–	0.165
Red vs. control	−1.124	0.584	3.076 (0.98, 9.662)	0.054
Yellow vs. control	−0.768	0.613	2.155 (0.648, 7.16)	0.210
Green vs. control	−0.225	0.671	1.252 (0.336, 4.662)	0.738
Among control and shapes[a]	–	–	–	0.010
Circle vs. control	0.303	0.764	0.738 (0.165, 3.300)	0.691[b]
Stripe vs. control	−1.321	0.563	3.746 (1.243, 11.293)	0.049[b]
Mesh vs. control	−0.383	0.646	1.466 (414, 5.195)	0.594[b]
Circle vs. stripe	1.624	0.633	0.197 (0.057, 0.681)	0.048[b]
Circle vs. mesh	0.686	0.707	0.504 (0.126, 2.014)	0.595[b]
Stripe vs. mesh	−0.938	0.483	2.556 (0.991, 6.588)	0.102[b]
Among control and forms[a]	–	–	–	<0.001
Weevil vs. control	−1.940	0.544	6.958 (2.398, 20.193)	<0.001[b]
Cricket vs. control	−2.895	0.523	18.075 (6.481, 50.41)	<0.001[b]
Weevil vs. cricket	0.955	0.263	0.385 (0.23, 0.645)	<0.001[b]

[a]Joint hypothesis tested by using a likelihood ratio test
[b]*p*-value adjusted by Westfall's logically constraint multiplicity adjustment to control false error rate

(Huang, 2007). Furthermore, field-caught lizards may have encountered weevils before, and their behavior supports the theory that males and larger lizards encounter more aposematic prey, resulting in hesitation to attack them.

Although all of the bright markings on *P. tobafolius* on Green Island possess metallic blue-green, our findings showed that weevils of different colors were not attacked more frequently than the control group prey, which can be explained in two ways. First, the lizards have attacked other red or yellow prey, such as the unpalatable *Menochilus sexmaculatus* (Coleoptera: Coccinellidae) and *Physopelta gutta*

(Hemiptera: Heteroptera) (Chao et al., 2008). Ladybugs and true bugs have been shown to have chemical defenses against predators, and their distinctive coloration can be attributed to aposematism (Dolenská et al., 2009; HotovVá Svádová et al., 2010). Second, lizard responses are influenced not only by obvious coloration but also by weevil visual signals. Several studies have found a significant effect of additional visual signals in mimic prey (Kauppinen & Mappes, 2003; Dolenská et al., 2009; Veselý et al., 2013).

According to field observations, some *P. tobafolius* spots have worn out and become indistinct in the field. Several studies have suggested that spots are a key feature of avian predators used to identify mimics, and thus large warning signals should benefit mimics (Forsman & Merilaita, 1999; Prokopová et al., 2010; Remmel & Tammaru, 2011). Previous research, for example, has shown that the presence of spots increases ladybug protection (Prokopová et al., 2010). However, current findings indicate that differently sized spots have no effect on the warning effect. This could be because the spot sizes of the sympatric two other weevils, *P. sarcitis* and *Kashotonus multipunctatus*, are larger and smaller, respectively, than those of *P. tobafolius*, neither of which is attacked by lizards (Tseng et al., 2014). As a result, weevils were similarly attacked by lizards when all of the bright colors of *P. tobafolius* were masked. The small population of *P. tobafolius* on Green Island may be due to the scarcity of host plants (Starr & Wang, 1992). Thus, the response of *D. swinhonis* to these weevils may be influenced by prior experience with local prey communities or habitat overlap. Attack hesitation times in lizards were not significantly different between controls and unspotted prey, implying that body shape may provide additional cues that would be ignored in the absence of aposematic signals.

Simple pattern components like spots or stripes can help with detection and avoidance (Stevens & Ruxton, 2012). For example, studies on blue tits (*Cyanistes caeruleus*) have revealed that avoid prey with stripes more than those without (Aronsson & Gamberale-Stille, 2013). Similarly, research on dragonflies as predators has revealed that they avoid prey with black and yellow stripes, similar to wasps (Kauppinen & Mappes, 2003). In the current study, lizards attacked stripe-patterned weevils more frequently than control, circular, or reticulated prey, implying that spots or similar patterns (circles) may be sufficient to explain weevil discrimination.

We show that the general appearance of *P. tobafolius* is an important visual cue for predators to recognize them as unpalatable prey. Experiments with similar avian predators have revealed that when the whorls of ladybugs are removed, predators are willing to attack them (Dolenská et al., 2009). The appearance of *P. tobafolius* was altered, which increased the attack rate of Swinhoe tree lizards and decreased the attack latency period (Figs. 11.13 and 11.14). Furthermore, while the spot pattern of *P. tobafolius* remained, the lizard attacked the cricket-attack-elytra of *P. tobafolius* by more than 50%. As a result, the general appearance of *P. tobafolius* in antipredation is important. *P. tobafolius* has a distinctive body shape that is shared by most species, regardless of color or pattern(Starr & Wang, 1992). When this shape changes, predators become perplexed and frequently attack the weevil.

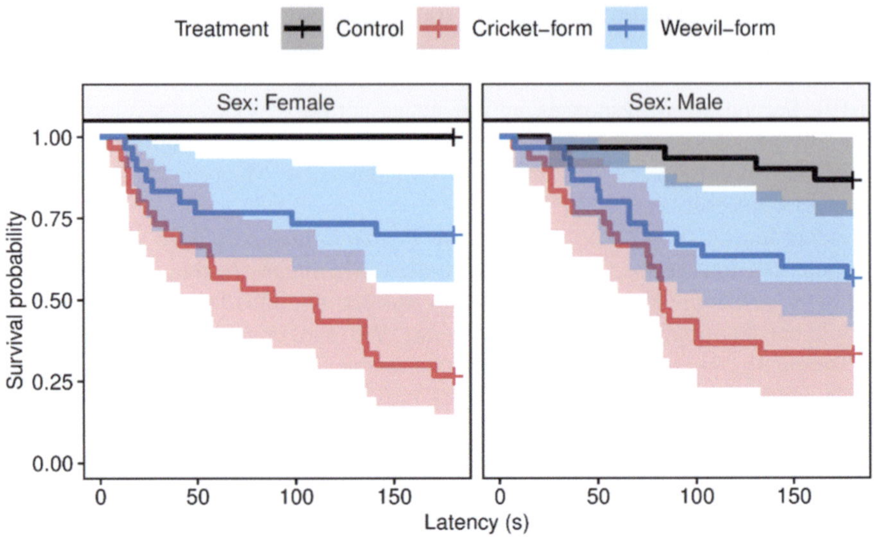

Fig. 11.14 Prey survival curves of female (left panel) and male (right panel) lizards as predators in control and treatments that differed from control. The attack latencies to the weevil-form and cricket-form prey were all shorter than the control prey, respectively. (Redrawn from Lee et al. (2018), Journal of Zoology)

11.3 The Functional Significance of Aposematic Colors: Geographic Variation in Responses to Colored Invertebrate Prey by a Wide Range of Lizard Predators

Conspicuous coloration is widely recognized as an antipredator strategy that signals prey unprofitability and reduces attack risk. Whether such warning signals are effective depends strongly on geographic overlap between predators and aposematic prey. We experimentally examined the role of conspicuous coloration in deterring visually hunting predators using Old World pachyrhynchid weevils (*Pachyrhynchus tobafolius* and *Kashotonus multipunctatus*) and Swinhoe's tree lizard (Swinhoe tree lizard; Agamidae, *Diploderma swinhonis*). In controlled encounters, sympatric lizards attacked weevils lacking conspicuous patterns more frequently than individuals with full patterns. In contrast, allopatric lizards attacked full-patterned weevils at higher rates than did sympatric lizards. The attack rate of sympatric lizards on masked weevils was also lower, implying that other characteristics of weevils (size/shape/odor) may aid in identification. Although initially more prone to attack, allopatric lizards rapidly learned to avoid weevils after a single encounter and retained this aversive response for over 3 weeks. Our results suggest that visual predators exhibit imperfect recognition of unpalatable prey, whether aposematic signals are present or they are experimentally altered. This limitation in predator perception may facilitate the persistence and diversification of aposematic and mimic strategies by exploiting biases and learning constraints within predator visual systems.

Predators that are able to identify and avoidprey items as unprofitable can achieve substantial fitness benefits. Accordingly, many poisonous or distasteful organisms have evolved conspicuous coloration that functions as an early-warning signal to deter predation (Smith, 1977), a phenomenon commonly referred to as aposematism. By recognizing such warning signals at an early stage, predators reduce unnecessary time and energy expenditure, while prey benefit from increased survival (Sillén-Tullberg, 1985; Kuchta et al., 2008) and predators avoid inefficient attack attempts (Ruxton et al., 2004). As a result, visually distinct signals have independently evolved across a wide array of animal groups, including Lepidoptera [e.g., moths (Lindstedt et al., 2008)], Coleoptera [e.g., ladybugs (Dolenská et al., 2009)], Hemiptera [e.g., true bugs (Gamberale-Stille & Tullberg, 1999, Veselý et al., 2013)], and Squamata [e.g., coral snakes (Kikuchi & Pfennig, 2010)], Anurans [e.g., poisonous frogs (Maan & Cummings, 2012)], and Teleostei [e.g., catfishes (Alexandrou et al., 2011)]. However, not all brightly colored organisms are aposematic. In many cases, conspicuous coloration serves alternative functions such as attracting prey or mates, or assessing the competitive ability of conspecific rivals (Endler, 1980; Kodric-Brown, 1998; Heiling et al., 2003; Tso et al., 2006; Cogliati et al., 2010). Clarifying the selective pressures underlying the evolution of bright patterns across taxa is therefore essential for understanding their functional significance.

Aposematic colors and patterns are generally readily detected, learned, and avoided by both vertebrate and invertebrate predators (Ruxton et al., 2004; Prudic et al., 2007; Kuchta et al., 2008), as demonstrated in laboratory (Langham, 2004; Ham et al., 2006; Exnerová et al., 2007; Prudic et al., 2007; Svádová et al., 2009) and field studies (Kapan, 2001; Borer et al., 2010). Color elements that stand out can help predators identify unpalatable prey (Terrick et al., 1995; Prudic et al., 2007). In particular, strong chromatic contrast and high brightness can accelerate avoidance learning and prolong aversive response memory retention (Gittleman & Harvey, 1980; Prudic et al., 2007). For instance, predators such as mantids maintain avoidance responses toward conspicuously colored prey for longer periods than toward covert prey (Prudic et al., 2007). Memory retention is influenced not only by visual salience but also by interactions among different sensory modalities within a signal (e.g., smell, shape, behavior) and the intensity of the aversive experience (Speed, 2000). These elements can hasten avoidance learning and memory retention. Given the complexity of memory formation, the duration of aversive responses may differ among predator taxa, yet empirical studies addressing long-term retention of aposematic signals remain limited (Speed, 2000; Ruxton et al., 2004).

Pachyrhynchid weevils (Insecta: Coleoptera: Curculionidae) are the most colorful group of Old World tropics island beetles (Wallace, 1895; Schultze, 1923). They frequently have brightly colored stripes, circles, and/or spots on the prothorax, dorsum, and legs that contrast with the dark body's high contrast (Fig. 11.15a). Alfred Russel Wallace proposed that weevils' bright colors function as a warning signal to predators (Wallace, 1889; Welch et al., 2007).

Fig. 11.15 Pachyrhynchid weevils used in our experiments. (**a**) *Pachyrhynchus tobafolius*; (**b**) *Kashotonus multipunctatus*; (**c**) and (**d**) show *P. tobafolius* and *K. multipunctatus*, respectively, with their aposematic markings experimentally masked with a black marker pen. (Adapted from Tseng et al. (2014), PLOS ONE, licensed under CC BY 4.0. Photos modified)

Because of their extremely hard exoskeletons, weevils may be difficult for some predators to eat [the chemical defenses of these insects are unknown (Schultze, 1923; Pasteels et al., 1983). Despite their prominence and diversity across the Old World tropics, the adaptive significance of weevils' bright colors remains unresolved after more than 120 years. Moreover, their color structure and iridescence mechanisms are still poorly understood (Welch et al., 2007; Seago et al., 2009). Many insects that co-occur with pachyrhynchid weevils, including other weevil genera (Schultze, 1923), other Coleoptera [e.g., longhorned beetles (Wallace, 1889; Linsley, 1959; Starr & Wang, 1992), and even Orthoptera (Wallace, 1889)], display similar colors and patterns, suggesting the possibility of mimicry. Many predators generalize avoidance responses from aposematic species to harmless mimics, as documented in taxa such as coral snakes (Pfennig et al., 2001) and monarch butterflies (Jeffords et al., 1979).

In this study, we experimentally examined the biological function and adaptive significance of pachyrhynchid weevils' conspicuous coloration. First, we investigated the aposematic function of bright patterns by comparing predator responses to conspicuously colored weevils with intact or experimentally masked colors (Fig. 11.15a, b).

Fig. 11.16 Swinhoe's tree lizards (*Diploderma swinhonis*) often inhabit the same trees with pachyrhynchid weevils, but they rarely attack weevils. (Redrawn from Tseng et al. (2014), PLOS ONE, licensed under CC BY 4.0. Photo modified)

Second, by comparing the attack frequencies between sympatric and allopatric predator populations, we looked for geographic variation in predator behaviors to conspicuously colored weevils (Fig. 11.16). Because the composition of individual weevil species varies between islands, widespread predator species may encounter different weevil species within their range, and some predator populations may have no exposure to weevils. We therefore predicted that allopatric predators would prey on weevils at a higher rate due to a lack of prior experience. Finally, we looked into whether allopatric predators were capable of learning to avoid weevils after initial encounter, as well as whether such avoidance response persisted over time. To provide a robust test of these predictions, we conducted controlled experiments with predator populations from various sources.

The majority of the 798 lizards tested attempted to consume the mealworms after presentation with a weevil. Cases where neither prey item was eaten were removed from analyses (2.6–6.3% of individuals in each population; n = 4, 15, 9, and 8 individuals).

11.3.1 Responses of Predators to Intact and Masked *P. tobafolius*

Only 26.0% (20/77) of sympatric lizards from Orchid Island attacked intact *P. tobafolius*, whereas 48.0% (36/75) attacked masked weevils ($\chi^2 = 7.92$, df = 1, $p = 0.01$) (Fig. 11.17a). Green Island sympatric lizards, on the other hand, demonstrated comparable attack rates ($\chi^2 = 0.01$, df = 1, $p = 0.92$) against intact (51.8%, 43/83) and masked (52.5%, 42/80) *P. tobafolius* (Fig. 11.17a).

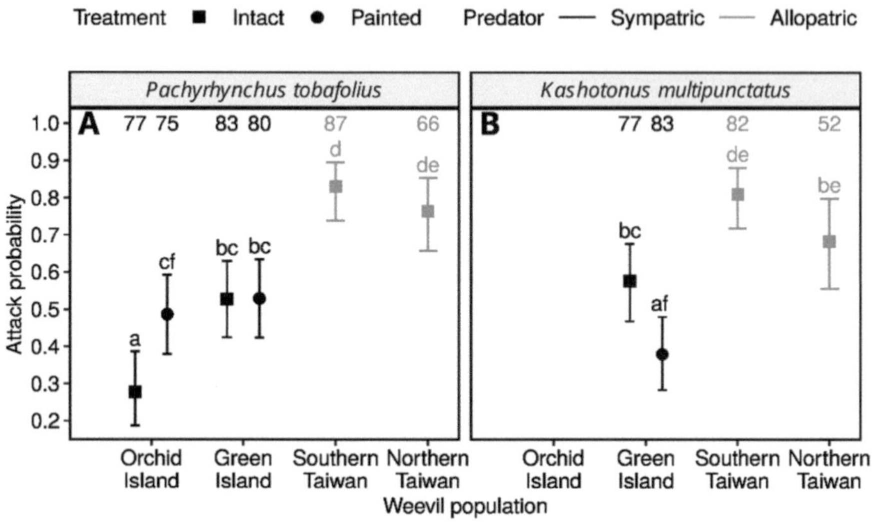

Fig. 11.17 The percentage of Swinhoe's tree lizards (*Diploderma swinhonis*) that exhibited different predatory behavior to two weevil species (presented by panels), shown for sympatric (black bars) and allopatric (gray bars) predator populations. Numbers above bars indicate the sample sizes. Different letters indicate statistically significant differences. Numbers above bars indicate the sample sizes. Note that *K. multipunctatus* does not occur on Orchid Island. (Redrawn from Tseng et al. (2014), PLOS ONE, licensed under CC BY 4.0)

11.3.2 Predatory Responses Toward Intact and Masked *K. multipunctatus*

Green Island sympatric lizards attacked masked *K. multipunctatus* more frequently (57.1%, 44/77) than intact *K. multipunctatus* (36.1%, 30/83) ($\chi^2 = 7.08$, df $= 1$, $p = 0.01$). The masked weevils were attacked at similar rates by Orchid Island lizards for *P. tobafolius* and those from Green Island lizards for *K. multipunctatus* (Fig. 11.17b).

11.3.3 Predation Responses of Sympatric and Allopatric Lizards to *P. tobafolius*

The attack rates of sympatric lizards from southern ($n = 87$) and northern Taiwan ($n = 66$) and sympatric lizards from Orchid Island ($n = 83$) were significantly higher than those from Orchid Island ($n = 77$) (southern Taiwan vs. northern Taiwan vs. Orchid Island, $\chi^2 = 37.41$, df $= 1$, $p < 0.01$; Green Island vs. Orchid Island, $\chi^2 = 11.17$, df $= 1$, $p < 0.01$) (Fig. 11.17a). Similar predation responses were observed between the two allopatric lizards from southern ($n = 82$) and northern Taiwan ($n = 52$).

11.3.4 Predation Responses of Sympatric and Allopatric Lizards to *K. multipunctatus*

The rates of attacks on *K. multipunctatus* were also higher in these allopatric populations than sympatric lizards on Green Island ($n = 83$) (southern Taiwan vs. Green Island, $\chi^2 = 37.43$, df $= 1$, $p < 0.01$; northern Taiwan vs. Green Island, $\chi^2 = 14.01, p < 0.01$). (Fig. 11.17b). Overall, most of the lizards that attacked weevils spat them out (97.1%), with only a few chewing, crushing, or eating the weevils (2.9%; $n = 13/444$ lizards that attacked weevils).

11.3.5 Memory Retention for Prey Avoidance

Conspecific lizards avoided eating subsequent weevils for up to 23 days after the initial weevil attack (Fig. 11.18). Individual lizards' propensity to reattack weevils after different time intervals showed no significant differences between treatments ($\chi^2 = 1.22$, df $= 3$, $p = 0.748$). Only 14.3% of lizards were willing to attack the weevil after 1 day, and this strong avoidance behavior was persisted for at least 23 days after the initial attack (6.7% attack rate; (Fig. 11.18).

11.3.6 Discussion

Our findings show that the conspicuous coloration of pachyrhynchid weevils can act as an aversive signal to reduce predation by sympatric predators. The conspicuous patterns of *P. tobafolius* and *K. multipunctatus* significantly reduced the attack rate of sympatric Swinhoe's tree lizards from Orchid Island and Green Island, respectively. Despite the fact that allopatric predators were more likely to initially attack fully colored weevils, the lizards quickly learned to avoid weevils that persisted for over 23 days after an encounter. These predators' (possibly learned) avoidance of prey can lead to effective mimicry of aposematic signals by other taxa. This is

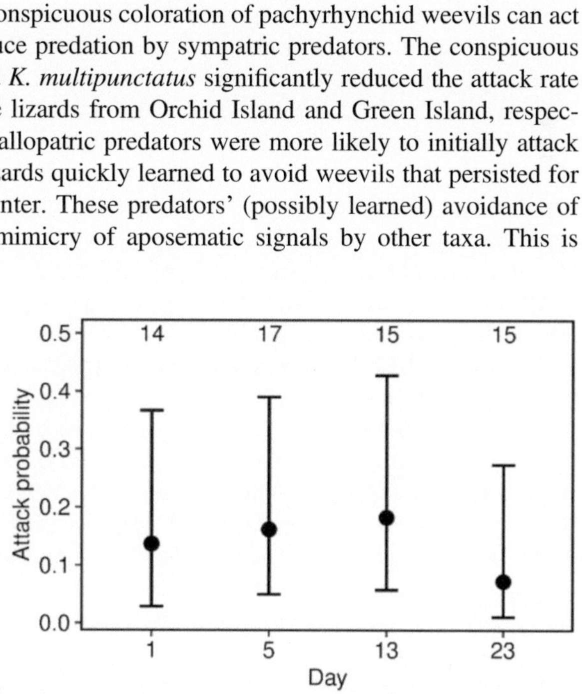

Fig. 11.18 Memory retention tests of the predatory responses of allopatric Swinhoe's tree lizards (*Diploderma swinhonis*) to colorful weevil prey (*K. multipunctatus*). Numbers above bars indicate the sample sizes. (Redrawn from Tseng et al. (2014), PLOS ONE, licensed under CC BY 4.0)

powerful evidence for the operation of the aposematism in nature, and it is consistent with other predators' responses to organisms with distinct coloration (Hasegawa & Taniguchi, 1994; Langham, 2004; Dolenská et al., 2009; Chouteau & Angers, 2011; Finkbeiner et al., 2012; Merrill et al., 2012; Veselý et al., 2013). The Old World tropics host a rich diversity of weevils as well as other taxa with similar appearances, such as other weevil genera (Schultze, 1923) and other insect families like Coleoptera and Orthoptera (Wallace, 1889, Linsley, 1959, Starr & Wang, 1992), which may gain protection from reduced predation by imitating the aposematic pattern of weevils.

Many mimic animals emit or possess chemicals such as alkaloids (Holloway et al., 1991), myristic acid (Reichstein et al., 1968), or formic acid (Kanehisa & Murase, 1977). However, previous research did not identify any chemical defense or secretion organs in pachyrhynchid weevils (Schultze, 1923, Pasteels et al., 1983). The hard exoskeleton of these weevils may be their primary defense mechanism (Wallace, 1889, Welch et al., 2007). In our study, only a few of the hundreds of lizards that attacked the weevils were willing to eat the weevils. Aposematism presents an evolutionary paradox, as conspicuous colors are thought to be a selective flaw that may enhance prey detectability and reducing fitness. Nonetheless, aposematism can be established when predators are capable of learning and remembering their association with unprofitable prey. The hard exoskeleton in our study system may have functioned as an effective defense and aided in the evolution of conspicuous coloration in weevils.

The variable responses in sympatric lizard to these weevils may be due to individual predators' prior experience, which may be shaped by geographic and habitat overlap, as well as local abundance (Endler & Rojas, 2009). On Green Island, for example, sympatric lizards exhibit similar attack rates to both intact and masked *P. tobafolius*. *P. tobafolius* was less abundant on Green Island than *K. multipunctatus*, possibly due to a lack of host plants (Starr & Wang, 1992). When prey abundance is low, predators like these lizards may have limited opportunities to encounter and learn specific mimicry patterns and thus attack unpalatable species. These types of "errors" may result in selection on recognition of co-evolved with pair signals, such as body size or shape,

Overall, sympatric lizard attack rates were significantly higher for masked weevils than for intact weevils, but these rates were lower than those of allopatric lizards attacking intact weevils (Fig. 11.17). These findings suggest that lizards can detect the disinterest of masked weevils using cues beyond conspicuous colors. Aposematic signal is not limited to coloration; it can also include complex patterns (Kikuchi & Pfennig, 2010; Aronsson & Gamberale-Stille, 2012), shapes (Dolenská et al., 2009; Hunter, 2009; Valkonen et al., 2011; Veselý et al., 2013), sizes (Pekár & Jarab, 2011), and even odors (Siddall & Marples, 2008). In our study, nearly half of the sympatric lizards (42.9–52.0%, depending on the species) avoided attacking masked weevils, indicating that predators may rely on multiple cues, such as physical shape, size, behavior, or scent, to facilitate avoidance in the absence of aposematic signals.

Our findings provide the first evidence of geographic variation in prey recognition in squamates. Similar patterns have been observed in other taxa, most notably avian predators (Noonan & Comeault, 2009; Chouteau & Angers, 2011). Different selective pressures on predators arising from different local prey communities may drive the geographic variation in prey recognition even in adjacent geographic areas (Mallet & Joron, 1999; Chouteau & Angers, 2011). Allopatric lizards were much more likely to attack aposematic weevils, because they may lack prior experience with these prey species' patterns. Similar occurrences are fairly common in many predator-prey systems (Mallet & Joron, 1999; Noonan & Comeault, 2009; Chouteau & Angers, 2011). Poison frogs, for example, show significant geographic variation in coloration, and local predators identify local aposematic morphs better than other geographically distant morphs (Noonan & Comeault, 2009; Chouteau & Angers, 2011). If aposematic species do not vary locally in color or pattern, local predators favor novel, unfamiliar phenotypes over the local aposematic morphs (Langham, 2004; Noonan & Comeault, 2009). However, these predators may exhibit avoidance of prey with novel aposematic morphs due to neophobia or dietary conservatism in other cases (Marples et al., 1998; Ruxton et al., 2004; Exnerová et al., 2007).

We found that a significant proportion of the sympatric lizards in our study attacked and ate their native weevil species. Incomplete predator avoidance behavior reflects a continuous learning/continuous testing process (Noonan & Comeault, 2009), poor learning/forgetfulness (Lynn, 2005), or a lack of prior experience (Saporito et al., 2007). Many organisms, particularly birds, can learn to avoid unprofitable prey by responding to new color signals (Gamberale-Stille & Tullberg, 1999; Ham et al., 2006; Svádová et al., 2009). However, research on avoidance learning in reptiles is still limited (Terrick et al., 1995). Swinhoe's tree lizards can successfully learn to avoid weevils after only one encounter, and this avoidance behavior can be maintained for at least 23 days.

Unlike many avian studies where memory tests last under 7 days (Gamberale-Stille & Tullberg, 1999; Ham et al., 2006) or garter snakes (*Thamnophis radix*) whose avoidance decreases over 22 days (Terrick et al., 1995), Swinhoe's tree lizards maintained a higher and more consistent rate of sustained prey avoidance behavior (Fig. 11.18). demonstrating that aposematic coloration not only influences immediate predation but also strengthens long-term predator memory. Because predators retained their ability to identify unpalatable prey through the end of our 23-day trial, weevil mimicry signals not only altered predator behavior but also assisted predators in forming strong and persistent memory associations.

Rat snakes, mice, and kukri snakes all eat sea turtle eggs, but on Orchid Island, only the latter enters the turtle nest to eat the eggs. The former two do not, and kukri snakes are not natural enemies of rat snakes and mice. So what are they afraid of? The next chapter will tell an interesting story.

References

Alexandrou, M. A., Oliveira, C., Maillard, M., McGill, R. A. R., Newton, J., Creer, S., & Taylor, M. I. (2011). Competition and phylogeny determine community structure in Müllerian co-mimics. *Nature, 469*, 84–88.

Arenas, L. M., Troscianko, J., & Stevens, M. (2014). Color contrast and stability as key elements for effective warning signals. *Frontiers in Ecology and Evolution, 2*, 25.

Aronsson, M., & Gamberale-Stille, G. (2012). Colour and pattern similarity in mimicry: Evidence for a hierarchical discriminative learning of different components. *Animal Behaviour, 84*, 881–887.

Aronsson, M., & Gamberale-Stille, G. (2013). Evidence of signaling benefits to contrasting internal color boundaries in warning coloration. *Behavioral Ecology, 24*, 349–354.

Barbour, H. R., Archer, M. A., Hart, N. S., Thomas, N., & Dunlop, S. A. (2002). Retinal characteristics of the ornate dragon lizard, *Ctenophorus ornatus*. *Journal of Comparative Neurology, 450*, 334–344.

Barnett, C. A., Skelhorn, J., Bateson, M., & Rowe, C. (2011). Educated predators make strategic decisions to eat defended prey according to their toxin content. *Behavioral Ecology, 23*, 418–424.

Bogale, B. A., Sugawara, S., Sakano, K., Tsuda, S., & Sugita, S. (2012). Long-term memory of color stimuli in the jungle crow (Corvus macrorhynchos). *Animal Cognition, 15*(2), 285–291.

Borer, M., Van Noort, T., Rahier, M., & Naisbit, R. E. (2010). Positive frequency-dependent selection on warning color in alpine leaf beetles. *Evolution, 64*, 3629–3633.

Bowers, D. M. (1992). The evolution of unpalatability and the cost of chemical defense in insects. In B. D. Roitberg & M. B. Isman (Eds.), *Insect chemical ecology: An evolutionary approach* (pp. 216–244). Chapman and Hall.

Britson, C. A., & Gutzke, W. H. N. (1993). Antipredator mechanisms of hatchling freshwater turtles. *Copeia, 1993*, 435–440.

Burghardt, G. M., Wilcoxon, H. C., & Czaplicki, J. A. (1973). Conditioning in garter snakes: Aversion to palatable prey induced by delayed illness. *Animal Learning & Behavior, 1*(4), 317–320.

Caro, T. (2009). Contrasting coloration in terrestrial mammals. *Philosophical Transactions of the Royal Society B: Biological Sciences, 364*, 537–548.

Chao, R. F., Lien, J. C., Hsu, H. C., Fu, C. M., Tsai, W. C., Lin, T. C., Cheng, S. F., & Chang, C. H. (2008). *The survey of terrestrial invertebrate fauna in Green Island*. I-Shou University.

Chouteau, M., & Angers, B. (2011). The role of predators in maintaining the geographic organization of aposematic signals. *The American Naturalist, 178*, 810–817.

Cogliati, K. M., Corkum, L. D., & Doucet, S. M. (2010). Bluegill coloration as a sexual ornament: Evidence from ontogeny, sexual dichromatism, and condition dependence. *Ethology, 116*, 416–428.

Cooper, J. W. E. (1994). Chemical discrimination by tongue-flicking in lizards: A review with hypotheses on its origin and its ecological and phylogenetic relationships. *Journal of Chemical Ecology, 20*, 439–487.

Cooper, J. W. E., & Burghardt, G. M. (1990). A comparative analysis of scoring methods for chemical discrimination of prey by squamate reptiles. *Journal of Chemical Ecology, 16*, 45–65.

Cooper, W. E., & Greenberg, N. (1992). Reptilian coloration and behavior. In C. Gans & D. Crews (Eds.), *Biology of the Reptilia: Vol. 18. Physiology E. Hormones, brain, and behavior* (pp. 298–422). London, UK: University of Chicago Press.

Coppinger, R. P. (1970). The effect of experience and novelty on avian feeding behavior with reference to the evolution of warning coloration in butterflies. II. Reactions of naive birds to novel insects. *The American Naturalist, 104*, 323–335.

Cott, H. B. (1940). *Adaptive coloration in animals*. Methuen.

Darst, C. R., & Cummings, M. E. (2006). Predator learning favours mimicry of a less-toxic model in poison frogs. *Nature, 440*, 208–211.

Dolenská, M., Nedvěd, O., Veselý, P., Tesařová, M., & Fuchs, R. (2009). What constitutes optical warning signals of ladybirds (Coleoptera: Coccinellidae) towards bird predators: Colour, pattern or general look? *Biological Journal of the Linnean Society, 98*, 234–242.

Doucet, S. M., & Meadows, M. G. (2009). Iridescence: A functional perspective. *Journal of the Royal Society Interface, 6*, S115–S132.

Dumbacher, J. P., Beehler, B. M., Spande, T. F., Garraffo, H. M., & Daly, J. W. (1992). Homobatrachotoxin in the genus Pitohui: Chemical defense in birds? *Science, 258*, 799.

Edmunds, M. (1974). *Defence in animals: A survey of anti-predator defences*. Longman Group Limited.

Eisner, T., & Grant, R. P. (1981). Toxicity, odor aversion, and "olfactory aposematism". *Science, 213*, 476.

Endler, J. A. (1980). Natural selection on color patterns in *Poecilia reticulata*. *Evolution, 34*, 76–91.

Endler, J. . A., & Mappes, J. (2004). Predator mixes and the conspicuousness of aposematic signals. *The American Naturalist, 163*, 532–547.

Endler, J. . A., & Rojas, B. (2009). The spatial pattern of natural selection when selection depends on experience. *The American Naturalist, 173*, E62–E78.

Evans, D. L., & Schmidt, J. O. (1990). *Insect defenses: Adaptive mechanisms and strategies of prey and predators*. State University of New York Press.

Exnerová, A., Štys, P., Fučíková, E., Veselá, S., Svádová, K., Prokopová, M., Jarošík, V., Fuchs, R., & Landová, E. (2007). Avoidance of aposematic prey in European tits (Paridae): Learned or innate? *Behavioral Ecology, 18*, 148–156.

Fabricant, S. A., & Smith, C. L. (2014). Is the hibiscus harlequin bug aposematic? The importance of testing multiple predators. *Ecology and Evolution, 4*, 113–120.

Fabricant, S. A., Exnerová, A., Ježová, D., & Štys, P. (2014). Scared by shiny? The value of iridescence in aposematic signalling of the hibiscus harlequin bug. *Animal Behaviour, 90*, 315–325.

Finkbeiner, S. D., Briscoe, A. D., & Reed, R. D. (2012). The benefit of being a social butterfly: Communal roosting deters predation. *Proceedings of the Royal Society B: Biological Sciences, 279*, 2769–2776.

Forsman, A., & Herrström, J. (2004). Asymmetry in size, shape, and color impairs the protective value of conspicuous color patterns. *Behavioral Ecology, 15*, 141–147.

Forsman, A., & Merilaita, S. (1999). Fearful symmetry: Pattern size and asymmetry affects aposematic signal efficacy. *Evolutionary Ecology, 13*, 131–140.

Gamberale-Stille, G., & Guilford, T. (2003). Contrast versus colour in aposematic signals. *Animal Behaviour, 65*, 1021–1026.

Gamberale-Stille, G., & Tullberg, B. S. (1999). Experienced chicks show biased avoidance of stronger signals: An experiment with natural colour variation in live aposematic prey. *Evolutionary Ecology, 13*, 579–589.

Gittleman, J. L., & Harvey, P. H. (1980). Why are distasteful prey not cryptic? *Nature, 286*, 149–150.

Gittleman, J. L., Harvey, P. H., & Greenwood, P. J. (1980). The evolution of conspicuous coloration: Some experiments in bad taste. *Animal Behaviour, 28*(3), 897–899.

Gosliner, T. M. (2001). Aposematic coloration and mimicry in opisthobranch mollusks: New phylogenetic and experimental data. *Bollettino Malacologico, 37*, 163–170.

Guilford, T. (1990). The evolution of aposematism. In D. L. Evans & J. O. Schmidt (Eds.), *Insect defenses: Adaptive mechanisms and strategies of prey and predators* (pp. 23–61). State University of New York Press.

Ham, A. D., Ihalainen, E., Lindström, L., & Mappes, J. (2006). Does colour matter? The importance of colour in avoidance learning, memorability and generalisation. *Behavioral Ecology and Sociobiology, 60*, 482–491.

Hasegawa, M., & Taniguchi, Y. (1994). Visual avoidance of a conspicuously colored carabid beetle *Dischissus mirandus* by the Lizard *Eumeces okadae*. *Journal of Ethology, 12*, 9–14.

Heiling, A. M., Herberstein, M. E., & Chittka, L. (2003). Crab-spiders manipulate flower signals. *Nature, 421*, 334–334.

Ho, H.-Y., Hsu, Y.-C., Ho, J.-Z., Lo, L.-W., & Chuang, Y.-C. (2006). Volatiles in the dorsal abdominal glands and exuviae of *Leptocoris abdominalis* and *Leptocoris augur* (Heteroptera: Rhopalidae). *Annals of the Entomological Society of America, 99*(5), 859–864.

Holloway, G. J., de Jong, P. W., Brakefield, P. M., & de Vos, H. (1991). Chemical defence in ladybird beetles (Coccinellidae). I. Distribution of coccineline and individual variation in defence in 7-spot ladybirds (*Coccinella septempunctala*). *Chemoecology, 2*, 7–14.

HotovVá Svádová, K., Exnerová, A., Kopečková, M., & Štys, P. (2010). Predator dependent mimetic complexes: Do passerine birds avoid central European red-and-black Heteroptera? *European Journal of Entomology, 107*, 349–355.

Howard, R. R., & Brodie, E. D. (1973). A Batesian mimetic complex in salamanders: Responses of avian predators. *Herpetologica, 29*, 33–41.

Huang, W. S. (2007). Ecology and reproductive patterns of the agamid lizard, *Japalura swinhonis*, on an East Asian Island, with comments on the small clutch sizes of Island lizrds. *Zoological Science, 24*, 181–188.

Hunter, J. S. (2009). Familiarity breeds contempt: Effects of striped skunk color, shape, and abundance on wild carnivore behavior. *Behavioral Ecology, 20*, 1315–1322.

Jeffords, M. R., Sternburg, J. G., & Waldbauer, G. P. (1979). Batesian mimicry: Field demonstration of the survival value of pipevine swallowtail and monarch color patterns. *Evolution, 33*, 275–286.

Kanehisa, K., & Murase, M. (1977). Comparative study of the pygidial defensive systems of carabid beetles. *Applied Entomology and Zoology, 12*, 225–235.

Kapan, D. D. (2001). Three-butterfly system provides a field test of müllerian mimicry. *Nature, 409*, 338–340.

Kauppinen, J., & Mappes, J. (2003). Why are wasps so intimidating: Field experiments on hunting dragonflies (Odonata: *Aeshna grandis*). *Animal Behaviour, 66*, 505–511.

Kazemi, B., Gamberale-Stille, G., Tullberg, B. S., & Leimar, O. (2014). Stimulus salience as an explanation for imperfect mimicry. *Current Biology, 24*, 965–969.

Kikuchi, D. W., & Pfennig, D. W. (2010). Predator cognition permits imperfect coral snake mimicry. *The American Naturalist, 176*, 830–834.

Ko, Y. W., Liao, C. P., Clark, R. W., Hsu, J. Y., Tseng, H. Y., & Huang, W. S. (2020). Aposematic coloration of prey enhances memory retention in an agamid lizard. *Animal Behaviour, 161*, 1–13.

Kodric-Brown, A. (1998). Sexual dichromatism and temporary color changes in the reproduction of fishes. *American Zoologist, 38*, 70–81.

Kuchta, S. R., Krakauer, A. H., & Sinervo, B. (2008). Why does the yellow-eyed ensatina have yellow eyes? Batesian mimicry of Pacific newts (genus Taricha) by the salamander *Ensatina eschscholtzii xanthoptica*. *Evolution, 62*, 984–990.

Kuo, C. Y., Lin, Y. S., & Lin, Y. K. (2009). Sexual size and shape dimorphism in an agamid lizard, *Japalura swinhonis* (Squamata: Lacertilia: Agamidae). *Zoological Studies, 48*, 351–361.

Langham, G. M. (2004). Specialized avian predators repeatedly attack novel color morphs of *Heliconius* butterflies. *Evolution, 58*, 2783–2787.

Leal, M., & Powell, B. J. (2012). Behavioural flexibility and problem-solving in a tropical lizard. *Biology Letters, 8*, 28–30.

LeBas, N. R., & Marshall, N. J. (2000). The role of colour in signalling and male choice in the agamid lizard *Ctenophorus ornatus*. *Proceedings of the Royal Society B: Biological Sciences, 267*, 445–452.

Lee, C. Y., Yo, S. P., Clark, R. W., Hsu, J. Y., Liao, C. P., Tseng, H. Y., & Huang, W. S. (2018). The role of different visual characters of weevils signalling aposematism to sympatric lizard predators. *Journal of Zoology, 306*, 36–47.

Lin, Y. S., & Lu, K. H. (1982). Population ecology of the lizard *Japalura swinhonis formosensis* (Sauria: Agamidae) in Taiwan. *Copeia, 1982*, 425–434.

Lindstedt, C., Lindström, L., & Mappes, J. (2008). Hairiness and warning colours as components of antipredator defence: Additive or interactive benefits? *Animal Behaviour, 75*, 1703–1713.
Lindström, L., Alatalo, R. V., Lyytinen, A., & Mappes, J. (2001). Predator experience on cryptic prey affects the survival of conspicuous aposematic prey. *Proceedings of the Royal Society B: Biological Sciences, 268*, 357–361.
Linsley, E. G. (1959). Ecology of Cerambycidae. *Annual Review of Entomology, 4*, 99–138.
Lynn, S. K. (2005). Learning to avoid aposematic prey. *Animal Behaviour, 70*, 1221–1226.
Maan, M. E., & Cummings, M. E. (2012). Poison frog colors are honest signals of toxicity, particularly for bird predators. *The American Naturalist, 179*, E1–E14.
Mallet, J., & Joron, M. (1999). Evolution of diversity in warning color and mimicry: Polymorphisms, shifting balance, and speciation. *Annual Review of Ecology and Systematics, 30*, 201–233.
Manrod, J. D., Hartdegen, R., & Burghardt, G. M. (2008). Rapid solving of a problem apparatus by juvenile black-throated monitor lizards (*Varanus albigularis albigularis*). *Animal Cognition, 11*, 267–273.
Marples, N. M., Roper, T. J., & Harper, D. G. C. (1998). Responses of wild birds to novel prey: Evidence of dietary conservatism. *Oikos, 83*, 161–165.
Merrill, R. M., Wallbank, R. W. R., Bull, V., Salazar, P. C. A., Mallet, J., Stevens, M., & Jiggins, C. D. (2012). Disruptive ecological selection on a mating cue. *Proceedings of the Royal Society B: Biological Sciences, 279*, 4907–4913.
Noonan, B. P., & Comeault, A. A. (2009). The role of predator selection on polymorphic aposematic poison frogs. *Biology Letters, 5*, 51–54.
Pasteels, J. M., Grégoire, J. C., & Rowell-Rahier, M. (1983). The chemical ecology of defense in arthropods. *Annual Review of Entomology, 28*, 263–289.
Pegram, K. V., & Rutowski, R. L. (2014). Relative effectiveness of blue and orange warning colours in the contexts of innate avoidance, learning and generalization. *Animal Behaviour, 92*, 1–8.
Pegram, K. V., Lillo, M. J., & Rutowski, R. L. (2013). Iridescent blue and orange components contribute to the recognition of a multicomponent warning signal. *Behaviour, 150*, 321–336.
Pekár, S., & Jarab, M. (2011). Assessment of color and behavioral resemblance to models by inaccurate myrmecomorphic spiders (Araneae). *Invertebrate Biology, 130*, 83–90.
Pfennig, D. W., Harcombe, W. R., & Pfennig, K. S. (2001). Frequency-dependent Batesian mimicry. *Nature, 410*, 323–323.
Polo-Cavia, N., Oliveira, J. M., Redondo Villa, A. J., & Márquez, R. (2016). Background colour matching in a wild population of *Alytes obstetricans*. *Amphibia-Reptilia, 37*, 253–260.
Poulton, E. B. (1890). *The colours of animals: Their meaning and use, especially considered in the case of insects*. New York, NY: D. Appleton & Company.
Prokopová, M., Veselý, P., Fuchs, R., & Zrzavý, J. (2010). The role of size and colour pattern in protection of developmental stages of the red firebug (*Pyrrhocoris apterus*) against avian predators. *Biological Journal of the Linnean Society, 100*, 890–898.
Prudic, K. L., Skemp, A. K., & Papaj, D. R. (2007). Aposematic coloration, luminance contrast, and the benefits of conspicuousness. *Behavioral Ecology, 18*, 41–46.
Reichstein, T., von Euw, J., Parsons, J. A., & Rothschild, M. (1968). Heart poisons in monarch butterfly: Some aposematic butterflies obtain protection from cardenolides present in their food plants. *Science, 161*, 861–866.
Remmel, T., & Tammaru, T. (2011). Evidence for the higher importance of signal size over body size in aposematic signaling in insects. *Journal of Insect Science, 11*, 4.
Roper, T. J. (1990). Responses of domestic chicks to artificially coloured insect prey: Effects of previous experience and background colour. *Animal Behaviour, 39*, 466–473.
Roper, T. J. (1994). Conspicuousness of prey retards reversal of learned avoidance. *Oikos, 69*, 115–118.

Roper, T. J., & Redston, S. (1987). Conspicuousness of distasteful prey affects the strength and durability of one-trial avoidance learning. *Animal Behaviour, 35*, 739–747.
Roper, T. J., & Wistow, R. (1986). Aposematic colouration and avoidance learning in chicks. *The Quarterly Journal of Experimental Psychology Section B, 38*, 141–149.
Ruxton, G. D., Sherratt, T. N., & Speed, M. P. (2004). *Avoiding attack: The evolutionary ecology of crypsis, warning signals and mimicry* (1st ed.). Oxford University Press.
Saporito, R. A., Zuercher, R., Roberts, M., Gerow, K. G., & Donnelly, M. A. (2007). Experimental evidence for aposematism in the dendrobatid poison frog *Oophaga pumilio*. *Copeia, 2007*, 1006–1011.
Schmidt, V., & Schaefer, H. M. (2004). Unlearned preference for red may facilitate recognition of palatable food in young omnivorous birds. *Evolutionary Ecology Research, 6*(6), 919–925.
Schultze, W. (1923). A monograph of the pachyrrhynchid group of the Brachyderinae, Curculionidae: Part I. The genus *Pachyrrhynchus* Germar. *The Philippine Journal of Science, 23*, 609–673.
Seago, A. E., Brady, P., Vigneron, J.-P., & Schultz, T. D. (2009). Gold bugs and beyond: A review of iridescence and structural colour mechanisms in beetles (Coleoptera). *Journal of the Royal Society Interface, 6*, S165–S184.
Servedio, M. R. (2000). The effects of predator learning, forgetting, and recognition errors on the evolution of warning coloration. *Evolution, 54*, 751–763.
Shen, T.-J. (2009). *The roles of chemical and visual senses in predatory behavior of Swinhoe's japalura (Japalura swinhonis Gunther)*. National Dong Hwa University.
Siddall, E. C., & Marples, N. M. (2008). Better to be bimodal: The interaction of color and odor on learning and memory. *Behavioral Ecology, 19*, 425–432.
Sillén-Tullberg, B. (1985). Higher survival of an aposematic than of a cryptic form of a distasteful bug. *Oecologia, 67*, 411–415.
Smith, S. M. (1975). Innate recognition of coral snake pattern by a possible avian predator. *Science, 187*, 759.
Smith, S. M. (1977). Coral-snake pattern recognition and stimulus generalisation by naive great kiskadees (Aves: Tyrannidae). *Nature, 265*, 535–536.
Speed, M. P. (2000). Warning signals, receiver psychology and predator memory. *Animal Behaviour, 60*, 269–278.
Starr, C. K., & Wang, H.-Y. (1992). Pachyrhynchine weevils (Coleoptera: Curculionidae) of the islands fringing Taiwan. *Journal of Taiwan Museum, 45*, 5–13.
Stevens, M., & Ruxton, G. D. (2012). Linking the evolution and form of warning coloration in nature. *Proceedings of the Royal Society B: Biological Sciences, 279*, 417–426.
Stevens, M., Castor-Perry, S. A., & Price, J. R. F. (2009). The protective value of conspicuous signals is not impaired by shape, size, or position asymmetry. *Behavioral Ecology, 20*, 96–102.
Summers, K., Cronin, T. W., & Kennedy, T. (2003). Variation in spectral reflectance among populations of *Dendrobates pumilio*, the strawberry poison frog, in the Bocas del Toro Archipelago, Panama. *Journal of Biogeography, 30*, 35–53.
Svádová, K., Exnerová, A., Štys, P., Landová, E., Valenta, J., Fučíková, A., & Socha, R. (2009). Role of different colours of aposematic insects in learning, memory and generalization of naïve bird predators. *Animal Behaviour, 77*, 327–336.
Terrick, T. D., Mumme, R. L., & Burghardt, G. M. (1995). Aposematic coloration enhances chemosensory recognition of noxious prey in the garter snake *Thamnophis radix*. *Animal Behaviour, 49*, 857–866.
Toledo, L. F., & Haddad, C. F. B. (2009). Colors and some morphological traits as defensive mechanisms in anurans. *International Journal of Zoology, 2009*, 910892.
Tseng, H.-Y., Lin, C.-P., Hsu, J.-Y., Pike, D. A., & Huang, W.-S. (2014). The functional significance of aposematic signals: Geographic variation in the responses of widespread lizard predators to colourful invertebrate prey. *PLoS One, 9*, e91777.
Tso, I.-M., Liao, C.-P., Huang, R.-P., & Yang, E.-C. (2006). Function of being colorful in web spiders: Attracting prey or camouflaging oneself? *Behavioral Ecology, 17*, 606–613.

Umbers, K. D. L. (2013). On the perception, production and function of blue colouration in animals. *Journal of Zoology, 289*, 229–242.
Valkonen, J. K., Nokelainen, O., & Mappes, J. (2011). Antipredatory function of head shape for vipers and their mimics. *PLoS One, 6*, e22272.
Veselý, P., Luhanová, D., Prášková, M., & Fuchs, R. (2013). Generalization of mimics imperfect in colour patterns: The point of view of wild avian predators. *Ethology, 119*, 138–145.
Vitt, L. J., Pianka, E. R., Cooper, W. E. J., & Schwenk, K. (2003). History and the global ecology of squamate reptiles. *The American Naturalist, 162*, 44–60.
Wallace, A. R. (1889). *Darwinism: An exposition of the theory of natural selection, with some of its applications*. Macmillan and.
Wallace, A. R. (1895). *Natural selection and tropical nature: Essays on descriptive and theoretical biology*. Macmillan and.
Ward-Fear, G., Pearson, D. J., Brown, G. P., Rangers, B., & Shine, R. (2016). Ecological immunization: In situ training of free-ranging predatory lizards reduces their vulnerability to invasive toxic prey. *Biology Letters, 12*, 20150863.
Webb, J. K., Du, W.-G., Pike, D. A., & Shine, R. (2010). Generalization of predator recognition: Velvet geckos display anti-predator behaviours in response to chemicals from non-dangerous elapid snakes. *Current Zoology, 56*, 337–342.
Welch, V., Lousse, V., Deparis, O., Parker, A., & Vigneron, J. P. (2007). Orange reflection from a three-dimensional photonic crystal in the scales of the weevil *Pachyrrhynchus congestus pavonius* (Curculionidae). *Physical Review E—Statistical, Nonlinear, and Soft Matter Physics, 75*, 041919.
Whitman, D. W., & Vincent, S. (2008). Large size as an antipredator defense in an insect. *Journal of Orthoptera Research, 17*, 353–371.
Woo, K. L., Burke, D., & Peters, R. A. (2009). Motion sensitivity of the jacky dragon, Amphibolurus muricatus: Random-dot kinematograms reveal the importance of motion noise for sigal detection. *Animal Behaviour, 77*, 307–315.
Woo, K. L., Rieucau, G., & Burke, D. (2017). Computer-animated stimuli to measure motion sensitivity: Contraints on signal design in the Jacky dragon. *Current Zoology, 63*, 75–84.

The Sum of Intraguild Predators' Fears Drives the Survival of Green Sea Turtle (*Chelonia mydas*) Eggs

12

Abstract

Top predators are thought to stabilize food webs by protecting basal resources, yet the behavioral mechanisms remain unclear. We combined long-term field observations with experiments to examine interactions among predators of sea turtle eggs (*Chelonia mydas*): a top predator (*Elaphe carinata*) and two intermediate predators (*Rattus norvegicus* and *Oligodon formosanus*). Despite all readily consuming eggs, survival remained high. Results show that fear of intraguild predation, cannibalism, habitat structure, and territorial behavior interact to limit exploitation. These findings highlight how behaviorally mediated interactions, particularly fear effects, can stabilize trophic systems and enhance persistence of vulnerable prey in complex natural ecosystems.

Keywords

Food chain · Intraguild fears · Sea turtle eggs · Snakes · Mouse

There is a rhyming saying in Taiwan: Who fears whom? People fear fame, pigs fear fatness, turtles fear hammers, and cockroaches fear slippers. Another story goes like this: A powerful nation is strong because there are many unmarried women. More women mean more cats, more cats mean fewer mice, fewer mice mean more grain, more grain means a stronger nation, a stronger nation means more wars, more wars mean more soldiers needed, more soldiers mean fewer men of marriageable age domestically, and fewer men of marriageable age mean more unmarried women. This is somewhat like the concept of a food chain. This chapter uses the concept of a food chain and the principle of "who fears whom" to tell an interesting story.

W.-S. Huang, *Reptiles in the Changing Climate*, Fascinating Life Sciences,
https://doi.org/10.1007/978-981-92-1887-5_12

12.1 Introduction

Top-level predators strongly influence organisms at lower trophic levels and frequently regulate prey abundance through a combination of direct and indirect density-dependent pathways (Lima, 2002; Ripple & Beschta, 2012; Kohl et al., 2018). Through competition and intra-group predation, their effects can also shape populations of intermediate predators (Estes et al., 2011; Ripple et al. 2014). Such pressures can drive evolutionary changes in prey behavior, natural history, and morphology that reduce vulnerability to predation, and these adaptations cause changes across additional trophic levels (Lima & Dill, 1990). As a result, predators build communities through multiple mechanisms and influence the natural processes of ecosystem functioning via trait-mediated effects (Carey & Wahl, 2010; Greeney et al., 2015). The indirect effects of predator-caused "fear landscapes" can exert effects that equal or exceed those caused by direct killing, according to current theoretical and empirical research on trophic cascades (Schmitz et al., 1997; Preisser et al., 2005; le Roux et al., 2018).

However, assessing trophic cascades in natural ecosystems necessitates the collection of difficult data as well as reliable causal inferences (Peterson et al., 2014). For example, only one of three potential predators of sea turtle eggs on tropical islands, direct egg consumption was observed only for one species (reptile-egg-eating snake, *Oligodon formosanus*). We couldn't figure out why so many brown rats (*Rattus norvegicus*) and stinky ratsnakes (*Elaphe carinata*) don't eat these nutritious and defenseless eggs. According to classical optimal foraging theory, foragers should maximize energetic returns by minimizing search and handling costs, leading to intensive exploitation of stationary, energy-rich prey with minimal defenses (Werner & Hall, 1974; Belovsky & Schmitz, 1991). Sea turtle eggs, therefore, appear to represent an ideal food source, yet their survival rate during the immobile and undefended incubation phase is unexpectedly high.

Turtle eggs, as a basic food web resource, experience indirect protection by interactions between consumers of higher nutritional classes, such as competition or intraguild predation. Theoretical studies suggest that food webs incorporating intraguild predation can exhibit multiple alternative steady states (Polis & Holt, 1992). Depending on the relative effects of direct increased consumption, competition, intraguild predation, and indirect predator effects, the net effect of multiple consumers can generate suppression or facilitation on the underlying resource in a food web (Duffy et al., 2007; Vance-Chalcraft et al., 2007). According to both models and intermediate experiments, the food web stability and basal resource use depend on interactions between the number of species within a trophic level (horizontal diversity) and the number of trophic levels (vertical diversity) (McCann et al., 1998; Zhao et al., 2019). In addition, habitat structure can modify susceptibility to predation by altering predator foraging behavior (Grabowski, 2004) and by offering species-specific refugia (Eklöv & Persson, 1995). Despite all potential egg predators being able to dig up sea turtle eggs easily, even this limited sheltering can have significant effects on consumer interactions. The ultimate impact of a shelter is determined by its role in mediating the likelihood and outcome of encounters between interacting species (Savino & Stein, 1989; Finke & Denno, 2006).

Although many studies demonstrate that fear of predators can strongly affect community function, wildlife population dynamics (e.g., Lima & Dill, 1990), significant gaps in our understanding remain that turtle egg-based food webs can help resolve. The majority of empirical food webs research relies either on highly controlled intermediate chambers for small organisms (Schmitz, 2010; Wise & Lensing, 2019) or on observational field studies focused on megafaunal species (Laundre et al., 2001; Surad et al., 2016; le Roux et al., 2018). Experimental studies that integrate natural observations with behavioral quantification of species interactions are comparatively uncommon, and more research is essential for determining whether the predicted population-level patterns were caused by the potential effects of fearful landscapes in natural systems.

Furthermore, our understanding of the influence of predator-prey behavioral interactions on food web structure is hampered by a number of critical factors: (1) studies have focused on a relatively limited range of the taxonomic diversity; (2) predators are frequently viewed as static risk factors instead of interactive behavioral actors (Lima, 2002); and (3) predator-prey behavior is frequently investigated in dyads, despite the fact that their interactions are embedded in a more complex ecosystem involving simultaneous interaction among multiple species [i.e., the multi-predator effect (Schmitz, 2007)]. Foraging strategies are principally governed by behavioral trade-offs between predation risk and foraging efficiency (Lima, 1985; Schmitz et al., 1997; Guillemain et al., 2007). Foraging and vigilance are frequently mutually in conflict, as individuals must look down to forage and/or chase prey, interfering with predator detection (e.g., Lima & Dill, 1990; Guillemain et al., 2007). In addition, obtaining, handling, or digesting food often requires exposure foragers themselves in open areas, increasing their risk of predation relative to individuals hiding in more protected areas (e.g., Lima, 1985). The majority of these studies, however, have concentrated on the responses of highly visual species (i.e., birds and insects) to isolated presentations of artificial predator cues (i.e., predator models, chemical cues, human surrogates). Less common are studies incorporating species that depend on other sensory mechanisms (e.g., olfaction) to detect food or predators, or that combine the dynamic behavioral interactions among multiple predator and prey species (Barbour & Clark, 2012; Whitford et al., 2019).

To fill some of the gaps in our understanding of food web dynamics and predator–prey interactions, we integrated field observations with controlled behavioral experiments. Our objective was to test the hypothesis that faunal predation on Orchid Island leads to relatively high egg survival rates. Importantly, our research system includes multiple predators whose common prey is a sessile food resource, and allowing us to compare the behavioral responses of these species to each other not only on Orchid Island, but also on Taiwan's mainland, where the island population originated. We first monitored multiple free-ranging predator species over an extended period while they foraged around sea turtle nests to record arrival times (Fig. 12.1), abundance, and behavioral interactions among the three predators.

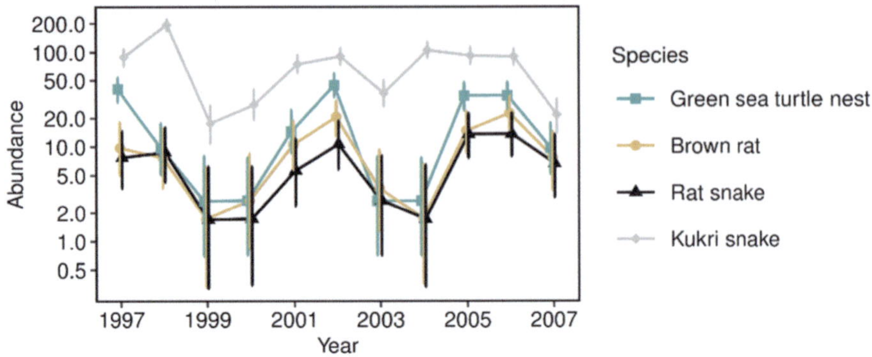

Fig. 12.1 Number of observed sea turtle (*C. mydas*) nests, brown rats (*R. norvegicus*), ratsnakes (*E. carinata*), and kukri snakes (*O. formosanus*) from 1997 to 2007

Second, because direct interactions among predators in the field are uncommon, we conducted manipulative experiments under controlled conditions to gather information on interactions between pairs and multispecies assemblages. Third, we repeated the laboratory experiments with conspecifics from the mainland Taiwan population to assess how behavioral responses differed from the derived Orchid Island population, where the turtles nest together (green sea turtles do not nest in mainland Taiwan).

We used the results of these experiments to assess whether different predators exert asymmetric effects on their competitors' foraging success and behavior, and to construct a model that emphasizes the importance of intra-faunal predation in buffering high mortality at low nutritional levels.

12.2 Field Observations

Between 1997 and 2007, surveys conducted at Little Paiday on Orchid Island documented a total of 203 green sea turtle nests, along with 107 brown rats, 78 ratsnakes, and 842 kukri snakes (Fig. 12.2).Turtle nests, brown rats, and ratsnakes all had positive correlations (turtle nests with brown rats, Spearman's correlation $r = 0.897$, Padj $= 0.001$; turtle nests with ratsnakes, Spearman's correlation $r = 0.837$, Padj $= 0.005$; brown rats with ratsnakes, Spearman's correlation $r = 0.905$, Padj $= 0.\ 001$). In contrast, kukri snake abundance was not significantly correlated with any of the other three variables (turtle nest and kukri snake, Spearman's correlation $r = 0.369$, Padj $= 0.521$; brown rat and kukri snake, Spearman's correlation r $= 0.338$, Padj $= 0.527$; ratsnake and kukri snake, Spearman's correlation $r = 0.506$, Padj $= 0.337$). Brown rats were the earliest egg predators to arrive at the nest (typically within an hour of the turtle's departure). However, once the kukri snake arrived (within 0.5–2 h), the rats quickly retreated; thus, the rats' excavation of turtle eggs was never observed. Kukri snakes, on the other hand, attempted to enter turtle nests as soon as they reached the beach.

Fig. 12.2 Order of arrival of egg predators at Little Paiday, Orchid Island. (**a**) Green sea turtles (*Chelonia mydas*) nest from July to September. (**b**) Brown rats (*Rattus norvegicus*) always appeared at the turtle nests immediately after turtles left. (**c**) Male kukri snakes (*Oligodon formosanus*) arrived within 0.5–2 h, earlier than females. (**d**) If female kukri snakes never entered a nest, males would consume all the eggs in less than 5 days. (**e**) A female kukri snake entering a nest would eject all the males. (**f**) A female kukri snake would continue to defend the whole nest from males and consume some of the eggs. (**g**) Nests with a female kukri snake would hatch some turtles after 50–60 days, and hatchlings would leave the nests to travel toward the sea at night. (**h**) After all the eggs have hatched, female kukri snakes dig a hole near the surface of the sand to avoid the high temperatures of the beach during the day. (**i**) Stink ratsnakes (*Elaphe carinata*) were present around the turtle nests and frequently consumed kukri snakes. (All pictures photographed by W.S. Huang)

12.3 Predator–Prey Pair Tests

We discovered that brown rats and ratsnakes from the Orchid Island population never excavate eggs from the sand (0% consumption), whereas kukri snakes always did (100% consumption; Fig. 12.3a). In contrast, when exposed to turtle eggs, all predators consumed them (100% consumption; Fig. 12.3b). In experiments using the same species in continental populations, all three readily excavated eggs: regardless of whether the eggs were buried or exposed, all predators almost always consumed turtle eggs (90–100% consumption) (Fig. 12.3a, b).

Ratsnakes always consumed mice in trials with different predator pairs (100% consumption; Fig. 12.3c), whereas kukri snakes never did, irrespective of their source population (0% consumption; Fig. 12.3c). When exposed to the sand, ratsnakes always consumed kukri snakes (100% consumption; Fig. 12.3c). In experiments with two conspecifics, No cannibalism was recorded among rats or among kukri snakes (0% consumption; Fig. 12.3c). In contrast, the larger ratsnake easily swallowed the smaller conspecifics (90% consumption; Fig. 12.3c) in both locations.

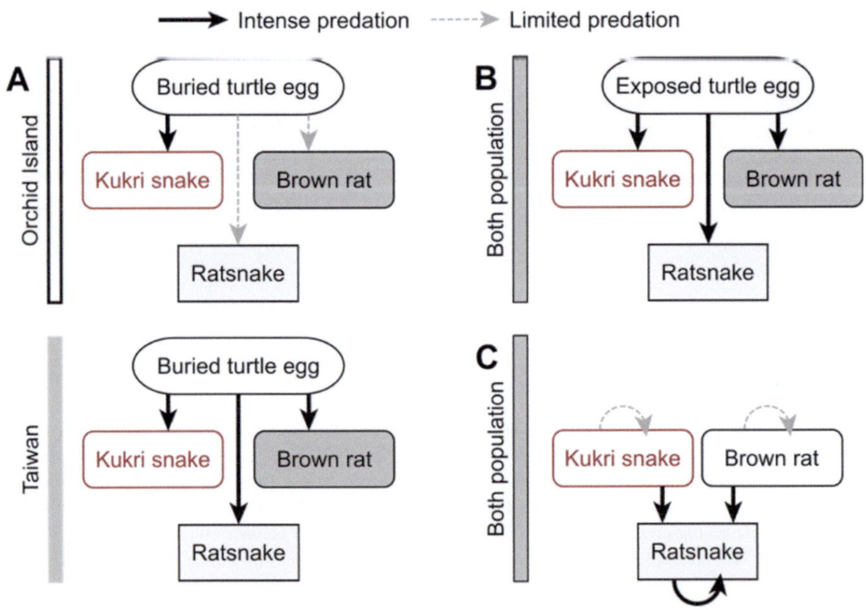

Fig. 12.3 Prey consumed in paired trials using Orchid Island and Taiwan populations. Experiments included buried turtle eggs paired with brown rat, kukri snake, and ratsnake (**a**); exposed turtle eggs paired with brown rat, kukri snake, and ratsnake (**b**); intraspecific and interspecific pairings of brown rat, kukri snake, and ratsnake (**c**). Trials in each experiment were replicated ten times. One arrow corresponds to one type of experiment (a pair in a specific population). (Redrawn from Liao et al. (2021), Proceedings of the Royal Society B: Biological Sciences)

12.4 Multiple Predator Test

Brown rats from Orchid Island did not excavate buried eggs (0% consumption; Fig. 12.4a), but always consumed exposed eggs (100% consumption; Fig. 12.4b), whereas brown rats from Taiwan readily consumed both buried and exposed eggs (91% consumption; 100% consumption; Fig. 12.4a, b). Kukri snakes from Taiwan and Orchid Island consumed both buried and exposed eggs (Fig. 12.4c).

In contrast to the results of the paired predator-turtle egg trial, the kukri snake never consumed turtle eggs under any multiple-predator condition (0% consumption; Fig. 12.4d), implying that the presence of rats or ratsnakes suppressed egg predation by the kukri snake.

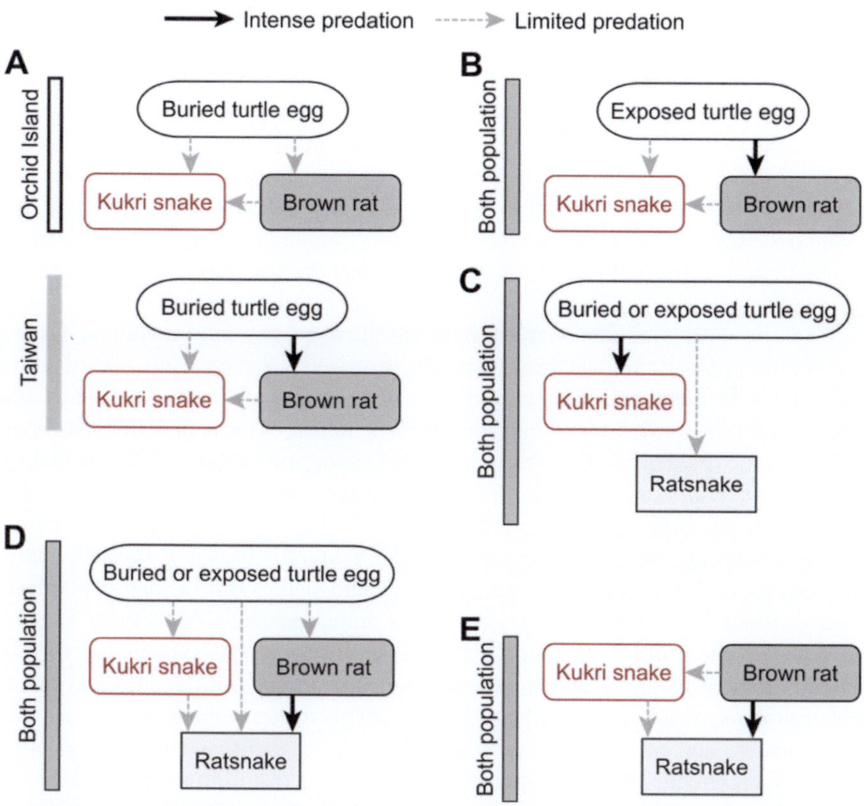

Fig. 12.4 Different prey items consumed in multiple predator trials using Orchid Island and Taiwan populations. Buried (**a**) and exposed (**b**) turtle eggs placed in enclosures with brown rat and kukri snake; buried or exposed (**c**) turtle eggs with kukri snake and ratsnake; buried or exposed (**d**) turtle eggs with all three predators; and all three predators alone (**e**). Arrows with same color indicate the same experimental treatment (an experiment set in a specific population). (Redrawn from Liao et al. (2021), Proceedings of the Royal Society B: Biological Sciences)

Ratsnakes preyed heavily on brown rats (85–100% consumption, Fig. 12.4d, e) but rarely consumed kukri snake (0–9% consumption, Fig. 12.4d, e) in trials with all three predators present. In these experiments, if eggs were present (either buried or exposed), none of the three predators consumed them (0% consumption, Fig. 12.4d). Both mainland and island populations displayed similar patterns.

Finally, regardless of population, in the absence of brown rats, ratsnakes almost always consumed kukri snake (90–100% consumption), but ignored eggs regardless of whether they were buried or exposed (0% consumption; Fig. 12.4c, d). Thus, ratsnakes preferred brown rats most, kukri snakes next, and turtle eggs least.

12.5 Discussion

Turtle eggs were abundant, immobile, undefended, nutritious, and co-present with three abundant predators at Orchid Island under experimental conditions, even when buried beneath sand. Under these conditions, we discovered that sea turtle eggs have an unexpectedly high survival rate in nature. How do they stay alive? Our findings suggest that depending on the native population, the presence of other predators, and the exposure of the eggs, the three predator species (brown rat, kukri snake, and ratsnake) behave differently towards each other and towards sea turtle eggs. The combined outcome of these interacting factors, together with the kukri snake's female-biased territoriality (Pike et al., 2016), has the net effect of protecting sea turtle eggs.

When no other predators were present, all three species from mainland Taiwan (the source populations for Orchid Island) consumed the turtle eggs placed in the enclosure with them, even though the eggs were buried under 66 cm of sand (the typical depth of a turtle nest). This demonstrates that all of these predators easily ate the eggs and were able to find them when they were buried (Fig. 12.5a). However, neither in the wild nor in captivity, rats and ratsnakes on Orchid Island attempted to excavate buried turtle eggs (Fig. 12.5b).

These findings suggest a strong interaction among predator species, native populations, and turtle egg exposure, with rats and snakes from island populations being cautious about digging up eggs, whereas individuals from continental relatives are not. We use multiple predator experiments to suggest an explanation for this derived caution: Orchid Island's buried turtle eggs attract kukri snakes and rats in large numbers, which attract a large number of ratsnakes, which readily eat rats, kukri snakes, and other ratsnakes (Fig. 12.1). As a result, sea turtle eggs were excavated on Orchid Island to increase in the predation risk from ratsnakes. Because cannibalism is common in this species, even ratsnakes are vulnerable to attacks by larger conspecifics, and individuals with its head buried in sand may be at a significant disadvantage if conspecifics attack.

Our field observations support the hypothesis that turtle eggs are protected by fear of rats. Rats are always present on nesting beaches, and they are the first predators seen around turtle nests following egg deposition and after the turtles have left. Orchid Island rats, on the other hand, will pick up eggs on the surface but avoid

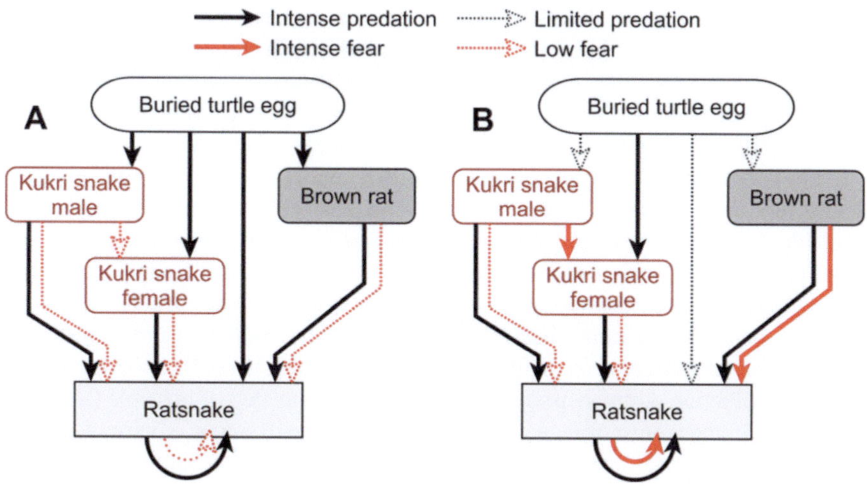

Fig. 12.5 The altered trophic relationships between the organisms in both Taiwan (**a**) and on Orchid Island (**b**). Red and black arrows indicate the direction of fear and consumption, respectively. Males expelled by female kukri snakes had been reported by Lee et al. (2019). (Redrawn from Liao et al. (2021), Proceedings of the Royal Society B: Biological Sciences)

digging up turtle nests. This could be because kukri snakes, which are quickly attracted to the beach and nest in large numbers within 0.5 h of egg laying, and ratsnakes arrive afterward. . Rats are the ratsnake's favorite prey (Fig. 12.4), which may explain the low rate of successful rat predation by ratsnakes; we have never seen free-ranging ratsnakes capture rats. Ratsnakes, on the other hand, are very successful at catching and eating kukri snakes, they prefer turtle eggs (Fig. 12.4). The availability of prey for kukri snakes, together with the risk of conspecific predation, offers a plausible explanation for ratsnakes' reluctance to dig turtle nests (Fig. 12.3). Only Kukri snakes are willing to dig turtle nests.

Although ratsnakes prey on kukri snakes, the energetic payoff of egg consumption likely outweighs the risk of digging a turtle nest, especially for female kukri snakes. Individual kukri snakes are capable of defending nests from conspecifics, providing them with food many times their size. Moreover, kukri snakes with buried eggs are relatively safe because rats and ratsnakes will not dig the turtle egg nests. Although kukri snakes will consume as many eggs as possible, the large clutch size produced and the individual snake's small size have a protective effect on the nest: the individual snake's territorial behavior excludes other kukri snakes from entering the nest, and the fear of ratsnakes deters the other mesopredators. The fear expressed by predators of each turtle egg species combined to promote the survival of abundant, sessile, palatable eggs, with the majority of eggs hatching [i.e., 53–94% hatching rate (Cheng et al., 2009)].

Given the diversity of food availability and fear of predators, it is not surprising that carnivores frequently balance these two behaviors. The non-lethal costs of these predations are just as important to overall prey dynamics as the lethal costs (Preisser et al., 2005; Creel et al., 2009). Orchid Island's ecosystem may be unique in that even top predators are constrained by this trade-off. Ratsnakes on Mainland Taiwan excavate sand to eat turtle eggs or snakes, whereas rats on Orchid Island do not, which could be because turtle eggs on Orchid Island are buried underground, where there is a risk of conspecific predation as well as more appealing prey.

Although mainland Taiwan rats are capable of digging for deeply buried eggs and brown rats are well documented to burrow extensively if a suitable substrate is available, we have never observed rats excavating for turtle eggs on Orchid Island, either in the field or in the laboratory (Boice, 1977). The first rats to colonize Orchid Island (presumably in collaboration with human colonists) most likely maintained this trait and excavated turtle nests, but the intense predation pressure from Orchid Island ratsnakes (attracted to the beach in search of kukri snake) resulted in a rapid change in behavior. Animals preferentially forage in safe areas if possible (Verdolin, 2006), but if they must forage in risky areas, they will adopt a variety of risk-management strategies, such as reducing foraging time, increasing alert periods, or simply abandoning foraging in risky areas (Lima & Dill, 1990). Predation risk has a significant sublethal behavioral effect on prey, causing them to change where they go and how long they stay at a foraging site (Laundre et al., 2010). Thus, natural selection on Orchid Island may favor rats that maintain high levels of vigilance, as at Little Paiday, would prevent them from burying their heads in the sand to dig for eggs.

However, rats can still be seen on the beach because of the opportunity to find and eat eggs accidentally exposed from nests by later arriving green turtles—a foraging activity that is more consistent with avoiding ratsnake predation.

The different responses of rats and ratsnakes to buried eggs highlight the importance of habitat structure in mitigating the indirect effects of predators. When wolves are nearby, moose preferentially feed on lower quality food close to forest safety (Hernandez & Laundre, 2005; Creel et al., 2009), and rats and ratsnakes appear hesitant to engage in dangerous digging activities when their own predators are plentiful. Sand provides a safe haven not only for eggs, but also for kukri snakes eager to dig eggs, because once a kukri snake enters a nest, it is protected from ratsnake predation. This finding is consistent with the effects commonly observed in mesocosm studies, in which increased habitat complexity can either reduce or increase consumption of basal resources, depending on how the structure is used by different consumers in specific situations (Eklöv & Persson, 1995; Zhao et al., 2019).

The reptilian egg-eating snake (*O. formosanus*) has a very different foraging strategy than the other two predators. Regardless of the number of sea turtle nests or the presence of ratsnakes, Kukri snakes arrived on the beach within 0.5–8 h after laying eggs (Huang et al., 2011); Fig. 12.1a) and immediately attempt to enter the nest. Why don't they appear to be afraid of top predators? This phenomenon can be explained by three non-exclusive mechanisms: low behavioral plasticity, dilution effects, and the relative security of the nest.

As dietary specialists that primarily consume reptile eggs, kukri snakes may exhibit low foraging plasticity; on Orchid Island, green sea turtle eggs and long-tailed skink (*Eutropis longicaudata*) are the two best food resources for kukri snakes (Huang et al., 2011). Turtle eggs are both more plentiful and predictable, whereas skinks aggressively guard the eggs, making them more expensive to consume (Huang, 2006; Huang & Wang, 2009). Thus, sea turtle eggs are an important food source worth pursuing for kukri snakes despite the risk of predation. Furthermore, turtle nests function as safe areas for kukri snakes; as previously stated, ratsnakes on Orchid Island do not enter turtle nests, shielding kukri snakes from ratsnakes once buried.

Finally, kukri snakes are much more numerous than ratsnakes. In 1998, for example, we counted 195 Kukri snakes on the beach, compared to only 9 ratsnakes. Ratsnakes can only consume one prey item at a time and, if successful, will cease foraging for an extended period of time during digestion. Thus, if kukri snakes invade the beach in large numbers within a short period of time, they may benefit from a dilution effect whereby individual predation risk declines as group size increases. Similar phenomena have been documented in other taxa. Pitcher butterflies, for example, can form conspicuous swarms (Molleman, 2011). Although these flocks are likely to be discovered by butterfly-eating birds, their visibility is offset by a dilution effect that decreases per capita mortality (Turner & Pitcher, 1986; Burger & Gochfeld, 2001). Another important factor in the protective effect of these interacting predators on turtle eggs is the territorial behavior of female kukri snakes (Huang et al., 2011)). In the absence of female defense, a male kukri snake would be capable of consuming an entire nest (with an average of 102 eggs) in only 5 days (Huang et al., 2011). However, the fear of being killed by the female's tail bite drives male snakes away from the nest, exposing the nest to predation by only one snake, and the eggs' hatching rates can reach up to 80% (Cheng et al., 2009). Turtle reproduction was successful in this case primarily because "the enemy of the enemy is a friend," and rats and male kukri snakes as enemies were suppressed by ratsnakes and female kukri snakes, respectively.

Our results are based on long-term observations of Orchid Island's sole remaining turtle nesting beaches. Although studying this system at only one location has obvious limitations, the trophic dynamics of natural systems frequently depend on site-specific environmental and ecological conditions (Lawton, 1999). Nonetheless, such natural system case studies are required to assess predictions derived from models and experimental studies conducted in artificial intermediate laboratories.

The factors influencing animal nest site selection are never the result of a single factor. As discussed in Chap. 7, long-tailed skinks lay their eggs in concrete holes and interact with kukri snakes. However, due to the global warming effect, these eggs will also face the impact of high temperatures. How significant will this impact be? The next chapter will address this issue.

References

Barbour, M. A., & Clark, R. W. (2012). Ground squirrel tail-flag displays alter both predatory strike and ambush site selection behaviours of rattlesnakes. *Proceedings of the Royal Society B: Biological Sciences, 279*, 3827–3833.

Belovsky, G. E., & Schmitz, O. J. (1991). Mammalian herbivore optimal foraging and the role of plant defenses. In R. T. Palo & C. T. Robbins (Eds.), *Plant defenses against mammalian herbivory* (pp. 1–28). CRC Press.

Boice, R. (1977). Burrows of wild and albino rats: Effects of domestication, outdoor raising, age, experience, and maternal state. *Journal of Comparative and Physiological Psychology, 91*, 649–661.

Burger, J., & Gochfeld, M. (2001). Smooth-billed ani (*Crotophaga ani*) predation on butterflies in Mato Grosso, Brazil: Risk decreases with increased group size. *Behavioral Ecology and Sociobiology, 49*, 482–492.

Carey, M. P., & Wahl, D. H. (2010). Interactions of multiple predators with different foraging modes in an aquatic food web. *Oecologia, 162*, 443–452.

Cheng, I. J., Huang, C.-T., Hung, P.-Y., Ke, B.-Z., Kuo, C.-W., & Fong, C.-L. (2009). Ten years of monitoring the nesting ecology of the Green turtle, *Chelonia mydas*, on Lanyu (Orchid Island), Taiwan. *Zoological Studies, 48*, 83–94.

Creel, S., Winnie, J. A., & Christianson, D. (2009). Glucocorticoid stress hormones and the effect of predation risk on elk reproduction. *Proceedings of the Natioaal Academy of Sciences of the United States of America, 106*, 12388–12393.

Duffy, J. E., Cardinale, B. J., France, K. E., McIntyre, P. B., Thébault, E., & Loreau, M. (2007). The functional role of biodiversity in ecosystems: Incorporating trophic complexity. *Ecology Letters, 10*, 522–538.

Eklöv, P., & Persson, L. (1995). Species-specific antipredator capacities and prey refuges: Interactions between piscivorous perch (*Perca fluviatilis*) and juvenile perch and roach (*Rutilus rutilus*). *Behavioral Ecology and Sociobiology, 37*, 169–178.

Estes, J. A., Terborgh, J., Brashares, J. S., Power, M. E., Berger, J., Bond, W. J., Carpenter, S. R., Essington, T. E., Holt, R. D., Jackson, J. B. C., Marquis, R. J., Oksanen, L., Oksanen, T., Paine, R. T., Pikitch, E. K., Ripple, W. J., Sandin, S. A., Scheffer, M., Schoener, T. W., Shurin, J. B., Sinclair, A. R. E., Soule, M. E., Virtanen, R., & Wardle, D. A. (2011). Trophic downgrading of planet earth. *Science, 333*, 301–306.

Finke, D. L., & Denno, R. F. (2006). Spatial refuge from intraguild predation: Implications for prey suppression and trophic cascades. *Oecologia, 149*, 265–275.

Grabowski, J. H. (2004). Habitat complexity disrupts predator-prey interactions but not the trophic cascade on oyster reefs. *Ecology, 85*, 995–1004.

Greeney, H. F., Meneses, M. R., Hamilton, C. E., Lichter-Marck, E., Mannan, R. W., Snyder, N., Snyder, H., Wethington, S. M., & Dyer, L. A. (2015). Trait-mediated trophic cascade creates enemy-free space for nesting hummingbirds. *Science Advances, 1*, e1500310.

Guillemain, M., Arzel, C., Legagneux, P., Elmberg, J., Fritz, H., Lepley, M., Pin, C., Arnaud, A., & Massez, G. (2007). Predation risk constrains the plasticity of foraging behaviour in teals, *Anas crecca*: A flyway-level circumannual approach. *Animal Behaviour, 73*, 845–854.

Hernandez, L., & Laundre, J. W. (2005). Foraging in the "landscape of fear" and its implications for habitat use and diet quality of elk *Cervus elaphus* and bison *Bison bison*. *Wildlife Biology, 11*, 215–220.

Huang, W. S. (2006). Parental care in the long-tailed skink, *Mabuya longicaudata,* on a tropical Asian Island. *Animal Behaviour, 72*, 791–795.

Huang, W. S., & Wang, H. Y. (2009). Predation risks and anti-predation parental care behavior: An experimental study in a tropical skink. *Ethology, 115*, 273–279.

Huang, W. S., Greene, H. W., Chang, T. J., & Shine, R. (2011). Territorial behavior in Taiwanese kukrisnakes *(Oligodon formosanus)*. *Proceedings of the National Academy of Science, 108*, 7455–7459.

Kohl, M. T., Stahler, D. R., Metz, M. C., Forester, J. D., Kauffman, M. J., Varley, N., White, P. J., Smith, D. W., & MacNulty, D. R. (2018). Diel predator activity drives a dynamic landscape of fear. *Ecological Monographs, 88*, 638–652.

Laundre, J. W., Hernandez, L., & Altendorf, K. B. (2001). Wolves, elk, and bison: Reestablishing the "landscape of fear" in Yellowstone National Park, USA. *Canadian Journal of Zoology, 79*, 1401–1409.

Laundre, J. W., Hernandez, L., & Ripple, W. J. (2010). The landscape of fear: Ecological implication of being afraid. *Open Ecology Journal, 3*, 1–7.

Lawton, J. H. (1999). Are there general laws in ecology? *Oikos, 84*, 177–192.

le Roux, E., Kerley, G. I. H., & Cromsigt, J. P. G. M. (2018). Megaherbivores modify trophic cascades triggered by fear of predation in an African savanna ecosystem. *Current Biology, 28*, 2493–2499.

Lee, C.-Y., Pike, D. A., Tseng, H.-Y., Hsu, J.-Y., Huang, S.-L., Shaner, L. P.-J., Liao, C.-P., Manica, A., & Huang, W.-S. (2019). When males live longer: Resource-driven territorial behavior drives sex-specific survival in snakes. *Science Advances, 5*, eaar5478.

Liao, C.-P., Hsu, J.-Y., Huang, S.-P., Clark, R. W., Lin, J.-W., Tseng, H.-Y., & Huang, W.-S. (2021). Sum of fears among intraguild predators drives the survival of green sea turtle (*Chelonia mydas*) eggs. *Proceedings of the Royal Society B: Biological Sciences, 288*, 20202631.

Lima, S. L. (1985). Maximizing feeding efficiency and minimizing time exposed to predators: A trade-off in the back-capped chickadee. *Oecologia, 66*, 60–67.

Lima, S. L. (2002). Putting predators back into behavioral predator-prey interactions. *Trends in Ecology and Evolution, 17*, 70–75.

Lima, S., & Dill, L. M. (1990). Behavioral decisions made under the risk of predation: A review and prospectus. *Canadian Journal of Zoology, 68*, 619–640.

McCann, K., Hastings, A., & Huxel, G. R. (1998). Weak trophic interactions and the balance of nature. *Nature, 395*, 794–798.

Molleman, F. (2011). Puddling: From natural history to understanding how it affects fitness. *Entomologia Experimentalis et Applicata, 134*, 107–113.

Peterson, R. O., Vucetich, J. A., Bump, J. M., & Smith, D. W. (2014). Trophic cascades in a multicausal world: Isle Royale and Yellowstone. *Annual Review of Ecology, Evolution, and Systematics, 45*, 325–345.

Pike, D. A., Clark, R. W., Manica, A., Tseng, H. Y., Hsu, J. Y., & Huang, W. S. (2016). Surf and turf: Predation by egg-eating snakes has led to the evolution of parental care in a terrestrial lizard. *Scientific Reports, 6*, 22207.

Polis, G. A., & Holt, R. D. (1992). Intraguild predation: The dynamics of complex trophic interactions. *Trends in Ecology & Evolution, 7*, 151–154.

Preisser, E. L., Bolnick, D. I., & Benard, M. F. (2005). Scared to death? The effects of intimidation and consumption in predator-prey interactions. *Ecology, 86*, 501–509.

Ripple, W. J., & Beschta, R. L. (2012). Trophic cascades in Yellowstone: The first 15 years after wolf reintroduction. *Biological Conservation, 145*, 205–213.

Ripple, W. J., Estes, J. A., Beschta, R. L., Wilmers, C. C., Ritchie, E. G., Hebblewhite, M., Berger, J., Elmberg, J., & Letnic, M. (2014). Status and ecological effects of the world's largest carnivores. *Science, 343*, 1241484.

Savino, J. F., & Stein, R. A. (1989). Behavioural interactions between fish predators and their prey: Effects of plant density. *Animal Behaviour, 37*, 311–321.

Schmitz, O. J. (2007). Predator diversity and trophic interactions. *Ecology, 88*, 2415–2426.

Schmitz, O. J. (2010). *Resolving ecosystem complexity*. Princeton University.

Schmitz, O. J., Beckerman, A. P., & O'Brien, K. M. (1997). Behaviorally mediated trophic cascades: Effects of predation risk on food web interactions. *Ecology, 78*, 1388–1399.

Surad, J. P., Clinchy, M., Dill, L. M., Roberts, D., & Zanette, L. Y. (2016). Fear of large carnivores cause a trophic cascade. *Nature Communications, 7*, 10698.

Turner, G. F., & Pitcher, T. J. (1986). Attack abatement: A model for group protection by combined avoidance and dilution. *The American Naturalist, 128*, 228–240.

Vance-Chalcraft, H. D., Rosenheim, J. A., Vonesh, J. R., Osenberg, C. W., & Sih, A. (2007). The influence of intraguild predation on prey suppression and prey release: A meta-analysis. *Ecology, 88*, 2689–2696.

Verdolin, J. L. (2006). Meta-analysis of foraging and predation risk tradeoffs in terrestrial systems. *Behavioral Ecology and Sociobiology, 60*, 457–464.

Werner, E. E., & Hall, D. J. (1974). Optimal foraging and the size selection of prey by the bluegill sunfish (*Lepomis macrochirus*). *Ecology, 55*, 1042–1052.

Whitford, M. D., Freymiller, G. A., Higham, T. E., & Clark, R. W. (2019). Determinants of predation success: How to survive an attack from a rattlesnake. *Functional Ecology, 33*, 1099–1109.

Wise, D. H., & Lensing, J. R. (2019). Impacts of rainfall extremes predicted by climate-change models on major trophic groups in the leaf litter arthropod community. *Journal of Animal Ecology, 88*, 1486–1497.

Zhao, Q., Van den Brink, P. J., Carpentier, C., Wang, Y. X. G., Rodríguez-Sánchez, P., Xu, C., Vollbrecht, S., Gillissen, F., Vollebregt, M., Wang, S., & De Laender, F. (2019). Horizontal and vertical diversity jointly shape food web stability against small and large perturbations. *Ecology, 22*, 1152–1162.

The Effects of Climate Change on Lizard Fitness Are Dependent on Nesting Habitat 13

Abstract

Maternal nest-site selection shapes incubation conditions and offspring fitness, but climate change may alter its benefits. In the long-tailed skink (*Eutropis longicaudata*), females nest either under rocks (natural habitat) or in concrete walls (artificial habitat), where temperatures are higher. Previously, artificial sites produced larger, faster-maturing hatchlings with higher survival. However, over the past decade, warming has disproportionately increased temperatures in artificial nests (~1.5 °C vs. 0.5 °C in natural sites), reversing this advantage. Now, artificial sites reduce offspring viability, while natural sites enhance it. Laboratory experiments confirm temperature as the key driver. These results show how human modifications can initially benefit organisms but, under climate change, create ecological traps that undermine adaptive maternal behavior.

Keywords

Long-tailed skink · Enviromental effects · Offspring-fitness · Survival · Global warming

Global warming due to the greenhouse effect has been known to affect the lives of living organisms on Earth. For example, the Arctic ice sheet has dissolved due to rising temperatures, and the ice used to be thick enough to support the weight of polar bears and thus provide an excellent hunting ground for bears to kill seals. In the long run, wild polar bears will not be able to find food to feed their cubs or themselves, so the day of extinction is not far away! Similarly, as the temperature rises, plants that bloom in spring are forced to bloom earlier, so many insects that live on plants must also appear earlier to collect food so as not to run out of food to eat; however, many migratory birds suffer because they feed on insects to replenish the energy needed for their journey. In recent years, there have been many exotic species

W.-S. Huang, *Reptiles in the Changing Climate*, Fascinating Life Sciences,
https://doi.org/10.1007/978-981-92-1887-5_13

of lizards in Taiwan, such as the Polynesian lizard, which originally lived in the hotter climates of Southeast Asia, but the temperature in Taiwan has been rising year after year, probably due to the greenhouse effect, so Taiwan has become a perfect place for the Polynesian lizard to survive. In less than 20 years, almost all habitats were occupied by them, and the less competitive native lizards (e.g., squamates) had no place to stand.

The author has been following the Orchid Island long-tailed skink for more than two decades and found that the survival rate and hatching rate of Orchid Island long-tailed skink on man-made concrete walls are much lower than the performance of long-tailed skink living in the natural environment due to global warming caused by the greenhouse effect, which further killed the Orchid Island long-tailed skink. The survival rate and hatching rate of long-tailed skinks in artificial concrete walls are much lower than the performance of long-tailed skinks in the natural environment. This is another example of human-induced environmental change affecting the survival of animals.

Lizards will lay their eggs in a warmer environment so that their young will grow larger and faster to increase their survival rate. About 22 years ago, the author found that long-tailed skink on Orchid Island had this phenomenon, but because the female lizard laid eggs in a concrete hole, it would expose its smell and attract the kukri snake to eat the eggs. When the kukri snake, a snake that specializes in eating reptiles' eggs, crawls into the nest, the mother lizard will immediately attack the snake and push it out of the nest. However, developing eggs are sensitive to environmental temperature, so mother lizards usually choose their nests carefully to ensure egg incubation and survival. The temperature or humidity in the environment can have a significant impact on the external morphology of the hatchlings, which in turn affects their future foraging, avoidance of enemies, and survival. If the incubating eggs experience temperatures that they are not capable of withstanding, the external morphology of the hatchlings will be affected, for example, if they hatch with deformed hatchlings, their fitness will be naturally reduced, or if the embryos do not survive at all and die before they hatch.

In this chapter, we found that some pregnant females laid their eggs under rocks in the natural environment or in concrete holes. After comparing the long-term ecological data (including survival rate, hatching rate and other external morphology over 10 years), we found that the hatching rate of eggs laid in concrete caves was higher (82%) than in the natural environment (55%). The survival rate of hatchlings hatched in concrete caves is higher because they can avoid enemies more quickly and have better foraging speed than hatchlings hatched in the natural environment. For example, hatchlings hatched earlier have a longer time to feed before winter comes, so they naturally have more energy to store to survive the winter; while hatchlings born in cement holes in late summer have larger heads and can feed on a wider variety of invertebrates, so they can safely survive the time when food is scarce from September to January. Naturally, its survival rate is more advantageous than that of smaller lizards born in the natural environment that eat small amounts of food.

However, in recent years, due to global warming, the average temperature of the natural environment of long-tailed skink has increased by 0.5 °C (the record of the Central Weather Bureau's Orchid Island weather station for 70 years), but the average temperature of the concrete hole has increased by 1.7 °C, and the maximum temperature has even reached over 40 °C, which has exceeded the loadable range of eggs. This not only affects the choice of nesting sites by female lizards, but also significantly reduces their reproductive success rate and egg hatching rate to less than 20%, which in turn has an impact on the reproduction of long-tailed skinks in the wild. In contrast, the survival rate of juveniles in natural environments is higher than that of juveniles in concrete caves; this reversal is causally related to global warming.

The man-made environment plays an important role in the ecology, evolution, and behavior of certain animals. For example, the concrete holes in this study affected temperature changes, and these changes attracted animals to live here, and sometimes became an ecological trap for animals (ecological traps: human activities to change the environment to attract animals to live, but when this place has become unsuitable for living, these animals still choose this place to move, thus reducing their fitness). In this study, for example, when the temperature of the man-made concrete hole was raised, the long-tailed skink used it as the best place to lay eggs to increase suitability, but after global warming, the temperature here had risen to the extent that the eggs could not be loaded, and some lizards still used it as a place to lay eggs and were trapped in it, which reduced their hatching rate and survival rate.

Field data from this study suggest that the continued increase in global temperature will change the egg-laying nest choice of Orchid Island long-tailed skinks, and that when temperatures reach a certain level, long-tailed skinks will no longer choose to lay eggs in artificial concrete holes. The long-tailed skink's unique egg-protection behavior around the world will also disappear. Even a small change in temperature can cause a huge survival crisis and adaptation change for the long-tailed skink, and bring about a wider ecological impact.

In egg-living species with minimal or no parental care, choosing nest sites with suitable conditions for embryo development can improve egg survival and thus viability. However, substantial variation in offspring phenotype can arise in response to environmental factors such as temperature and humidity within the range of incubation environments that permit successful hatching, (Blouin-Demers et al., 2004; Brown & Shine, 2004; Shine, 2004). This phenotypic variation can affect a range of traits, including morphology, gender, and behavior (Elphick & Shine, 1998; Bragg et al., 2000; Qualls & Shine, 2000; Morjan, 2003). For example, studies of natural nests in the field have demonstrated that maternal nest site selection shapes the hatchling phenotypes (Muth, 1980; Packard et al., 1980; Brown & Shine, 2004) and can affect neonate survival (Kolbe & Janzen, 2001). This evidence suggests that nesting females assess the environmental conditions and seek out nest sites that maximize their own and their offspring's fitness (Goldsbrough et al., 2004). As a result, current climatic changes (e.g., temperature, rainfall) may potentially reshape the abundance and distribution of suitable nesting sites, as well as the phenotype of nestlings from these sites.

Temperature changes in the microhabitat can have an impact on incubating eggs, either positively or negatively. Potentially beneficial effects of temperature on hatchling phenotype can occur only within the temperature range that permits successful egg incubation (Packard & Packard, 1988; David et al., 1998). Temperatures above the species-specific tolerable range can significantly reduce embryo viability (e.g., Packard & Packard, 1988; Löwenborg et al., 2010). As a result, if the current temperature of the nest is at the low end of the viable incubation range, current climate change may enhance fitness parameters until the temperature exceeds this threshold. Fitness will be greatly decreased once this threshold is reached. Recent research in this area has concentrated on the sex ratio of species with temperature-dependent sex determination (e.g. Mitchell et al., 2008; Telemeco et al., 2009).

Despite the pressing need to understand how climate affects fitness, it is inherently challenging to estimate the survival of eggs incubated under various temperature and humidity regimes; thus, such studies are uncommon (Brown & Shine, 2004). One reason is that most ectotherms' nesting habits are unknown (e.g., Doody et al., 2009). Furthermore, logistical difficulties in quantifying environmental variation among female nest site conditions and understanding how this variation affects both eggs and resulting offspring survival over longer time periods complicate the situation (e.g. Pike et al., 2008; Löwenborg et al., 2010). Another limitation is that ambient temperature (for which historical data is widely available) does not necessarily match the microhabitat conditions, such as those experienced by incubating eggs (Kearney et al., 2009). This makes it difficult to collect biologically relevant data to answer these questions. Recent advances in mechanistic ecological niche modeling overcome the latter limitation by analogizing nest temperature and predicting the effects of environmental temperature on offspring characteristics such as sex [e.g., sex; (Mitchell et al., 2008; Kearney & Porter, 2009)]. Despite recent advances, empirical data are still required to understand how temperature affects the life history traits of interest. This highlights the importance of long-term field studies of maternal nest site selection, including a thorough understanding of the effects of nest site selection on juvenile phenotypes, growth rates, and, most importantly, survival rates.

Using the long-term field data from tropical lizards, the long-tailed skinks (*Eutropis* [*Mabuya*] *longicaudata*), we investigated how nest site selection affects fitness and whether current climate change affects the availability of optimal nest sites (Fig. 13.1). Female long-tailed skinks lay their eggs in two types of microhabitats: under rocks in natural habitats and in concrete retaining walls in human-modified habitats. We first demonstrate that the thermal and hydric regimes of these two microhabitats differ, which affects female reproductive success (incubation time, egg survival, hatchling phenotype) and offspring fitness (growth, maturation age, and survival). Our findings then indicate that modern climate change disproportionately affects human-modified habitats, such as increasing nest temperatures faster than nests in their natural habitat. This, in turn, alters the location and availability of optimal nesting sites across the landscape.

Fig. 13.1 A long-tailed skink (*Eutropis longicaudata*) on Orchid Island, Taiwan. (Photograph by W.-S.H)

13.1 Effect of Maternal Attendance

Female attendance time (i.e., or > 1 week) did not significantly affect clutch size and quality, hatching time, egg survival, or hatchling size (all $p > 0.34$). Similarly, nest temperature and moisture regimes has no differences with maternal attendance time (temperature: mean 30.2 1.2 °C vs. 30.1 1.4 °C, respectively; nest type, $F_{1,43} = 1.16$, $p = 0.29$, nest type*time interaction, $F_{3,129} = 1.00$, $p = 0.39$; moisture: mean 92.8 0.9% vs. 92.1 0.6%, respectively; nest type, $F_{1,43} = 1.75$, $p = 0.19$, nest type*time interaction $F_{3,129} = 0.50$, $p = 0.68$). As a result, maternal attendance had no effect on any of the variables studied, and any subsequent differences between natural and artificial nest sites are unlikely to be due to maternal behavior.

13.1.1 Does Nest Site Location Affect Fitness?

In 2001, we collected data from egg laying to hatching for 25 artificial nests and 25 natural nests. Although nest sizes were comparable across habitat types, eggs in artificial habitats were larger, hatched nearly 5 days earlier, and had 18.9% higher survival rates than eggs in natural habitats (Table 13.1). Mean incubation

Table 13.1 Comparisons of reproductive success and hatchling body size attributes between *Eutropis longicaudata* nests laid beneath rocks in natural habitat ($n = 25$) vs. artificial habitat ($n = 25$). Egg survival is the percentage of eggs hatching from each clutch. Values are for 2001 only and are presented as ranges followed by means ± SE in parentheses

Variable	Natural habitat	Artificial habitat	*p*-value
Clutch size (n)	2–11 (6.1 ± 1.4)	1–13 (6.5 ± 1.2)	0.44
Egg mass (g)	0.7–1.4 (1.0 ± 0.4)	1.0–1.4 (1.3 ± 0.3)	<0.001
Incubation duration (days)	27–35 (31.6 ± 0.39)	24–30 (26.9 ± 0.4)	<0.001
Egg survival (%)	33.3–100 (67.1 ± 3.1)	66.7–100 (86.0 ± 1.5)	<0.001
Hatchling SVL (mm)	32.0–35.9 (34.5 ± 0.2)	35.1–39.2 (37.4 ± 0.2)	<0.001
Hatchling body mass (g)	0.60–1.28 (0.92 ± 0.03)	0.89–1.29 (1.22 ± 0.03)	<0.001
Hatchling tail length (mm)	51.0–66.0 (62.3 ± 1.1)	51.0–71.0 (65.2 ± 1.1)	<0.001
Hatchling head length (mm)	9.0–10.3 (10.3 ± 0.1)	8.9–11.3 (11.1 ± 0.1)	<0.001

temperatures in nests from artificial versus natural habitats were significantly higher (habitat type, $F_{1,68} = 17.10, p = 0.0001$; habitat type*time interaction, $F_{3,204} = 1.29$, $p = 0.28$; Fig. 13.2a). During the hottest part of the day, the mean temperature for natural nests was 30.5 0.3 °C and 31.4 0.3 °C for artificial nests. Furthermore, nest temperatures in artificial habitats ranged from an overall minimum ($F_{1,68} = 459.12$, $p < 0.0001$) to a maximum ($F_{1,68} = 1876.45$, $p < 0.0001$) and mean ($F_{1,68} = 1222.07$, $p < 0.0001$), with thermal variation ($F_{1,68} = 636.93$, $p < 0.0001$; Fig. 13.2b). Similarly, humidity was significantly higher in artificial nests than in natural nests (habitat type, $F_{1,68} = 5.99$, $p = 0.017$; habitat type*time interaction, $F_{3,204} = 2.37$, $p = 0.07$; Fig. 13.2c). Humidity decreased as nest temperature increased, with total body humidity averaging 89.7 0.4% in natural nests and 91.6 0.6% in artificial nests during the hottest part of the day. In terms of total minimum ($F_{1,68} = 5.64$, $p = 0.020$), maximum ($F_{1,68} = 3.57$, $p = 0.063$), and mean ($F_{1,68} = 5.99$, $p = 0.017$) temperatures, nests in artificial habitats had significantly higher humidity than nests in natural habitats, but there was no difference in water level variation ($F_{1,68} = 0.001$, $p = 0.98$; Fig. 13.2d). We excluded humidity from further analyses due to the small diurnal variation (generally 4%).

13.1.1.1 Offspring Body Size

Eggs hatched from artificial habitats hatched earlier than those from natural habitats, and hatchlings from artificial habitats were larger in terms of body length, body weight, relative tail length, and relative head length than hatchlings in natural habitats (Table 13.1).

13.1.1.2 Growth Rate and Maturity Age

From 2002 to 2006, we recaptured at least 68 hatchlings from artificial habitats and 86 hatchlings from natural habitats out of 100 hatchlings released per nest type. Growth rates significantly negatively correlated with SVL midpoint, indicating that as hatchling size increased, growth rate decreased (artificial nests, $R^2 = -0.39$, $p < 0.001$; natural nests, $R^2 = -0.32$, $p < 0.001$). Juveniles from both habitats had the same growth rate in relation to their body size (ANOVA, $F_{1,151} = 0.01$,

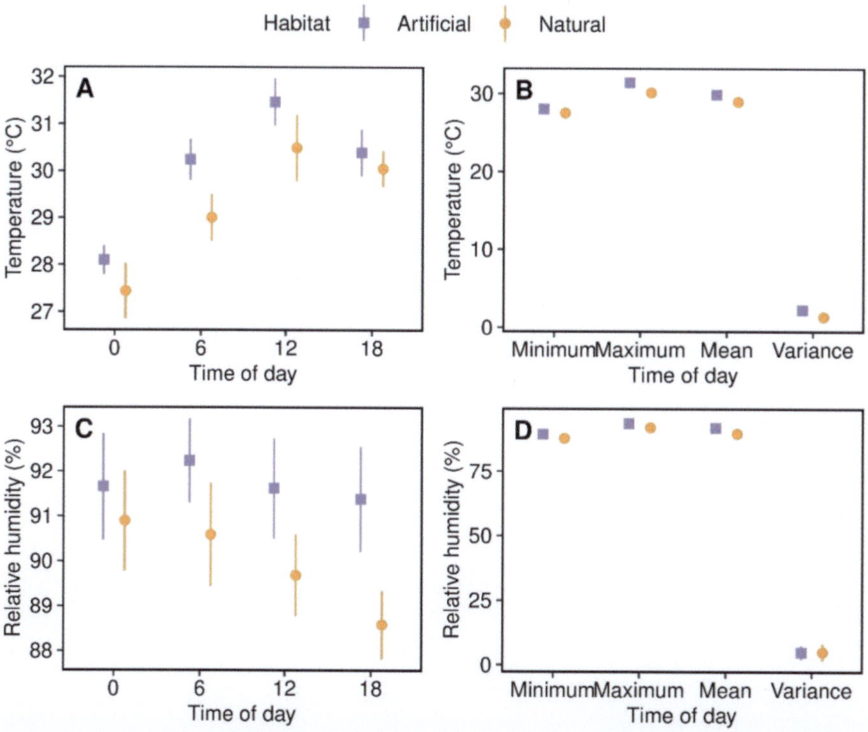

Fig. 13.2 Mean daily (**a**) thermal regimes, (**b**) thermal variation, (**c**) hydric regimes, and (**d**) hydric variation inside *Eutropis longicaudata* nests laid in artificial vs. natural habitats. Daily regimes are averaged over 6-h intervals from May to August 2001. (Redrawn from Huang and Pike (2011a, b), Functional Ecology)

$p = 0.99$). Despite this, juveniles born in artificial habitats each year were consistently larger than those born in natural habitats (two-way ANOVA; nest type. $F_{1,342} = 217.02$, $p < 0.0001$; year. $F_{5,342} = 86{,}188.90$, $p < 0.0001$; nest*year interaction. $F_{5,342} = 14.61$, $p < 0.0001$) (Fig. 13.3a). The von Bertalanffy growth model predicts that juveniles born in artificial habitats will reach sexual maturity 4 months earlier than those born in natural habitats, based on the minimum size of sexually mature females (i.e., 1.58 years or 19 months versus 1.9 years or 23 months; Fig. 13.3b). More female hatchlings nested in their birth habitat than in the opposite habitat ($n = 52$ females; 35 hatchlings born in artificial habitat nested there, 8 hatchlings born in natural habitat nested there, 6 hatchlings born in artificial habitat nested in natural habitat, and 3 hatchlings born in natural habitat nested in artificial habitat; comparing individuals nesting at birth to those nesting at maturity).

13.1.1.3 Offspring Survival

The model with the highest level of support from our mark-recapture data (87%) included differences in juvenile survival over time for various habitat types (i.e.,

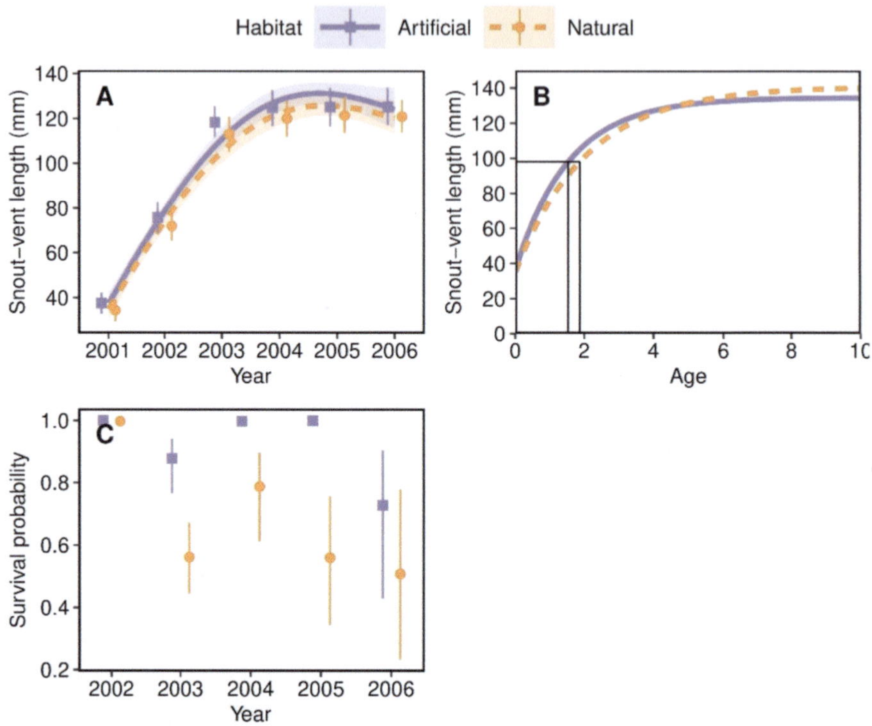

Fig. 13.3 (**a**) Mean body size (snout-vent length, SVL; ± 95% CIs) of *Eutropis longicaudata* emerging from nests laid in artificial vs. natural habitats. Data for 2001 represent body size at hatching, and data for 2002–2006 show how these hatchlings subsequently grew. (**b**) Von Bertalanffy growth curves for *E. longicaudata* hatchlings incubated in artificial habitats (solid line) vs. natural habitats (dashed line). The horizontal line indicates the minimum size (98.1 mm) of sexually mature females and the vertical lines show the estimated age at maturity for artificial vs. natural nests. (**c**) Annual survival estimates (± 95% CIs) of *E. longicaudata* hatching from artificial vs. natural habitats in 2001. Note that in some cases, error bars are too small to be visible. (Redrawn from Huang and Pike (2011a, b), Functional Ecology)

2002–2006; Table 13.2). The model with the second-highest ranking was one in which survival varied over time but not with habitat type (Table 13.2). The weights of the other candidate models were all low (Table 13.2). Because the difference between the two highest-ranked models was significant (QAICc = 3.92), the highest-ranked model was selected for estimating survival parameters. The mean annual survival rate for juveniles born in artificial habitats was significantly higher (paired $t = 3.30$, $p = 0.03$; Fig. 13.3c). During the period 2002–2006, hatchlings hatched in artificial habitats had a 24% higher survival rate than those in natural habitats (Fig. 13.3c).

Table 13.2 Candidate models used to assess the influence of nest type (artificial vs. natural) on hatchling *Eutropis longicaudata* survival rates. Hatchlings were born and marked in 2001 and recaptured from 2002 to 2006. We ranked models according to Akaike's Information Criterion (AIC) with adjustment for over dispersion ($QAIC_c$). The best-supported models have $\Delta QAIC_c$ values <2

Survival	Recapture	$\Delta QAIC_c$	$QAIC_c$ weight	Model likelihood	Parameters
Time × nest	Constant × Nest	0	0.87	1.00	12
Time	Constant × Nest	3.92	0.12	0.14	7
Time	Constant	10.21	0.01	0.01	6
Constant × nest	Constant × Nest	14.19	0	0	4
Time × nest	Constant	16.73	0	0	11
Constant	Constant × Nest	21.91	0	0	3
Constant	Constant	24.79	0	0	2
Constant × nest	Constant	26.32	0	0	3

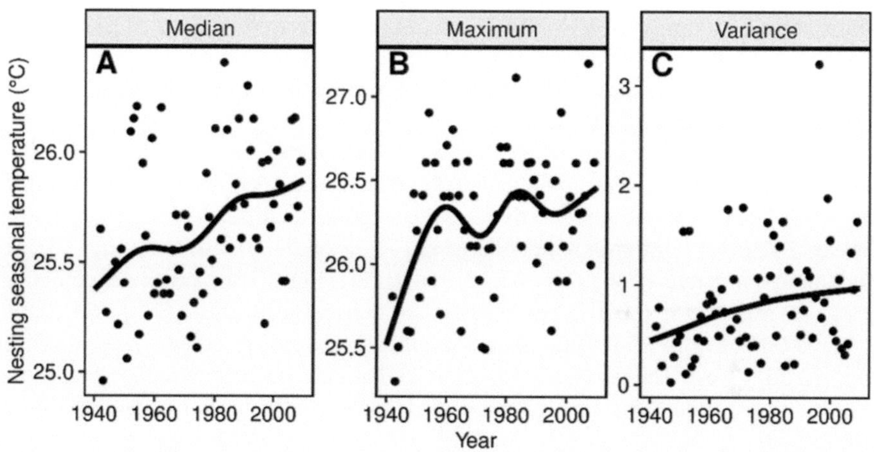

Fig. 13.4 Ambient temperatures averaged throughout the *Eutropis longicaudata* nesting season (May–August) on Orchid Island, Taiwan from 1940 to 2009. Shown are (**a**) monthly median temperatures, (**b**) mean monthly maximum temperatures, and (**c**) mean monthly thermal variance (right panel). Data for 1945 and 1946 were unavailable. (Redrawn from Huang and Pike (2011a, b), Functional Ecology)

13.1.2 Has Contemporary Climate Change Altered Nest Quality?

Temperature in the surroundings—during the lizard nesting season, the median ambient temperature at Orchid Island increased significantly from 1940 to 2009 ($N = 66$ years, no data available for 1945 and 1946; $R^2 = 0.13$, $p = 0.003$; Fig. 13.4a). During this time, ambient temperature increased by 0.065 °C per decade, for a total average increase of 0.455 °C. Similarly, the mean monthly maximum temperature ($R^2 = 0.09$, $p = 0.014$; Fig. 13.4b) and monthly thermal difference ($R^2 = 0.08$, $p = 0.022$; Fig. 13.4c) increased during this time period.

Table 13.3 Coefficient of determination (R^2) between hatchling *Eutropis longicaudata* phenotypes and nest temperatures ($n = 75$ nests in artificial habitat and 25 in natural habitat). Statistical tests are from individual regressions between phenotype and nest temperature, using data from both artificial and natural habitats. Significant results are indicated in bold

Hatchling phenotype	Minimum nest temperature	Maximum nest temperature	Mean nest temperature	Nest thermal variance
Egg hatching success	0.01	**0.09***	**0.06***	**0.18*****
Snout-vent length	**0.47*****	**0.28*****	**0.31*****	**0.08****
Body mass	0.04	0.02	0.02	0.01

* $p < 0.05$; ** $p < 0.01$; *** $p < 0.001$

We found significant differences in the relationship between ambient and nest temperatures across habitat types for minimum ($F_{1,211} = 16.68$, $p = 0.0001$) and maximum ($F_{1,211} = 7.53$, $p = 0.007$) temperatures, as well as thermal differences ($F_{1,211} = 23.06$, $p = 0.0001$), but not for mean temperatures ($F_{1,211} = 2.23$, $p = 0.14$). Nest temperature influenced egg hatching success and hatchling length but not hatchling quality (Table 13.3).

Mean nest temperature increased significantly in artificial habitats ($R^2 = 0.27$, $p < 0.0001$) but not in natural habitats (excluding a statistical anomaly, $R^2 = 0.05$, $p = 0.30$) from 2001 to 2009. We discovered a significant difference between habitat type and year for minimum temperatures ($F_{1,211} = 68.53$, $p < 0.0001$; Fig. 13.5a), maximum temperatures ($F_{1,211} = 43.17$, $p < 0.0001$; Fig. 13.5b), mean temperatures ($F_{1,211} = 50.79$, $p < 0.0001$; Fig. 13.5c), and thermal differences ($F_{1,211} = 40.26$, $p < 0.0001$; Fig. 13.5d). Significant interactions were observed ($p < 0.0001$; Fig. 13.5d). Nest Temperatures in artificial habitats became warmer and more variable over time, and this increase was faster than that in natural habitats (Fig. 13.5).

Beginning in 2006, these thermal variables increased in artificial habitats, but this pattern did not appear in natural habitats (Fig. 13.5). We found significant differences in mean nest temperatures across nest type and time period when we divided these years into two time periods (2001–2005 vs. 2006–2009) (habitat type. $F_{1,211} = 1630.90$, $p < 0.0001$; time period. $F_{1,211} = 314.70$, $p < 0.001$; Fig. 13.6a). Nest temperatures have increased in both artificial and natural habitats in recent years (Fig. 13.6a), but the relative increase in artificial habitats was three times greater than in natural habitats (i.e., 1.5 °C vs. 0.5 °C, respectively; Fig. 13.6b).

Changes in hatchling phenotype—we discovered a significant interaction between habitat type and year on egg hatching success ($F_{1,211} = 68.53$, $p < 0.0001$; Fig. 13.7a) and hatchling SVL ($F_{1,211} = 43.17$, $p < 0.0001$; Fig. 13.7b), but not on hatchling body mass ($F_{1,211} = 50.79$, $p < 0.0001$; Fig. 13.7c).

In our study, egg hatching success declined in artificial habitats while increasing in natural habitats (Fig. 13.7a). SVL of hatchlings remained constant in artificial

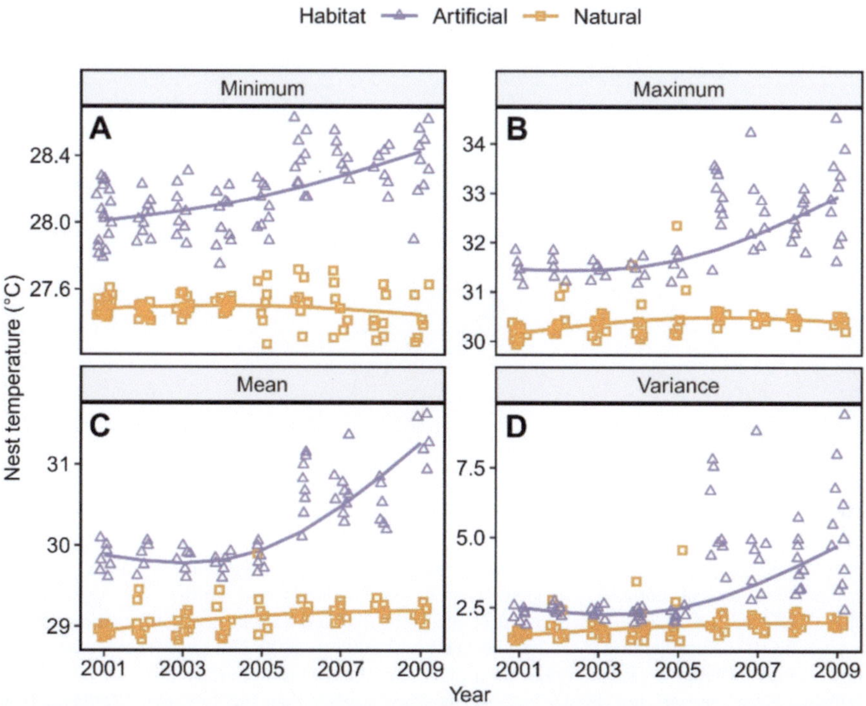

Fig. 13.5 *Eutropis longicaudata* nest temperatures in artificial vs. natural habitats from 2001 to 2009. Shown are overall (**a**) minimum, (**b**) maximum, and (**c**) mean temperatures along with (D) thermal variance. The black trend line is for artificial habitat and the gray trend line is for natural habitat. (Redrawn from Huang and Pike (2011a, b), Functional Ecology)

habitats but increased in natural habitats (Fig. 13.7b). Body weight followed the same relative pattern, but there was no significant variation over time (Fig. 13.7c).

We discovered that hatching success varied by habitat type and time period (habitat type. $F_{1,96} = 0.73, p = 0.39$; time period, $F_{1,96} = 2.23, p = 0.14$; interaction, $F_{1,96} = 21.27, p < 0.0001$); hatching success in artificial habitats decreased by 28% in recent years, while increased by 14% in natural habitats (Fig. 13.8a).

SVL, on the other hand, displayed the opposite pattern (habitat type. $F_{1,92} = 165.26$, $p < 0.0001$; time period; $F_{1,92} = 6.56$, $p = 0.012$; interaction; $F_{1,92} = 10.68, p = 0.002$; Fig. 13.8b), with hatchlings in natural habitats exhibiting an increase in SVL in recent years but no change in artificial habitats (Fig. 13.8b).

Body mass did not differ significantly between time periods ($F_{1,92} = 2.94$, $p = 0.09$; time period. Interaction; $F_{1,92} = 2.47$, $p = 0.12$; Fig. 13.8c). In recent years (2006–2009), a direct comparison between habitat types showed similiar hatching success (ANOVA, $F_{1,97} = 1.94, p = 0.17$; Fig. 13.8a) and hatchling quality ($F_{1,94} = 3.62, p = 0.06$; Fig. 13.8c), but hatchlings from artificial habitats had greater body length than those from natural habitats ($F_{1,94} = 162.99, p < 0.0001$; Fig. 13.8b, c).

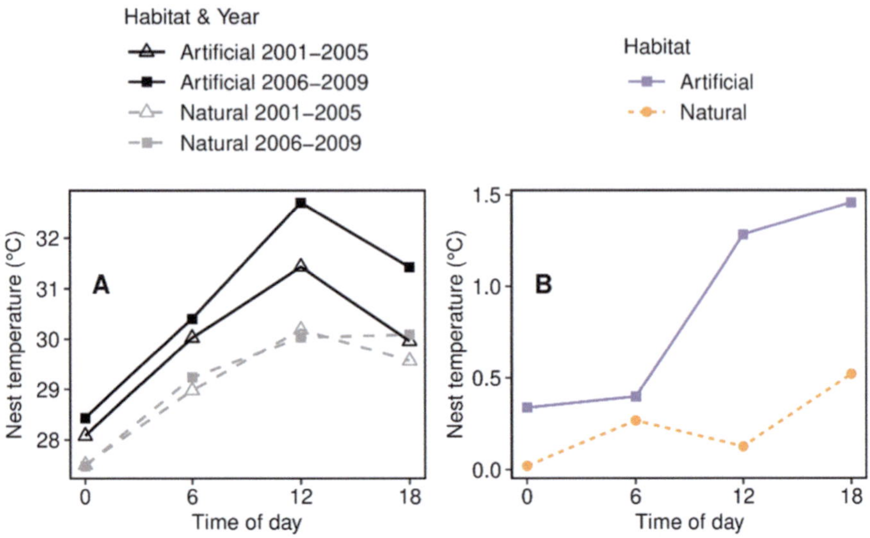

Fig. 13.6 (**a**) Mean nest temperatures inside *Eutropis longicaudata* nests laid in artificial vs. natural habitat, averaged over 6-hour intervals and shown for May–August 2001–2005 and 2006–2009. (**b**) Relative increases in *E. longicaudata* nest temperatures located in artificial vs. natural habitat from 2001 to 2005 and 2006 to 2009. Note that values at 0 hrs are shown twice to better depict the entire 24 h cycle, but these data were only included once in the analyses. Error bars are not shown because they are smaller than the symbols. (Redrawn from Huang and Pike (2011a, b), Functional Ecology)

Finally, our laboratory experiments revealed that hatching success was strongly influenced by incubation temperature ($\chi^2 = 10.30$, df $= 2$, $p < 0.0001$; Fig. 13.9a). Egg survival was highest at moderate incubation temperatures (mimicking nest temperatures in artificial habitats relative to cool years) and lowest at high incubation temperatures (mimicking nest temperatures in artificial habitats relative to warm years; Fig. 13.9a). Hatching time ($F_{2,43} = 17.55$, $p < 0.0001$; Fig. 13.9b), body length ($F_{2,43} = 52.41$, $p < 0.0001$; Fig. 13.9c), and body weight ($F_{2,43} = 3.73$, $p = 0.03$; Fig. 13.9d) were also significantly affected by incubation temperature. Hatchlings from the coldest incubation treatment had the longest hatching time and the smallest body size, whereas hatchlings from the two hotter treatments had a shorter hatching time, a larger body size, and nearly equal body size (Fig. 13.9b, c). Post hoc tests revealed that the low temperature treatment differed significantly from the other two treatments in all cases ($p < 0.05$), whereas the medium and high temperature treatments were similar in all cases ($p > 0.31$).

13.1.3 Discussion

Our findings show that nest site selection by tropical long-tailed skinks affects both offspring and maternal fitness, but that current climate change has modified the

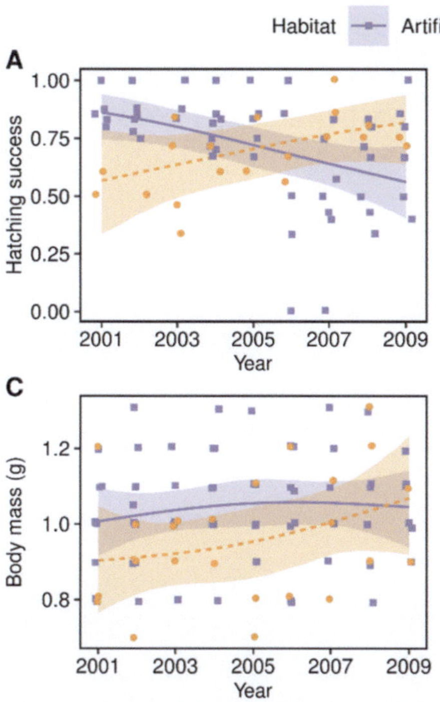

Fig. 13.7 Mean (**a**) egg hatching success, (**b**) hatchling snout-vent length (SVL), and (**c**) body mass (all ±SE) of *Eutropis longicaudata* from 2001 to 2009 shown for nests incubating in artificial vs. natural habitat. The solid trend line is for artificial habitat, and the dashed trend line is for natural habitat. (Redrawn from Huang and Pike (2011a, b), Functional Ecology)

location and availability of "optimal" nest sites. Long-tailed skink females nested in two different microhabitats (under rocks or within retaining walls in natural habitats). Female long-tailed skinks in artificial habitats improved egg survival and produced larger offspring that hatched earlier (Table 13.1) , matured (Fig. 13.3b), and survived longer (Fig. 13.3c) than female long-tailed skinks in their natural habitat. Thus, female lizards could improve their fitness levels in the early years of our study by nesting in human-modified habitats. However, as our research progressed, the temperature of the man-made environment began to rise above that of the natural habitat, reversing this pattern of nest site quality. As a result, the quality of artificial habitats is rapidly declining while natural habitats are improving. Because natural habitats are changing at an alarming rate around the world, these findings have far-reaching implications for the survival of oviposition species in human-dominated landscapes.

Available nests differ in two ways, which appear to explain the mechanisms underlying the fitness disparity between the two nesting habitats. (1) Females nesting within artificial walls stay with their eggs during incubation, and (2) artificial nests are warmer and more humid than natural nests (and may differ in other unmeasured

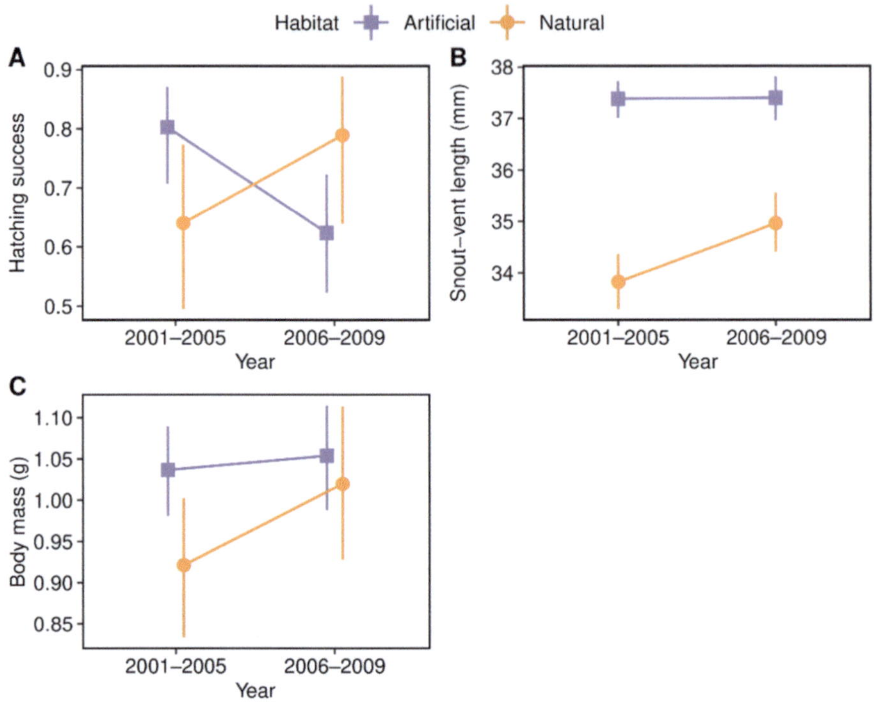

Fig. 13.8 Changes in (**a**) hatching success (percentage of eggs hatching), (**b**) hatchling snout-vent length, and (**c**) hatchling body mass for *Eutropis longicaudata* hatching from artificial or natural nests in the early (2001–2005) vs. late (2006–2009) years of our study. Shown are mean values ±SE. (Redrawn from Huang and Pike (2011a, b), Functional Ecology)

variables). Although nest attendance by long-tailed skink can deter whole-nest predation from egg-eating snakes (Huang, 2006), in the current study, none of the clutches were fully preyed upon. Furthermore, we discovered no evidence that the timing of maternal attendance had effects on incubation conditions (such as incubation time, temperature, and humidity) or offspring phenotype. As a result, the distinct maternal attendance behavior displayed by females nesting in artificial habitats appears unlikely to serve any purpose other than to deter predators from consuming eggs (in contrast to some other nest-keeping lizard species; reviewed in Shine (1988)). Thus, the fitness differences we observed may be a direct result of the incubation environment, and this claim was supported by our laboratory experiments (Fig. 13.9). Other studies have found that warmer and/or more humid incubation environments result in larger and faster-growing offspring (Blouin-Demers et al., 2004; Brown & Shine, 2004; Radder & Shine, 2007). Because the humidity exhibited minimal diel variation, we suspect that temperature is the more important factor (Fig. 13.2c, d). In the future, it would be worthwhile to investigate the interaction of temperature and humidity on offspring phenotypes.

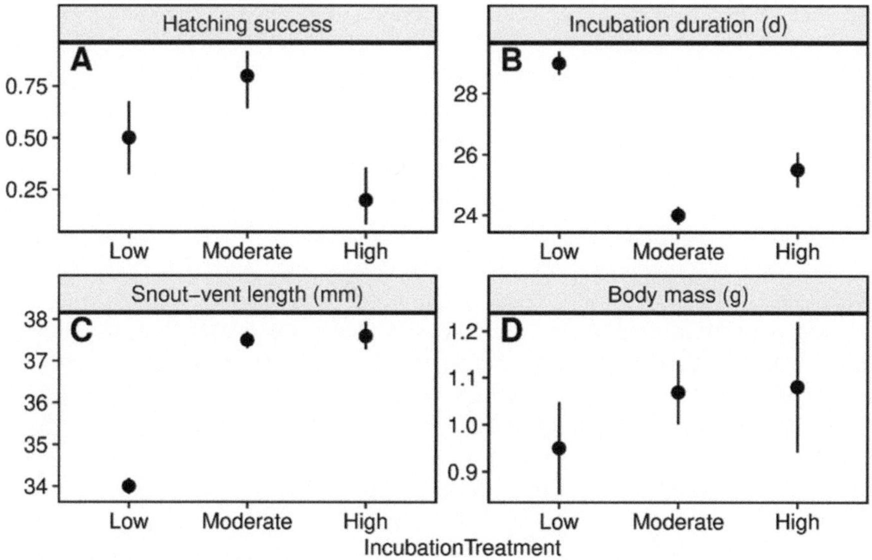

Fig. 13.9 Hatchling phenotype of *Eutropis longicaudata* eggs incubated under laboratory conditions at low (mean temperature: 29.7 °C, range: 25–37 °C), moderate (30.6 °C, range: 25–39 °C), or high temperatures (31.1 °C, range: 25–41 °C) for (**a**) hatching success of eggs ($N = 30$/treatment), (**b**) incubation duration, (**c**) snout-vent length, and (**d**) body mass. (Redrawn from Huang and Pike (2011a, b), Functional Ecology)

Individuals who hatch earlier in the season have a significant advantage over those who hatch later, primarily because early-hatching individuals have a longer period of activity before the onset of cold weather (Olsson & Shine, 1997; Shine & Elphick, 2001; Warner & Shine, 2005). For example, during the late activity season in our study area, the availability of small invertebrates (the primary food source for hatchlings) was low (September–January; Huang, 2004). Because head size is correlated with the size of prey that lizards can consume (Schoener, 1967), smaller juvenile lizards may be at a disadvantage when foraging for prey during the late season. Despite recent studies demonstrating "catch-up growth," in which smaller hatchlings grow faster than larger counterparts (thus compensating for initially smaller size) (Radder et al., 2007), hatchlings from both nest types exhibited the same growth rate relative to their size in our study. As a result, individuals from natural nests remained consistently smaller than those from artificial nests at any given time (Fig. 13.3a). Because reptiles typically reach sexual maturation based on body size (rather than age), this persistent body size advantage explains the difference in age at sexual maturity (Fig. 13.3b) and may increase lifetime reproductions (Warner & Shine, 2007).

In our study, what factors or mechanisms caused the temperature inside the concrete wall to steadily rise, but not in nests found in their natural habitat? Concrete walls are placed in an area with an open canopy and receive direct sunlight exposure, whereas long-tailed skinks in natural habitats laid eggs at the edge of the forest and are thus more shaded. As a result, the effect of ambient temperature on nest temperature varied depending on habitat type. Larger structures like concrete walls absorb and store heat more effectively than smaller stones (particularly those located on soil, which acts as a heat sink). It is not surprising that these two microhabitats react differently to changes in ambient temperature.

Human-modified habitats can have a significant impact on species ecology, evolution, and behavior. In some cases, despite the fact that individuals may utilize both habitat types in the same way (e.g., Kolbe & Janzen, 2002), modified habitats support higher population densities than natural habitats (e.g., Germaine & Wakeling, 2000).

Artificial habitat modification can alter the nesting microenvironment, allowing some species to nest at higher temperatures (e.g., Sartorius et al., 1999; Angilletta et al., 2009; Löwenborg et al., 2010). Some habitats' nesting temperatures may be lower than those of other species (Mrosovsky et al., 1995; Kolbe & Janzen, 2002). Some habitats may become ecological traps as a result of this, i.e., an attractive habitat with reduced quality, but animals continue to select that habitat because it was previously beneficial (e.g., Kolbe & Janzen, 2002). The available dichotomous nesting sites in our study arose as a result of anthropogenic habitat modification, specifically the construction of an artificial wall in 1997 to prevent soil erosion. Long-tailed skinks quickly colonized this wall, and some of them later developed distinct maternal nest protection behaviors (Huang, 2006, 2007a, b). Despite the quality of artificial nest sites had declined, the majority of hatched female long-tailed skinks continuously returned to nest in their natal habitat after maturation. This implies that human habitat modification can provide advantages for animals and thus improve their survival; however, climate change can disproportionately change the microhabitats of these habitat types, making them unsuitable for nesting through lowered egg survival.

The presence of artificial walls, we hypothesize, has increased population size over the last decade since 1997 by providing a large number of new nesting sites. However, further temperature increases may ultimately eliminate this important nesting habitat (in terms of producing viable offspring), which could have a population-wide cascading effect by slowing population growth rates and potentially weakening the distinctive nest protection behavior displayed by long-tailed skinks nesting in artificial habitats. Furthermore, females breeding early in the season (when temperatures are cooler) may continue to utilize artificial walls for nesting, but as temperatures rise seasonally, they may switch to nesting under rocks in their natural habitat. Our research focuses on the interactive effects of multiple human environmental activities on animals and shows how modern climate change can transform human-modified habitats into ecological traps.

In addition to the global greenhouse effect discussed in Chap. 13, climate change will also have a comprehensive impact on the reptile populations of Orchid Island. When frequent and powerful typhoons pass through this small island, they destroy the sandy beaches where sea turtles lay their eggs. In other words, once the nesting habitats are destroyed, sea turtles will no longer come ashore to lay eggs. How will the kukri snake, which primarily feeds on eggs, adapt to this situation? It may instead begin preying on other lizard eggs, which could have significant implications for lizard populations. The next chapter will address this issue.

References

Angilletta, M. J. J., Sears, M. W., & Pringle, R. M. (2009). Spatial dynamics of nesting behavior: Lizards shift microhabitats to construct nests with beneficial thermal properties. *Ecology, 90*, 2933–2939.

Blouin-Demers, G., Weatherhead, P. J., & Row, J. R. (2004). Phenotypic consequences of nest-site selection in black rat snakes (*Elaphe obsoleta*). *Canadian Journal of Zoology, 82*, 449–456.

Bragg, W. K., Fawcett, J. D., & Bragg, T. B. (2000). Nest-site selection in two eublepharid geckospecies with temperature-dependent sex determination and one with genotypic sex determination. *Biological Journal of the Linnean Society, 69*, 319–332.

Brown, G. P., & Shine, R. (2004). Maternal nest-site choice and offspring fitness in a tropical snake (*Ttropidonophis mairii,* Colubridae). *Ecology, 85*, 1627–1634.

David, J. R., Gibert, P., Gravot, E., Petavy, G., Morin, J.-P., Karan, D., & Moreteau, B. (1998). Phenotypic plasticity and developmental temperature in Drosophila analysis and significance of reaction norms of morphometrical traits. *Journal of Thermal Biology, 22*, 441–451.

Doody, J. S., Freedberg, S., & Keogh, J. S. (2009). Communal egg-laying in reptiles and amphibians: Evolutionary patterns and hypotheses. *The Quarterly Review of Biology, 84*, 229–252.

Elphick, M. J., & Shine, R. (1998). Longterm effects of incubation temperatures on the morphology and locomotor performance of hatchling lizards (*Bassiana duperreyi* , Scincidae). *Biological Journal of the Linnean Society, 63*, 429–447.

Germaine, S. S., & Wakeling, B. F. (2000). Lizard species distributions and habitat occupation along an urban gradient in Tucson, Arizona, USA. *Biological Conservation, 97*, 229–237.

Goldsbrough, C. L., Hochuli, D. F., & Shine, R. (2004). Fitness benefits of retreat-site selection: Spiders, rocks, and thermal cues. *Ecology, 85*, 1635–1641.

Huang, W. S. (2004). *Reptile ecology and the evolution of parental care on a tropical Asian Island.* PH. D. Cornell University.

Huang, W. S. (2006). Ecology and reproductive patterns of the grass lizard, *Takydromus sauteri*, in a tropical rain forest of an East Asian Island. *Journal of Herpetology, 40*, 267–273.

Huang, W. S. (2007a). Costs of egg caring in the skink, *Mabuya longicaudata*. *Ecological Research, 22*, 659–664.

Huang, W. S. (2007b). Ecology and reproductive patterns of the agamid lizard, *Japalura swinhonis*, on an East Asian Island, with comments on the small clutch sizes of Island lizrds. *Zoological Science, 24*, 181–188.

Huang, W. S., & Pike, D. A. (2011a). Climate change impacts on fitness depend on nesting habitat. *Functional Ecology, 25*, 1125–1136.

Huang, W. S., & Pike, D. A. (2011b). Determinants of homing in nest-guarding females: Balancing risks while travelling through unfamiliar landscapes. *Animal Behaviour, 82*, 263–270.

Kearney, M. R., & Porter, W. P. (2009). Mechanistic niche modelling: Combining physiological and spatial data to predict species' ranges. *Ecology Letters, 12*, 334–350.

Kearney, M. R., Porter, W., & Shine, R. (2009). The potential for behavioral thermoregulation to buffer "cold-blooded" animals against climate warming. *Proceedings of the National Academy of Science, 106*, 3835–3840.

Kolbe, J. J., & Janzen, F. J. (2001). The influence of propagule size and maternal nest-site selection on survival and behaviour of neonate turtles. *Functional Ecology, 15*, 772–781.

Kolbe, J. J., & Janzen, F. J. (2002). Impact of nest-site selection on nest success and nest temperature in natural and disturbed habitats. *Ecology, 83*, 269–281.

Löwenborg, K., Shine, R., Kärvemo, S., & Hagman, M. (2010). Grass snakes exploit anthropogenic heat sources to overcome distributional limits imposed by oviparity. *Functional Ecology, 24*, 1095–1102.

Mitchell, N., Kearney, M. R., Nelson, N. J., & Porter, W. P. (2008). Predicting the fate of a living fossil: How will global warming affect sex determination and hatching phenology in tuatara? *Proceedings of the Royal Society B: Biological Sciences, 275*, 2185–2193.

Morjan, C. L. (2003). Variation in nesting patterns affecting nest temperatures in two populations of painted turtles (*Chrysemys picta*) with temperature-dependent sex determination. *Behavioral Ecology and Sociobiology, 53*, 254–261.

Mrosovsky, N., Lavin, C., & Godfrey, M. H. (1995). Thermal effects of condominiums on a turtle beach in Florida. *Biological Conservation, 74*, 151–156.

Muth, A. (1980). Physiological ecology of desert iguana (*Dipsosaurus dorsalis*) eggs: Temperature and water relations. *Ecology, 61*, 1335–1343.

Olsson, M., & Shine, R. (1997). The seasonal timing of oviposition in sand lizards (*Lacerta agilis*): Why early clutches are better. *Journal of Evolutionary Biology, 10*, 369–381.

Packard, G. C., & Packard, M. J. (1988). The physiological ecology of reptilian eggs and embryos. In C. Gans & R. B. Huey (Eds.), *Biology of the reptilia. Volume 16. Ecology B. Defense and life history* (pp. 523–605). Alan R. Liss.

Packard, M. J., Packard, G. C., & Boardman, T. J. (1980). Water balance of the eggs of a desert lizard (*Callisaurus draconoides*). *Canadian Journal of Zoology, 58*, 2051–2058.

Pike, D. A., Pizzatto, L., Pike, B. A., & Shine, R. (2008). Estimating survival rates of uncatchable animals: The myth of high juvenile mortality in reptiles. *Ecology, 89*, 607–611.

Qualls, F. J., & Shine, R. (2000). Post-hatching environment contributes greatly to phenotypic variation between two populations of the australian garden skink, *Lampropholis guichenoti*. *Biological Journal of the Linnean Society, 71*, 315–341.

Radder, R. S., & Shine, R. (2007). Why do female lizards lay their eggs in communal nests? *Journal of Animal Ecology, 76*, 881–887.

Radder, R. S., Warner, D. A., & Shine, R. (2007). Compensating for a bad start: Catch-up growth in hatchling lizards (*Amphibolurus muricatus*, Agamidae). *Journal of Experimental Zoology Part A: Ecological Genetics and Physiology, 307A*, 500–508.

Sartorius, S. S., Vitt, L. J., & Colli, G. R. (1999). Use of naturally and anthropogenically disturbed habitats in Amazonian rainforest by the teiid lizard *Ameiva ameiva*. *Biological Conservation, 90*, 91–101.

Schoener, T. W. (1967). The ecological significance of sexual dimorphism in size in the lizard *Anolis conspersus*. *Science, 155*, 474–477.

Shine, R. (1988). Parental care in reptiles. In C. Gans & R. B. Huey (Eds.), *Biology of the Reptilia Vol. 16 Ecology B. Defense and life history* (pp. 275–330). Alan R. Liss.

Shine, R. (2004). Seasonal shifts in nest temperature can modify the phenotypes of hatchling lizards, regardless of overall mean incubation temperature. *Ecology, 18*, 43–49.

Shine, R., & Elphick, M. J. (2001). The effect of short-term weather fluctuations on temperatures inside lizard nests, and on the phenotypic traits of hatchling lizards. *Biological Journal of the Linnean Society, 72*, 555–565.

Telemeco, R., Elphick, M. J., & Shine, R. (2009). Nesting lizards (*Bassiana duperreyi*) compensate partly, but not completely, for climate change. *Ecology, 90*, 17–22.

Warner, D. A., & Shine, R. (2005). The adaptive significance of temperature-dependent sex determination: Experimental tests with a short-lived lizard. *Evolution, 59*, 2209–2221.

Warner, D. A., & Shine, R. (2007). Fitness of juvenile lizards depends on seasonal timing of hatching, not offspring body size. *Oecologia, 154*, 65–73.

Raging Ocean and Fragile Island: Loss of Sea Turtle Eggs Leads to the Collapse of Island Reptile Communities

14

Abstract

Terrestrial ecosystems often rely on marine subsidies, particularly on low-productivity islands, yet the consequences of losing such inputs are difficult to predict. Using 23 years of survey data from Orchid Island, we examined the effects of an abrupt loss of sea turtle eggs due to climate-driven beach erosion and predator exclusion. Following this loss, predatory snakes shifted their foraging inland, increasing predation pressure on terrestrial reptiles. As a result, most reptile species declined. In contrast, comparable turtle-free regions showed stable lizard populations. These findings demonstrate that removing a single marine subsidy can trigger cascading food-web effects, highlighting the importance of considering indirect interactions and unintended consequences in island conservation management.

Keywords

Orchid Island · Reptiles collapse · Climate change · Sea turtle eggs

Have you not seen the waters of the Yellow River flowing down from the heavens, rushing to the sea never to return? Have you not seen the bright mirror in the hall reflecting white hair, morning like black silk, evening like snow? When life is good, one must enjoy it to the fullest; do not let the golden cup stand empty before the moon. Heaven has given me talents that must be useful; even if I spend all my money, it will come back again. Let us cook sheep and slaughter cattle for our enjoyment; we must drink three hundred cups. Cen the scholar, Danqiu the sage, pour the wine, do not stop. I will sing you a song; please listen carefully. Bells, drums, and jade dishes are not precious; I only wish to stay drunk and never wake up. Throughout history, sages and saints have all been lonely, but only drinkers have left their names behind. Chen Wang once feasted at Pingle, spending ten thousand

W.-S. Huang, *Reptiles in the Changing Climate*, Fascinating Life Sciences,
https://doi.org/10.1007/978-981-92-1887-5_14

coins on wine and revelry. Why does the host say there is little money? Go and buy more wine to drink with you. Five-flower horses, thousand-coin robes, Call the boy to bring them out and exchange them for fine wine, to share with you and dispel the sorrows of the ages!

The poem "Will You Drink?" by Li Bai, a poet of the Tang Dynasty, is a poem that has been passed down through the ages. I would like to emphasize only the first line, "The waters of the Yellow River come from the heavens," while this chapter emphasizes "Island resources come from the sea." What kind of impact would the disruption of marine resources have on the island's reptiles? Although Li Bai's poem mentioned above is unrelated to this chapter, Chap. 14 is the last chapter of this book. If you can have a drink while reading this, you might gain some insight!

14.1 Introduction

Supplies of nutrients and energy are crucial for shaping ecosystem community dynamics, and a portion of energy inputs comes from neighboring or nearby ecosystems (Polis et al., 1997; Graham et al., 2018). These subsidies often arise through nutrient transport by organisms that move between habitats, such as seabirds (Graham et al., 2018; Bokhorst et al., 2019), seals (Bokhorst et al., 2019), salmon (Hocking & Reynolds, 2011), and sea turtles (Bouchard & Bjorndal, 2000). By affecting the species distribution and abundance (Rose & Polis, 1998; Hocking & Reynolds, 2011; Graham et al., 2018; Roffler et al., 2023) as well as food web interactions (Polis & Hurd, 1995; Spiller et al., 2010; Piovia-Scott et al., 2019; Roffler et al., 2023), nutrient subsidies can have a significant impact on the receiving system, particularly in ecosystems with low primary productivity (Polis & Hurd, 1995; Polis & Hurd, 1996). If such cross-ecosystem inputs are abundant and temporally and spatially predictable, then they may sustain high densities of keystone species despite the system with relatively low primary productivity (Polis & Hurd, 1995; Polis & Hurd, 1996), which in turn affects interactions among multiple species in the receiving ecosystem (Pike et al., 2016). Because the transfer of subsidies across the system may be mediated by one or a few keystone species, human activities can easily disrupt these inputs when these keystone organisms are affected (Moss, 2017; Schindler & Smits, 2017; Graham et al., 2018). Such species-level disruptions can have profound effects on the entire food web associated with the subsidy, similar to the loss of dominant species in a bottom-up process or the disruption of keystone species in a top-down process, ultimately resulting in severe ecosystem-wide impacts.

Sea turtles are an important link between marine and terrestrial ecosystems, building nests in their beach habitats and transferring nutrients and energy from the ocean to the land (Bouchard & Bjorndal, 2000; Moss, 2017; Lovich et al., 2018). Their nests contain large amounts of biomass, nutrients, and energy, much of which is retained in the terrestrial ecosystem. For example, in 1996, loggerhead sea turtles had nests containing approximately 1.6 million eggs on a 21-kilometre-long beach, but only about 27% of the total energetic content from these nests was returned to the ocean after the hatchlings had hatched, with the majority of the resources remaining onshore and entering the local terrestrial ecosystem (Bouchard & Bjorndal, 2000). Nutrients provided by sea turtles play a significant role in the health and functioning of inland ecosystems, supporting fauna and flora communities, particularly in island ecosystems where nutrient availability from other sources is more limited (Lovich et al., 2018). Unfortunately, this critical input is currently being disrupted by human activities through two different pathways. The first is beach erosion, which is a growing problem due to surging coastal populations, global climate change-induced sea level rise, and an increase in the frequency and severity of extreme storms (Schlacher et al., 2007; Defeo et al., 2009). For example, on eight monitored Caribbean islands, 33–93% of the beaches showed clear signs of erosion, with a mea erosion rate of 0.5 m/year recorded since 1985(Cambers, 2009). This global erosion has substantially reduced available nesting areas for sea turtles. Another major impact of human activities on marine subsidies to sea turtles comes unexpectedly from conservation efforts. More specifically, predator exclusion, a common management strategy for sea turtle conservation, restricts the transfer of nutrients from turtles to terrestrial ecosystems (Riley & Litzgus, 2013; Sainsbury et al., 2021). While predator exclusion can protect effectively vulnerable sea turtle nests from predators of sea turtle eggs (often invasive species), this practice also restricts native predators that have coevolved with the turtles from consuming the eggs, thus interrupting the import of marine nutrients into the terrestrial community. The disappearance of this critical resource could have cascading effects on other members of the inland food web, especially on small, low-productivity islands where dense native predator populations rely heavily on high levels of marine nutrient supplementation. Surprisingly, no research has yet assessed the impact of this loss on terrestrial communities, despite sea turtles playing a crucial role in inland ecosystems and predator exclusion measures being broadly practiced in sea turtle conservation.

Here we report a case study based on 23 years of long-term ecosystem monitoring in southeastern Taiwan, including a site where we were able to collect sea turtle eggs as information on key members of the terrestrial community before and after the disappearance of nutrient subsidies. On Orchid Island, the reptile community is connected to green turtle (*Chelonia mydas*) eggs through two main predators: the reptilian egg-feeding Kukri snake (*Oligodon formosanus*) and the generalist predator, the stink ratsnake (*Elaphe carinata*, the apex predator within the island community) (Fig. 14.1a). Historically, sea turtle nests provided a rich food source for the thriving kukri snake population on this small island, promoting significant behavioral evolution of both prey and predator species (Huang, 2006; Huang et al., 2011; Pike et al., 2016; Lee et al., 2019). However, due to beach erosion and predator exclusion measures to protect sea turtle nests, neither the kukri snakes nor the stink ratsnakes were able to access this important resource. The gradual erosion of beaches, together with several unusually strong storms in 2001, has led to the loss of larger beaches that no longer support sea turtle nesting (Fig. 14.2). As a result, sea turtles nesting is now restricted to a single, smaller beach, Bardai Beach(Cheng et al., 2009). In addition, sea turtle nests at Bardai Beach have been protected since 1997

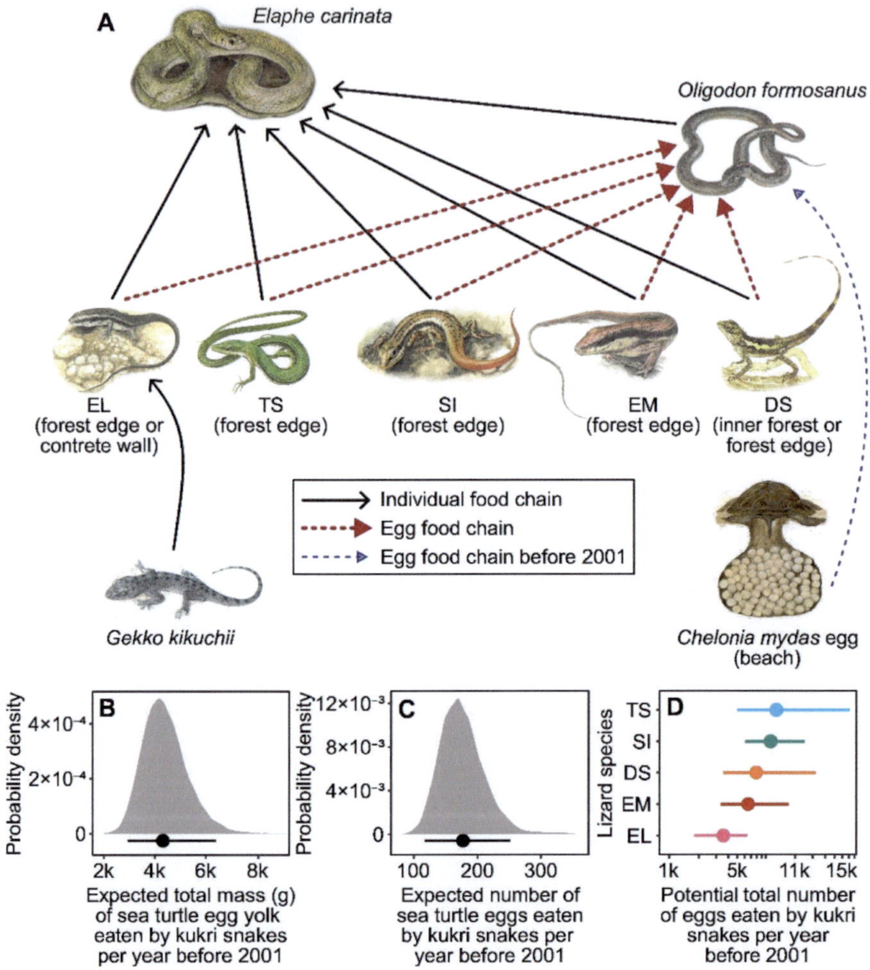

Fig. 14.1 The reptile food web on Orchid Island and the importance of sea turtle eggs in the context of egg predation by kukri snakes (*Oligodon formosanus*). (**a**) Trophic links between reptile species, with egg predation shown by dashed arrows and predation on hatched individuals shown by solid arrows. For species that nest in particular habitats, nesting habitat is given parenthetically after the species name. (**b**) Estimated mass and (**c**) number of sea turtle eggs consumed annually by kukri snakes from 1997 to 2000 at Tungching Beach; (**d**) sea turtle egg mass consumed by kukri snake expressed as the equivalent number of lizard eggs for five lizard species. DS, *Diploderma swinhonis*, Swinhoe's tree lizard; *EL Eutropis longicaudata*, long-tailed sun skink; *EM Eutropis multicarinata*, multi-keeled sun skink; *SI Sphenomorphus incognitus*, brown forest skink; *TS Takydromus sauteri*, Sauter's grass lizard. (Adapted from Lin et al. (2023), Science Advances. Originally published under CC BY-NC 4.0; used with permission. Illustrations rearranged and modified; plots redrawn)

Fig. 14.2 From 1997 to 2016, the height of the sea sand dropped sharply by about 1.2 m and the beach receded by about 11 m. After 2002, the remaining sea sand (depth of about 20 to 40 cm) was insufficient for sea turtles to lay eggs

by means of predator exclusion. Over a period of several years, several fence designs were tested and placed, culminating in the use of a fine plastic fence structure around each nest, which proved to effectively prevent snakes from entering the nests. As a result, sea turtle eggs have virtually disappeared from the terrestrial food web since 2001, despite being historically integrated into the island's ecosystem.

To evaluate how the removal of this major marine-derived subsidy influenced the terrestrial reptile communities, we employed a combination of experimental manipulations and field observations. Egg-feeding experiments and short-term mark-recapture data were used to estimate the quantity of sea turtle eggs biomass consumed by the kukri snake prior to the loss of nesting beaches. We subsequently analyzed long-term survey data to examine whether the disappearance of sea turtle eggs led to shifts in the spatial distribution and foraging behavior of snake predators. These same datasets were also used to assess the population dynamics of all reptile species on the island before and after sea turtle eggs disappeared. Finally, to determine whether the population trajectory was unique to Orchid Island, we compared population data for two formerly abundant lizard species with data from two ecologically similar sites where sea turtle eggs had never occurred (another island and a site on mainland Taiwan).

14.2 Number of Eggs Eaten by Kukri on Orchid Island

A total of 28 kukri snakes that had been recaptured on the sea turtle nesting beach exhibited weight loss (i.e., had not fed) since they were first captured 1–6 days earlier. Using these data, we estimated free-ranging kukri snakes lost weight at an average daily rate of 1.34% (0.92–1.77%; Bayesian $R^2 = 0.276$ [0.150–0.387]). Using the weight increases of 10 captive kukri-fed sea turtle eggs, we calculate an egg mass conversion rate of 32.2% (27.0–37.4%). Applying these values, we calculated that each snake consumed on average 0.321 g of sea turtle egg quality per day prior to 2001 (0.270–3.734), equivalent to 0.081 (0.060–0.105) sea turtle eggs.

Combined with our estimates of the annual abundance of kukri snakes and the annual duration of the sea turtle breeding season found on Tungching Beach before 2001, we calculated that kukri snakes consumed approximately 5.40 kg (3.6–7.8; Fig. 14.1b) of sea turtle eggs per year during 1997–2000, equivalent to 120.1 sea turtle eggs (79.8–174.2; Fig. 14.1c). If the Kukri Snake population had consumed the same number of eggs from the five species of soft-shelled lizards present on Orchid Island after 2001, this would have been equivalent to an annual consumption of approximately 5000–18,000 lizard eggs (4903 long-tailed skink eggs, 8992 Multi-keeled sun skink eggs, 15,369 Swinhoe's tree lizard eggs, 10,684 brown forest skink eggs, or 17,589 Sauter's grass lizard eggs) (Fig. 14.1d).

14.3 Changes in Prey and Foraging Habitat of Snakes on Orchid Island

Segmented Poisson model estimates indicated a significant increase in the frequency of lizard eggs attacked by kukri snakes in 2001, coinciding with the disappearance of sea turtle nesting beaches (3- to 19-fold, depending on species), followed by a slow decline in subsequent years (Fig. 14.3). This change in behavior was accompanied by changes in habitat use (Fig. 14.4). Since 1997, the abundance of kukri snakes observed on beaches has declined significantly (26%, [−23 to −30%] per year) and has remained consistently low in subsequent years (Fig. 14.4a). Although the estimated breakpoint didn't occur in 2001, the snake number on the beach declined from about 20 individuals prior to 2001 to less than 5 after 2001 (Fig. 14.4a). During the same period, the model showed the abundance of kukri snakes significantly increased in concrete retaining walls (28% [17–40%] per year before and after 2003; Fig. 14.4b) and in forested habitats (61% [17–120%] per year before and after 2003; Fig. 14.4c). Following this increase, kukri snake populations remained stable in forested habitats (−2.4% [−5.9–1.2%] per year; Fig. 14.4c), but declined at retaining walls (−37% [−30 to 43%] per year; Fig. 14.4b). Stink ratsnake population also decreased in beach after 2003 (beach, −42% [−25 to 56%] per year; Fig. 14.4d) and in concrete retaining wall habitats in 2006 (concrete, −50% [−22 to 68%] per year; Fig. 14.4e), but remained unchanged in forested habitats (+2% [−2 to 6.1%] per year; Fig. 14.4f).

14.4 Population Trends of Snakes and Lizards on Orchid Island

Population models revealed broad declines among snakes and lizards on Orchid Island across the food web. From 1997 to 2020, kukri snake populations were estimated to decrease by −12% (−10 to 13%) annually, while stink ratsnakes declined by −8.3% (−5.7 to 11%) per year (Fig. 14.5a, b). Except for geckos, model estimates also showed that substantial reductions in most Orchid Island lizard species, with annual declines of −11 to −25% per year (Fig. 14.5c–g). Population models indicated that Sauter's grass lizard and the long-tailed sunstone lizard have a steady decline similar to that of snakes, while the Swinhoe's tree lizard, the brown forest stone lizard, and the polydactyl sunstone lizard showed population numbers declined sharply at first, followed by stabilization at low population levels (Fig. 14.5c–g).

Based on our Poisson regression model, the populations of kukri snakes, stink ratsnakes, Sauter's grass lizards, long-tailed sun skink, and multi-keeled sun skink may continue to decline and reach zero within several decades (see estimated boundaries for predicted observed declines below 5 and 1, Fig. 14.5a–d, g). Populations of Swinhoe's tree lizards and brown forest skinks are predicted to remain low (Fig. 14.5e, f). Unlike the other reptile species, Botel's geckos showed a significant increase in population from 1997 to 2020. Their hard-shelled eggs prevent predation by kukri snakes, and their population exhibited a positive yearly growth rate of 9.3% (6.8–12%)(Fig. 14.5h).

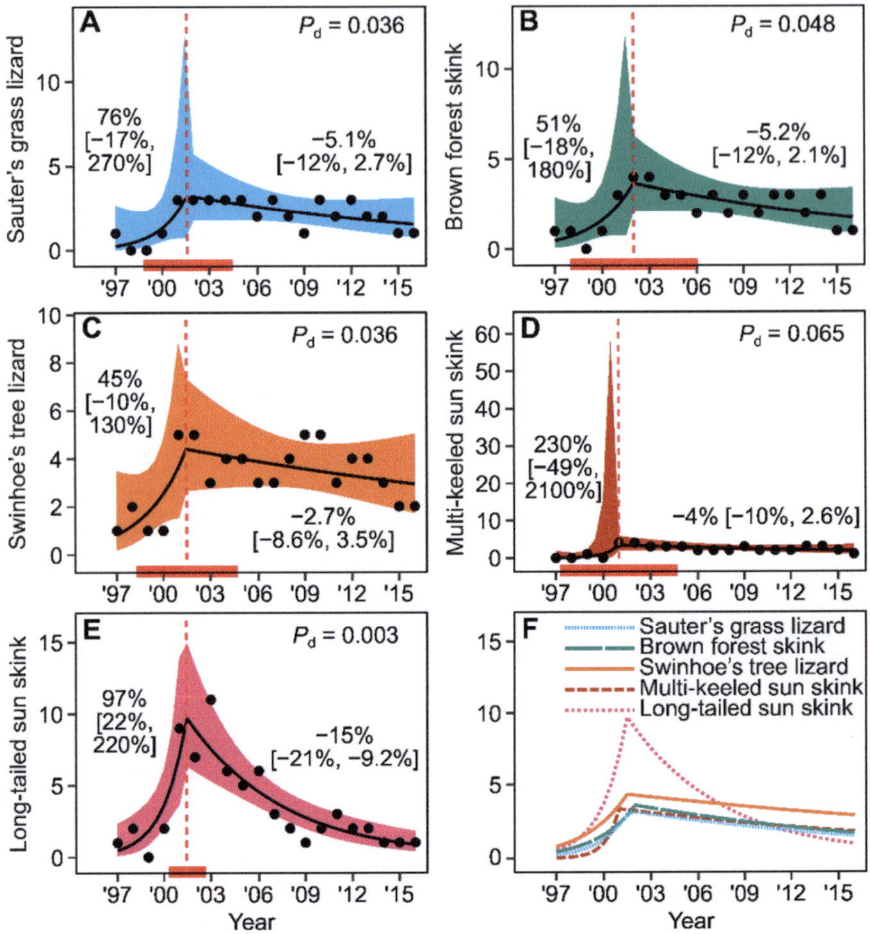

Fig. 14.3 The frequency of lizard clutches attacked by kukri snakes. Panels represent the number of attacks observed on egg clutches of (**a**) Sauter's grass lizard (*T. sauteri*), (**b**) brown forest skink (*S. incognitus*), (**c**) Swinhoe's tree lizard (*D. swinhonis*), (**d**) multi-keeled sun skink (*E. multicarinata*), and (**e**) long-tailed sun skink (*E. longicaudata*) and (**f**) predictions for above five species, respectively. Percentages indicate the estimated annual rate of increase of attack and the corresponding 95% confidence intervals. Vertical dashed red lines and horizontal red lines indicate breakpoints and 95% confidence intervals estimated by segmented Poisson regressions. p_d represents the *p* value corresponding to the comparison of attack rates before and after breakpoints. (Adapted from Lin et al. (2023), Science Advances. Originally published under CC BY-NC 4.0; used with permission. Plots rearranged)

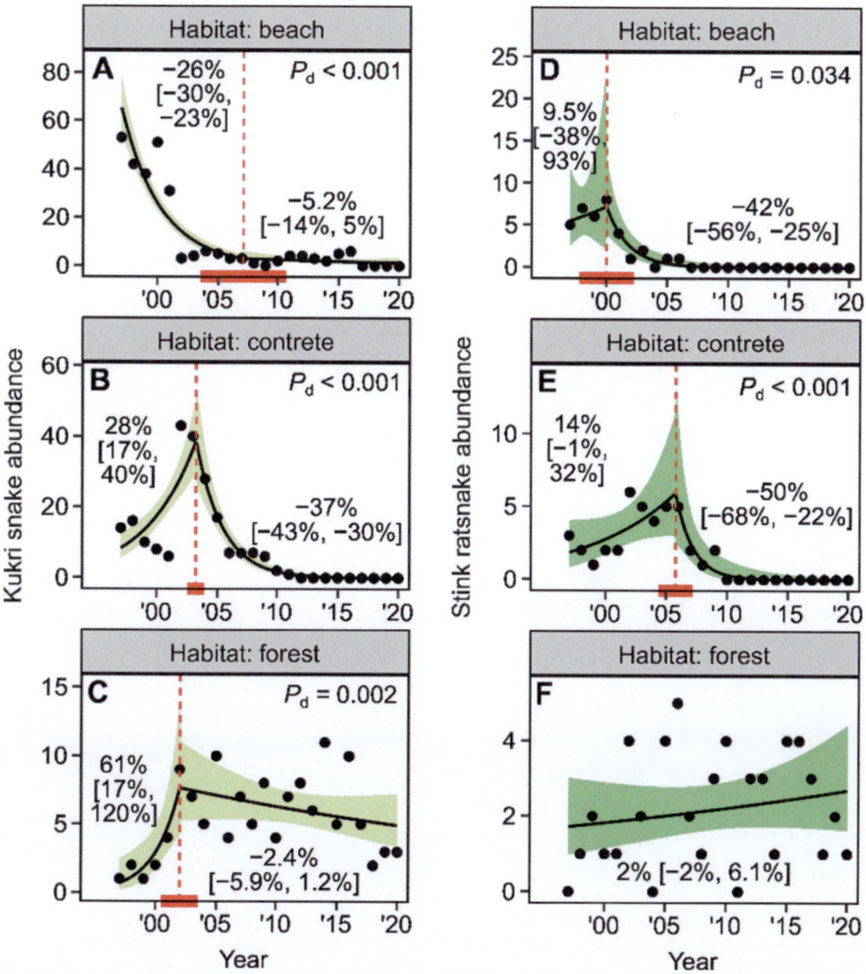

Fig. 14.4 Abundance of kukri snake (*O. formosanus*) and stink ratsnake (*E. carinata* in beach, concrete, and forest habitats. (**a** to **c**) and (**d** to **f**) present the abundances of kukri snakes (*O. formosanus*) and stink ratsnakes (*E. carinata*) in beach (**a**) and (**d**), concrete (**b**) and (**e**), and forest (**c**) and (**f**) habitats. Percentages indicate the annual growth rate and the corresponding 95% confidence intervals. Vertical dashed red lines and horizontal red lines indicate breakpoints and their 95% confidence intervals, respectively. p_d represents the p value corresponding to the comparison between growth rates before and after breakpoints. A panel without red lines indicates that no breakpoint was favored by Bayesian information criteria. (Adapted from Lin et al. (2023), Science Advances. Originally published under CC BY-NC 4.0; used with permission. Plots rearranged)

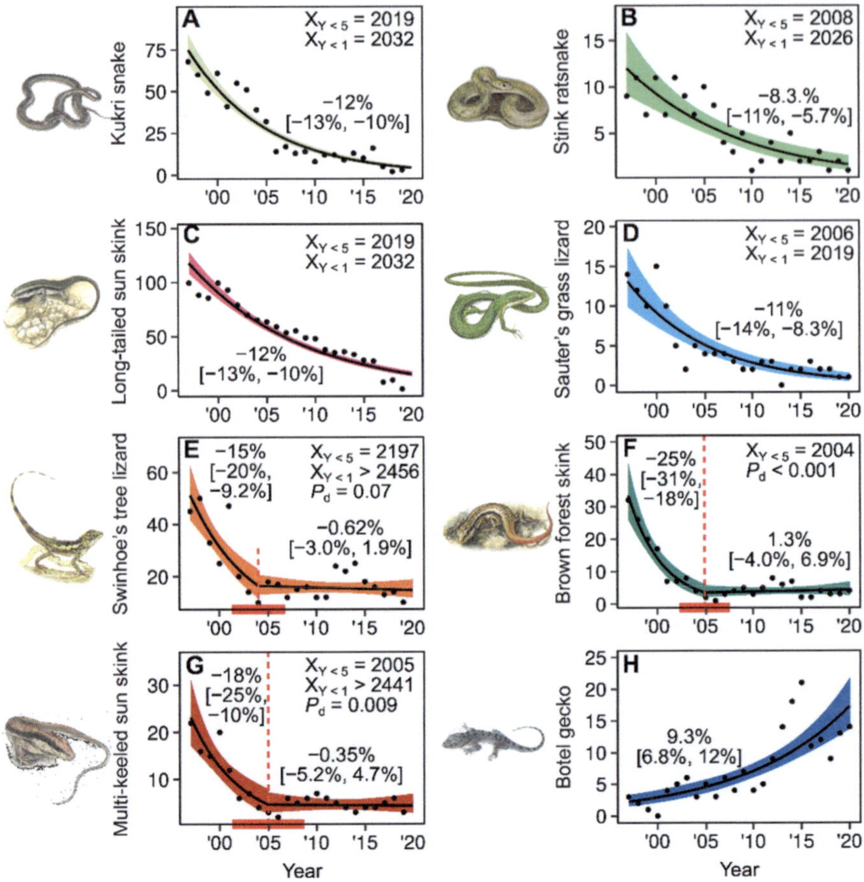

Fig. 14.5 Population trends of the eight terrestrial reptile species native to Orchid Island. Annual changes in abundance of eight reptile species, (**a**) kukri snake (*O. formosanus*), (**b**) stink ratsnake (*E. carinata*), (**c**) long-tailed sun skink (*E. longicaudata*), (**d**) Sauter's grass lizard (*T. sauteri*), (**e**) Swinhoe's tree lizard (*D. swinhonis*), (**f**) brown forest skink (*S. incognitus*), (**g**) multi-keeled sun skink (*E. multicarinata*), and (**h**) Botel gecko (*G. kikuchii*). Percentages indicate the estimated annual growth rates and the corresponding 95% confidence intervals. Vertical dashed red lines and horizontal red lines indicate breakpoints and their 95% confidence intervals, respectively. p_d represents the p value corresponding to the comparison between growth rates before and after breakpoints. A panel without red lines indicates no breakpoint favored by the Bayesian information criteria. $X_{Y<5}$ and $X_{Y<1}$ indicate the boundaries when the observed counts will decrease to fewer than 5 and 1, respectively, as predicted by the model. (Adapted from Lin et al. (2023), Science Advances. Originally published under CC BY-NC 4.0; used with permission. Illustrations modified; plots rearranged)

14.5 Lizard Population Trends at Nearby Sites

To provide background information on Orchid Island, we compared long-tailed sun skink and Swinhoe's tree lizard population trends with survey data from Orchid Island, neighbouring Green Island and Pingtung County (Taiwan). Analyses were limited to the habitat types shared among the three sites (concrete retaining walls for skinks and forested habitat for tree lizards). Prior to 2002, population changes of long-tailed sun skinks and tree lizards were observed at all three sites. Prior to 2002, the trend in population size of the long-tailed skinks at the three sites was generally stable (−3% to 1.9%; Fig. 14.6a–c). After 2002, however, Orchid Island populations of the long-tailed skinks declined sharply, with a significant negative annual growth rate estimated by the model (−32% [−27 to 36%] per year; Fig. 14.6a). There were no breakpoints in population growth rates of long-tailed sunshine skinks at Green Island and Pingtung, with a slight decline estimated by the model (−2.1% [−0.087 to 4.1%] p.a. at Green Island, and −3% [−1.1 to 4.8%] p.a. at Pingtung; Fig. 14.6b, c).

Swinhoe's tree lizards exhibited a similar pattern. In Orchid Island, population growth rate was strongly negative from 1997 to 2004 (−16% [−11 to 21%] per year), followed by stabilization at low abundance (−2.8% [−1.5 to 7.3%]; Fig. 14.6d). The population estimates of Swinhoe's tree lizards in Green Island and Pingtung remained stable throughout the study period without a break point (−0.11% [−1.4 to 1.2%], −0.78% [−2.3 to 0.74%] per year, respectively; Fig. 14.6e, f).

14.6 Discussion

In this chapter, we reveal the ecological consequences of losing sea turtle eggs, an important marine subsidy in a small island reptile community. From 1997 to 2001, we recorded significant changes in the foraging behavior and habitat use of Kukri's snakes, as large nesting beach used by sea turtles disappeared and their nests on smaller beach were protected. Fewer kukri snakes were observed on beaches, but more were observed in forested habitats and concrete retaining walls (Fig. 14.4), where kukri snakes were observed feeding more frequently on lizard eggs (Fig. 14.3). Habitat preferences of rat snakes showed a similar change as those of kukri snakes (Fig. 14.4d, e).

Although the trends in the abundance of lizard species we documented on the islands were complex (Fig. 14.5), the overall pattern of decline was consistent with their relative susceptibility to kukri snakes. Geckos lay eggs that are not preyed upon by kukri snakes, and their populations continued to increase throughout the study period. The Swinhoe's tree lizard, the brown forest skink, and the multi-keeled sun skink which are the lizard species most susceptible to predation by kukri snakes (Fig. 14.7), declined sharply during the period when sea turtle eggs were disappearing, and populations continued to decline in subsequent years. These species produced relatively large numbers of eggs, depositing them in open, bare

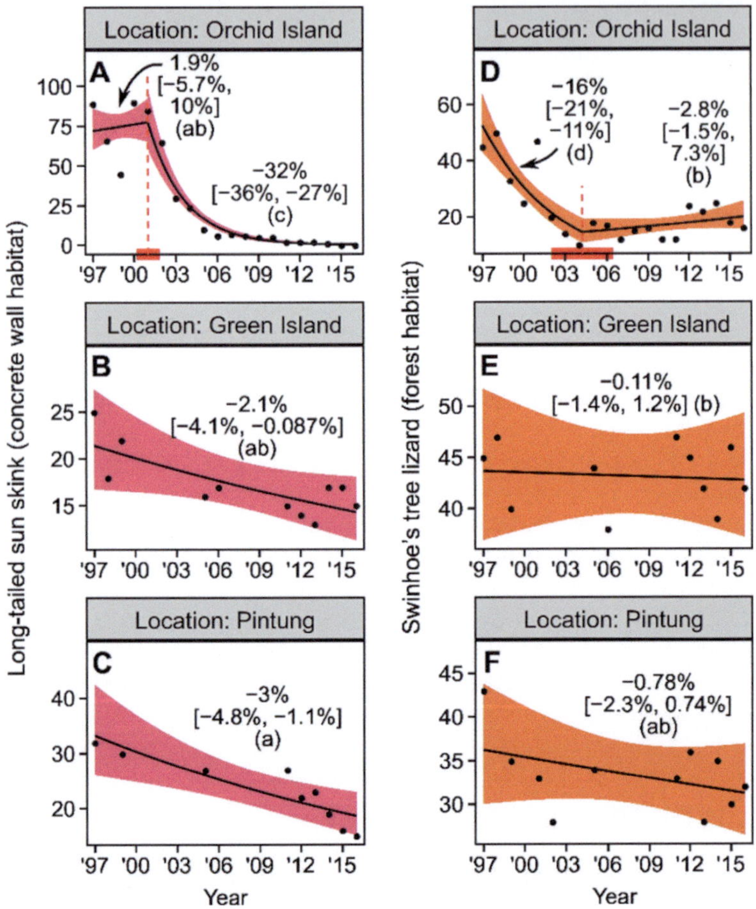

Fig. 14.6 Comparisons of population trends of two lizard species common to three regional localities. (**a** to **c**) Abundance of long-tailed sun skink (*E. longicaudata*) found in concrete wall habitats and (**d** to **f**) Swinhoe's tree lizard (*D. swinhonis*) in forest habitats on (**a**) and (**d**) Orchid Island, (**b**) and (**e**) Green Island, and (**c**) and (**f**) Pingtung County. Percentages indicate annual growth rates and the corresponding 95% confidence intervals. Vertical dashed red lines and horizontal red lines indicate breakpoints and their 95% confidence intervals, respectively. Lower-case letters in parentheses indicate the results of the post hoc comparisons among growth rates, with the absence of a common letter for growth rates indicating a significant difference. A panel without red lines indicates that no breakpoint was favored by Bayesian information criteria. (Adapted from Lin et al. (2023), Science Advances. Originally published under CC BY-NC 4.0; used with permission. Plots rearranged)

soil under the forest canopy, and lacked any form of parental care behavior. Populations of long-tailed skinks and grass lizards declined more slowly throughout the study period (Fig. 14.5). The lower susceptibility of these species to kukri snake egg predation may result from different reproductive strategies: long-tailed skinks displayed parental care behaviors specifically designed to guard clutches against

Fig. 14.7 Photos of kukri snakes (*Oligodon formosanus*) entering holes to consume eggs in various types of habitats on Orchid Island (**a**) beach, (**b**) concrete retaining wall, (**c**) forest edge, and (**d**) inner forest

Kukri snakes(Huang, 2006), whereas grass lizards typically laid only two small eggs in dense grasses and in soil between grass roots.

Although our results provide strong evidence that the loss of sea turtle eggs has led to changes in snake predation behavior, which in turn has led to a dramatic decline in lizard populations, the data are only correlative and multiple additional factors may have contributed to the patterns we have recorded. The area is vulnerable to human development (Cheng et al., 2009), increasingly frequent and severe windstorms due to climate change (Mei & Xie, 2016), sea level rise (Church et al., 2006), and invasive species such as rats (Liao et al., 2021), so more intensive monitoring or additional empirical studies are necessary to definitively confirm a causal relationship between Kukri snakes' predation on lizard eggs and the decline of lizard populations. Nevertheless, the fact that lizard populations remained stable over the same time period in nearby areas that lacked significant Kukri snake populations (Fig. 14.6) supports the hypothesis that the main reason for the decline in lizard populations on Orchid Island was the loss of an important marine food resource that sustained a large Kukri Snake population.

14.7 Marine Subsidies: Predator Effects and Trade-Offs in Sea Turtle Conservation

Conservation strategies often prioritize the protection of individual species, but our study highlights the importance of adopting a more holistic approach that considers how conservation measures will influence relationships among species, particularly where cross-ecosystem interactions are involved. On Orchid Island, our findings suggest that protecting sea turtle eggs from egg-eating snakes may have led to the breakdown of the island's lizard community. The conservation of green sea turtles is important because the population of this species has declined so severely (Cheng et al., 2009) that there is only one small beach left on Orchid Island (Bardai Beach) that is suitable for nesting. However, predator exclusion activities have been taking place on this beach for over 20 years, which has effectively eliminated the turtles from land-based island communities for prolonged periods of time. As a result, when a series of storms and sea level rise caused another nesting beach (Tungching Beach) unavailable to the turtles, marine subsidies sustaining the Kukri and Stink snakes disappeared within a few years, with serious consequences for the lizard community across the island.

The dramatic decline in the populations across almost all members of the Orchid lizard community underscores the possibility that conservation efforts targeting a single species may lead to unforeseen ecological consequences. Similar examples can be found in other systems (Ripple & Beschta, 2012; Ripple et al. 2014; Roffler et al., 2023). For example, protection of wolves in Yellowstone National Park has facilitated the restoration of willows and the return of beavers, which in turn has greatly reshaped stream ecosystems (Ripple & Beschta, 2012). In island systems, such as the sea otter–wolf relationship (Roffler et al., 2023), these types of interactions can have serious impacts, bringing relatively small local populations to the brink of extinction, as observed in our study. Therefore, conservation measures targeting specific species need to be carefully evaluated to determine their potential impacts on interactions with other species (Simberloff, 1998). Despite sea turtle eggs have long been recognized as a significant nutrient supplement (Bouchard & Bjorndal, 2000; Moss, 2017; Lovich et al., 2018), surprisingly few research examined the ecological consequences of disrupting this connection between ecosystems. Our study suggests that protection of the flagellar sea turtle (*Chelonia mydas*) may have serious cascading effects across terrestrial communities.

On Orchid Island, the populations of green turtle remains threatened by beach erosion and human disturbance (Cheng et al., 2009). Balancing the protection of both the sea turtles and the sensitive terrestrial reptile community is a conservation dilemma, especially given the vulnerable status of both species on this small island. Nevertheless, sea turtle nesting populations at Bardai Beach, the remaining nesting site, have remained stable for the past two decades (Nozawa et al., 2019). At the same time, our data indicate that many species are at risk of extinction, showing the terrestrial reptile community requires urgent protection.

Most sea turtle conservation efforts have used predator exclusion to prevent predation pressure from invasive species such as foxes, dogs, or rats (Sainsbury et al., 2021), but no evidence currently indicates that invasive species prey on sea turtle eggs on Orchid Island(Liao et al., 2021). We suggest that avoiding the exclusion of kukri from nesting beaches while conducting long-term monitoring of all species populations, which may provide a more balanced approach for protecting both sea turtles and the local reptile community. Allowing Kukri snakes to re-enter sea turtle nests will help restore the historical connection between the marine and terrestrial ecosystems on this small island and reduce predation pressure on the island's lizard community. As an alternative to the fine-meshed plastic fencing currently in use, isolation fences could be modified to include openings large enough to permit passage of the relatively slender kukri snakes while still restricting other predators. The effect of the kukri on sea turtle populations is probably limited. Kukri's snakes exclude conspecifics from sea turtle nests, and each individual can only eat a small portion of the dozens of eggs in a nest before hatching (Liao et al., 2021). Moreover, sea turtles display typical "Type III" survival rates, with extremely high egg and hatchling mortality rates, and previous studies have demonstrated that conservation measures targeting adult turtles are more effective than those targeting eggs (Heppell et al., 1996). Ultimately, maintaining the delicate balance between marine and island ecosystems is critical for the persistence of individual species and the broader ecological communities associated with these managed species.

References

Bokhorst, S., Convey, P., & Aerts, R. (2019). Nitrogen inputs by marine vertebrates drive abundance and richness in Antarctic terrestrial ecosystems. *Current Biology, 29*, 1721–1727.

Bouchard, S. S., & Bjorndal, K. A. (2000). Sea turtles as biological transporters of nutrients and energy from marine to terrestrial ecosystems. *Ecology, 81*, 2305–2313.

Cambers, G. (2009). Caribbean beach changes and climate change adaptation. *Aquatic Ecosystem Health & Management, 12*, 168–176.

Cheng, I. J., Huang, C.-T., Hung, P.-Y., Ke, B.-Z., Kuo, C.-W., & Fong, C.-L. (2009). Ten years of monitoring the nesting ecology of the Green turtle, *Chelonia mydas*, on Lanyu (Orchid Island), Taiwan. *Zoological Studies, 48*, 83–94.

Church, J. A., White, N. J., & Hunter, J. R. (2006). Sea-level rise at tropical Pacific and Indian Ocean islands. *Global and Planetary Change, 53*, 155–168.

Defeo, O., McLachlan, A., Schoeman, D. S., Schlacher, T. A., Dugan, J., Jones, A., Lastra, M., & Scapini, F. (2009). Threats to sandy beach ecosystems: A review. *Estuarine, Coastal and Shelf Science, 81*, 1–12.

Graham, N. A. J., Wilson, S. K., Carr, P., Hoey, A. S., Jennings, S., & MacNeil, M. A. (2018). Seabirds enhance coral reef productivity and functioning in the absence of invasive rats. *Nature, 559*, 250–253.

Heppell, S. S., Crowder, L. B., & Crouse, D. T. (1996). Models to evaluate headstarting as a management tool for long-lived turtles. *Ecological Applications, 6*, 556–565.

Hocking, M. D., & Reynolds, J. D. (2011). Impacts of salmon on riparian plant diversity. *Science, 331*, 1609–1612.

Huang, W. S. (2006). Parental care in the long-tailed skink, *Mabuya longicaudata,* on a tropical Asian Island. *Animal Behaviour, 72*, 791–795.

Huang, W. S., Greene, H. W., Chang, T. J., & Shine, R. (2011). Territorial behavior in Taiwanese kukrisnakes *(Oligodon formosanus). Proceedings of the National Academy of Science, 108*, 7455–7459.

Lee, C.-Y., Pike, D. A., Tseng, H.-Y., Hsu, J.-Y., Huang, S.-L., Shaner, L. P.-J., Liao, C.-P., Manica, A., & Huang, W.-S. (2019). When males live longer: Resource-driven territorial behavior drives sex-specific survival in snakes. *Science Advances, 5*, eaar5478.

Liao, C.-P., Hsu, J.-Y., Huang, S.-P., Clark, R. W., Lin, J.-W., Tseng, H.-Y., & Huang, W.-S. (2021). Sum of fears among intraguild predators drives the survival of green sea turtle (*Chelonia mydas*) eggs. *Proceedings of the Royal Society B: Biological Sciences, 288*, 20202631.

Lin, J.-W., Liao, C.-P., Chou, C.-C., Clark, R. W., Tseng, H.-Y., Hsu, J.-Y., & Huang, W.-S. (2023). Loss of sea turtle eggs drives the collapse of an insular reptile community. Science. *Advances, 9*, eadj7052.

Lovich, J. E., Ennen, J. R., Agha, M., & Gibbons, J. W. (2018). Where have all the turtles gone, and why does it matter? *Bioscience, 68*, 771–781.

Mei, W., & Xie, S. P. (2016). Intensification of landfalling typhoons over the Northwest Pacific since the late 1970s. *Nature Geoscience, 9*, 753–757.

Moss, B. (2017). Marine reptiles, birds and mammals and nutrient transfers among the seas and the land: An appraisal of current knowledge. *Journal of Experimental Marine Biology and Ecology, 492*, 63–80.

Nozawa, Y., Fong, C.-L., He, Z.-W., Chen, P.-Y., & Su, H. (2019). *Report of population survey project of sea turtle around Taiwan*. Academia Sinica.

Pike, D. A., Clark, R. W., Manica, A., Tseng, H. Y., Hsu, J. Y., & Huang, W. S. (2016). Surf and turf: Predation by egg-eating snakes has led to the evolution of parental care in a terrestrial lizard. *Scientific Reports, 6*, 22207.

Piovia-Scott, J., Yang, L. H., Wright, A. N., Spiller, D. A., & Schoener, T. W. (2019). Pulsed seaweed subsidies drive sequential shifts in the effects of lizard predators on Island food webs. *Ecology Letters, 22*, 1850–1859.

Polis, G. A., & Hurd, S. D. (1995). Extraordinarily high spider densities on islands: Flow of energy from the marine to terrestrial food webs and the absence of predation. *Proceedings of the National Academy of Sciences, 92*, 4382–4386.

Polis, G. A., & Hurd, S. D. (1996). Linking marine and terrestrial food webs: Allochthonous input from the ocean supports high secondary productivity on small islands and coastal land communities. *The American Naturalist, 147*, 396–423.

Polis, G. A., Anderson, W. B., & Holt, R. D. (1997). Toward an integration of landscape and food web ecology: The dynamics of spatially subsidized food webs. *Annual Review of Ecology and Systematics, 28*, 289–316.

Riley, J., & Litzgus, J. J. W. R. (2013). Evaluation of predator-exclusion cages used in turtle conservation: Cost analysis and effects on nest environment and proxies of hatchling fitness. *Wildlife Research, 40*, 499–511.

Ripple, W. J., & Beschta, R. L. (2012). Trophic cascades in Yellowstone: The first 15 years after wolf reintroduction. *Biological Conservation, 145*, 205–213.

Ripple, W. J., Beschta, R. L., Fortin, J. K., & Robbins, C. T. (2014). Trophic cascades from wolves to grizzly bears in Yellowstone. *Journal of Animal Ecology, 83*, 223–233.

Roffler, G. H., Eriksson, C. E., Allen, J. M., & Levi, T. (2023). Recovery of a marine keystone predator transforms terrestrial predator–prey dynamics. *Proceedings of the National Academy of Sciences, 120*, e2209037120.

Rose, M. D., & Polis, G. A. (1998). The distribution and abundance of coyotes: The effects of allochthonous food subsidies from the sea. *Ecology, 79*, 998–1007.

Sainsbury, K., Morgan, W., Watson, M., Rotem, G., Bouskila, A., Smith, R., & Sutherland, W. (2021). *Reptile conservation: Global evidence for the effects of interventions for reptiles*. University of Cambridge.

Schindler, D. E., & Smits, A. P. (2017). Subsidies of aquatic resources in terrestrial ecosystems. *Ecosystems, 20*, 78–93.

Schlacher, T. A., Dugan, J., Schoeman, D. S., Lastra, M., Jones, A., Scapini, F., McLachlan, A., & Defeo, O. (2007). Sandy beaches at the brink. *Diversity and Distributions, 13*, 556–560.

Simberloff, D. (1998). Flagships, umbrellas, and keystones: Is single-species management passé in the landscape era? *Biological Conservation, 83*, 247–257.

Spiller, D. A., Piovia-Scott, J., Wright, A. N., Yang, L. H., Takimoto, G., Schoener, T. W., & Iwata, T. (2010). Marine subsidies have multiple effects on coastal food webs. *Ecology, 91*, 1424–1434.

Afterword

The Importance of Balance in Nature

When I finished the book, a thought came to me: Oh my God! What have I found? This thought immediately gave me goose bumps all over my body. Is this what the Taoists call "Tao"? Or is it what religionists call [Truth] or what Buddhists call [Prajna]? From the very beginning, in 1997, I was ignorantly collecting information on the diet, habitat, number of eggs laid, reproductive cycle, etc., of various reptiles, although when I first started my research, one of my colleagues called me a stamp collector, meaning that I was doing some boring and not very creative research. As time passed, I continued to study the behavior of snakes and lizards, and it was only in recent years, when studies have been conducted and published, that climate change has affected the reptile community throughout Orchid Island, that the importance of what was initially thought to be a stamp-collection behavior has now become apparent. This is because without the integration of the most basic ecological data and the understanding of biological behavior, it is absolutely impossible to see the whole picture of the operation and mechanism of the reptile community.

This feeling is just like Su Dongpo, a famous Literary Scholar of the Northern Song Dynasty, who happened to travel to LuShan, Jiangxi Province, more than a thousand years ago and wrote a poem titled "Inscription on LuShan" because of the marvelous beauty of the mountain: "Looking across the mountains, the peaks become peaks on the sides, and the heights are different from each other near and far. Why was it impossible to see the true face of Mountain Lushan? Because the author was in the midst of Mountain Lushan, his vision was limited by the peaks in front of him, and what he saw was only a part of Mountain Lushan. This is true of what one sees when looking at a mountain, and it is also true of analyzing the nature of things and distinguishing between good and evil.

This state of mind is also similar to what the Buddhists say:[The green bamboos are all Dharma, and the yellow flowers are all prajna]. If you can realize this truth, then the green bamboos and yellow flowers are all Dharma; if you can't, then even though the Buddha's path of Zen is still there, you can't correspond to it. This is what is called [actual prajna]. Whether one is wise or not, actual prajna is there in the first

W.-S. Huang, *Reptiles in the Changing Climate*, Fascinating Life Sciences,
https://doi.org/10.1007/978-981-92-1887-5

place; only the foolish mortal does not know it, because it has no definite appearance and no action, and when one opens the eyes of wisdom, one realizes that actual prajna is already there.

I, an ecologist, have spent my life in pursuit of ecological meaning, expression, and purpose. After collecting and re-collecting fragments, like assembling a jigsaw puzzle, it was only after assembling the last piece of the puzzle that I suddenly realized that the operation of the ecology is actually a trade-off. It is like a religious person who is pursuing truth and suddenly sees the road ahead in the misty fog. It seems to be what the ancients called "Tao."

In Lao Tzu's book [Tao Te Ching], there is a saying: "Man follows the earth, the earth follows the sky, the sky follows the Tao, and the Tao follows nature" It means that man should follow the example of the earth, which is selfless and carries everything; the earth should follow the example of the sky, which is selfless and illuminates everything; the sky should follow the example of the great road, which is selfless and pervades everything; and the road should follow the example of nature, which is selfless and produces all things. In fact, according to Lao Tzu, Tao is the way of nature. To say that "the Tao follows nature" is that the Tao follows itself. The reason why Lao Tzu took the trouble to emphasize that "the Tao follows nature" is to point out that the Tao is nature. It is just that Tao is a universal principle in form, which is more abstract and not easy to grasp, whereas nature is more concrete and traceable.

The [Zhong Yong], as described by Confucius, means "too much, not enough," and the [Zhong Yong] is the core of Confucius' method of thinking. "Zhong" means "no bias," and "Yong" means "no change," and "zhongyong" means "neutral and constant," that is to say, appropriate and in line with the [Zhong Yong]. Zhuang Zi's Wu Wei Zhi Zi (govern by doing nothing that goes against nature) respects nature and human nature, and does not arbitrarily or deliberately interfere with the rules of personal life and living. We should respect the "Tao of Heaven" and the "Tao of Man" and realize the "unity of the Tao of Man" and the "unity of man and nature."

The endless destruction of nature by human beings in order to capture natural resources for their own use has, in fact, jeopardized the balance of nature. For example, extracting large quantities of oil stored in the ground and refining it into gasoline for use in cars and other fuels, dispersing large amounts of CO_2 into the atmosphere and causing the greenhouse effect, which in turn causes global climate change, stronger typhoons, hurricanes, or high temperatures around the globe, are not uncommon. There is also an imbalance in the distribution of rainfall, which in turn affects crop yields and causes famine. These are often the precursors of nature's imbalances. In more serious cases, it can lead to human extinction. Therefore, humans' respect for nature should be emphasized. In short, the way nature works is not to interfere, and the mechanism for doing so is balance.

Of course, there are some follow-up studies that should be continued; for example, I am not an expert in genetic evolution, so I have not studied the genetic mechanism of oviparous-to-viviparous evolution. Although it is known that the tree lizard (*Diploderma swinhole*) is moving towards the path of viviparous evolution, I have not been able to do so due to my lack of knowledge. Other mammals have not been studied either. For example, the Masked palm civet (*Paguma larvata*) was seen at night in the early days of Orchid Island, and it should have interacted with reptiles, but because the Dawu preyed on them, they have not been seen much in recent years. All I can do is focus on the reptiles themselves and look to future generations for further research.

Since 1997, I have been doing research in Orchid Island, trying to catch lizards on the island during the day and snakes on the beach at night for microchips and other numerical data collection. When I first arrived in Orchid Island, I could see lizards running around in the forest, on the roadside, or on the beach, or basking in the sun. For example, at least 50 nests of long-tailed lizards were recorded every year in the concrete caves, grass lizards could be seen foraging on the grass, not to mention the competition for the territory of tree lizards could be seen everywhere. However, after 2020, these observations have almost disappeared, and even the nests of long-tailed lizards are only seen in one nest at most. Beaches are thinning or disappearing due to typhoons, so naturally turtles come ashore and find no suitable nesting sites and give up laying their eggs and go back to the sea, and *Oligodon formosanus* does not appear to be looking for turtle eggs to eat. All these phenomena make me feel frustrated as an ecologist. The apparent collapse of these reptile populations is the result of climate change affecting the energy flow in the trophic cascade, prompting *Oligodon formosanus* to forage for other lizard eggs, as described in the book.

Ecologists who do ecological research on individuals, communities, and behaviors in the wild will eventually see the entire habitat change over time due to a variety of factors. These ecologists will always see that the number of animals living in nature will decrease or even disappear over time. Of course, there are many factors involved, such as human destruction of habitats, environmental pollution, chemical pollution, and so on. Eventually, ecologists will become conservationists. For example, the living organisms on Orchid Island are basically free from the abovementioned factors, but the reptile population is still decreasing year by year, and it is obviously disappearing. Why? It may be related to climate change in recent years. The warming of the Earth due to the greenhouse effect has caused the melting of the polar ice sheets and glaciers around the world, and the influx of fresh water into the oceans has changed the salinity of the oceans, and the ocean currents have also changed, resulting in the disruption of the originally stable climate. The sea level has risen and the number of severe typhoons has increased, resulting in the loss of sea sand. The habitat suitable for sea turtles to lay eggs is lost, and the disappearance of energy from the sea to Orchid Island destroys the originally stable food chain, as the reptile-egg-eating snakes turn to eat the eggs of inland lizards, which destroys the balance of the reptile community in Orchid Island, leading to the disappearance or even extinction of the lizard community. Some conservation ecologists have proposed transporting sand from other places to Orchid Island to make up for the loss of sand so that sea turtles can come ashore to lay eggs again. Although the intention is good, before the warming of the environment improves, these time-consuming and labor-intensive transportation of sand will be swallowed up by the sea, and no effect will be seen. These silent extinctions are not unlike the disappearance of polar bears and penguins, which will happen all over the world. Although there is a strong call for zero carbon emissions by 2050 to curb climate change, is this likely to happen? I am pessimistic. Although I feel pessimistic in the short term, I believe that future scientific developments should be able to solve the problem of carbon sequestration in climate change.

Index

W.-S. Huang, *Reptiles in the Changing Climate*, Fascinating Life Sciences,
https://doi.org/10.1007/978-981-92-1887-5

P

MIX
Papier aus verantwortungsvollen Quellen
Paper from responsible sources
FSC® C105338

If you have any concerns about our products,
you can contact us on
ProductSafety@springernature.com

In case Publisher is established outside the EU,
the EU authorized representative is:
Springer Nature Customer Service Center GmbH
Europaplatz 3, 69115 Heidelberg, Germany

Printed by Libri Plureos GmbH
in Hamburg, Germany